'The definitive guide to the UK 2011 Census, covering the collection, access, visualisation and analysis of census data, providing a handy go-to reference point for all things Census 2011 related. A particular strength worth highlighting is its breadth of coverage of approaches to the analysis of census data. A worthy and welcome successor to similar academic guides to previous UK censuses.'

– **Paul Williamson**, *Senior Lecturer in Geography and Planning, University of Liverpool, UK.*

'The census remains a vital source of information for decision makers and researchers, despite the recent explosion of new 'Big Data' from both government administrative and commercial customer files. Its coverage of the whole population, range of topics and geographical detail cannot be equalled, and the Census Offices have led the open data movement by making it freely available and easily accessible. This book helps users to unlock this rich source of data, and to create further value with their own projects, by drawing on contributions from a remarkable gathering of census experts.'

– **Keith Dugmore MBE**, *Director, Demographic Decisions Ltd., and Honorary Professor, University College London, UK.*

'This book is a definitive and authoritative guide for anyone using the expanded range of data products from the UK 2011 Census. Ably edited by John Stillwell, a longstanding census expert and user/enhancer of census data, the book includes a wealth of information on the last census in its historical context, the mechanisms of data delivery and visualisation, and research studies based on census data of all types. The book will be of interest to both census takers and census data users globally.'

– **Emily Grundy**, *Professor of Demography, London School of Economics and Political Science, UK.*

The Routledge Handbook of Census Resources, Methods and Applications

The collection of reliable and comprehensive data on the magnitude, composition and distribution of a country's population is essential in order for governments to provide services, administer effectively and guide a country's development. The primary source of basic demographic statistics is frequently a population census, which provides hugely important datasets for policy makers, practitioners and researchers working in a wide range of different socio-demographic contexts.

The Routledge Handbook of Census Resources, Methods and Applications provides a comprehensive and authoritative guide to the collection, processing, quality assessment and delivery of the different data products that constitute the results of the population censuses conducted across the United Kingdom in 2011. It provides those interested in using census data with an introduction to the collection, processing and quality assessment of the 2011 Census, together with guidance on the various types of data resources that are available and how they can be accessed. It demonstrates how new methods and technologies, such as interactive infographics and web-based mapping, are now being used to visualise census data in new and exciting ways. Perhaps most importantly, it presents a collection of applications of census data in different social and health science research contexts that reveal key messages about the characteristics of the UK population and the ways in which society is changing. The operation of the 2011 Census and the use of its results are set in the context of census-taking around the world and its historical development in the UK over the last 200 years.

The results of the UK 2011 Census are a unique and reliable source of detailed information that are immensely important for users from a wide range of public and private sector organisations, as well as those working in Population Studies, Human Geography, Migration Studies and the Social Sciences more generally.

John Stillwell is Professor of Migration and Regional Development in the School of Geography at the University of Leeds. He is Director of the UK Data Service-Census Support and has worked on both the delivery of census data to users and their application in a research context throughout his academic career. He was the Co-ordinator of the ESRC-funded Understanding Population Trends and Processes (UPTAP) programme and has been the co-editor of Applied Spatial Analysis and Policy (ASAP) since its inception in 2008.

International Population Studies

Series Editor: Philip Rees

This book series provides an outlet for integrated and in-depth coverage of innovative research on population themes and techniques. International in scope, the books in the series cover topics such as migration and mobility, advanced population projection techniques, microsimulation modelling, lifecourse analysis, demographic estimation methods and relationship statistics. The series includes research monographs, edited collections, advanced level textbooks and reference works on both methods and substantive topics. Key to the series is the presentation of knowledge founded on social science analysis of hard demographic facts based on censuses, surveys, vital and migration statistics. All books in the series are subject to review.

The Routledge Handbook of Census Resources, Methods and Applications

Unlocking the UK 2011 Census

Edited by John Stillwell

Routledge
Taylor & Francis Group

LONDON AND NEW YORK

First published 2018
by Routledge

2 Park Square, Milton Park, Abingdon, Oxfordshire OX14 4RN
52 Vanderbilt Avenue, New York, NY 10017

Routledge is an imprint of the Taylor & Francis Group, an informa business

First issued in paperback 2020

British Library Cataloguing-in-Publication Data
A catalogue record for this book is available from the British Library.

Library of Congress Cataloging-in-Publication Data
Names: Stillwell, John C. H. (John Charles Harold), 1952– editor.
Title: The Routledge handbook of census resources, methods and applications : unlocking the UK 2011 census / edited by John Stillwell.
Description: Abingdon, Oxon ; New York, NY : Routledge, 2017.
Identifiers: LCCN 2016058394 | ISBN 9781472475886 (hardback) | ISBN 9781315564777 (ebook)
Subjects: LCSH: Great Britain—Census, 2011.
Classification: LCC HA1124 2011 .R68 2017 | DDC 314.1—dc23
LC record available at https://lccn.loc.gov/2016058394

ISBN: 978-1-4724-7588-6 (hbk)
ISBN: 978-0-367-66003-1 (pbk)

Typeset in Bembo
by Apex CoVantage, LLC

Foreword

The census is an enormous undertaking. No other single data collection exercise sets out to collect information from all individuals and households in the United Kingdom at a point in time. The nature of this undertaking will have seemed equally remarkable before, during and after each census in the UK dating back to 1801.

A successful census is not, of course, defined by how many people completed the form. That details of 94 per cent of the UK population (in excess of 59 million individuals) were provided on a census return is nevertheless an astonishing achievement for the 2011 Census. A successful census is defined by how it is used. As set out in the chapters of this book, the 2011 Census has provided an enormous opportunity to understand and make decisions in modern Britain.

While the temptation with such a rich source might be to jump in and get going, those looking to help realise the full benefit of census data should first understand what census data are. For example, it is important to understand that statistical methods are used to provide census estimates representing 100 per cent of the UK population rather than just on the 94 per cent who responded. An accessible guide to the collection and processing of census data is provided in this book.

Once the user has jumped in, they will find that census data are accessible as never before. The Census Offices have worked in collaboration with a wide range of organisations and users to develop tools to access data. I welcome the guidance provided in this book not just to accessing census data (aggregate, record level and longitudinal) but also to a number of the tools made available by third-party suppliers. Such tools were made possible by the 'open data' approach taken for the census release.

The majority of this book is devoted to what census data have already achieved. It is particularly striking how the consumption of statistical information has changed between 2001 and 2011. Social media has provided a platform for the dissemination of statistical stories and new ways of engaging with users. Online technologies have enabled third parties to make remarkably beautiful and insightful visualisations of census data.

It is the breadth of the topics covered that is most striking about the research examples provided in this book. However, while the coverage is broad (informal and elderly care, migration, ethnicity, social inequality, housing need), the real insight the census provides is in the detail. To have a single source which has been consistently collected which can be analysed at any level of geographic detail is what makes the census-type statistics special. Of course, this should not be a surprise. This is what the census is designed to do and is what the census has been doing for hundreds of years.

Importantly, the book also looks to the future of the census in the UK, to 2021 and beyond. In 2014, the Board of the UK Statistics Authority and the Government accepted and endorsed my predecessor's recommendation that a predominantly online census in 2021 should be supplemented by further use of administrative data. Alongside planning for 2021, the census-taking offices are conducting research into the potential of using administrative data and surveys to replace the ten-yearly censuses beyond 2021. A recommendation will be made on the future of population statistics in 2023.

We should not need reminding how valuable census-type statistics are. Indeed, greater use of administrative data in particular has the potential to provide this level of insight in a far more timely way than once every ten years.

John Pullinger
National Statistician
January 2017

Preface

This book sets out to provide a guide to the collection, processing, quality assessment and delivery of the different data products that constitute the results of the censuses conducted by the national statistical authorities across the United Kingdom in 2011. It also endeavours to demonstrate that, collectively, the different census datasets are hugely important for research in a wide range of different socio-demographic contexts from geodemographic profiling of small areas to understanding the geography of immigration and the changes in the ethnic structure of British cities. Thus, in addition to a detailed account of the types of data that are available from the census and how they can be accessed by users, it contains an anthology of applications that exemplify the way in which the data are being used together with a series of shorter synopses of contemporary methods of visualisation through maps and other forms of infographic. The operation of the 2011 Census and the use of its results are set in the context of census-taking around the world and its historical development in the UK over the last 200 years.

The book is particularly important in terms of the value that it provides for those wanting to understand what 2011 Census data are available and how and where different types of data can be accessed, for those interested in understanding the state-of-the-art in data visualisation, for those wanting to absorb some of the latest findings of contemporary research in social science in the UK, and for those concerned to understand more about the census and its utility at a time when its future existence is under threat. The most important conclusion from the book as a whole is that it would be folly to abandon such a vital resource of reliable and comprehensive data without putting in place an infrastructure for a much improved collection of data from other sources, be they administrative records, sample surveys or new digital devices.

John Stillwell
University of Leeds
January 2017

The Book and the Series

The Routledge Handbook of Census Resources, Methods and Applications appears simultaneously in both the Routledge Handbooks series and as the first book in the Routledge International Population Studies series. This dual identity is significant. The book informs the reader about the latest UK Census held in 2011 and how to use the huge collection of associated data outputs. The census is the most significant geographic set of official statistics on the United Kingdom and is key to answering research questions and in designing policy at scales from the nation to the neighbourhood. The book appears in the International Population Studies series because it serves as a world-leading example of how to produce and disseminate outputs from an enumeration of a population of 63 million people. UK census data have been made available to ever-growing communities of users. Data from the 2011 Census are available to users in the UK and also worldwide, via a few mouse clicks.

The book has been assembled by one of the most productive and skilled book editors in social science, my colleague John Stillwell. The topic focus is the 2011 Census of Population in the United Kingdom. Quantitative Population Studies as a field of research relies in large measure on a data infrastructure provided by official statistical organisations, both national and international, working in collaboration with constituencies of users in national and local government, in business and in academia. In the UK, the Economic and Social Research Council (ESRC) from the early 1990s to the present has been committed and generous in supporting the development and dissemination of the UK's collections of large social science datasets, with census data among the most important. This book represents the fruit of these investments, in close collaboration with the UK's National Statistical Offices.

Since 2012, the Editor of the book has been Co-ordinator of ESRC's Census Dissemination Programme, part of the UK Data Service hosted at the University of Essex. He is third in the sequence of Co-ordinators, preceded by David Martin (2002–12), the original designer of census output areas that all census users know and love. I was the first Co-ordinator (1992–2002). It is enormously pleasing to see ambitions that we had in the 1990s coming to fruition in the 2010s. Virtually all dissemination of census data is now via the internet; most of the datasets are freely downloadable under the Open Government Licence to interested researchers or citizens in the world (Chapter 1 and Part III of the book). The processes of census-taking and database creation by UK National Statistical Offices have benefitted from use of radically better statistical and quality assurance procedures (Part II). Sophisticated graphical and mapping software has improved our ability to visualise census data to obtain new insights and to generate new hypotheses about causation (Part IV). Analyses of census data (Part V), such as geodemographic classifications, exemplified on the book cover, are now fourth-generation products. Our knowledge of the UK population's ethnic composition, family makeup, health and care geographies, housing and household patterns, poverty and inequality, internal migration and commuting has improved substantially as better data have been produced and as analytical methods have improved. Part V also provides examples of fresh uses of census data in microsimulation modelling and in demographic accounting.

Although the book focuses on the latest UK Census and some key findings, international readers will find plenty to interest them, whether they are government statisticians, population analysts or university students. As you read the book, you may want to pose the following questions. How does your country handle the production, dissemination and analysis of census or equivalent data? What model of collaboration is there between data producers and data users? Does the UK experience provide guidance for your country? Do any of the examples of census data visualisation or statistical analysis have relevance for your country? Could you include in your international study census data from the UK? I hope you will be convinced that, despite the result of the June 2016 Referendum on the UK's membership of the European Union, UK government statisticians and researchers are committed to open collaboration with colleagues beyond our shores in order to better understand the condition of the human population.

Philip Rees
University of Leeds
January 2017

Acknowledgements

This volume has been produced on behalf of the UK Data Service–Census Support which is funded by the Economic and Social Research Council (ESRC) under grant ES/J02337X/1.

All UK census data mentioned in this book are Crown copyright and have been produced by three separate agencies (Office for National Statistics, National Records of Scotland, Northern Ireland Statistics and Research Agency) and licensed under the Open Government Licence v.3.0 (except where otherwise stated).

The image on the front cover is by *Chris Gale* (whilst at University College London) and *Oliver O'Brien* (at the Consumer Data Research Centre), and shows the 2011 Area Classification for Output Areas (or 2011 OAC) for the Brighton and Hove area. It contains National Statistics and OS data © Crown copyright and database right 2017, and is derived from a larger image that appears in Chapter 17.

The editor is immensely grateful to the following: *Alison Manson* at the University of Leeds for all her hard work in standardising, improving and, in many cases, redrawing the illustrations supplied by contributors; *Calum Carson*, UK Data Service–Census Support administrative assistant based at the University of Leeds, for his efforts in checking all the references and ensuring that, at the time of manuscript submission, all the web links were live; *Faye Lererinck, Priscilla Corbett* and *Jo Hardern* at Routledge, for their help at different stages of the commissioning and production processes; *Pam Smith* for reading and correcting the proofs so diligently; *Garnett Compton* at the ONS, for co-ordinating contributions to the book from the Census Offices and for his comments on the first two chapters; *Philip Rees*, the editor of the book series in which this volume is published, for his valuable comments on each chapter of the draft manuscript: and his encouragement throughout the duration of the project; and each of the *authors* for their willingness to deliver chapters and respond to requests for revision.

Contents

Contents

Contents

Contributors

Dimitris Ballas, Department of Geography, University of the Aegean, Lesvos Campus, University Hill, Mytilene 81100, Greece. Email: d.ballas@aegean.gr

Dimitris Ballas is Associate Professor in the Department of Geography at the University of the Aegean. He has previously worked as a Senior Lecturer at the University of Sheffield and has held visiting positions at Harvard University (USA), the International Institute for Applied Systems Analysis (Austria) and Ritsumeikan University (Japan).

Mark Birkin, Consumer Data Research Centre, School of Geography, University of Leeds, Leeds, LS2 9JT, United Kingdom. Email: m.h.birkin@leeds.ac.uk

Mark Birkin is Professor of Spatial Analysis and Policy and the Director of the Consumer Data Research Centre (CDRC). His major interests are in simulating social and demographic change within cities and regions, and in understanding the impact of these changes on the need for services like housing, roads and hospitals, using techniques of microsimulation, agent-based modelling and GIS.

Tom Birkin, 154 Caversham Road, Reading, RG1 8AZ, United Kingdom. Email: tom.birkin@live.co.uk

Tom Birkin is a Leeds native who graduated with first class honours in a Master of Mathematics degree from the University of Manchester in July 2015. He spent six months as a data science intern in the Consumer Data Research Centre at Leeds University and now works in Reading as a trainee actuarial consultant.

Lisa Buckner, School of Sociology and Social Policy, University of Leeds, Leeds, LS2 9JT, United Kingdom. Email: l.j.buckner@leeds.ac.uk

Lisa Buckner is Associate Professor of Social Statistics in the School of Sociology and Social Policy at the University of Leeds. Her research interests include the analysis of large and complex datasets, and the use of official statistics and surveys for policy-applied research.

Gemma Catney, Department of Geography and Planning, School of Environmental Sciences, Roxby Building, University of Liverpool, Liverpool, L69 7ZT, United Kingdom. Email: g.catney@liverpool.ac.uk

Gemma Catney is Lecturer in the Department of Geography and Planning at the University of Liverpool. She is a population geographer with research interests in internal migration, and ethnic residential segregation and mixing in Britain. Her research focuses on understanding the evolution of ethnically diverse neighbourhoods and on the existence, and persistence over time, of ethnic inequalities.

Tony Champion, Centre for Urban and Regional Development Studies (CURDS), Newcastle University, Claremont Bridge, Newcastle upon Tyne, NE1 7RU, United Kingdom. Email: tony.champion@ncl.ac.uk
Tony Champion, Emeritus Professor of Population Geography, is based in CURDS at Newcastle University. His research interests include change in population distribution and composition, with particular reference to counterurbanisation and population deconcentration in developed countries and the policy implications of changes in local population profiles.

Stephane Chatagnier, School of Geography, University of Leeds, Leeds, LS2 9JT, United Kingdom. Email: gy13sjhc@leeds.ac.uk
Stephane Chatagnier is a part-time PhD student at the University of Leeds researching how indicators of internal migration in the UK are influenced by scale and zonation effects. He is using the *IMAGE Studio* software to compare migration intensities and patterns using data from the 2001 and 2011 Censuses.

James Cheshire, Department of Geography, University College London, Gower Street, London, WC1E 6BT, United Kingdom. Email: james.cheshire@ucl.ac.uk
James Cheshire has a BSc in Physical Geography from the University of Southampton and a PhD from UCL on spatial analysis and visualisation of large surname databases. In 2014, he became a Lecturer in Quantitative Human Geography at UCL. A wide range of his maps and visualisations have been featured in the popular print press as well as online.

Stephen Clark, School of Geography, University of Leeds, Leeds, LS2 9JT, United Kingdom. Email: s.d.clark@leeds.ac.uk
Stephen Clark has recently completed his PhD on modelling local health care demand at the School of Geography, University of Leeds. He is currently working on projects to update local authority ethnic population projections in light of subsequent census releases and to forecast long-term water demand.

Garnett Compton, Office for National Statistics, Segensworth Road, Titchfield, Fareham, Hampshire, PO15 5RR, United Kingdom. Email: garnett.compton@ons.gsi.gov.uk
Garnett Compton was the Head of 2011 Census Quality and Outputs at the Office for National Statistics. He led the overall statistical design work for the 2011 Census, in particular the development of the methods and processes to quality assure the census results.

Mike Coombes, Centre for Urban and Regional Development Studies (CURDS), Newcastle University, Claremont Bridge, Newcastle upon Tyne, NE1 7RU, United Kingdom. Email: mike.coombes@ncl.ac.uk
Mike Coombes is Emeritus Professor of Geographic Information in CURDS. His analyses of social and economic data aim to map how different people in different areas live their lives. In particular, the analyses of commuting and migration flows that reveal the boundaries of labour and housing market areas include four decades of census analyses for official statisticians that define Travel To Work Areas.

James Crone, EDINA, Information Services, Argyle House, 3 Lady Lawson Street, Edinburgh, EH3 9DR, United Kingdom. Email: james.crone@ed.ac.uk
James Crone has over ten years' experience in delivering census outputs to a wide user base by supporting national services providing online access to census products. He has significant geospatial

data management, analysis and processing experience which supports and enhances the delivery of the UK national census services. He provides technical GIS support and is the principal user support contact for census geography, producing promotional and help/training materials.

Andreas Culora, Department of Geography, 1st Floor Martin Hall Building, Loughborough University, Leicestershire, LE11 3TU, United Kingdom. Email: a.culora@lboro.ac.uk
Andreas Culora is a doctoral student at Loughborough University. His research interests include population changes tied to the production of housing of multiple occupation away from university and coastal towns. He is examining the ways in which internal migration and the residential decision making of people living in multi-person households in 'off-the-beaten-track' places give rise to childless populations and homogenous communities.

Fran Darlington-Pollock, Department of Geography and Planning, University of Liverpool, Liverpool, L69 7ZT, United Kingdom. Email: f.darlington-pollock@liverpool.ac.uk
Fran Darlington-Pollock is a Lecturer at the University of Liverpool. She is a population and health geographer, with research interests in health, migration, and ethnic inequalities. Her research has looked at the nature of ethnic inequalities in health and explanations for changing health gradients rooted in migration, deprivation change and social mobility. Fran has also spent time as a visiting researcher at the University of Auckland, exploring the influence of migration and deprivation change on health and mortality differences between ethnic groups in New Zealand.

Chris Dibben, Geography Building, School of Geosciences, University of Edinburgh, Drummond Street, Edinburgh, EH8 9XP, Scotland, United Kingdom. Email: chris.dibben@ed.ac.uk
Chris Dibben is Director of the Longitudinal Studies Centre Scotland, the Scottish Longitudinal Study and the Administrative Data Research Centre – Scotland. He has worked on, amongst other subjects, epidemiological studies into recovery after heart attacks, the causes of low birth weight, the survival of drug misusers and the impact of air pollution. He also contributed to work on the UK NHS health funding formula and on measuring health inequalities, for example developing the Health Poverty Index for the Department of Health and Information Centre (NHS).

Danny Dorling, School of Geography and the Environment, University of Oxford, South Parks Road, Oxford, OX1 3QY, United Kingdom. Email: danny.dorling@ouce.ox.ac.uk
Danny Dorling is Halford Mackinder Chair of Geography at Oxford. He grew up in Oxford and went to university in Newcastle upon Tyne. He has worked in Newcastle, Bristol, Leeds, Sheffield and New Zealand. His work concerns issues of housing, health, employment, education, inequality and poverty. His recent books include *Population Ten Billion* (2013), *All That Is Solid* (2014), and *Injustice* (second edition 2015).

Oliver Duke-Williams, Department of Information Studies, Foster Court, University College London, London, WC1E 6BT, United Kingdom. Email: o.duke-williams@ucl.ac.uk
Oliver Duke-Williams is a Senior Lecturer in Digital Information Studies in the Centre for Digital Humanities at University College London. He is a co-applicant on the ESRC UK Data Service-Census Support project and has worked on promoting the use of census flow data and the development of querying systems for flow data for many years. He is also Co-Director of the CALLS Hub, which integrates knowledge, access and practice across the three census longitudinal studies in the UK, and is a Co-Investigator in the Centre for Longitudinal Study Information and User Support (CeLSIUS), providing access to and support for users of the Office for National Statistics (ONS) Longitudinal Study of England and Wales.

Rob Dymond-Green, Jisc, 6th Floor, Churchgate House, 56, Oxford Street, Manchester, M1 6EU, United Kingdom. Email: rob.dymond-green@jisc.ac.uk
Rob Dymond-Green is the technical co-ordinator of the UK Data Service-Census Support team and is the technical architect and lead developer for the *InFuse* service, which delivers access to census aggregate data. Part of his role involves engaging with the technical teams of the ONS and other national statistics institutions.

Meghan Elkin, Office for National Statistics, Segensworth Road, Titchfield, Fareham, Hampshire, PO15 5RR, United Kingdom. Email: meghan.elkin@ons.gsi.gov.uk
Meghan Elkin is working with the Admin Data Census Project in the Census Transformation Programme at the ONS. She has worked on assessing the quality of administrative sources and identifying the requirements for administrative data to meet user information needs. She is now leading on publishing the Admin Data Research Outputs.

Maria Evandrou, EPSRC Care Life Cycle, ESRC Centre for Population Change, and Centre for Research on Ageing, Social Sciences, University of Southampton, SO17 1BJ, United Kingdom. Email: maria.evandrou@soton.ac.uk
Maria Evandrou is Professor of Gerontology, Co-Director of the ESRC Centre for Population Change and Director of the Centre for Research on Ageing at the University of Southampton. Her research interests span three distinct but related areas of investigation: inequalities in later life; informal carers and employment; and the retirement prospects of future generations of elders.

Jane Falkingham, EPSRC Care Life Cycle and ESRC Centre for Population Change, Social Sciences, University of Southampton, SO17 1BJ, United Kingdom. Email: j.c.falkingham@soton.ac.uk
Jane Falkingham is Professor of Demography and International Social Policy, and Director of the ESRC Centre for Population Change at the University of Southampton. A consistent theme throughout her research is demographic change and its consequences for the distribution of social and economic welfare. Particular areas of work include ageing and intergenerational relations across the lifecourse, changing needs for social security across the lifecourse, and informal/formal care provision between different generations.

Nissa Finney, School of Geography and Sustainable Development, Irvine Building, University of St Andrews, North Street, St Andrews, KY16 9AL, Fife, Scotland, United Kingdom. Email: nissa.finney@st-andrews.ac.uk
Nissa Finney is Reader in Human Geography at the University of St Andrews, member of the ESRC Centre on Dynamics of Ethnicity (CoDE) and member of the ESRC Centre for Population Change (CPC). Her research focuses on internal migration, neighbourhood and ethnic inequalities; her current interests are in lifecourse approaches to migration and the relation between inequalities in housing and internal migration.

Orlaith Fraser, Office for National Statistics, Segensworth Road, Titchfield, Fareham, Hampshire, PO15 5RR, United Kingdom. Email: orlaith.fraser@ons.gsi.gov.uk
Orlaith Fraser is a Senior Research Officer at the Office for National Statistics and has spent the last two years producing specialist products from the 2011 Census, with a particular focus on microdata. Currently, she is working on research for the design of the 2021 Census.

Brian French, Census Office, Northern Ireland Statistical and Research Agency (NISRA), McAuley House, 2–14 Castle Street, Belfast, BT1 1SA, United Kingdom. Email: brian.french@dfpni.gov.uk
Brian French joined the Census Office of the Northern Ireland Statistics and Research Agency in September 2012 and oversaw the specification, production, analysis, dissemination and benefits realisation measurement of the 2011 Census outputs for Northern Ireland.

Robert Fry, Office for National Statistics, Segensworth Road, Titchfield, Fareham, Hampshire, PO15 5RR, United Kingdom. Email: robert.fry@ons.gsi.gov.uk
Robert Fry is Head of Data Visualisation at the Office for National Statistics. He has a degree in Maths and an MSc in Official Statistics. Prior to his current role exploring interactive graphics, he spent several years as a statistician at ONS working on a variety of official outputs.

Chris Gale, Social Sciences, University of Southampton, Southampton, SO17 1BJ, United Kingdom. Email: c.gale@soton.ac.uk
Chris Gale is a Research Fellow at the University of Southampton working as part of the Administrative Data Research Centre for England. Previously he was a Research Associate at University College London, where he was also awarded a PhD in 2014.

Myles Gould, School of Geography, University of Leeds, Leeds, LS2 9JT, United Kingdom. Email: m.i.gould@leeds.ac.uk
Myles Gould is a Lecturer in Human Geography at the University of Leeds. His research interests include analysis of complex data, multi-level modelling, residential movement and health geography.

Justin Hayes, Jisc, 6th Floor, Churchgate House, 56, Oxford Street, Manchester, M1 6EU, United Kingdom. Email: justin.hayes@jisc.ac.uk
Justin Hayes co-manages a UK Data Service-Census Support team at Jisc, which develops tools to provide quick, easy, comprehensive and integrated online delivery of aggregate statistics from the five most recent UK censuses conducted between 1971 and 2011. He has a background in GIS and has worked with UK census aggregate statistics as both a user and supplier for almost 20 years.

Spencer Hedger, *Nomis*, Department of Geography, Durham University, Lower Mountjoy, South Road, Durham, DH1 3LE, United Kingdom. Email: spencer.hedger@durham.ac.uk
Spencer Hedger is a data scientist with a BSc degree in Computing for Business. His experience spans over 16 years, having worked on a variety of software development projects for the *Nomis* website on behalf of the UK ONS. He has worked on billing systems, the development of complex web applications and APIs, and has been a key contributor to the ESRC-funded PARLER project.

Alison Heppenstall, School of Geography, University of Leeds, Leeds, LS2 9JT, United Kingdom. Email: a.j.heppenstall@leeds.ac.uk
Alison Heppenstall has research interests focused on the development of artificial intelligence methods (especially agent-based modelling) for understanding processes and evolution of geographical systems. She has a particular interest in methods for incorporating human behaviour in individual-level models and developing approaches to improve synthetic population generation.

Christopher D. Lloyd, Department of Geography and Planning, School of Environmental Sciences, University of Liverpool, Roxby Building, Chatham Street, Liverpool, L69 7ZT, United Kingdom. Email: c.d.lloyd@liverpool.ac.uk

Christopher D. Lloyd, Professor of Quantitative Geography at the University of Liverpool, has a particular interest in geographic inequalities and residential segregation and has published extensively on spatial analysis and population studies.

Nik Lomax, School of Geography, University of Leeds, Leeds, LS2 9JT, United Kingdom. Email: n.m.lomax@leeds.ac.uk
Nik Lomax is a University Academic Fellow in Spatial Data Analytics at the University of Leeds, where he obtained his PhD. He has expertise in collating and analysing large migration datasets and is a Co-Investigator on the ESRC-funded NewETHPOP project, tasked with projecting ethnic group populations for UK local authorities. He provides advice to national statistical agencies, to local government and to business on methods for producing population estimates and projections, and on the impact of population change.

Paul Longley, Room PB115, Department of Geography, University College London, Gower Street, London, WC1E 6BY, United Kingdom. Email: p.longley@ucl.ac.uk
Paul Longley is a Professor of Geographic Information Science at University College London, where he was appointed in 2000. He is a co-author of *Geographic Information Science and Systems* (fourth edition, 2015) and over 150 other refereed publications.

Robin Lovelace, Consumer Data Research Centre, School of Geography, University of Leeds, Leeds, LS2 9JT, United Kingdom. Email: r.lovelace@leeds.ac.uk
Robin Lovelace is a researcher at the Leeds Institute for Transport Studies (ITS) and the Consumer Data Research Centre (CDRC). He has wide ranging experience researching the energy impacts of transport systems, modelling sustainable transport uptake and visualizing transport futures. These skills have been applied on a number of projects with real-world applications, most recently the Propensity to Cycle Tool (www.pct.bike), a nationally scalable interactive online mapping application, and the stplanr package for sustainable transport planning.

Kitty Lymperopoulou, Cathie Marsh Institute for Social Research, University of Manchester, Humanities Bridgeford Street, Oxford Road, Manchester, M13 9PL, United Kingdom. Email: kitty.lymperopoulou@manchester.ac.uk
Kitty Lymperopoulou is a Research Associate at the Centre on Dynamics of Ethnicity (CoDE) at the University of Manchester. Her research interests are in new immigration, ethnic inequalities and neighbourhood inequalities. Her research has examined immigrant settlement patterns and neighbourhood experiences of immigration, spatial patterns of ethnic inequalities, and ethnic inequalities in education and employment.

David Martin, Geography and Environment, University of Southampton, Southampton, SO17 1BJ, United Kingdom. Email: d.j.martin@soton.ac.uk
David Martin is Professor of Geography at the University of Southampton and was a member of the 2011 Census High-Level Quality Assurance Panel. He is Deputy Director of the ESRC's UK Data Service and Administrative Data Research Centre for England and a Co-Director of the National Centre for Research Methods. He has worked extensively with the ONS and devised the system of output areas employed in the 2001 and 2011 Censuses.

Victoria Moody, Jisc, 6th floor, Churchgate House, 56, Oxford Street, Manchester, M1 6EU, United Kingdom. Email: victoria.moody@jisc.ac.uk

Victoria Moody leads the impact and communications function of the UK Data Service and supports the service in engaging with data users, data owners and other stakeholders internationally, developing opportunities for enhancing the impact of the UK Data Service.

Michelle Morris, Leeds Institute for Data Analytics (LIDA), School of Medicine, University of Leeds, LS2 9JT, United Kingdom. Email: m.morris@leeds.ac.uk
Michelle Morris is a University Academic Fellow in LIDA at the University of Leeds. Her primary research interests are in spatial and social variations in diet and health. An interdisciplinary researcher with a background spanning spatial analysis and policy, nutritional epidemiology and health economics, she currently uses quantitative methods to investigate geographies of consumption and how consumer behaviours are linked with health outcomes.

Thomas Murphy, Knight Frank, 55, Baker Street, London, W1U 8AN, United Kingdom. Email: thomas.murphy@knightfrank.com
Thomas Murphy is a quantitative analyst in the Global Residential Research Department of Knight Frank. His PhD from the University of Leeds in 2016 was focused on the analysis of UK census data on commuting to work and the changes taking place between 2001 and 2011.

James Nicholson, SMART Centre, School of Education, Durham University, Leazes Road, Durham, DH1 1TA, United Kingdom. Email: j.r.nicholson@durham.ac.uk
James Nicholson is Principal Research Fellow, Durham University. Administrative roles have included Chair of the British Congress of Mathematical Education; Council Member for the Institute for Mathematics and its Applications; Co-ordinator of the Royal Statistical Society schools workshop programme; Vice President of the International Association for Statistics Education; Head of Mathematics at the Belfast Royal Academy; consultant to NFER; and Ofqual subject expert.

Paul Norman, School of Geography, University of Leeds, LS2 9JT, United Kingdom. Email: p.d.norman@leeds.ac.uk
Paul Norman is a population and health geographer with particular expertise in time-series and longitudinal analyses of both area- and individual-level data derived from census, survey and administrative records. Research includes studies on health-selective migration and small area-based studies of UK demographic change.

Oliver O'Brien, Department of Geography, University College London, Gower Street, London, WC1E 6BT, United Kingdom. Email: o.obrien@ucl.ac.uk
Oliver O'Brien is a researcher and software developer at the UCL Department of Geography, where he investigates and implements new ways to visualise spatial data, particularly mapping of open demographic and socio-economic datasets, especially London-focused ones, using *OpenLayers*.

Lara Phelan, Office for National Statistics, Segensworth Road, Titchfield, PO15 5RR, United Kingdom. Email: lara.phelan@ons.gov.uk
Lara Phelan is Head of Stakeholder Engagement & Communications for the Census Transformation Programme at the ONS. As Head of the Census Benefits Realisation Team, she developed the overarching approach on how to realise and measure the benefits generated from the 2011 Census in England and Wales.

Philip Rees, School of Geography, University of Leeds, LS2 9JT, United Kingdom. Email: p.h.rees@leeds.ac.uk
Philip Rees is Emeritus Professor of Population Geography at the University of Leeds. His research interests focus on population forecasting, for which the census provides essential inputs. In 2016 he completed with colleagues a new set of projections of UK local authority ethnic populations to 2061. In 2017 these projections have been extended to 2101 as an input to long-term forecasts of domestic water consumption for Thames Water.

James Reid, EDINA, Information Services, Argyle House, 3 Lady Lawson Street, Edinburgh, EH3 9DR, Scotland, United Kingdom. Email: james.reid@ed.ac.uk
James Reid has over 20 years' experience in the geographical information industry, serving in both public sector and tertiary education sectors. He is currently a Co-Investigator for the UK Data Service-Census Support and for the Administrative Data Centre Scotland. He is also Council Member for the Association for Geographic Information Scotland and represents the UK on the European INSPIRE Metadata Group.

Jim Ridgway, SMART Centre, School of Education, Durham University, Leazes Road, Durham, DH1 1TA, United Kingdom. Email: jim.ridgway@durham.ac.uk
Jim Ridgway directs the SMART Centre. His professional roles have included membership of: the Royal Statistical Society Centre for Statistics Education Technical Advisory Group; Advisory Board on National Test Development for the Swedish National Agency; OECD Expert Panel for the International Assessment of Adult Competencies; Executive Committee of the European Association for Research on Learning and Instruction; and the ESRC Research Priorities Board. He is Vice President of the International Association for Statistics Education.

James Robards, ESRC National Centre for Research Methods (NCRM), Social Sciences, University of Southampton, SO17 1BJ, United Kingdom. Email: james.robards@soton.ac.uk
James Robards is a Research Fellow in the NCRM and Division of Social Statistics and Demography at the University of Southampton. He is currently researching automated zone design techniques for linked data and previously held a post within the EPSRC Care Life Cycle where he was engaged with research on the drivers of health and social care including research on transitions to different forms of caring arrangements and subsequent mortality in later life, using the ONS Longitudinal Study.

Vassilis Routsis, Department of Information Studies, University College London, Gower Street, London, WC1E 6BT, United Kingdom. Email: v.routsis@ucl.ac.uk
Vassilis Routsis is a Research Associate for the UK Data Service-Census Support where he collaborates with the ONS in the dissemination of 2011 Census interaction datasets via the *WICID* software system. He is also a PhD student at the UCL Centre for Digital Humanities, exploring modern-day questions revolving around the societal impact of technology in the areas of privacy and data protection.

Nikola Sander, Population Research Centre, University of Groningen, Landleven 1, 9747 AD Groningen, The Netherlands. Email: n.d.sander@rug.nl
Nikola Sander is an Assistant Professor in the Population Research Centre at the Faculty of Spatial Sciences, University of Groningen. Her research focuses on internal and international migration flows, urbanisation, population projections and interactive data visualisations.

Nicola Shelton, Health and Social Surveys, Department of Epidemiology and Public Health, University College London, 1–19 Torrington Place, London, WC1E 6BT, United Kingdom. Email: n.shelton@ucl.ac.uk
Nicola Shelton is a Senior Lecturer in the Department of Epidemiology and Public Health at UCL, and is Director of the Centre for Longitudinal Study Information and User Support (CeL-SIUS). Her research interests are on health and population geography, health surveillance and the outcomes that can be measured through large and complex health datasets.

Ian Shuttleworth, School of Geography, Archaeology and Palaeoecology, Queen's University Belfast, Belfast, BT7 1NN, United Kingdom. Email: i.shuttleworth@qub.ac.uk
Ian Shuttleworth is a Senior Lecturer in the School of Geography, Archaeology and Palaeoecology at Queen's University Belfast and Director of the Northern Ireland Longitudinal Study (NILS). His research interests lie in applied social and political geography and include census analysis, religious and national identities, internal migration within Northern Ireland and its implications for residential segregation, labour markets and labour mobility.

Ludi Simpson, Cathie Marsh Institute for Social Research, Humanities Bridgeford Street, University of Manchester, Manchester, M13 9PL, United Kingdom. Email: ludi.simpson@ manchester.ac.uk
Ludi Simpson works with population, census and survey statistics, aiming to extend their use by communities and governments. His current research focuses on local population dynamics within the UK and Latin America.

Alex Singleton, Department of Geography and Planning, University of Liverpool, Roxby Building, Liverpool, L69 7ZT, United Kingdom. Email: alex.singleton@liverpool.ac.uk
Alex Singleton is a Professor of Geographic Information Science at the University of Liverpool, where he was appointed as a Lecturer in 2010. Previously he held research positions at UCL, where he was also awarded a PhD in 2007.

Alan Smith, Financial Times, One Southwark Bridge Road, London SE1 9HL, United Kingdom. Email: alan.smith@ft.com
Alan Smith is Data Visualisation Editor at the Financial Times. Previously, he was Head of Digital Content at the ONS, where he established a dedicated centre for data visualisation. He was appointed OBE in the 2011 Queen's Birthday Honours List for services to official statistics.

Darren Smith, Martin Hall, Geography Department, Loughborough University, Loughborough, LE11 3TU, United Kingdom. Email: d.p.smith@lboro.ac.uk
Darren Smith is Professor of Geography at Loughborough University, UK. His research interests include: sub-national family migration, population change and housing markets, rural gentrification, and studentification. He is currently Co-editor of *Population Space and Place*, Chair of the Organising Committee of the International Conference on Population Geographies (ICPG) and Steering Group Member of the IGU Population Commission.

John Stillwell, School of Geography, University of Leeds, Leeds, LS2 9JT, United Kingdom. Email: j.c.stillwell@leeds.ac.uk
John Stillwell is Professor of Migration and Regional Development in the School of Geography at the University of Leeds. He is Director of the UK Data Service-Census Support project and

has worked on the delivery of census data to users and their application in a research context throughout his academic career.

Sinclair Sutherland, Nomis, Department of Geography, Durham University, Lower Mountjoy, South Road, Durham, DH1 3LE United Kingdom. Email: sinclair.sutherland@ durham.ac.uk
Sinclair Sutherland is Director of *Nomis*, the UK Office for National Statistics website for labour market data, and Deputy Director of the Centre for Health and Inequalities Research. He has extensive experience of developing web-based data dissemination and visualisation tools having worked in partnership with ONS for over 20 years to provide online access to their statistics.

Andy Teague, Office for National Statistics, Segensworth Road, Titchfield, Fareham, Hampshire, PO15 5RR, United Kingdom. Email: andy.teague@ons.gsi.gov.uk
Andy Teague is leading the Beyond 2021 Project in the Census Transformation Programme at the ONS in the UK on continued research into the potential of using administrative data and surveys for replacing the ten-yearly census beyond 2021. Andy led the statistical research into administrative data in the previous Beyond 2011 Programme, culminating in a suite of research evidence in some 30 published research papers.

Athina Vlachantoni, EPSRC Care Life Cycle, ESRC Centre for Population Change, and Centre for Research on Ageing, Social Sciences, University of Southampton, SO17 1BJ, United Kingdom. Email: a.vlachantoni@soton.ac.uk
Athina Vlachantoni is Associate Professor in Gerontology at the Centre for Research on Ageing and the ESRC Centre for Population Change, University of Southampton. Her research broadly combines the areas of ageing and social policy, incorporating a lifecourse perspective, for example changing need for and provision of informal and formal care, and pension protection among individuals from ethnic minorities in mid and late life.

Nigel Walford, Department of Geography and Geology, School of Natural and Built Environments, Kingston University, Penrhyn Road, Kingston upon Thames, KT1 2EE, United Kingdom. Email: n.walford@kingston.ac.uk
Nigel Walford, Professor of Applied GIS at Kingston University, undertakes research involving the application of GIS to the mapping and analysis of geodemographic and agri-environmental information. ESRC funding has allowed him to create retrospectively digital boundary data for the enumeration districts used in the 1981 and 1971 Censuses and to apply spatial analytic tools to estimate counts of census attributes for consistent spatial units spanning the 1971 to 2001 enumerations.

Paul Waruszynski, Office for National Statistics, Segensworth Road, Titchfield, Fareham, Hampshire, PO15 5RR, United Kingdom. Email: paul.waruszynski@ons.gsi.gov.uk
Paul Waruszynski is a Senior Research Officer at the ONS. He has worked on the 2011 Census since starting at ONS in 2008, which has included design and development of the questionnaire, implementing the processing of the data and designing and managing the outputs.

Jo Wathan, Cathie Marsh Institute for Social Research, School of Social Sciences, Humanities Bridgeford Street, University of Manchester, Manchester, M13 9PL, United Kingdom. Email: jo.wathan@manchester.ac.uk
Jo Wathan is a Research Fellow at the Cathie Marsh Institute where she started as a PhD student. Having been an early adopter of census microdata in 1994, she has worked in data support and

data enhancement roles relating to census microdata since 1999, including undertaking user consultations, user support, data recovery and training.

Amy Wilson, Ladywell House, Ladywell Road, Edinburgh, EH12 7TF, Scotland, United Kingdom. Email: amy.wilson@nrscotland.gov.uk
Amy Wilson was the Head of 2011 Census Statistics at the National Records of Scotland from May 2012 to April 2014. As the lead statistician on the programme, she was responsible for the processing of the data and the production of the 2011 Census outputs.

Pia Wohland, Supportive Care, Early Diagnosis and Advanced Disease Research Group, Centre for Health and Population Sciences, The Hull York Medical School, University of Hull, Hertford Building, Cottingham Road, Hull, HU6 7RX, United Kingdom. Email: pia.wohland@hyms.ac.uk
Pia Wohland is a Lecturer in Health Inequalities at the Hull York Medical School. She is a quantitative health geographer with expertise in: health inequalities – across geographical areas, time, age and population sub-groups; population projections and future composition of the UK population; and population ageing. She has worked on a wide range of projects, including research in ethnic mortality in the UK and health differences for local areas in the UK and across European countries.

Figures

Figures

Tables

Tables

Part I
Introducing the census

.

1

The 2011 Census in the United Kingdom

John Stillwell

1.1 Introduction

Most countries around the world recognise that, in order to provide services, administer effectively and guide development, it is essential to collect reliable and comprehensive data on the magnitude, composition and distribution of their populations. The primary source of basic demographic statistics is a population census which involves "the total process of collecting, compiling, evaluating, analysing and publishing or otherwise disseminating demographic, economic and social data pertaining, at a specified time, to all persons in a country or in a well-delimited part of the country" (United Nations Statistics Division (UNSD), 1997, p. 3). The socio-demographic data that are collected by this process not only support needs assessment, policy formulation and strategic planning, but they also underpin a wide range of research studies and are often used in combination with data from other sources or as a benchmark, a gold standard, against which data from elsewhere may be compared.

On 27 March 2011, censuses were taken by each of the national statistical offices (NSOs) in the United Kingdom (UK) – the Office for National Statistics (ONS) in England and Wales, National Records of Scotland (NRS) and the Northern Ireland Statistics and Research Agency (NISRA) – continuing a tradition that began in 1801 in England, Wales and Scotland and in 1821 in Northern Ireland. In a 'traditional' census, questionnaire forms are delivered to and collected from every household by designated enumerators, but the UK is one of a number of countries whose census forms in 2011 were posted out to 25 million addresses with the option of respondents being able to complete and submit their answers to the census questionnaire online or by post and with the additional phase of identification and follow-up of non-respondents.

This introductory chapter of the handbook has three aims. First, in Section 1.2, it provides some global context for the UK Census held in 2011 by reviewing the ways in which populations have been counted in countries across the world in what the UNSD (2013) refers to as the 2010 Census round. Second, given the evidence that many countries, particularly in Europe, have decided to replace traditional census collection methods with approaches that collect demographic data from administrative registers and sample surveys (Valente, 2010), Section 1.3 considers the arguments for and against the continuation of the census in the UK in its current form. This debate, leading to the National Statistician's recommendation in 2014 for a census in 2021, has underpinned ONS' Beyond 2011 programme[1] and the subsequent Census Transformation Programme.[2]

Due to the volume and detail of the data collected for processing, various statistical products have been released by the NSOs in stages commencing in July 2012, with ONS producing all statistics for the UK. In Section 1.4, the different channels of access to these statistics are summarised, and the role and structure of the UK Data Service-Census Support in providing a 'one stop shop' for users to access and download aggregate statistics, origin–destination flows, boundary data and cross-sectional microdata from the 2011 Census is explained. In the final section of the chapter, the rationale for the structure and content of the chapters that follow is outlined.

1.2   Counting populations around the world

The 2010 World Population and Housing Census Programme[3] of the UNSD reports that 214 out of 235 countries (or areas) conducted some form of population and housing census at least once during the 2010 Census round (between 2005 and 2014) where the term 'census' is used in its broadest sense to include traditional censuses, population registers, the use of administrative records, sample surveys and data from other surveys. The UK is one of 60 countries whose census was taken in 2011, in close temporal proximity to population censuses carried out in Portugal (21 March), the Czech Republic (25 March), Curaçao (26 March), Croatia and Poland (31 March). Subsequently, the UNSD (2013) reported the results of two surveys, the first based on a sample of 138 responding countries on how they have implemented their censuses, and a second of 126 countries on what lessons can be learned to inform the formulation of the *Third Revision of The Principles and Recommendations for Population and Housing Censuses* (UNSD, 2014). The results of the first survey (UNSD, 2011), summarised in Table 1.1, suggest that whilst 83 per cent of the 138 countries responding carried out a traditional census with full enumeration as their main methodology, many countries had developed alternative methods when compared to previous rounds.

All the countries in Africa, Oceania, Latin America and the Caribbean responding to the survey took a traditional census whilst all but four in Asia did so; Bahrain and Singapore used a register-based census, whilst Turkey and Israel, together with five European countries (Estonia, Italy, Latvia, Lithuania and Poland), used a combination of data sources. France is the only country in the world which counts its population using a rolling census, whereas nearly one-third of the responding countries in Europe now use administrative registers. The results of the UNSD survey align with those obtained from other studies. An inventory of internal migration data collections

Table 1.1 Main census methodology for the 2010 Census round, by geographical region

World region	Countries responding	Traditional census	Administrative register(s)	Rolling census	Other*
Africa	29	29			
Asia	39	35	2		2
Europe	39	21	12	1	5
North America	4	3	1		
Oceania	5	5			
Latin America and Caribbean	22	22			
Total	138	115	15	1	7

* Combination of data sources including registers with full or partial enumeration and/or sample surveys.

Source: UNSD (2011, Table 2.1, p. 10).

among the 193 UN member states built as part of the IMAGE project (Bell et al., 2015) indicates that 88 per cent of the 179 countries collecting migration data did so using a census, although this study confirms an increasing number of countries, particularly in Europe, are using 'register-based censuses' or 'combined censuses' which link data from registers and surveys. Similarly, the Ethnicity Counts? project (Kukutai et al., 2015), investigating how different countries count their ethnic populations between 1984 and 2014, created a database which provides evidence of an absolute decline in countries undertaking a traditional census between the 1990 and 2010 rounds, the growing use of administrative data, and the wider use of survey data.

In summary, whilst census-taking retains its importance in many countries, particularly in the less developed world, full enumeration using a traditional census is becoming much less popular in more developed countries, particularly in Europe where many countries use population registers as the source of their demographic statistics (Poulain and Herm, 2013) and where opportunities to generate population counts that combine data from different administrative sources using electronic data linkage techniques has gained ground in recent years. In view of these developments, Coleman (2013) refers to the 'twilight of the census' as many NSOs consider the future of their census-taking methodologies and explore alternative options that are available. In the UK, the debate on the future of the census is captured in the consultation and research activities involved in the ONS' Beyond 2011 which, since January 2015, has become the 'Census Transformation Programme'.

1.3   Census-taking in the UK: benefits and concerns

The question we address in this section is: what are the reasons that are driving the consideration of alternative approaches to a traditional census which has served the country pretty well for over 200 years and which generates a range of products that provide essential data for public administration, governance and research as well as for strategic planning by private enterprises, community groups and voluntary sector agencies?

In England and Wales, the aggregate data products derived from the 2011 Census and released subsequently by the ONS include: population and household estimates plus headcounts for post-codes; univariate Key Statistics (KS) presented in 35 tables and Quick Statistics (QS) which are more detailed data about a single topic available from 74 tables; Detailed Characteristics (DC) with more detailed multivariate data (218 tables) that are not available for the smallest areas (output areas) and Local Characteristics (LC) which are less detailed (184 tables) but are available for small areas; and populations with a base alternative to usual residence that include those with second addresses and short-term residents as well as workplace and workday populations. Whilst each of these products, after statistical disclosure controls have been applied, provides estimates of single area attributes, Origin–Destination Statistics (ODS), also known as flow data, include the counts of flows of migrants, those commuting to work, students and second home owners between two geographical areas. In addition, there are the cross-sectional samples of microdata derived from the full census and the longitudinal microdata that link a sample of individuals between censuses.

As well as the 100 per cent census aggregate and flow estimates and the various unadjusted sample census datasets, there are also data products that are derived from the primary data which are of value to different user communities. Among the examples of these products are the area classifications that are produced by the NSO, by researchers in collaboration with an NSO or by independent researchers in the public or private sectors. For example, the first 2011 Census geodemographic classification was produced for output areas (2011 Output Area Classification (OAC)) in partnership with University College London (Gale, 2014). This is a

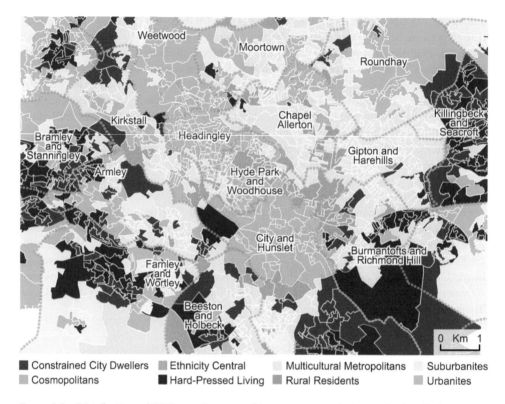

Figure 1.1 Distribution of 2011 geodemographic supergroups in inner city Leeds, by output area, 2011

three-tiered hierarchical geodemographic classification of the whole of the UK consisting of eight supergroups, 26 groups and 76 sub-groups.[4] Figure 1.1 illustrates the distribution of supergroups in the central and inner suburbs of Leeds, showing the 'Cosmopolitan' demographic of much of City and Hunslet, Hyde Park and Woodhouse and Headingley, all census wards with large student populations. To the east and south are found suburbs classified as 'Multicultural Metropolitan' (Chapel Allerton, Gipton and Harehills and Beeston and Holbeck), interwoven with areas classified as 'Hard-Pressed Living' (such as Burmantofts and Richmond Hill) and 'Constrained City Dwellers' (Armley) with increasing numbers of 'Suburbanites' and 'Urbanites' (e.g. Weetwood, Moortown, Roundhay) with distance from the city centre. More recently, a Classification Of Workplace Zones (WZs) for England and Wales (COWZ-EW) has been constructed at the University of Southampton in collaboration with ONS using k-means clustering based on 48 census variables selected for their ability to differentiate types of workers and workplaces to produce a two-tier WZ classification of seven supergroups and 29 groups[5] (Cockings et al., 2015).

The 2011 Census is therefore an unrivalled source of data that are exploited by researchers from a wide range of academic backgrounds. Geographers, in particular, benefit from the high level of coverage that makes it possible to analyse and understand the socio-demographic characteristics of populations in very small areas as shown in Figure 1.1. Output areas (OAs) were introduced across the UK in 2001 and are the lowest level of spatial unit for which census estimates are provided. In 2011, there were 171,372 OAs in England, 10,036 in Wales, 46,351 in Scotland

and 5,022 in Northern Ireland, although in the case of the latter, the OAs defined initially in 2001 were aggregated into 4,537 small areas (SAs) in 2011. The requirement to maintain consistency between 2001 and 2011 has meant that only 2.6 per cent of OAs in England and Wales, for example, have changed. Since these are the building blocks for higher-level geographies, this has facilitated spatial analysis of change between the two censuses. Comparisons of small areas between 2001 and 1991 were problematic because data in 1991 were released for enumeration districts, the census collection areas, with OAs being used for the first time in 2001.

In response to a public consultation organised by the ONS (2013) as part of the Beyond 2011 programme, the Royal Geographical Society with the Institute of British Geographers (2014) provided a submission consisting of a number of case studies that highlighted the value of small area census estimates and the importance of their flexibility in answering important societal questions. When detailed geography is less important, the cross-sectional and longitudinal microdata come into their own, providing users with detailed attributes for individuals or households that can be cross-classified according to the researcher's requirements, unconstrained by the limited combination of variables imposed on the aggregate and flow data to preserve confidentiality.

One of the underlying motivations for the ONS Beyond 2011 review was a concern over the rising costs of delivery aligned with tighter fiscal constraints, a view aired publicly by Francis Maude, Minister for the Cabinet Office in the Coalition Government, and reported in the popular press (Hope, 2010). Maude considered that the decennial census was an expensive and inaccurate method of measuring the population of the UK and revealed that the Government was looking at alternative and less expensive ways to count the population more regularly and make use of existing registers and administrative data sources. The debate on the future form of the census in the UK has been paralleled by similar debates reported by the press in other countries around the world including, for example, the USA (Singer, 2010), Canada (Underhay, 2011) and Australia (Hutchens and Martin, 2015).

The need to control cost is by no means the only concern of those who consider the days of the traditional census to be numbered. The list of concerns of those commenting on the census debate around the world includes issues of privacy, frequency, accuracy and apathy, as well as the pressure to make better use of information available elsewhere (Valente, 2010; Fienberg and Prewitt, 2010; Coleman, 2013).

In the UK, as in other countries like the USA and Canada, there is tension between the need for the Government, through its NSOs, to collect information on which to formulate policy and to base legislation and the fundamental human rights of individual privacy and freedom from discrimination. Fienberg and Prewitt (2010) indicate that privacy was a major issue in the 2010 Census in the USA, with questions relating to age, gender and ethnicity, in particular, considered by some to be too invasive. In preparing for the 2011 Census in the UK, the usual approval by Parliament was required for secondary legislation to enable the census to be taken and parliamentary debates in the run-up reflected the conflict between the collection of information necessary for running the country and the privacy burden it imposes on the public. Despite the fact that the Statistics and Registration Service Act (SRSA) 2007 gives responsibility to the UK Statistics Authority (UKSA) to maintain confidentiality of census data and prohibit disclosure of personal information, there is increasing distrust of politicians and public servants and increasing uncertainty about what the data collected by the census will be used for (Coleman, 2013). This is partly responsible for lower public co-operation and the growing reluctance of people to participate (Valente, 2010). Mistrust of government and concerns about privacy are by no means confined to census data; these issues are equally if not more relevant in the context of the growing use of data from administrative and register-based sources. Singer et al. (2011) suggest that negative public opinion has been one of the main reasons why the US Census Bureau has been loath to implement the use of administrative data on a large scale.

One of the major longstanding criticisms of the traditional census in the UK has been its relative infrequency and the time taken for the results to be processed and released. Having only a once-a-decade snapshot of the population is considered insufficient when social and demographic changes are occurring at a relatively rapid rate. The UK population, estimated to be 63.2 million at the time of the 2011 Census, increased by 4.1 million or nearly 7 per cent over the preceding decade, the largest decadal growth since censuses began. Population development in terms of both size and demographic structure in the 2000s has been driven, in particular, by the process of ageing and by the influx of unprecedented numbers of international immigrants. The significant increase in net migration to the UK during the 2000s coincided with the enlargement of the European Union (EU) and the opening up of opportunities for migrants from the Accession 8 (A8) countries to work in the UK. It is estimated by the Migration Observatory that the average annual net migration gain was nearly a quarter of a million migrants between 2004 and 2014, compared with an annual average of 65,000 between 1991 and 1999 (Vargas-Silva and Markaki, 2015). The perceived inability of local authorities and other agencies to cope with this influx has been reflected in the rising popularity of the UK Independence Party (UKIP), which campaigned successfully for UK withdrawal from the EU.

The mid-year population estimates are rebased on the decennial census and consequently, as each year passes following the year of the census, the margin of error is likely to increase and the need to use existing, up-to-date data from administrative sources becomes more paramount. The value of data from each census in this context therefore deteriorates over time up until the release of data from the next census. This cycle is illustrated in Figure 1.2, which has been adapted from the ONS Beyond 2011 Consultation of User Requirements document (ONS, 2011, p. 6) where it was used for illustrative purposes. The graph indicates how the net benefit from the census over alternative administrative sources will gradually reduce over time as the quality of data collection using the latter matures.

In response to a proposal in 2015 by the Australian Bureau of Statistics to hold a census every ten years rather than the traditional five years, one senior academic from the Queensland Centre for Population Research stated that "It's the one benchmark we have for population dynamics

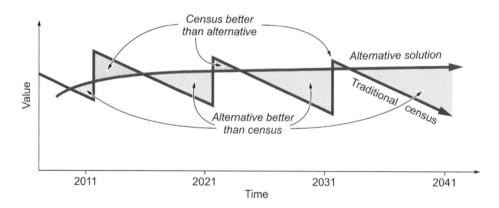

Figure 1.2 A comparison of the value of traditional and alternative (administrative) census-taking over time (adapted from ONS, 2011, p. 6)

in Australia, and 10 years is way too long to wait for a benchmark to enable one to put together effective, accurate information" (Martin Bell, quoted in *Sydney Morning Herald*, 2015). However, since there is no data source to compare with the census in the UK when it comes to rich socio-demographic data for small areas, its value tends to be preserved from one census to the next, at least as far as researchers are concerned. This is partly because of the time taken to release particular datasets and partly due to the access conditions surrounding their release. Thus, for example, different sets of 2011 origin–destination statistics classified as either open, safeguarded or secure have only become available in 2015 and will only be fully exploited in the second half of the inter-censal period.

Whilst the time delay between census data collection and release is a justifiable concern, especially for the policy community, there is the additional problem that the data are increasingly inappropriate for addressing the problems of contemporary society. Social and economic change makes conventional enumeration through a census more and more difficult because of the complexity of household arrangements and social circumstances in an increasingly mobile world (Sheller and Urry, 2006). The concept of a single place of usual residence is becoming increasingly less meaningful for a range of different groups including the children of divorced parents, for example, and the classifications used to categorise variables such as ethnicity and occupation are increasingly inappropriate.

Various negative reasons are therefore apparent to suggest that change in the traditional method of population counting is required but there is also the pressure to make much greater use of the information that is already available elsewhere in digital form. Various consultation exercises have been undertaken as part of the Beyond 2011 programme with eight alternative methods proposed initially for investigation and summarised as statistical options in the appendix of the ONS consultation document (ONS, 2011). ONS had eliminated six options and retained two options by September 2013: (i) an online census supplemented by the use of administrative data to take account of annual population change during the inter-censal period; and (ii) the use of administrative data already held by government departments to estimate mid-year populations, supplemented by a rolling annual survey to help estimate the socio-demographic characteristics of the population. ONS commissioned an independent report (Skinner et al., 2013) which supported the online census option subject to the inclusion of full follow-up for non-response, but which was less positive about the administrative data option, suggesting that more evidence (and research) was needed to support this alternative.

The National Statistician considered the findings of the Skinner Review alongside other evidence, including responses to the ONS consultation and research on public acceptability, and recommended a modernised census for England and Wales in 2021 and the "increased use of administrative data and surveys in order to enhance the statistics from the 2021 Census and improve annual statistics between censuses" (ONS, 2014, p. 2). The value and benefits of the 2011 Census and the options for collecting population data in the future were also considered by a House of Commons Public Administration Select Committee (PASC) whose report and formal minutes endorsed the recommendations of the Skinner Review and stipulated that the ONS "now scope and set out a more ambitious vision for the creative and full use of administrative data to provide rich and valuable population statistics" (PASC, 2014, p. 20). A statement of agreement between the National Statistician and the Registrars General in Scotland and Northern Ireland to conduct a census simultaneously across the UK was issued in October 2015. The programme to run the 2021 Census in England and Wales, to increase the use of administrative data and surveys in producing the 2021 Census results and improve inter-censal statistics, and to consider plans beyond 2021 are the three components of ONS' Census Transformation Programme.

1.4 Channels of access to census data: the UK Data Service-Census Support

In order to maximise the use of census data, effective dissemination is critical and the changing information environment presents ongoing challenges for those who collect, process, supply and provide access to census data. The NSOs continuously want to improve the efficiency of their operations and gain better value from their data production and dissemination mechanisms. Compared to previous censuses, 2011 Census outputs include data generated in new multi-dimensional forms and supplied using new transfer methods to devices such as mobile phones or iPads. On the demand side, census users expect more immediate access through delivery mechanisms that are more intuitive and user-friendly.

ONS' Web Data Access (WDA) project was established to deliver an enhanced website capability for organisations and individuals not only to use and explore census data but also other ONS statistics more effectively. Direct access to bulk census data in a machine-to-machine readable format was implemented through a public Application Programming Interface (API) and the data available from this open API were also accessible from the *Data Explorer (Beta)*.[6] This interface offers functionality for users to discover or find datasets by area or topic, view and manipulate datasets online, produce charts and maps as well as refine queries and download data. The other NSOs developed similar websites. In Scotland, the NRS *Census Data Explorer*[7] also provides access to a *Data Warehouse*[8] where all the standard tables are downloadable in bulk form as .csv files. The *Northern Ireland Neighbourhood Information Service (NINIS)* website[9] fulfils an equivalent function and has interactive content that can be filtered by subset, geography or year. The ONS' *Neighbourhood Statistics (NeSS)* website[10] is a similar online source, containing datasets that describe the characteristics of a neighbourhood, with a particular focus on deprivation and including results from the 2011 Census as well as the 2001 Census and other official government statistics. All the 2011 Census data for England and Wales, including 'open' origin–destination statistics for the UK, are available from the *Nomis* website,[11] the service provided by the ONS that gives users free access to the most detailed and up-to-date UK labour market statistics from official sources. Digital boundaries for 2011 Census areas and lookup tables are available from ONS' 'Open Geography' geoportal and from the NRS and NISRA websites.

It is clear from the examples above that each of the NSOs in the UK has taken advantage of the new digital technologies to develop online systems and interfaces that have facilitated access to the results of the 2011 Census as well as other data from official sources. This has been encouraged by much greater recognition of the importance of transparency in government and the belief outlined in the Open Data White Paper that "opening up data will empower citizens, foster innovation and reform public services" (HM Government, 2012, p. 5). Data produced by statistical offices such as the census and key socio-economic indicators are one type of open data, i.e. data that must be available as a whole and at no more than a reasonable reproduction cost, preferably by downloading over the internet in convenient and modifiable form, that are provided under terms that permit reuse and redistribution, and that anyone must be able to use, reuse and redistribute.

The last two decades have witnessed the development of a set of services designed to provide members of the academic community in the UK with quick and easy access to census datasets, primarily for research purposes but also for teaching. Independent units providing services for academic users to access the census aggregate statistics, boundary data, flow data and microdata from pre-2011 censuses were funded by the Economic and Social Research Council (ESRC) under the Census Programme up until August 2012. Thereafter, it was agreed that these hitherto rather disparate services should become part of a more integrated Census Support (CS) service which itself would become part of the ESRC-funded UK Data Service, aiming to allow researchers access to a wider range of data from census, administrative and survey sources through a single point of entry. Figure 1.3 shows the Census Support homepage of the UK Data Service website.

The UK Data Service-Census Support (UKDS-CS) is essentially composed of the four 'data support units' operating under the Census Programme that provided expert support and online access to different datasets (Stillwell et al., 2013). The first of these, providing support for the aggregate outputs, was the Census Dissemination Unit (CDU) based within Mimas at the University of Manchester until July 2014, at which point Mimas became part of Jisc, and moved to Churchgate House in central Manchester. The second unit is the former UKBORDERS service within EDINA at the University of Edinburgh, which provides access to census-related digital boundary data. The third unit (formerly the Centre for Interaction Data Estimation and Research, CIDER, at the University of Leeds) is based at University College London and provides access to census and related interaction datasets, whilst the fourth unit, providing advice and guidance to users of census microdata sets, which were known Samples of Anonymised Records

Figure 1.3   UK Data Service-Census Support homepage

Source: https://census.ukdataservice.ac.uk/.

(SARs) prior to 2011, is based at the Cathie Marsh Institute for Social Research (CMist) at the University of Manchester. The UKDS–CS Director is based at the University of Leeds.

Collectively, the CS units are referred to as 'value-added services' of the UK Data Service, receiving census data supplied by the NSOs (sometimes through a respective API), ingesting them for storage in various databases, and maintaining interfaces through which users can discover, query, retrieve, analyse and download the data they require. In the first 12 months since the census data became 'open' in January 2014, there were 25,600 data downloads from *InFuse*; the boundary data and postcode directories averaged around 9,500 downloads in the first two quarters after becoming open in 2015; and there were 2,650 downloads of the 2011 flow data during the first three quarters in 2015 after they became available in *WICID* (Web Interface to Census Interaction Data). One downside of the transition to open data access is that, once the need for authentication is removed, detailed monitoring of service usage becomes impossible. However, Google Analytics data on service providers does provide an indication of which academic institutions are the principal users. By way of example, Figure 1.4 illustrates usage of *InFuse* in 2015, as measured by the number of user sessions per higher education institution.

The online interfaces and their associated information systems are one of the unique selling points of the UKDS–CS. *InFuse*, the user interface for aggregate data from the 2011 Census, has been developed to avoid the conventional table structure of census aggregate statistics on

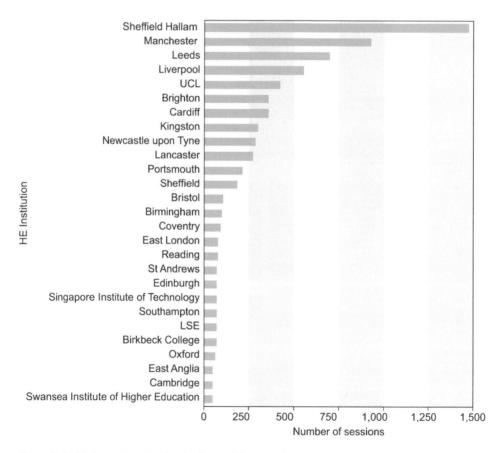

Figure 1.4 Higher education institutions with over 50 *InFuse* sessions during 2015

which its predecessor, *Casweb*, was based, to improve the consistency of the labelling of variable categories (such as age groups) in different tables and to provide clear metadata with each data download. The construction of this more flexible interface has required the reformatting of the underlying database into a multi-dimensional structure based on *Statistical Data and Metadata eXchange* (*SDMX*), a data model which has been adopted as the standard for data and metadata exchange for aggregate time-series data. Users of *InFuse* are able to locate data more easily and intuitively than with *Casweb* without hunting through a multitude of tables.

UKDS-CS supports a suite of services available for users to select and download the geography outputs of the last five decennial censuses as well as a range of historical census geographies, associated lookup tables, postcode directories and non-census geographical boundary data. *Easy Download* allows users to extract complete sets of the most regularly requested census boundaries available as ready-to-use national datasets in popular formats whilst the *Boundary Data Selector* allows the selection of boundaries that are required, for the area required, in the format required. Equivalent systems are available for access to current and historical postcode directories and sets of postcodes. *GeoConvert* is a geography matching and conversion tool that allows users to obtain and manipulate census geographical and postcode data in a straightforward way.

Census interaction data, which have been produced from modern censuses from 1981 onwards, tabulate information about flows within and between locations, and take the conceptual form of matrices of origins by destinations for migrants, students, commuters or details of those individuals with second homes. Sparsity becomes a problem when the zones are small and the numbers of origins and destinations are high, such as with a ward-to-ward matrix of migrants in England. The main access path to the flow data is via *WICID*, which allows users to flexibly build and execute a query which extracts the required data either for online analysis or direct download. *WICID* provides a number of routes into the data that aim to address the practical problems for users of sparse matrices.

Delivery of the microdata is a more collaborative affair. Safeguarded data and the census microdata which can be distributed under licence by the UK Data Service are archived with the UK Data Archive, a UK Data Service partner organisation. These data can be downloaded or accessed online in a manner similar to the other survey microdata that the UK Data Service provides. Once users have the data, they are then able to produce their own outputs in a flexible manner.

One of the long-term goals of the UK Data Service is to work with data owners to identify and remove all unnecessary barriers to accessing data. To this end, a generic access policy of three tiers – open, safeguarded and controlled (or secure) – has been implemented in collaboration with ONS. The large majority of datasets held by UKDS-CS are available without registration or authentication using an 'open data licence' for data that are not classified as personal. Data licensed for use in the 'safeguarded' category are not personal, but the data owner has identified a risk of disclosure resulting from linkage to other data, such as private databases. 'Safeguarded data' is the ONS preferred term for data which UKDS-CS provides under an End User Licence (EUL) and which requires user registration and specification of the purpose for which the data are required. Controlled data are personal data which may be identifiable and thus potentially disclosive and thus require registration and special user training to enable access.

The only means of access to the 'safeguarded' tables of flow data that have been produced from the 2011 Census is through the *WICID* interface and therefore users must be registered and abide by the stipulations of the EUL. A time series of inter-district migration flow estimates based on NHS patient re-registrations and supplied by ONS is also available from *WICID* together with estimated migration matrices from certain ESRC-funded research projects. Other non-census data are available from other UKDS-CS systems such as deprivation indicators from *GeoConvert* and there are a wide range of non-census geographical boundary datasets available from *Easy Download* or *Boundary Data Selector*. However, one of the key user attractions of the UKDS-CS resources is the ability to access data from previous censuses. *InFuse* provides access to 2001 as well as 2011 aggregate data, whereas *Casweb* holds UK aggregate data for 2001 and 1991 and GB data for 1981 and 1971. Migration and commuting data from censuses back to 1981 are available from *WICID* and census boundary data are available from 1991. Users of *InFuse* benefit from the effort that has been spent on integrating aggregate data supplied by each of the NSOs into a single system, and thereby facilitating UK-wide analysis and avoiding the need to extract data for different home nations from different online systems. Boundary data for the UK are available for a range of geographies which match exactly with the geographies used in *InFuse*.

It is important to recognise that the function of the UKDS-CS is not just to disseminate census data. The technical expertise that has accumulated in handling large and complex datasets and developing online database systems and interfaces has been valuable in collaborative initiatives with the census offices. UKDS-CS staff are involved in many of the census advisory groups and advice has been provided, in particular, to the data suppliers on the structure and content of the 2011 microdata and on the table structure, disclosure thresholding and dissemination of the 2011 flow data. Work is currently ongoing to prepare a trimmed-down version of *WICID* to support access to the secure flow data held in the ONS virtual microdata laboratory (VML). However, as well as providing users with seamless and flexible access to a wide range of data resources that facilitate high-quality social and economic research and education, it is also the role of the UKDS-CS to support members of the user community with advice and guidance on particular issues, to maintain a helpdesk that provides rapid response to issues as they arise, to provide regular training courses and webinars and to develop a bank of user support materials available online that can be used by researchers and students.

1.5   Rationale and structure of the handbook

UK census handbooks or user guides have been produced for previous censuses in 1981 by Rhind (1983), in 1991 by Dale and Marsh (1993) and Openshaw (1995) and in 2001 by Rees et al. (2002), providing details of the census process and the datasets that are produced. The 2011 Census Handbook is unique in endeavouring to demonstrate, through inclusion of an extensive set of case studies, how different census datasets are hugely important for research in academia in a wide range of different socio-demographic contexts from geodemographic profiling of small areas to understanding the geography of immigration and the changes in the ethnic structure of British cities.

The book is divided into six parts, the first of which is composed of two chapters that provide context. This introductory chapter has reviewed census-taking around the world, flagged up the trend apparent in many European countries and in developed countries elsewhere (USA, Canada, Australia) towards using hybrid methods involving data from administrative sources and surveys, and discussed both the benefits and shortcomings of the traditional census. The next chapter provides a synopsis of how the census in the UK has evolved throughout its history. Part II also contains two chapters that will provide readers with an understanding of the steps involved in the run-up to the 2011 Census, the actual data collection framework, methods of collection, problems, imputation, adjustment and evaluation. Attention is given to the devolved nature of the census and the different accountabilities. Both chapters have been written by individuals with first-hand involvement in the 2011 Census preparation, collection, adjustment and evaluation procedures. In Part III of the book, there are five chapters which spell out the types of data generated by the 2011 Census in detail and illustrate the means of user access and extraction of each type. The chapters provide guidance on using the web interface services of the UKDS-CS, in particular.

The chapters in Parts II and III are essentially focused on the supply side of the census process and the paths through which census data travel to the user whilst those in Parts IV and V are primarily concerned with the demand side or how the data are being used for visualisation and analysis, in this case almost entirely by members of the research community rather than users from other public or private sector organisations. Part IV contains seven relatively short chapters that introduce and exemplify some of the ways in which new technologies are being used to create innovative visualisations of census data through web-based infographics and mapping applications. Part V of the book, on the other hand, includes a series of chapters that cover the use of census data in a range of thematic areas, beginning with families, households and individuals, and then focusing on identity, ethnicity and religion, health care and deprivation, housing and inequality, migration and commuting, and finishing with issues of scale and long-term change. The chapters are all written by members of the academic research community. Finally, Part VI of the book contains a single chapter that takes a forward look at ONS' activities in planning for the 2021 Census and the future role of the census vis-à-vis alternative approaches to counting the population. Throughout the volume, we use Census with an upper case C when referring to a particular census (e.g. the 2011 Census) and lower case c otherwise.

In conclusion, this book is important in terms of the information that it provides for those wanting to understand how 2011 Census data were collected, what data products are available, and how and where different types of data can be accessed, for those interested in understanding the state-of-the-art in data visualisation, for those wanting to absorb some of the latest findings of contemporary research in social science in the UK, and for those concerned to understand more about the census and its utility at a time when its future existence is under threat. We hope you agree with the overall conclusion that the 2011 Census has been extremely valuable in underpinning a wealth of diverse academic research which has generated some fascinating results as well as some important messages for policy makers. It would be folly to abandon such a vital resource of reliable and comprehensive data without putting in place an infrastructure for a much improved collection of data from other sources, be they administrative records, sample surveys or big data in its various guises.

Acknowledgement

The author is grateful to Jo Wathan and Garnett Compton for comments on an initial version of this chapter.

Notes

1 http://www.ons.gov.uk/ons/about-ons/who-ons-are/programmes-and-projects/beyond-2011/reports-and-publications/index.html.
2 http://www.ons.gov.uk/ons/guide-method/census/2021-census/about-the-census-transformation-programme/index.html.
3 http://unstats.un.org/unsd/demographic/sources/census/census3.htm.
4 http://geogale.github.io/2011OAC/.
5 http://cowz.geodata.soton.ac.uk/.
6 http://www.ons.gov.uk/ons/data/web/explorer.
7 http://www.scotlandscensus.gov.uk/ods-web/home.html.
8 http://www.scotlandscensus.gov.uk/ods-web/data-warehouse.html.
9 http://www.ninis2.nisra.gov.uk/public/Home.aspx.
10 https://www.neighbourhood.statistics.gov.uk/dissemination/.
11 https://www.nomisweb.co.uk/census/2011.

References

Bell, M., Charles-Edwards, E., Kupiszewska, D., Kupiszewski, M., Stillwell, J. and Zhu, Y. (2015) Internal migration around the world: Assessing contemporary practice. *Population, Place and Space*, 21(1): 1–17.

Cockings, S., Martin, D. and Harfoot, D. (2015) A Classification of Workplace Zones for England and Wales (COWZ-EW) User Guide, Geography and Environment, University of Southampton. Available at: http://cowz.geodata.soton.ac.uk/download/files/COWZ-EW_UserGuide.pdf.

Coleman, D. (2013) The twilight of the census. *Population and Development Review*, 38(Supplement): 334–351.

Dale, A. and Marsh, C. (eds.) (1993) *The 1991 Census User's Guide*. HMSO, London.

Fienberg, S.E. and Prewitt, K. (2010) Save your census. *Nature*, 466(26): 1043.

Gale, C.G. (2014) Creating an open geodemographic classification using the UK Census of the Population. Doctoral Thesis, Department of Geography, University College London. Available at: http://discovery.ucl.ac.uk/1446924/.

HM Government (2012) *Open Data White Paper Unleashing the Potential*. Cmnd 8353. The Stationery Office, Norwich.

Hope, C. (2010) National census to be axed after 200 years. *The Telegraph*, 9 July. Available at: www.telegraph.co.uk/news/politics/7882774/National-census-to-be-axed-after-200-years.html.

Hutchens, G. and Martin, P. (2015) Researchers attack ABS plan to move census to 10 years. *The Canberra Times*, 20 February. Available at: www.canberratimes.com.au/federal-politics/political-news/researchers-attack-abs-plan-to-move-census-to-10-years-20150219-13jg59.

Kukutai, T., Thompson, V. and McMillan, R. (2015) Whither the census? Continuity and change in census methodologies worldwide, 1985–2014. *Journal of Population Research*, 32: 3–22.

ONS (2011) Beyond 2011 Consultation on User Requirements: 17 October 2011 – 20 January 2012 Consultation Document. Available at: www.ons.gov.uk/ons/about-ons/get-involved/consultations-and-user-surveys/archived-consultations/2012/beyond-2011-public-consultation/index.html.

ONS (2013) *The Census and Future Provision of Population Statistics in England and Wales Public Consultation*. Beyond 2011 C1 Consultation, ONS, Titchfield. Available at: www.ons.gov.uk/ons/media-centre/statements/census-consultation/index.html.

ONS (2014) *The Census and Future Provision of Population Statistics in England and Wales: Recommendation from the National Statistician and Chief Executive of the UK Statistics Authority*. Available at: http://webarchive.nationalarchives.gov.uk/20160105160709/http://www.ons.gov.uk/ons/about-ons/who-ons-are/programmes-and-projects/beyond-2011/beyond-2011-report-on-autumn-2013-consultation-and-recommendations/index.html.

Openshaw, S. (1995) *Census User's Handbook*. Wiley, Chichester.

PASC (2014) *Too soon to scrap the Census. Fifteenth Report of Session 2013–14*. Report, together with formal minutes relating to the Report, HC 1090, The Stationery Office Limited, London. Available at: www.publications.parliament.uk/pa/cm201314/cmselect/cmpubadm/1090/1090.pdf.

Poulain, M. and Herm, A. (2013) Central population registers as a source of demographic statistics in Europe. *Population-E*, 68(2): 182–212.

Rees, P., Martin, D. and Williamson, P. (eds.) (2002) *The Census Data System*. Wiley, Chichester.

Rhind, D. (ed.) (1983) *A Census User's Handbook*. Methuen, London.

Royal Geographical Society with the Institute of British Geographers (2014) *Small Area Data Looking Towards a 2021 Census*. Policy Briefing, RGS-IBG, London. Available at: www.rgs.org/NR/rdonlyres/80857592-3D48-4C2B-9CB9-52A107AE249F/21508/RGSIBGPolicyDocumentSmallAreaDataForweb1.pdf.

Sheller, M. and Urry, J. (2006) The new mobilities paradigm. *Environment and Planning A*, 38(2): 207–226.

Singer, A. (2010) Census 2010 can count on controversy. *CNN*, 4 January. Available at: http://edition.cnn.com/2010/OPINION/01/04/singer.census2010.challenges.controversy/.

Singer, E., Bates, N. and Van Hoewyk, J. (2011) Concerns about privacy, trust in government, and willingness to use administrative records to improve the decennial census. Paper presented at the 64th Annual Conference of the American Association for Public Opinion Research (AAPOR), 12–15 May, Phoenix. Available at: www.amstat.org/sections/srms/proceedings/y2011/Files/400168.pdf.

Skinner, C., Hollis, J. and Murphy, M. (2013) Beyond 2011: Independent Review of Methodology. Available from: www.ons.gov.uk/ons/about-ons/who-ons-are/programmes-and-projects/beyond-2011/beyond-2011-independent-review-of-methodlogy/index.html.

Stillwell, J., Hayes, J., Dymond-Green, R., Reid, J., Duke-Williams, O., Dennett, A. and Wathan, J. (2013) Access to UK census data for spatial analysis: Towards an integrated Census Support service. In Geertman, S., Toppen, F. and Stillwell, J. (eds.) *Planning Support Systems for Sustainable Urban Development*. Springer, Heidelberg, pp. 329–348.

Sydney Morning Herald (2015) Researchers attack ABS plan to move census to 10 years. 20 February. Available at: www.smh.com.au/federal-politics/political-news/researchers-attack-abs-plan-to-move-census-to-10-years-20150219-13jg59.html.

Underhay, J. (2011) The long and short of the census debate. *Skeptic North*, 17 November. Available at: www.skepticnorth.com/2011/11/the-long-and-the-short-of-the-census-debate/.

UNSD (1997) Principles and Recommendations for Population and Housing Censuses Revision 1. *Statistical Papers, Series M No. 67/Rev.1*, United Nations, New York.

UNSD (2011) Report on the Results of a Survey on Census Methods used by Countries in the 2010 Census Round. *Working Paper: UNSD/DSSB/1c*, United Nations, New York.

UNSD (2013) *2010 World Population and Housing Census Programme*. Available at: http://unstats.un.org/unsd/demographic/sources/census/wphc/default.htm.

UNSD (2014) Principles and Recommendations for a Vital Statistics System Revision 3. *Statistical Papers, Series M No. 19/Rev.3*, United Nations, New York.

Valente, P. (2010) Census taking in Europe: How are populations counted in 2010? *Population & Societies*, 467(May): 1–4.

Vargas-Silva, C. and Markaki, Y. (2015) Long-term international migration flows to and from the UK. *Briefing*, The Migration Observatory, University of Oxford, Oxford. Available at: www.migrationobservatory.ox.ac.uk/briefings/long-term-international-migration-flows-and-uk.

A history of census-taking in the UK

Oliver Duke-Williams

2.1 Introduction

The first census in the UK was taken in 1801, and censuses have continued to be taken at ten-year intervals, with the most recent one being in 2011. The first four censuses were in fact aggregate collections of data, and it was not until the 1841 Census that the collection methodology shifted to individual-level data collection as recognised today. The exceptions to this ten-yearly pattern have been in 1941 when no census was taken due to the war, and in 1966 when an additional sample census was taken. The 1941 Census can be partly seen as being replaced with national registration; whilst similar data were gathered, the purpose and nature of civil registration was different from that of a census. Both the number and range of topics which have been included in UK censuses have grown considerably over time, with the 2011 Census being the most detailed in terms of topic coverage yet conducted, although the differences between the censuses conducted by statistical agencies in different parts of the UK have also grown.

A census is a collection of data across the whole population of a country. In the context of this book we are interested in population censuses, but other censuses may gather data about land use, production, employment, physical or intangible assets or particular population subgroups, with a defining common characteristic that censuses generally attempt to be comprehensive in their coverage. As described in Chapter 1 of this volume, the United Nations offers a definition of a population census (United Nations Statistics Division, 1997) which makes clear that a census is not just a set of numeric outputs, but a broader process which encompasses the collection and processing of data relating to the population of a country. In offering this definition, it is clear that a population census can be understood not just as a set of resulting count observations, but also the entire administrative process that has led up to the publication of those counts.

This chapter reviews the history of census-taking in this country (the scope of 'this country' is itself something not fixed over a two-century period), and the motivating factors that led to the first census in 1801. This historical synopsis can be divided into a number of distinct phases: the period leading up to the first census and the censuses of 1801 to 1831; the censuses during the remainder of the nineteenth century and the early twentieth century; the post-war twentieth-century censuses; and, more latterly, the censuses of the current century. The development of census methodology does not, of course, strictly follow temporal distinctions quite so easily, and is intertwined with the history of data processing technology, from data being collated entirely by hand, through mechanical processing of punched cards and early digital computers, to the current era in which data are not just processed by computer, but increasingly collected in a digital form.

The first part of this chapter works through this history of census-taking, identifying developments of note in census methodology: the points at which innovations were introduced and when certain questions were first asked. The second part of the chapter looks in more detail at the inclusion of questions on different topics in each census and presents a tabular summary of topics included in each census.

2.2 Early census-taking in the UK

Descriptions of census-taking in the UK typically start out with reference to Domesday as an early example of an attempt to collect data on a national (in this case, England and parts of Wales) basis. However, whilst it did refer to people, Domesday's remit, as ordered by William the Conqueror in 1085, was a survey of landholdings and their value for the purpose of taxation. It was, however, a very large data gathering exercise, and in that context continues to serve as a useful starting point for any history of census-taking. In fitting with the UN definition of a census as the process of collecting, processing and publishing data, it is notable that Domesday was the process of collecting data, and the results were collated and published as the Domesday Book (see Martin, 2003, for a modern translation), although popular history has tended to focus on the latter rather than the former.

Whilst Domesday is celebrated, data gathering continued to be carried out in subsequent periods, but is less remarked upon. The '1279–80 Hundred Rolls' were commissioned by Edward I in 1279 as a national survey of landholdings in England and were so called as data were collected within county divisions known as Hundreds. Few of the rolls survive and it is not known whether other results have been lost, or whether full data collection did not actually take place (Raban, 2004). The data collected in these 'Hundred Rolls' were more extensive than those collected by Domesday, and it is possible that it is primarily because the data were not edited and collated that they did not survive. In the past, as well as in modern census-taking, it is important to see data collection, processing and publication as equally vital parts of a production chain.

Data collection continued to be carried out by various stakeholders: by the Government and agents of the Crown, by private landholders and in parish registers. A bill proposing an annual enumeration of the population was put forward in Parliament in 1753, and passed through both Houses, but did not complete all committee stages before the end of the parliamentary session, and thus lapsed (Higgs, 1989). This enumeration was to be carried out by Overseers of the Poor, officers who would also be instrumental in the census.

The Census Act of 1800 paved the way for the 1801 Census. Census-taking had become common in a number of European countries in the eighteenth century, and had started in the USA with the Census of 1790; a quinquennial series of censuses in France started in 1801 (Durr, 2005). A variety of pressures faced England at this time (Taylor, 1951), which could all be informed by an assessment of the size of the population, a number which was then not known. It was a period of war and thus there was interest in determining the size of any army that could be raised. At the same time, there was pressure to increase agricultural production in order to produce food. Thomas Malthus' *Essay on the Principle of Population* in 1798 had drawn this into sharper focus. Malthus had argued that population and food production grew at separate rates:

that food production grew in an arithmetic progression, increasing at a constant rate, whereas population was subject to geometric growth, doubling and then re-doubling. A logical consequence was that, at some stage, the population would grow beyond the agricultural capability of the country to feed itself. In order to contemplate when this might occur, it was necessary to have an idea of both the current agricultural production and the current population size. There was interplay between the influence of war and of food production, in that military campaigns would divert labour from agricultural work to military service.

The 1801 Census – and those of 1811, 1821 and 1831 – were far removed from what we now consider to be typical census-taking. Higgs (1989) describes in detail both the content of these early censuses and the administrative machinery involved in their collection, which was based around overseers of the poor and the clergy. A set of six questions were posed in the schedule of the 1800 Census Act, and these sought to identify for each area (parish, township or place):

- the number of inhabited houses and the number of families inhabiting them; and the number of uninhabited houses;
- the number of people, by sex, excluding men on active military service;
- the number of people occupied in agriculture, in trade, manufacture or handicraft, or not occupied in those classes;
- the number of baptisms, by sex, in various years over the preceding century;
- the number of marriages, annually, since 1754; and
- any other remarks.

Thus, rather than collecting data about individuals, the data were collected on an aggregate basis. In that respect, the results of the early censuses resemble later census outputs, and we can see them in a continuity of censuses stretching across more than two centuries. However, there are no base individual records for demographers or family historians to examine. The next three censuses after 1801 were of a similar design, and also collected aggregate observations, albeit with modifications to the question schedules. The 1811 Census altered the focus of the question about occupations from persons occupied in different categories to the number of families employed or maintained by work in a given sector. Whilst counts of persons had always distinguished between men and women, the 1821 Census was the first to address the age of people as well as their gender, with counts gathered for five-year age groups up to 20 and ten-year age groups beyond that age. The 1831 Census saw significant expansion of the number of different occupations recognised, with additional sub-division of those employed in agriculture to separately tally those who occupied land and employed labourers, those who occupied land without employing others, and those who were employed as labourers.

The 1841 Census was the first to be based around household schedules, in which details of all individuals in households were recorded. The use of these schedules was authorised after pilot testing and allowed an assessment to be made about the size of the field force that would be required, given concerns that literacy levels would be a problem with household enumeration (Higgs, 1989). The 1841 enumeration gathered considerably more detail about individuals than had been permitted by the aggregate returns of 1801–31. For each person, sex and age in years was recorded, as was 'profession, trade (or) employment' or whether of independent means. Additional questions were included on the schedule which addressed a topic not included on the preceding censuses: where a person had been born. Whilst earlier censuses had (perhaps enforced by the collection design) viewed people as being connected to one location only, the 1841 Census recognised that populations were mobile. The question structure also recognised that geography and mobility could simultaneously be viewed at more than one scale. Thus, the first of these new questions asked whether or not a person had been born in the same county (as they were now being enumerated) and the second asked about country of birth.

As the nineteenth century progressed, so the census schedule continued to expand. Some question topics were introduced and later dropped out of use, but for the most part the story of the census has been one of steady accretion of question topics. This is considered in more detail in Section 2.7. Higgs (1989) argued that the mid-Victorian censuses of 1851–71, in their administration under the epidemiologist William Farr, and in their expansion of content, can be seen as rooted in medical surveys. Whilst the process of taking a census developed in the nineteenth century, there was not a permanently established Census Office, but, rather, a set of administrative processes which were established every ten years, with each census being carried out under the authority of a separate Census Act. Lawton's edited collection of essays (Lawton, 1978) shows how the nineteenth-century censuses were essential documents for the analysis of a range of demographic, social and economic variables, including the first comprehensive study of migration in the UK (Ravenstein, 1885).

2.3 Early twentieth-century censuses

The 1911 Census featured the first use in the UK census of automated technology (Campbell-Kelly, 1996) in the form of automated tabulating machines for processing the returns collected by enumerators. These machines – similar to the Hollerith machine developed for the 1890 US Census – stored and processed data on punched cards, improving data processing capabilities, but incurring additional costs for the acquisition of equipment. The census was also notable for an organised boycott of the census by suffragette organisations (Liddington and Crawford, 2011), and the dual recording of the address on census night of leading campaigner, Emily Davison, both at home in Russell Square and in a broom cupboard in the Houses of Parliament (UK Parliament, n.d.), thus highlighting her claim for equal political rights as men. Further developments followed in methodology and form design, in the technology used to process the census, and in the administrative and legal underpinnings of the census. The 1920 Census Act made provision for the regular taking of a census, removing the need for a separate Act for each census. The 1920 Act also formally introduced ideas of confidentiality and restrictions on release of individual data.

Census questions have always reflected the changing interests of the day. The 1921 Census was no different in this regard; following the First World War, it included a question on orphanhood (whether either or both parents were dead). Such a question had not been included before and has not been included since. Reflecting increasing population mobility, the 1931 Census was the first to include a question on usual residence.

2.4 The census and the Second World War

Were it not for the Second World War, it might have been easy to see the twentieth-century sequence of censuses as one of gradual change – of a slow, steady increase in the number and range of questions asked, and a change in the way that censuses were carried out and the way in which data were processed. However, the war enforced a mental division that permits us to view that history as pre- and post-war. The expected 1941 Census did not take place due to the war; the only time since 1801 that a decennial census was not conducted. However, under the 1939 National Registration Act, registration took place from September 1939, and this registration can be viewed as very much census-like. The Act mandated the collection of the following information for individuals:

1   names;
2   sex;
3   age;
4   occupation, profession, trade or employment;
5   residence;
6   condition as to marriage; and
7   membership of Naval, Military or Air Force Reserves or Auxiliary Forces or of Civil Defence Services or Reserves.

Aside from the final one of these 'particulars', this can readily be seen as a simplified version of the preceding censuses. The National Registration Act explicitly referenced the 1920 Census Act, and required "persons who have undertaken to perform duties in connection with the taking of a census under the Census Act, 1920, to perform in lieu thereof similar duties . . ." (www. histpop.org). However, in other respects, national registration should not be seen as a census. A census is conceptualised as being a snapshot count of people and their characteristics at a particular point in time. In contrast, national registration was a continuing process, covering all persons in the UK at the time of the Act and all those entering or born in the UK afterwards. There is also a significant difference in purpose: censuses are carried out specifically as a count of the population, whereas national registration served a wider set of purposes, including the issuance of identity cards.

A further occurrence during the war emphasises the distinction between the pre- and postwar periods in the census, when viewed as a series of datasets that can be analysed. In December 1942, a fire occurred at an Office of Works store in Hayes, Middlesex. Amongst the material destroyed were the schedules and enumeration books containing the raw data of the 1931 Census. The fire was not related to war. In a letter reporting the fire, W.A. Derrick (National Archives, RG 20/109 – Derrick, 1942) indicated that there was no salvageable material. The same letter also referred to the 1921 Census materials, and noted that some schedules (stored in a separate location) had previously sustained water damage but had since been dried out. An overall effect of the 1942 fire and the cancellation of the 1941 Census is that the sequence of original individual data ends with the 1921 Census, and does not pick up again until the 1951 Census.

2.5   The post-war censuses

The 1951 Census can be seen as a transitional step in the sequence of historical censuses. In terms of data collection and processing, it bore more similarity with earlier censuses than it would with later ones. However, in terms of subject coverage, it can be seen as heralding a new era of taking censuses which would all add new areas of interest. The 1951 Census was of considerable significance given the cancellation of the 1941 Census; it was the first full census in 20 years. The subsequent 1961 Census was the first to be processed by computer, opening the door to a wider range of possible outputs. A lengthy phase of data entry from hand-completed forms was still required. The 1961 Census was the first in the UK to use a two-form design: all households received a standard 'short' form, with every tenth household (thus a 10 per cent sample of households) also receiving an alternative 'long' form, with a wider set of questions for individuals. In the case of communal establishments, extra questions were asked of every tenth person (General Register Office, 1961). This long form was, however, broadly similar in length to the form used in 1951. More detailed outputs were made available from the 1961 Census than had previously been possible: the Small Area Statistics (SAS) product was developed, which published data from enumeration districts (EDs) for the first time (Dewdney, 1983).

The 1920 Census Act had allowed for the taking of a census every five years, although, until 1961, this had only been carried out every ten years in practice. A five-year census was conducted in 1966, and so far remains the only such census conducted in the UK. The 1966 Census was only collected for a 10 per cent sample of the population. Unlike the main decennial censuses, little information about or description of the 1966 Census is available. However, information about its content is reflected in documents such as the magnetic tape layout document held by the UK Data Service (Office of Population Censuses and Surveys, General Register Office (Scotland). Census Branch, 1966). Two questions introduced in the 1966 Census, which have been asked in subsequent censuses, were about the mode of transport used to travel to work, and about the availability of cars in households.

In 1961, a question about internal migration had been included, asking whether people had changed their usual residence in the year preceding the census. For the 1966 Census this question was repeated and paired with an additional question asking about change of usual residence in the preceding five years – thus linking back to the time of the 1961 Census. This five-year question was repeated in the 1971 Census (again, alongside a one-year transition question), providing a conceptual link to the 1966 Census. This implicit reference to earlier censuses has not since been repeated in the questions asked. The 1971 Census shared the parallel sample design of the 1961 Census, in that there was a 10 per cent sample of data as well as a full 100 per cent enumeration. However, unlike the 1961 two-form approach, all households in 1971 were given the long form. In the data entry phase, the answers to a number of 'hard to code' questions were only recorded for a 10 per cent random sample of households (or 10 per cent of persons in communal establishments). The 1971 Census Small Area Statistics were produced for all EDs, greatly increasing the volume of possible outputs. A further innovation of the 1971 Census was the geocoding of data, allowing re-aggregation of data into new boundaries, including a new 1 km grid square based set of outputs. A number of questions included in the 1971 Census explored areas such as occupation one year ago and country of birth of parents. The 1971 Census was not the first census to include a question on children born to married women. There was a question in 1951 asking about children born in marriage. However, the 1971 Census asked for details on women under 60 who were married, widowed or divorced and birth of each child in marriage.

Subsequent censuses continued to expand the range of outputs, and made growing use of new technologies for dissemination of data, but with a broadly similar methodology, with paper forms being delivered and collected by enumeration officers, and with the continued practice of 10 per cent coding of some hard to code fields in the 1981 and 1991 Censuses. These two censuses were not without innovation. An approximately 1 per cent sample of individuals had been drawn from the 1971 Census for England and Wales, which formed the basis of a new form of output: the Longitudinal Study (see Chapter 9 of this volume). The 1981 Census presented the first opportunity to add another wave of census data to this sample, permitting the same individuals to be observed over a ten-year transition period. In terms of census methodology, the 1981 Census was the first to feature a post-enumeration follow-up coverage survey (Britton and Birch, 1985), a practice which would be repeated and developed with subsequent censuses.

A number of the questions that had been featured in the 1971 Census were dropped for the 1981 Census and although a question on ethnicity was trialled, it was not included on the final census form (Dale, 2000). However, a question on ethnic group was included for the first time in the 1991 Census. The 1991 Census also saw the introduction of a new microdata product: the Samples of Anonymised Records (SARs) (Marsh and Teague, 1992). The 1991 SARs consisted of two files: a 1 per cent sample of households (and the individuals in them) and a 2 per cent sample of individuals. These files permit the flexible creation of any cross-tabulation of component variables, as long as the sample size gives sufficient values in each cell to make the resulting table usable or meaningful. Similar studies (with different sample sizes) would follow in Scotland (first census data from 1991) and Northern Ireland (first census data from 2001, with earlier waves added subsequently).

A considerable cause of concern arising from the 1991 Census was the level of enumeration, leading to a so-called 'missing million' (Simpson and Dorling, 1994): persons – often young adult men – who had not been captured in the census data. Much commentary identified concerns related to the Community Charge (or Poll Tax) as a possible reason for deliberate avoidance of the census. As a result of this, revised counts were produced with missing persons imputed (Office of Population Censuses and Surveys, 1994), and additional estimates produced by academic projects such as 'Estimating with Confidence' (Norman et al., 2008). Imputation and disclosure control have now become key steps in the 'census data system' (Rees et al., 2002). In the case of the interaction data from the 1991 Census – two sets of Special Migration Statistics (SMS) and one set of Special Workplace Statistics (SWS) – the suppression of counts under ten was used with potentially disclosive tables, necessitating innovative recovery methods to be devised by researchers requiring a complete picture of migration (Rees and Duke-Williams, 1997).

2.6   Twenty-first-century censuses

The 2001 Census saw a number of methodological innovations, as well as new questions included in the census questionnaires. Following the problems with enumeration in 1991 and the revision of counts, a strategy known as the 'One Number Census' was adopted (Brown et al., 1999), with the aim that outputs would have a 'final' set of counts and would not need revision. The process included imputation based on findings from a detailed Census Coverage Survey (CCS). The 2001 Census was the first to detach the census output geography from that used for enumeration for the main body of small area outputs. A new set of reporting units known as output areas (OAs) were produced for England and Wales after the census data were collected (Martin, 1998, 2002); these areas were designed to satisfy requirements relating to minimum population counts, to reflect postcode boundaries where possible and to maximise social homogeneity within zones.

Whereas the preceding censuses had used 10 per cent coding for some questions, the 2001 Census data were fully coded, with 100 per cent counts published for all observations. A further methodological innovation was the widespread use of 'post-back' return of census forms, with field officers then concentrating on collecting non-returned forms. The 2001 Census forms included a question on religion, with variations in wording across parts of the UK (Weller, 2004), the first time outside Northern Ireland that a question on religion had been included since 1851. In contrast to all other questions, the religion question was explicitly identified as being optional. In terms of statistical disclosure control in the 2001 Census, one of the methods used was the 'small cell adjustment method' (SCAM), essentially removing all counts of one or two individuals from any census tables and replacing them with values of zero and three and, in so doing, adding considerable uncertainty to the counts for small areas, particularly for flows between output areas (Stillwell and Duke-Williams, 2007).

The most recent round of censuses took place on 27 March 2011, again seeing methodological developments (White, 2009) as documented by Compton et al. in the next chapter of this book. Following the use of 'post-back' for the 2001 Census forms, one extension of methodology in 2011 was 'post-out' of forms, with questionnaires being delivered by mail rather than by an enumeration officer. In order to operationalise this, a comprehensive address register was constructed prior to the census and the register was used to underpin a new questionnaire tracking system. A major shift from previous practice was the development of an option to allow completion of the census form via the internet; around 16 per cent of census returns in England and Wales used this route in 2011 (ONS, 2012a). The availability of an internet channel has consequent implications for data editing and processing, as it is possible that individuals may return details both on paper and via the internet, with reconciliation thus required.

Following the introduction of OAs with the 2001 Census (revised where required for the 2011 outputs), an additional change was made to the reporting geographies in England and Wales (with similar geographies planned elsewhere) with the introduction of new workplace zones (WZs) (Martin et al., 2013) – small reporting areas for workplace-related data, permitting OAs to be split where a large number of people work, or merged in places where few people work.

A significant number of new questions were included in the 2011 Censuses, including ones on main language and English language proficiency, date of most recent arrival, passports held and national identity, and, in England and Wales, on whether or not an individual uses a second residence.

2.7 The questions asked in censuses

Since 1801, the number of questions asked in UK censuses has grown, and the subject areas covered have diversified. The inclusion of a question on a particular subject (or, in some cases, a set of subjects) is marked in Table 2.1, but it must be noted that the actual question asked may vary between different parts of the UK and may also vary over time. For example, a question on 'occupation' is ever-present, yet this has varied from a simple classification ('Agriculture', 'Trade, Manufacture and Handicraft' and 'Other') in early censuses, to the detailed classification of occupations with which we are now familiar. Similarly, the scope of questions about birthplace has changed over time from capturing detail of place of birth within Britain in the nineteenth century, to capturing country of birth for those born outside the UK in more recent censuses. Table 2.1 shows a general increase in the number of questions asked over time, and also suggests shifting perceptions as to what is considered to be of interest. Thus, for example, under household amenities, questions were asked between 1951 and 1991 about the availability of a flush toilet, and whether the entrance was inside or outside (with a related question about exclusive use of a toilet in 2001 in Northern Ireland). This question is no longer asked, but rather, questions about central heating are included. The changing pattern of questions over time may perhaps indicate something about issues that are or have been considered to be of political or social interest.

Table 2.1 is adapted from ONS (n.d.), which shows the variation in census questions from 1801 to 2011, and from a similar table showing topics from 1801 to 1991 by Diamond (1999). Documentation relating to the 1966 Census is less widely available than for other censuses. Information on the questions asked in different parts of the UK was assembled from a number of sources, including the UK Data Service user guide and Vidler (2001). The questions asked include some – such as name – for which responses are not retained in processed data, but which are noted here for completeness.

The table is simplified, and indicates whether questions on a particular subject were asked in at least one of the UK censuses. The first four columns refer to the 1801–31 Censuses and, as previously discussed, as aggregate observations of the total number of persons in an area, they are rather different in nature to later censuses. If the 1841 Census is also regarded as being somewhat transitional in nature, and the main sequence of censuses seen as running from 1851 onwards, we can see that there are a number of questions that have always been asked. These are name, age (or date of birth), sex, relationship to the head of household, marital status, occupation, economic

Table 2.1 Questions asked in censuses from 1801 to 2011

	1801	1811	1821	1831	1841	1851	1861	1871	1881	1891	1901	1911	1921	1931	1951	1961	1966	1971	1981	1991	2001	2011
Individual characteristics																						
Names					*	*	*	*	*	*	*	*	*	*	*	*	*	*	*	*	*	*
Sex	*	*	*	*	*	*	*	*	*	*	*	*	*	*	*	*	*	*	*	*	*	*
Age				*	*	*	*	*	*	*	*	*	*	*	*	*						
Date of birth																	*	*	*	*	*	*
Relationship to head of household						*	*	*	*	*	*	*	*	*	*	*	*	*	*	*	*	*
Relationships to others in household																					*	*
Whereabouts on census night															*	*	*	*	*			
Usual address													*	*	*	*		*	*	*	*	*
Absent persons																*		*	*	*		
Second address: use of; type																						*
Student term-time address																				*	*	*
Marriage and fertility																						
Marital status						*	*	*	*	*	*	*	*	*	*	*	*	*	*	*	*	*
Civil partnership status																						*
Year and month of liveborn children in marriage																	*					
Number of liveborn children in marriage												*			*	*						
Liveborn children in last 12 months															*	*						
Duration of marriage												*										
Year and month of marriage															*	*						
Education etc.																						
Student/scholar status						*	*	*	*	*	*	*			*						*	*
Age at which full-time education ceased															*	*						
Types of qualification																*		*	*	*	*	*
Subject of study																	*					
Awarding institution																	*					
Transport to place of study																					*	*
Address of place of study																					*	*
Employment & work																						
Economic activity						*	*	*	*	*	*	*	*	*	*	*	*	*	*	*	*	*
Students of working age														*	*	*		*	*	*	*	*
Full time/part time														*	*	*			*	*	*	*
Hours worked																*			*	*	*	*
Occupation (etc.)	*	*	*	*	*	*	*	*	*	*	*	*	*	*	*	*	*	*	*	*	*	*
Occupation one year prior to census																	*					
Apprentice/trainee														*	*			*	*	*		
Employer/employed				*		*	*	*	*	*	*	*	*	*	*	*		*	*	*	*	*
Industry													*	*	*	*		*	*	*	*	*
Place of work													*	*	*	*		*	*	*	*	*
Transport to work																	*	*	*	*	*	*

	1801	1811	1821	1831	1841	1851	1861	1871	1881	1891	1901	1911	1921	1931	1951	1961	1966	1971	1981	1991	2001	2011
Time since last worked																					*	*
Voluntary work																					*	
Migration & nationality																						
Birthplace					*	*	*	*	*	*	*	*	*	*	*	*	*	*	*	*	*	*
Parents' country of birth																		*				
Nationality												*	*	*	*	*						
National identity																						*
Passports held																						*
Year/month of entry																		*				*
Intended length of stay																						*
Address one year before																*	*	*	*	*	*	*
Address five years before																	*	*				
Ethnicity & religion																						
Religion						*																*
Religion (NI)															*	*		*	*	*	*	*
Religion brought up in																						*
Ethnic group																				*	*	*
Health & dependency																						
Ages of children under 16												*										
Orphanhood												*										
Infirmity					*	*	*	*	*	*	*											
General health																					*	*
Limiting long-term illness																				*	*	*
Long-term health condition																						*
Provision of care																					*	*
Language																						
Main language used																						*
Welsh language use										*	*	*	*	*	*	*		*	*	*	*	*
Irish language use																					*	*
Scottish language use									*	*	*	*	*	*	*	*		*	*	*	*	*
Ability in English																						*
Household characteristics																						
Number of rooms										*	*	*	*	*	*	*	*	*	*	*	*	*
Number of rooms with one or more window				*	*	*	*	*	*	*	*											
Number of bedrooms																						*
Sharing accommodation																*		*	*	*	*	*
Tenure																*	*	*	*	*	*	*
Furnished/unfurnished																			*			
Lowest floor level																			*			
Accommodation on more than one floor																			*			
Landlord																					*	*

(Continued)

Table 2.1 (Continued)

	1801	1811	1821	1831	1841	1851	1861	1871	1881	1891	1901	1911	1921	1931	1951	1961	1966	1971	1981	1991	2001	2011
Adaptation to accommodation																						*
Household amenities																						
Cooking stove; kitchen sink; water supply															*	*	*					
Fixed bath or shower															*	*	*	*	*	*	*	
Inside/outside WC															*	*	*	*	*	*	*	
Central heating: presence, type																				*	*	*
Car availability																	*	*	*	*	*	*
Returns by enumerators																						
Baptisms & burials	*	*	*																			
Marriages	*	*	*																			
Illegitimate children				*																		
Number of houses	*	*	*	*	*	*	*	*	*	*	*	*	*	*	*							
Families per house	*	*	*	*							*	*	*	*	*							
Vacant houses	*	*	*	*	*	*	*	*	*	*	*	*	*	*	*	*	*	*	*	*	*	*
Shared access																			*	*	*	*

status, whether an employer or employed and birthplace – broadly: who we are, who we live with, and what work we do.

2.8   The cost of censuses

Censuses are expensive operations to undertake; they require the recruitment or contracting of a large number of data entry and processing staff, of a field enumeration force (especially prior to more recent developments in postal delivery and return), as well as management and supervision staff. They also require the production and handling of a large number of physical documents, and the IT costs for data handling, storage and processing.

ONS have put the cost of the 2011 Census at £480 million (ONS, 2012b). Campbell-Kelly (1996) produced a table of costs of the censuses from 1841 to 1901, which provides some opportunity for comparison. The quoted costs rise from £66,727 in 1841, to £148,921 in 1901. The website MeasuringWorth.com (Officer and Williamson, 2016) provides a variety of time-adjusted prices, including inflation-adjusted prices and other measures that take into account changes in the relative purchasing power of money. The inflation-adjusted prices (in 2011 terms) thus range from £5 million for the 1841 Census to £13 million for the 1901 Census, whilst the economy cost (taking into account the cost of a project as a share of GDP) is calculated as ranging from £214 million in 1841 to £128 million in 1901. Whilst these Victorian censuses were expensive operations, they do not appear to have had the same relative cost as the 2011 Census; recent censuses, of course, include a larger number of questions, and lead to a wider range of outputs.

A gross comparison of inflation-adjusted 'cost' only addresses a limited part of the financial interpretation of a census: for a fuller analysis – beyond the scope of this chapter – it would also be necessary to consider the ways in which census data are used, costs that would be required to gather information were census data not available, and efficiency savings made through setting planning and policy decisions in a context informed by accurate data. Cope (2015) discussed the results of a cost–benefit analysis conducted by ONS, which suggested that benefits from use of census data rapidly outweighed the initial cost.

2.9 Conclusions

Many aspects of the 2011 Census are explored in the following chapters of this book; it is suggested here that the 2011 Census should not be considered in isolation, but that each modern census should be seen as building upon, extending and developing out of its predecessors. An overriding design issue for topic selection, for question phrasing and for response categories is to provide harmony where possible with earlier censuses. This is not always possible; an obvious recent example is to consider the response categories of the 'marital status' topic: whilst the *topic* remains the same, the definition of marriage has been expanded to include same-sex couples and thus the response categories used will change in line with changes in legislation such as those under the Civil Partnership Act and the Marriage (Same Sex Couples) Act.

A more subtle reading may be made of the purposes of taking a census: from early assessment of population stock, to assessment of medical and public health conditions, to a more wide-ranging collection of population statistics. A nuanced understanding of the 2011 Census is important, as it will in turn act as an important element for the design of the 2021 Census. Several topics were newly introduced in the 2011 Census, largely reflecting interest in the role or perceived role of immigration in shaping our society: in part, these questions should be seen as setting up follow-up questions on the same topics in future censuses or equivalent data collection instruments.

References

Britton, M. and Birch, F. (1985) *1981 Census, Post-enumeration Survey: An Enquiry into the Coverage and Quality of the 1981 Census in England and Wales.* OPCS, Social Survey Division, HMSO, London.

Brown, J.J., Diamond, I.D., Chambers, R.L., Buckner, L.J. and Teague, A.D. (1999) A methodological strategy for a one-number census in the UK. *Journal of the Royal Statistical Society: Series A (Statistics in Society),* 162(2): 247–267.

Campbell-Kelly, M. (1996) Information technology and organizational change in the British census, 1801–1911. *Information Systems Research,* 7(1): 22–36.

Cope, I (2015) The value of census statistics. Paper presented at Census Applications: Using the UK population census data, 16–17 July, University of Manchester. Available at: www.ukdataservice.ac.uk/media/455474/cope.pdf.

Dale, A. (2000) Developments in census taking in the last 25 years. *Population Trends,* 100: 40–46.

Derrick, W. (1942) Letter to Census National Registration Office, part of RG 20/109, The National Archives.

Dewdney, J. (1983) Censuses past and present. In Rhind D. (ed.) *A Census User's Handbook.* Methuen, London.

Diamond, I. (1999) The Census. In Dorling, D. and Simpson, S. (eds.) *Statistics in Society: The Arithmetic of Politics.* Arnold, London.

Durr, J.M. (2005) The French new rolling census. *Statistical Journal of the United Nations Economic Commission for Europe,* 22(1): 3–12.

General Register Office (1961) *Census 1961, England and Wales Preliminary Report.* HMSO, London.

Higgs, E. (1989) *Making Sense of the Census: The Manuscript Returns for England and Wales 1801–1901.* HMSO, London.

Lawton, R. (ed.) (1978) *The Census and Social Structure: An Interpretive Guide to 19th Century Censuses in England and Wales.* Frank Cass, Abingdon.

Liddington, J. and Crawford, E. (2011) 'Women do not count, neither shall they be counted': Suffrage, citizenship and the battle for the 1911 Census. *History Workshop Journal,* 71(1): 98–127.

Malthus, T. (1798) *An Essay on the Principle of Population.* J. Johnson, in St Paul's Church-yard, London.

Marsh, C. and Teague, A. (1992) Samples of anonymised records from the 1991 Census. *Population Trends,* 69: 17–26.

Martin, D. (1998) Optimizing census geography: The separation of collection and output geographies. *International Journal of Geographical Information Science*, 12(7): 673–685.

Martin, D. (2002) Geography for the 2001 Census in England and Wales. *Population Trends*, 108: 7–15.

Martin, D., Cockings, S. and Harfoot, A. (2013) Development of a geographical framework for census workplace data. *Journal of the Royal Statistical Society: Series A (Statistics in Society)*, 176(2): 585–602.

Martin, G. (2003) *Domesday Book: A Complete Translation*. Penguin Books, London.

Norman, P., Simpson, L. and Sabater, A. (2008) 'Estimating with Confidence' and hindsight: New UK small-area population estimates for 1991. *Population, Space and Place*, 14(5): 449–472.

Officer, L. and Williamson, S. (2016) Five ways to compute the relative value of a UK pound amount, 1270 to present. MeasuringWorth. Available at: www.measuringworth.com/calculators/ukcompare/index.php.

Office of Population Censuses and Surveys (1994) Undercoverage in Great Britain. *OPCS 1991 Census User Guide No. 58*, HMSO, London.

Office of Population Censuses and Surveys, General Register Office (Scotland). Census Branch (1966) Census Small Area Statistics (Ward Library), [data collection], UK Data Service. Accessed 18 May 2016, SN: 1488. Available at: http://dx.doi.org/10.5255/UKDA-SN-1488-1.

ONS (n.d.) Comparing census topics over time. Available at: www.ons.gov.uk/ons/guide-method/census/1991-and-earlier-censuses/guide-to-earlier-census-data/comparing-census-topics-over-time.pdf.

ONS (2012a) Providing the online census. Census Update February 2012. Available at: www.ons.gov.uk/ons/guide-method/census/2011/the-2011-census/the-2011-census-project/2011-census-updates-and-evaluation-reports/2011-census-update--providing-the-online-census.pdf.

ONS (2012b) Census 2011 Frequently Asked Questions. Available at: www.ons.gov.uk/ons/guide-method/census/2011/census-data/faq/2011-census-frequently-asked-questions.pdf.

Raban, S. (2004) *A Second Domesday? The Hundred Rolls of 1279–80*. Oxford University Press, Oxford.

Ravenstein, E.G. (1885) The laws of migration. *Journal of the Statistical Society*, XLVIII(II): 167–227

Rees, P. and Duke-Williams, O. (1997) Methods of estimating missing data on migrants in the 1991 British Census. *International Journal of Population Geography*, 3: 323–368.

Rees, P., Martin, D. and Williamson, P. (eds.) (2002) *The Census Data System*. Wiley, Chichester.

Simpson, S. and Dorling, D. (1994) Those missing millions: Implications for social statistics of non-response to the 1991 Census. *Journal of Social Policy*, 23(4): 543–567.

Stillwell, J. and Duke-Williams, O. (2007) Understanding the 2001 Census interaction data: The impact of small cell adjustment and problems of comparison with 1991. *Journal of the Royal Statistical Society: Series A (General)*, 170(2): 1–21.

Taylor, A.J. (1951) Taking of the census, 1801–1951. *British Medical Journal*, 1(4709): 715–720.

UK Parliament (n.d.) Emily Wilding Davison and Parliament. Available at: https://www.parliament.uk/about/living-heritage/transformingsociety/electionsvoting/womenvote/case-studies-women-parliament/ewd/.

United Nations Statistics Division (1997) Principles and Recommendations for Population and Housing Censuses, Revision 1. *Statistical Papers, Series M No. 67/Rev. 1*, United Nations, New York.

Vidler, G. (2001) The 2001 Census of Population. *Research Paper 01/21*, House of Commons Library Social and General Statistics Section. Available at: http://researchbriefings.files.parliament.uk/documents/RP01-21/RP01-21.pdf.

Weller, P. (2004) Identity, politics, and the future(s) of religion in the UK: The case of the religion questions in the 2001 decennial census. *Journal of Contemporary Religion*, 19(1): 3–21.

White, I. (2009) The 2011 Census taking shape: Methodological and technological developments. *Population Trends*, 136: 64.

Part II

Taking the 2011 Census and assuring the quality of the data

3

The 2011 Census

From preparation to publication

Garnett Compton, Amy Wilson and Brian French

3.1 Introduction

The planning and execution of the decennial population census is a vast and complex undertaking often described as the largest peacetime operation carried out in the UK. It is certainly the largest statistical exercise, aiming to collect socio-demographic information from every individual and household in the four home nations. Census-taking is a complex exercise because it involves the deployment of thousands of staff at peak times, the co-ordination of services across a number of different suppliers, the management of people and teams remotely, engagement with a wide range of stakeholders with a wide variety of interests and, crucially, connecting with every person and household in the country. The most recent census was undertaken in 2011 and was one of the most successful censuses of modern times.

Census-taking is a devolved responsibility in the UK. The Office for National Statistics (ONS) is responsible for the census in England and Wales, whereas in Scotland responsibility lies with the National Records of Scotland (NRS) and in Northern Ireland with the Northern Ireland Statistics and Research Agency (NISRA). The ONS is the department responsible for UK outputs and delivering UK statistics to Eurostat as part of the European Council (Framework) Census Regulation (EC No. 763/2008) (Eurostat, 2008). As a result of devolved responsibilities, the design and operation of the three censuses will differ in some aspects to meet local requirements and/or operational needs. However, a key aim of each of the Census Offices in the 2011 Census was to harmonise the census operation and most importantly the census outputs as far as possible. This was achieved through close liaison and co-operation between the three UK Census Offices. The National Statistician and the Registrars General (ONS, 2005) have a formal agreement to work together to achieve consistent and comparable census outputs – both to meet domestic users' requirements and to fulfil the UK's international obligations.

This chapter covers the main aspects involved in the taking of the 2011 Census, the vast majority of which are common between the three censuses. However, within the operation, there are areas where it was designed or implemented differently by the three UK Census Offices and, where these are significant to the story of how the 2011 Census was undertaken, these differences will be highlighted. The chapter seeks to give a flavour of some of the new approaches and methods that were introduced to ensure the delivery of a good-quality census that would meet the needs of users and address the challenges of a changing society. Given those complexities, only the key areas are explored; a more in-depth explanation of the planning, implementation and evaluation of the 2011 Censuses is available in each of the General Reports for England and Wales (ONS, 2015a), Scotland (NRS, 2015) and Northern Ireland (NISRA, 2015).

The key areas explored in the next five sections of this chapter are as follows: Section 3.2 summarises the planning and preparation for the 2011 Census, including testing and user engagement; Section 3.3 outlines the development of the questionnaire; Section 3.4 explains the data collection operation and key supporting processes such as publicity and stakeholder engagement; the data processing operation is documented in Section 3.5 and the production and dissemination of outputs are documented in Section 3.6. The chapter ends with some conclusions in Section 3.7.

3.2   Planning and preparation for the 2011 Census

Recommendations from the 2001 Census and strategic aims

The design of the 2011 Census took into account the lessons learned from the 2001 Census, as assessed by the Census Offices' own evaluations, and also the changes in society that were expected between 2001 and 2011. In addition, in England and Wales, recommendations made by external bodies, such as the Treasury Select Committee (House of Commons Treasury Committee, 2002), the National Audit Office (2002), the Public Accounts Committee (House of Commons Committee on Public Accounts, 2003) and the Local Government Association (2003) informed the design. For England and Wales, the key issues raised in these reviews covered the need to:

- select external suppliers of outsourced census operations early, using rigorous procurement procedures, and test their systems before the census;
- increase the efficiency of census questionnaire delivery by developing a high-quality and up-to-date address list;
- enable better central control of field processes and activities by developing robust field management and questionnaire-tracking systems;
- have earlier and more detailed engagement with stakeholders, particularly local authorities, and review consultation processes to ensure the disabled community's needs were taken into account;
- ensure that the views of people in Wales and the Welsh Government are better reflected in census planning, by reviewing consultation processes;
- review whether or not the coverage survey's design is sufficient to identify under-enumeration in the hardest-to-count areas;
- review the need to collect information on income;
- review the cost–benefit trade-offs in aiming to produce more timely outputs that are consistent and harmonised across the UK; and
- review the mechanisms to protect statistical confidentiality without eroding the utility of the data.

The separate internal reviews carried out in Scotland and Northern Ireland, whilst not so detailed, made recommendations in the same general areas.

Taking account of these and many other comments arising from the 2001 Census, the design of the 2011 Census in England and Wales was based on a number of broad strategic aims:

- to give the highest priority to estimating correctly the national and local populations;
- to build effective partnerships with other organisations, particularly local authorities, in planning and executing the field operation;
- to provide high-quality, value-for-money, fit-for-purpose statistics that meet user needs, inspire user confidence, and are consistent, comparable and accessible across the UK as far as possible;
- to maximise overall response rates and minimise differences in response rates in specific areas and among particular population sub-groups; and
- to protect, and be seen to protect, confidential personal census information.

The strategic aims informing the census design for Scotland and Northern Ireland were similar to those for England and Wales.

Key elements and innovations of the 2011 Census design

To achieve these aims, and to respond to changes in society since 2001, the design of the 2011 Census was significantly different from its predecessors. The societal changes included: an increasingly ageing population; a more mobile population with more complex living arrangements; increasing numbers of migrant communities, particularly from eastern European countries; and greater numbers of people in both single-person households and in dwellings with multiple household occupation.

The key elements in the design were that:

- the census would aim to cover everyone usually resident in the UK on census night, with a subset of information also collected from visitors present in households on census night;
- questionnaires would primarily be delivered by post, using a purpose-built address register;[1]
- field staff resources would be focused in areas which were particularly hard to enumerate, and from which initial response rates were low;
- the public could return completed questionnaires either by post or online;
- help would be available to anyone who had difficulty in completing the census questionnaire;
- there would be a slight increase in the number of questions compared with the 2001 Census, although the questionnaire would be re-designed to make it easier to complete;
- stringent confidentiality and security procedures would protect the information gathered in the census and would conform to the requirements of census confidentiality, data protection and freedom of information legislation, as well as to the provisions of the Statistics and Registration Service Act 2007;
- to help achieve the public co-operation that a census relies on, there would be publicity to convey the purpose and value of the census, and to give assurances about the confidentiality with which information is treated;
- initiatives would be put into place to maximise, and measure effectively, the quality of the information collected; in particular, census coverage and quality surveys would be carried out to measure the number of people not counted by the census and the quality of the responses given, quality assurance panels would review the outputs prior to publication to ensure differences with other sources were explained (thereby increasing users' confidence in the estimates and realising the benefits of the census); and
- the statistical outputs from the census would be designed to meet user requirements, and dissemination would be to a timetable.

These initiatives and other key elements of the census design are shown in Figure 3.1.

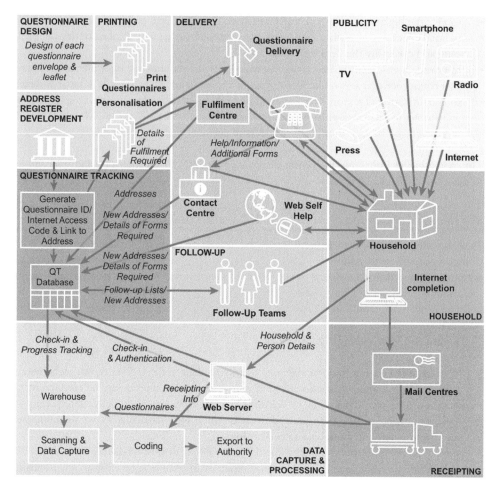

Figure 3.1   Key elements of 2011 Census design

Source: Adapted from ONS (2003, Figure 1, p. 6).

3.3   Meeting the information needs of users: the census questionnaire

The investment of time and resource in a national census can only be justified if the results are accessible to users and meet their information needs. To determine these needs, in accordance with the principles and practices in the Code of Practice for Official Statistics (UKSA, 2009), the Census Offices consulted widely with users of census data. In order to understand user needs, formal public consultations were held, supported by a number of national open meetings about particular issues, including equality-related questions on ethnicity, identity, religion and language. As part of this, the views and needs of government departments, local authorities, the health service, the academic community, the business sector, voluntary sector and local communities were collected.

The topics and questions chosen for inclusion on the 2011 Census questionnaires were those with the most demonstrable need by users, and for which census questions could be devised to produce reliable and accurate data. In each case, no other comparable and accessible source of information was available that would enable multivariate analysis with other census topics. As in the 2001 Census, the demand for questions was much more than was feasible and/or appropriate in a single census questionnaire. This demand led to challenging decisions for the Census Offices in deciding on questions and balancing the need for change against the advantages of continuity. There was a slight increase in the number of questions compared with the 2001 Census, and a re-designed questionnaire to make it easier to complete. In particular, these were:

- new questions on national identity and citizenship;[2]
- additional response categories in the ethnicity question, which differed between countries;
- new questions on second residences;[3]
- a new question on language;
- the inclusion of a civil partnership category in the marital status question;
- questions on date of entry into the UK for immigrants and intention to stay;[4]
- the omission of questions on access to toilet and bath/shower;
- in both Scotland and Northern Ireland there were questions on long-term health conditions and travel to place of study; and
- in Northern Ireland, new questions on adaptation of accommodation, Ulster-Scots and volunteering.

Developing and finalising harmonised questionnaires across the UK was particularly challenging given the devolved responsibilities for many policy and service delivery functions. Inevitably, there were some differences in the content and design of the three census questionnaires. The comparability of these questions and the subsequent results derived from the information collected are set out in the 2011 Census UK Comparability Report (ONS, 2015b).

Developing the questionnaire

A large programme of qualitative and quantitative question testing was undertaken to develop the questions. The testing consisted of five different activities: (i) cognitive testing, i.e. one-to-one interviews designed to ascertain whether or not a question was acceptable and worked as intended; (ii) a number of small-scale postal tests were conducted to collect sufficient quantitative information about questionnaire design and in particular the success of new questions; (iii) an opinions survey was used a number of times to test understanding and acceptability of new or changed questions (e.g. the migration questions) or towards colour terminology with the ethnicity question; (iv) a number of focus groups were held to understand the acceptability of particularly challenging topics and questions, including ethnicity, national identity and migration; and (v) large-scale tests in 2006 (Scotland) and in 2007 (in England, Wales and Northern Ireland) were used to test the acceptability, quality and impact on response through the inclusion of a question on income.

The near-final questionnaire was used in census rehearsals across the UK in 2009 and the final topics and questionnaire were endorsed by the respective legislatures, with the passing of the respective Census Orders and Census Regulations.

3.4 Data collection operation and key supporting processes

Using an address list as the spine of the census operation

As set out earlier, a key objective of the 2011 Census was to maximise overall response and minimise variation in response between local authorities. A critical element of this was the introduction of an up-to-date address list that in England, Wales and Northern Ireland would be crucial to the delivery of questionnaires – using Royal Mail to post questionnaires to the vast majority of households. Hand delivery of questionnaires by enumerators had been the design for each of the previous censuses. However, moving to a post-out methodology provided savings in enumerator time that could be focused more on supporting and encouraging responses from those that did not respond. It was also found that in the 2001 Census more than a third of households visited by the enumerator failed to make contact and ended up effectively posting a questionnaire anyway.

Questionnaire tracking and deployment of a flexible field force ended up effectively enabling the Census Offices to track each questionnaire through a unique identification number assigned to each dwelling at an address. This unique ID was then associated with the paper questionnaires as a barcode which was scanned when questionnaires were returned to the processing site. This enabled the Census Offices to have daily information on which addresses had not responded and prioritise field staff to areas where the response was lower than expected. This was a significant finding from the 2001 Census where there was insufficient information to accurately assess and manage the progress of the field enumeration.

Therefore, the design hinged around having an address list that had high levels of coverage to reduce the risk of under-coverage in the census operations. If households were excluded from the list, they were unlikely to receive a questionnaire and therefore had a higher propensity not to respond. However, the list also needed low over-coverage (such as commercial addresses mis-classified as residential, duplication, demolished properties and addressing errors) as these would lead to wasted postage, wasted field staff time resolving these in the field and the risk of irritating households who had already completed and returned their census questionnaires. Although hand delivery of questionnaires was still used for the vast majority of addresses in Scotland, the printed questionnaires were personalised to the address. Therefore, the accuracy of the address register was still vitally important to ensure that the correct questionnaire was delivered to the correct household.

There was no single national source of addresses that provided sufficiently recognised quality. Therefore, the Census Offices developed a single source through matching the main national address source Royal Mail Postal Address file (PAF), the local government maintained National Land and Property Gazeteer (NLPG) and the Ordnance Survey/Ordnance Survey NI address layer products and on-the-ground verification of address anomalies. More information on the development of address lists is set out in the General Reports of the three Census Offices.

In evaluating the 2011 Census, it was estimated that the address registers had under-coverage rates of about 1 per cent and over-coverage rates of nearly 4 per cent. Using information from the Census Coverage Surveys (CCS), the Census Offices were able to estimate and adjust for those households who did not complete the census, including those who were not on the original address list. The over-coverage rates were similarly estimated and resolved through field staff visits to these addresses and the CCS. However, these did have an impact on the field force and the public, in particular when the over-coverage was due to duplication of addresses.

Managing and deploying field staff

The field operation was designed to maximise the likelihood of achieving the strategic aims of the census. The main field operation comprised four distinct activities: delivery, collection, follow-up of non-responders and special enumeration.

The delivery of questionnaires to households was described earlier with the vast majority of households in England, Wales and Northern Ireland receiving their questionnaires through the post and the vast majority in Scotland receiving theirs via hand delivery. However, communal establishments and managed residential accommodation were enumerated in a more traditional manner with special enumerators visiting each communal establishment to identify the type of establishment and the number of residents, leaving the appropriate number of questionnaires. The public could respond to the census in several ways: (i) completing the paper questionnaire and posting it back to the processing site in the prepaid envelope; (ii) completing an online questionnaire (each paper questionnaire had a unique internet access code that let the householder enter the census website securely and complete their questionnaire online; this code linked their online return to their address); and (iii) handing their completed paper questionnaire to census staff on the doorstep.

The use of post-out and post-back, and online completion, were very cost effective ways to enable households to make a return with minimal effort. As a result of these changes, the 2011 Census in England and Wales was able to halve the overall size of the field force employed from about 70,000 employed in 2001 to just 35,000. At the same time, the amount of resource put into follow-up was three times more than in 2001, enabling field resources to be targeted at those unwilling or unable to make a return without support and/ or encouragement.[5] In effect, the entire field operation was designed to focus effort and resources on non-responders by: changing the field staff roles (and training) to concentrate primarily on collection; and having flexible workloads so that field staff were not assigned to one specific area.

This shift in approach required a different field staff allocation model. Rather than assigning a collector to a particular area with the sole responsibility of collecting questionnaires from non-responding households in their area (previous census models), collectors were assigned larger areas but all had the same workload (similar numbers of non-responding households). To do this, a model was developed to estimate the amount of follow-up resource required to achieve the minimum response rate thresholds for each lower super output area (LSOA). Inputs to the model included: estimates of how many households would return a questionnaire without follow-up (the initial return rate); how successful a collector would be in securing a completed questionnaire from each visit during follow-up; and how long each visit would take given the type of area (its hard-to-count classification). These inputs were based on evidence from other social surveys, the 2009 rehearsal and the 2007 test.

Generally, the right amount of resource was allocated to each area. As with any model, there will be errors in the various input parameters and to account for this some caution was used in those areas where the model was expected to have more variability. There were compensating errors in various assumptions of the model. The initial return rate was higher than expected and more visits per hour were achieved. These were offset by follow-up success being much lower than expected. However, the flexible allocation of field staff and being able to move them around during the operation and focus more resource on areas with lower-than-expected response rates (using the daily information from the questionnaire tracking system) meant that the field operation met its overall targets of maximising response and that all local authorities exceeded the minimum levels of response.

Online collection

For the first time, the 2011 Census offered households and individuals the opportunity to complete their return online, as an alternative to the traditional paper questionnaire. About 16 per cent of returns were completed through the secure online census; the majority of completed questionnaires were returned by post, as had been the case in 2001. Although the level of online response was lower than expected, the online service was regarded as a success, providing a number of benefits to the data collection operation. It:

- met the expectations of both the public and census stakeholders for an online questionnaire;
- provided an environment in which the security of the census information could be better protected;
- improved overall responses by offering an alternative to householders who may have been less inclined to complete a paper questionnaire;
- delivered a more accessible census for the disabled community; and
- avoided the need to scan and capture a significant proportion of the returns, thereby speeding up, and reducing the cost of, data processing.

Most importantly, the online service was easy to use, improved data quality by prompting for missing responses, limited the scope for incorrect responses, did not fail at any time during the process and had no security breaches. Indeed, security was the highest priority requirement in developing the online system. Confidentiality of personal information is a cornerstone of the Census Offices' assurance to the public, and any breach of data security would not be tolerated.

The daily volume of online returns for England and Wales from 21 March to 22 May is shown in Figure 3.2. Although these data are for England and Wales, the patterns of response are

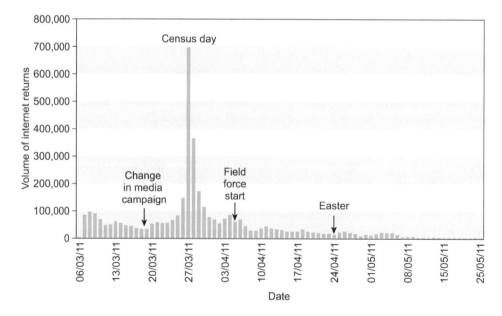

Figure 3.2 Daily volume of online census returns, England and Wales, 2011

Source: ONS (2012f, Figure 1, p. 2).

consistent across the UK. As expected, census day (27 March) had the highest volume of daily returns with about 19 per cent of the total online household returns. Apart from census day, the daily volumes of returns throughout each week peaked on a Monday.

Other notable influences on the volume of returns included: the increase from Saturday 19 March to census day following the change of media message to 'Fill it in now'; an increase on Wednesday 6 April, which coincided with the start of collectors visiting the homes of non-responders (there was a rise on Tuesday 26 April – after the Easter weekend – which probably reflected people returning from the weekend and finding reminder cards left by a collector); and rises during the week of 2 May, which reflected the last week of collectors' visits before the start of non-compliance. The great majority of respondents took between 10 and 20 minutes to complete the online questionnaire. This varied of course depending on the number of people in the household. The average time taken to complete the question-naire was 22 minutes.

The experience of running an online questionnaire, the first time by UK statistics offices, will be invaluable for informing the 2021 Census and other data collection operations. More information about the timings, location and characteristics of those that did respond online are available in the ONS report on 'Providing the Online Census' (ONS, 2012a).

Online help and support

In addition to the online completion facility, an online help system was available in English and Welsh, and Gaelic in Scotland. The system provided general information about the census, infor-mation about census questions, and answers to questions that the public might have. Answers to specific questions such as who to include on the questionnaire, how to obtain replacements, additional questionnaires or large-print versions of questions and translation leaflets proved very popular. The Help website received nearly 5 million visits. The system was especially effective at helping people decide how to complete certain parts of the questionnaire. The online help was also a popular method for requesting materials such as audio cassettes, Braille booklets or British Sign Language DVDs.

The census help-line handled over a million telephone calls during the operation. A large number of queries related to address register 'anomalies', most of which related to: houses that had been split into flats and had received only one questionnaire; houses that used to be flats and were no longer sub-divided which had received multiple questionnaires; properties that were incorrectly addressed (such as 'Basement flat' instead of 'Lower flat', or 'Flat A' instead of 'Flat 1'); and Welsh road names that were more commonly known in their English form, and *vice versa*.

Publicising the 2011 Census

The success of the census depended on making contact with every UK household. This presented a unique challenge to marketing communications as it necessitated engaging with every household in the UK and motivating them to fill in their census questionnaire. The increasing diversity of households made such contact difficult for key population groups such as ethnic minorities, migrants and young adults – which are some of the very groups for which census information is critical. It was impossible to reach all census audiences through one communications channel. Therefore, various communications channels and activities, including paid-for, owned and earned (free) media channels, were employed depending on the target audience.

The paid-for media channels for the campaign were: TV advertising; outdoor/out-of-home advertising; digital and social media; Black and minority ethnic (BME) radio and print advertising; and magazine advertising targeting young adults. Pro-active media relations, for example news releases, worked alongside advertising and paid-for communications to generate free coverage on radio, TV and print.

Local authority and community engagement – key partnerships

A census – which encompasses the whole population – has an exceptionally large number of stakeholders with varying degrees of influence and interest. For the 2011 Census, the UK Census Offices took a more systematic and consistent approach to communicating with stakeholders. In particular, extensive programmes of local authority and community liaison were initiated that were appropriate and relevant to each of the Census Offices. In particular, the Census Offices developed programmes that were aimed at working with local authorities and community representatives to utilise their knowledge and understanding of their local areas and communities.

To help support the strategic aims of the 2011 Census, the Census Offices initiated a programme of local authority liaison and engagement with the aims of: raising awareness and understanding of the 2011 Census; explaining the role that local authorities could play in participating and supporting the census; and building trust and confidence in census methodology.

Previous censuses have shown that certain population groups are less likely to complete and return their questionnaires. These include young men, certain black and minority ethnic groups, the very elderly, low-income families, non-English speakers and disabled people. The reasons for low response vary from an unwillingness to complete the questionnaire (because of concerns or misunderstandings of how the information is used), through to potential barriers such as a lack of English language skills. Community liaison activities were therefore designed and established to work with community representatives to:

- identify the main groups at risk of not responding and understand barriers to participation;
- initiate contacts with groups to communicate the purposes of the census and how data from it can be used;
- identify resources to interact with all community planning partnerships across the UK on census-related issues; and
- implement a system to identify 'champions' to promote the census from the largest identifiable group.

Some of the specific activities that flowed from these partnerships were:

- access to current, established, and effective networks and communication channels set up by such communities;
- access to translators (for example, contact with Northern Ireland Centre for Ethnic Minorities);
- designing tailored approaches to what best suits different groups (for the traveller community, for example, invitation to present and interact at a traveller-related event organised by Belfast City Council);
- partnership with influential and strategic organisations, including the Royal National Institute of Blind People (RNIB) and the Royal National Institute of the Deaf (RNID), that

resulted in targeted information campaigns and sign-posting to help-lines and facilities provided; and

- hints and tips on best methods of engagement.

Engagement with these local authority and community partners started before the rehearsal in 2009 and continued through the publication of outputs.

Confidentiality, security and privacy

The Census Offices recognised that the public needs to be confident that personal information collected in the census will be held securely. As in previous censuses, assurances were given to the public that all the information provided would be treated in the strictest confidence. An independent information assurance review (ONS, 2011a) was carried out prior to and during the census operation, covering a wide range of planning, management and implementation activities. The review team noted that, from the outset, ensuring the protection of personal information provided by the public had been a core objective in planning for the 2011 Census. They concluded: "As a result of our review, we are very satisfied that the three Census Offices are managing Information Assurance pragmatically, appropriately and cost-effectively. We are, therefore, confident that they are capable of delivering their IA [information assurance] objectives and that information will be handled in line with best practice and Government standards. The public can be assured that the information they provide to the 2011 Census will be well protected" (ONS, 2011a, p. 5).

The information collected in the 2011 Census was used solely for the production of statistics and statistical research. The Census Offices applied a statistical disclosure control process that modified some of the data before the statistics were released. The method employed was record swapping, which always introduces some uncertainty as to whether the value of any given small count is the true value. These measures proved satisfactory for protecting statistical confidentiality within the published census outputs.

3.5 Processing the results and data quality

In all, more than 27 million census questionnaires required data processing, which began by scanning the paper questionnaires and automatically capturing their data which were then merged with data captured through the online questionnaires. The data were validated to ensure that the values for each question were within the range specified in the relevant coding frame and that there were no duplicate responses. Coders assigned numerical values to written text and ticked boxes, applying coding rules and standardised national coding frames, such as SIC07 (Standard Industrial Classification 2007) and SOC2010 (Standard Occupational Classification 2010). This processing was carried out at specially commissioned and secure sites in Trafford Park, Manchester, for England, Wales and Northern Ireland returns, and in Livingston for Scottish returns.

After capture and coding, the data were passed to the UK Census Offices for further processing to validate, adjust and prepare for producing and disseminating results. This involved a number of key stages that are outlined in Table 3.1.

Overall, the methods and data sources used to capture, clean, validate and quality assure the census results were transparent and gave users confidence in the process and hence the census results.

Table 3.1 Key stages in the processing of 2011 Census results

Processing activity	Description
Range checks	This process checked that the value of each variable was within the valid range for that variable. For example, there were four valid values for the sex variable: male, female, missing or multi-tick. The 'range checks' process verified that all values for the sex variable were one of these four valid values. If an invalid value was found it was set to 'missing'. Missing values were then imputed as part of the edit and imputation process.
Removing false person records	A person record was created during the recognition phase each time at least one mark was detected in any of the person questions. Records could be created in error if, for example, there was dust on the scanners that was incorrectly interpreted as a mark. This process identified genuine person records and those found not to be genuine were flagged as an invalid person record.
	For a person record to be counted as a genuine response and kept in the data, the following information had to be present on the record:
	• name (from individual questions or household members table) or date of birth; and • at least one other item, different from the above filter, from: name (from individual questions), date of birth, sex, marital status, or name (from household members table).
Resolving multiple responses	This process resolved both household and individual multiple responses at the same address. There was an increased likelihood of multiple responses from the same household occurring in the 2011 Census because of the introduction of online completion. A multiple household response was created if a paper and an online response were returned for the same address. A multiple individual response was created if a person was included on the same questionnaire more than once.
	Multiple responses were identified by looking for more than one response for an address ID or by looking at all of the individuals within a household for the same individual more than once. Name, date of birth, and sex were the variables used for the primary matching.
	When multiple responses relating to the same household or individual were identified, the records were merged to leave just one record for the household or individual. The most complete response was kept, with any missing variables being filled in from the other response(s) if possible. In the case of multiple individual responses, a response on an individual questionnaire was given priority over a response on a household questionnaire.
	Across the UK, the numbers identified and removed as multiple responses were relatively low. For example, in England and Wales, 237,000 individual records (0.4 per cent) and 181,000 household records (0.8 per cent) were identified and removed as being multiple responses. More information on these early processes is set out in the ONS report, 'Data Capture, Cleaning and Coding' (ONS, 2012b).
Edit and imputation	Respondents to any census sometimes make mistakes in their answers. This results in missing data or invalid responses which are inconsistent with other values on the questionnaire. An edit and imputation method was used to correct inconsistencies and estimate missing data whilst preserving the relationships between census characteristics (ONS, 2012c).

Processing activity	Description

Coverage assessment and adjustment

The coverage assessment and adjustment operation helped the Census Offices to adjust for the number of people and households not counted in the 2011 Census. The extent of this under counting was identified using the post-enumeration Census Coverage Survey (CCS covered approximately 400,000 households in the UK). Standard statistical estimation techniques were then used to produce an adjusted database from which the final census results were produced. These results also formed the new 2011 base for the mid-year population estimates in each country. More information on coverage assessment and adjustment is detailed in Chapter 4 of this book and in ONS (2012d).

In Northern Ireland, the CCS was supplemented by the Census Under Enumeration (CUE) Project. The CUE Project was initiated to augment the coverage of the census enumeration (that is, completed questionnaires) by using activity-based administrative data from the medical system to supply core information on non-responding households. Activities such as the collection of a prescription, changes to registration details and treatment by a dentist or optician were considered to provide good evidence of residence. The administrative data were considered an additional source of information along with the CCS. The CUE methodology was piloted and refined during both the 2007 test and the 2009 rehearsal.

The results of the CUE/CCS were then matched, at the individual level, to the corresponding 2011 Census data, identifying the number and characteristics of those missed in the census. The combined census and CCS information, along with statistical models, were used to produce an estimate of the number of people missed by the census. The people and households estimated to have been missed were then added to the database using similar techniques and processes to the edit and imputation stage.

In approximate terms, 94 per cent of households were enumerated through completed questionnaires, 4 per cent through the CUE and 2 per cent through the CCS. More information can be found in the Northern Ireland General Report of the 2011 Census (NISRA, 2015).

It is expected that this approach will be further developed by the UK Census Offices, particularly given the anticipated increased use of administrative data to augment and quality assure the 2021 Census.

Quality assurance (QA)

QA procedures were built into all stages of data processing, and the 2011 Census estimates were subject to a rigorous QA process prior to their release (see ONS, 2011b). The overall aim was to provide confidence in the estimates by using comparator datasets and by conducting a series of vital checks.

The Census Offices carried out their own QA, but adopted the same general approach which is detailed in Chapter 4. Key steps in the process were as follows:

- a range of quality assurance panels reviewed estimates at varying levels of detail, including different geographic levels;
- a range of evidence was considered, including comparison with administrative data sources;
- the quality assurance process checked persons and their key characteristics (for example, students, armed forces, ethnicity);
- estimates of households occupied by usual residents were also quality assured;
- identifying issues which were adjusted for in the data processing; and
- further analysis to explain inconsistencies with the comparator data against which census estimates were evaluated.

(Continued)

Table 3.1 (Continued)

Processing activity	Description
Assigning output geographies	This activity assigns each person and household record a number of geographies based on the address information collected in the census (such as usual residence, workplace address, second address). These are then used to allocate the records to any particular output geography, such as output area or workplace zone.
Statistical disclosure control	This activity applies the statistical disclosure control measure, record swapping, to protect confidentiality and minimise the risk of accidentally disclosing information about an individual or household in the main census outputs (see ONS, 2012e).
Creating derived variables	Some of the census results use variables derived from more than one census question; for example, age is derived from date of birth, and distance travelled to work is estimated from the location of the addresses of the place of usual residence and the place of work. This activity seeks to create all of the required derived variables.

3.6 Outputs

The 2011 Census provides the most complex and comprehensive set of information about the population ever produced. The recent growth in the demand for information, especially through the internet and social media, has encouraged a high user expectation regarding content and delivery that includes collaboration and user participation. The ultimate benefits of the census are only realised when the users of census data make use of the published outputs. Therefore, the investment of time and resources in a census can be justified only if the results are made accessible and the outputs produced meet users' needs. Other chapters in Part V of this book provide excellent examples of the wide range of research uses that census results have been put to.

The Census Offices used their experience of the 2001 Census, user feedback and newly developed technologies to maximise census benefits, delivering a suite of products and services which included:

- the web as the primary dissemination route, with minimum paper products;
- increased ability to analyse census data online, for example through data explorer functionality and interactive graphics;
- comprehensive metadata delivered alongside the data;
- utilities to enable bulk download of census results via the web;
- larger set of products, ensuring maximum analytical use;
- microdata products (provided via secure mechanisms as appropriate);
- the provision of updated data for inclusion in the three UK Longitudinal Studies;
- licensed access for trusted users to more complex outputs that do not satisfy disclosure control requirements for public availability; and
- the provision of outputs to meet the requirements of EU regulations.

From the first release of 2001 Census results onwards, the Census Offices sought feedback on all aspects of census outputs. User forums such as the long-running Census Advisory Groups for England and Wales and Northern Ireland, and the Population and Migration Statistics Committee for Scotland, and specific interest groups such as the Microdata Working Group, provided the Census Offices with mechanisms to understand ongoing and evolving needs. The 2011 Census outputs teams also carried out focused consultations using mechanisms such as web surveys, blogs and wikis. The outcomes from all these consultations were collated and used to influence the decisions taken in defining the 2011 Census output programmes.

As far as possible, a UK-wide approach was taken to understanding users' high-level output requirements, with ONS, NRS and NISRA collaborating on different aspects of the user consultation programme on planned output. Users from all sectors – academic, commercial, central government, local government, health and others – confirmed that the planned design of outputs reflected their interests in, and needs for, both the retention of comparability with 2001 and information on new topics to be collected in 2011.

The main outcomes from the consultations that influenced the design and development of the census outputs were:

- for comparability and continuity purposes, only minor or no changes were made to the majority of existing 2001 Census outputs, other than those necessary due to changes in questions, or for statistical disclosure control purposes;
- to preserve stability, some outputs had additional age breakdowns incorporated and the statistical disclosure control method of targeted record swapping was developed to address the additivity and consistency problems arising from the post-tabular application of small cell adjustment in 2001;
- comparability documentation was published, highlighting areas of change between 2001 and 2011 and describing how outputs may be affected;
- the outputs reflect the width and breadth of the data collected on those topics included for the first time; and
- the generation, for the first time, of alternative population bases was a particularly popular innovation, including the bases for short-term migrants, workplace, workday/daytime, and out-of-term populations.

Output geography

Geography is a key element of census outputs. Every statistic produced from the 2011 Census is available for at least one of the various administrative or statistical geographies in the UK. An overall aim for the 2011 Census was to provide outputs in line with the National Statistics (NS) geography policy (ONS, 2015). The policy sets out the principles for using geographic information to produce and disseminate statistics. Its principles are driven by the objectives to:

- reference statistical events accurately, consistently and at as low a level of geographical referencing as possible;
- maximise the comparability of 'National Statistics';
- minimise the impact of changing area boundaries on National Statistics outputs; and
- provide the framework for defining and standardising how geographies and associated information are defined, used and presented in the production of statistics.

The 2011 Census results for output geographies are aggregations of whole output areas, which have been best-fitted to the higher geographies that were current at 31 December 2011. This is the method used to produce all 2011 Census and National Statistics, so that statistics produced on the same geography are consistent, comparable and non-disclosive. The exceptions to this are the exact-fit estimates for local authority areas (to which whole super output areas (SOAs) and thus output areas (OAs) are constrained), workplace zones (which have been created by merging or splitting whole OAs (Ralphs, 2011) and national parks (because best-fit estimates were considered to be inappropriate for this largely rural geography).

In Northern Ireland, the main outputs were based on a fully-nested hierarchical geographic structure of Small Areas (based largely on 2001 OAs) within SOAs within Electoral Wards within Local Government Districts. The main outputs were based on the administrative geography in place in 2011, namely the local government boundaries in place since a review held in 1992.

Dissemination

The vision for 2011 was that the web would be the primary dissemination route, and would offer users easy navigation and functionality to customise outputs, charting and thematic mapping. To achieve this, the Census Offices' existing web services were enhanced. Both Northern Ireland and Scotland provided enhanced access and functionality to the 2011 outputs through their websites and a Web Data Access programme was initiated to provide new functionality on the ONS website for the dissemination of England and Wales, and UK results. These channels were the primary vehicle for the publishing of the census standard products.

Technological developments have enabled the Census Offices to present data in more innovative ways, including infographics and data visualisations. Infographics, in particular, are an effective way to summarise census data and highlight key insights as indicated in Chapter 10 of this book. Similarly, data visualisations allow users to explore different variables and have more control over what they want to see. As a result, 2011 Census data are more relevant to a wider audience, as evidenced by the large numbers viewing the data and using the online tools.

The Census Offices strove to meet users' requirements for statistics at varying levels of detail, for a number of geographies, subject to the overriding requirement to protect statistical confidentiality. The main products disseminated through the Census Offices' websites include univariate and multivariate statistics, supported by statistical bulletins, short stories and data visualisations, including infographics and interactive maps. The key products made available from the Census Offices include standard and specialist census products. The former include:

- *Key Statistics* – largely percentages of selected key variables, designed to enable easy comparison across the geographies for which they were produced (for example local authorities).
- *Quick Statistics* – provided more detail on the breakdowns, or classifications, within a single census topic or variable, for output areas and higher (for example, a full breakdown of the ethnic group categories or single year of age population for a given geography).
- *Local Characteristics* – a series of more complex datasets designed to be available at output areas and all higher geographies. Often contained cross-classifications of more than one topic, such as age, sex, general health and ethnicity.
- *Detailed Characteristics* – similar to the local characteristics, but provided more detail within the topics (for example, full ethnic group classification) and available at a higher geography – mostly at middle layer super output area (MSOA) level and above.

As in previous censuses, the Census Offices have produced more specialist products aimed at specific audiences, such as the expert user. These include:

- *Small population groups* – subject to disclosure control restrictions, the detailed characteristics of some small population sub-groups, such as the Ravidassian and Nepalese communities who had made strong representations to ONS prior to the census. The threshold for table production is 50 or more qualifying people in any given MSOA. Separate sets of outputs have been developed for areas where there are 100 or more, and 200 or more, people from the same small population group.

- *Microdata files* – (often referred to as Samples of Anonymised Records (SARs)) have been produced from each census since 1991. These datasets comprise files containing a sample of individuals or households drawn from the census database that have been anonymised and made available through different channels. Available are: (i) a public 1 per cent sample of persons; (ii) a safeguarded file, with a sample size of about 5 per cent, through the UK Data Service (UKDS) and after application, accessible via users' desktops; and (iii) two secure files available through the ONS' virtual microdata laboratory (VML) and made available only to approved researchers. Containing maximum sample sizes of 10 per cent, these are similar to the 2001 Controlled Access Microdata Samples (CAMS) for both households and individuals.

- *Origin–destination data* – (also known as flow data) comprised the travel-to-work and migration patterns of individuals, cross-tabulated by key variables of interest (for example, occupation). New products for the 2011 Census, however, provided the flow patterns separately for those living at a student address one year before the census, and also provided data on the movement of people between their usual address and any second address on which information was collected for the first time. A large number of the origin–destination outputs are at UK level, providing flows for usual residents within and between England, Wales, Scotland and Northern Ireland. However, for the 2011 Census, the UK statistical disclosure control policy required that the disclosure protection of the most detailed origin–destination tables should be controlled, in the main, through access only via ONS' secure environment, the VML. This is a change from the 2001 Census, where the protection for similar outputs came from the post-tabular small cell adjustment that still allowed wide and easy access, but which also adversely affected the utility of the outputs. There are, however, a small number of less detailed tables that are available publicly.

- *Alternative population bases* – The main output base for the 2011 Census results is usual residents. Some basic demographic outputs using population bases other than usual residents are available from the 2011 Census. These use information from a combination of different census questions (such as workplace address) to focus on alternative population bases, including:
  - *workplace population*: figures for a given geography during standard working hours, taking account of the number of people who, for example, travel into a city to work (effectively a geographical redistribution of the usually resident population who are in work, allocated to their place of work);
  - *workday population*:[6] figures for persons present in any given geography during the day, including non-residents with a workplace in the area, but excluding residents with a workplace outside the area;
  - *out-of-term population*: figures for a given geography, including students counted at their non-term-time address (that may or may not be the same as their term-time address).

UK-based statistics

In addition to the information produced and disseminated by each of the three Census Offices, the same statistics for the UK as a whole were produced (such as those required to fulfil international obligations as well as meet domestic users' requirements):

- population figures by single year of age and sex; for the UK and all local authorities (or equivalent) in the UK, along with UK historic population pyramids (1911 to 2011);

- Key Statistics and Quick Statistics for all local authorities;
- Key Statistics and Quick Statistics for OAs, SOAs and equivalents in the UK; and
- EU outputs – a set of 60 multivariate tables consistent across the UK to meet UK require-ments under EC regulation 763/2008. This information is available through the Eurostat Census Hub[7] and provides comparable information across all member states.

These statistics, together with interactive maps and tools, are available on the 2011 UK Census web pages.

Making the 2011 Census data accessible was paramount. However, widening the census user base is possible only when potential users are educated about how they can benefit from the available data; potential new users also require help on how to access the data. Recognising the importance of these activities in maximising the benefits of the census, the Census Offices imple-mented a number of activities to promote and enhance the use of census data.

Case studies are a good and easy way to showcase different uses and benefits of census data to potential new users. The chapters in Part V of the book provide evidence of how census data are used by the academic research community. The Census Offices used their websites to illustrate how people/organisations can benefit from the 2011 Census. These pages show examples of the many ways of using census data and how different organisations from the private, public and voluntary sectors also benefit. Furthermore, toolkits, factsheets and instructions to help get people started with different census products were also made available to download. Some examples of uses and promotion outside the academic sector include:

- The Northern Ireland Housing Executive (NIHE) commissioned research to develop a census-based model of future housing need in Northern Ireland. In addition, on behalf of the NIHE, 2011 Census data have been used to model three key housing quality indicators at local government district level in relation to strategies on fuel poverty, decent homes and dwelling unfitness. This research, which benefited from additional work by Census Office staff in relation to the full 2011 Census dataset, enabled a reduction in the sample size of a recent Northern Ireland House Condition Survey and realised cost savings in the region of £250,000.
- In Scotland, NRS has supported a number of training programmes for prospective and practising school teachers and helped them develop practical skills in accessing and inter-preting the statistics. The training has been delivered to trainee teachers in mathematics and social subjects through Continuous Professional Development, and at a range of conferences including the Scottish Mathematical Council, the Scottish Association of Geography Teach-ers and the Royal Statistical Society.
- As part of the Curriculum for Excellence in Scotland, all teachers have a responsibility for developing numeracy, both within the early stages of a broad general education and within the senior phase of school education. In addition, the Curriculum for Excellence emphasis on developing learners' capacities as 'responsible citizens' and 'effective contributors' lends itself well to accessing and interpreting statistics about society. Studying Scotland's people is already a feature of Education Scotland's support materials but there was a clear opportunity for NRS to more widely publicise the use of the census outputs in schools and communities.
- Bristol City Council works with its partners through the Children and Young People's Trust to deliver a wide range of services. From education, health and welfare to youth services and play, these services support the 92,000 children and young people in Bristol and their families. Census data are vital for delivering these council services effectively. For example, the Council's Early Years, Children and Young People's Services uses lower super output area

(LSOA) data from the 2011 Census together with the Income Deprivation Affecting Children Index (IDACI), a deprivation index that measures the local area's proportion of children under age 16 that are living in low-income household, to prioritise initiatives and funding. More specifically, these services use: (i) LSOA census population estimates with IDACI to allocate around 85 per cent of all Children Centre funding in a targeted way; (ii) IDACI by LSOA for the allocation of deprivation supplements to providers of the free early education entitlement for three and four year olds; and (iii) LSOA census estimates to ensure sufficient childcare is available, assess take-up of Children Centre services, aid in developing and managing the childcare market, and plan for the implementation of the free entitlement for two year olds, all of which are statutory duties.

3.7 Conclusions

The census is a unique source of data that has a wide range of uses involving an extensive number of users. The 2011 Census was considered a success and the underlying reasons for that success were:

- the planning undertaken with the end-user in mind and being clear about the benefits that had to be realised – each Census Office shared a vision and an understanding about how the results were used and collectively worked towards fulfilling that vision;
- the programmes had agreed success criteria which were used to drive the design and decision making – continuous monitoring of progress towards goals, extensive testing and rehearsing of procedures and methods, responding to lessons learned and introducing new innovations to address lessons learned and changes in society and technology, as well as having sufficient resources and people with the right skills at the right time;
- recognition of the importance of stakeholder engagement and establishing partnerships with local authorities and community organisations and representatives; and
- ensuring that the confidentiality and security of the information collected in the census was a matter of the highest priority.

All of the Census Offices have undertaken a range of activities with users to promote and identify the benefits which flow from the census. These activities have been successful in broadening the use of the outputs and improving the Census Offices' understanding of the uses of census data. For the first time, the Census Offices attempted to measure the scale of the benefits generated by the census and having an understanding of the wide range of users and uses of census data was critical to quantifying the benefits. Measuring the benefits of official statistics such as those collected by the census is difficult, as users of census information find it hard to put a value on their use. Therefore, it was important for the Census Offices to work closely with individual users and representative bodies (e.g. trade associations) to understand the use made and to explore what alternative data might be used.

ONS concluded an assessment of the 2011 Census benefits in England and Wales in January 2014. ONS identified and quantified the benefits for government departments, the wider public sector, local authorities, businesses and genealogy/family history. Many examples of uses of census data were gathered from other sectors but, because of the diverse and fragmented nature of some of these sectors, it was not possible to estimate total benefits for each sector. More detail on the methods used in estimating the benefits is detailed in Cope (2015). This assessment concluded that over a ten-year period, the costs for the 2011 Census were estimated at £482 million in cash prices. The re-valued 2011 Census benefits were estimated at £489.5 million each year, which equates to a payback period of around 14 months. The total benefits from the 2011 Census turned out to be significantly higher than the £720 million estimated in the 2007 business case. However, it was also noted that quantifying the benefits to central government in areas such as policy development and evaluation was particularly difficult.

Notes

1 In Scotland, the vast majority of census questionnaires were hand delivered by enumerators. Postal delivery was only used in around 5 per cent of cases covering remote areas.
2 The question on citizenship was not asked in Scotland.
3 The question on second residence was only asked in England and Wales.
4 The question on intention to stay was not asked in Scotland.
5 In Scotland, where hand delivery by enumerators was used in both 2001 and 2011, the size of field force reduced from around 8,000 to just under 7,000. Northern Ireland adopted postal delivery of census questionnaires for the vast majority of addresses (other than in rural Fermanagh), but maintained the model of an enumerator taking responsibility for a given enumeration district; its number of enumerators fell from around 2,500 in 2001 to around 1,500 in 2011.
6 Daytime populations in Scotland and Northern Ireland.
7 http://ec.europa.eu/eurostat/web/population-and-housing-census/census-data/2011-census.

References

Cope, I. (2015) The Value of Census Statistics in England and Wales. *United Nations Economic and Social Council ECE/CES/GE.41/2015/16.* Available at: www.unece.org/fileadmin/DAM/stats/documents/ece/ces/ge.41/2015/mtg1/CES_GE.41_2015_16_-_UK.pdf.

Eurostat (2008) Regulation (EC) No. 763/2008 of the European Parliament and of the Council of 9 July 2008 on Population and Housing Censuses. *Official Journal of the European Union* L218/14. Available at: http://eur-lex.europa.eu/legal-content/EN/TXT/?qid=1412688957286&uri=CELEX:32008R0763.

House of Commons Committee on Public Accounts (2003) The Office for National Statistics: Outsourcing the 2001 Census. Ninth Report of Session 2002–03. HC 543. The Stationery Office, London. Available at: www.publications.parliament.uk/pa/cm200203/cmselect/cmpubacc/543/54302.htm.

House of Commons Treasury Committee (2002) The 2001 Census in England and Wales. First Report of Session 2001–02. HC 310. The Stationery Office, London.

Local Government Association (2003) 2001 One Number Census and its Quality Assurance: A Review. *LGA Research Briefing 6.03.* Local Government Association, London.

National Audit Office (2002) Outsourcing the 2001 Census. Report by the Controller and Auditor General. HC 1211, Session 2001/02. The Stationery Office, London. Available at: www.nao.org.uk/report/office-for-national-statistics-outsourcing-the-2001-census/.

NISRA (2015) Northern Ireland Census 2011: General Report. NRS (2015) Available at: https://www.nisra.gov.uk/publications/2011-census-general-report.

NRS (2015) Scotland's Census 2011: General Report. October, NRS. Available at: http://www.scotlandscensus.gov.uk/documents/censusresults/Scotland's_Census_2011_General_Report.pdf.

ONS (2003) *The 2011 Census: A Proposed Design for England and Wales.* Discussion Paper, ONS, Titchfield.

ONS (2005) The Conduct of the 2011 Censuses in the UK – Statement of Agreement of the National Statistician and the Registrars General for Scotland and Northern Ireland. Available at: www.ons.gov.uk/guide-method/census/2011/how-our-census-works/how-we-planned-the-2011-census/index.html.

ONS (2011a) 2011 Census Security: Report of the Independent Review Team. Available at: https://www.ons.gov.uk/file?uri=/census/2011census/confidentiality/assessingourmeasurestoprotectyourconfidentiality/iia_tcm77-229707.pdf.

ONS (2011b) 2011 Census – Methodology for Quality Assuring the Census Population Estimates. Available at: http://www.ons.gov.uk/ons/guide-method/census/2011/the-2011-census/processing-the-information/data-quality-assurance/2011-census---methodology-for-quality-assuring-the-census-population-estimates.pdf.

ONS (2012a) Providing the Online Census. Update. Available at: http://www.ons.gov.uk/ons/guide-method/census/2011/how-our-census-works/how-did-we-do-in-2011-/index.html.

ONS (2012b) Data Capture, Coding and Cleaning for the 2011 Census. Available at: http://www.ons.gov. uk/ons/guide-method/census/2011/census-data/2011-census-user-guide/quality-and-methods/qual- ity/quality-measures/data-capture--coding-and-cleaning/report--data-capture--coding-and-cleaning. pdf.

ONS (2012c) 2011 Census Item Edit and Imputation Process. 2011 Census Methods and Quality Report. Available at: http://www.ons.gov.uk/ons/guide-method/census/2011/census-data/2011-cen- sus-user-guide/quality-and-methods/quality/quality-measures/response-and-imputation-rates/ item-edit-and-imputation-process.pdf.

ONS (2012d) The 2011 Census Coverage Assessment and Adjustment Process. 2011 Census Meth- ods and Quality Report. Available at: http://www.ons.gov.uk/ons/guide-method/census/2011/ census-data/2011-census-user-guide/quality-and-methods/quality/quality-measures/response-and- imputation-rates/item-edit-and-imputation-process.pdf.

ONS (2012e) Statistical Disclosure Control for 2011 Census. Available at: http://www.ons.gov.uk/ons/ guide-method/census/2011/the-2011-census/processing-the-information/statistical-methodology/ statistical-disclosure-control-for-2011-census.pdf.

ONS (2012f) Proving the online census. *2001 Census Update*, ONS, Titchfield.

ONS (2015a) 2011 Census General Report for England and Wales. Available at: http://www.ons.gov. uk/ons/guide-method/census/2011/how-our-census-works/how-did-we-do-in-2011-/2011-census- general-report/index.html.

ONS (2015b) 2011 Census UK Comparability. Available at: http://www.ons.gov.uk/ons/guide-method/ census/2011/uk-census/index.html.

ONS (2015c) GSS Geography Policy. Available at: http://gss.civilservice.gov.uk/wp-content/uploads/2012/12/ GSS-Geography-Policy-is-now-available.pdf.

Ralphs, M. (2011) Exploring the Performance of Best Fitting to Provide ONS Data for Non-standard Geographical Areas. Available at: www.ons.gov.uk/ons/guide-method/geography/geographic-policy/ best-fit-policy/index.html.

UKSA (2009) Code of Practice for Official Statistics. Available at: www.statisticsauthority.gov.uk/monitoring- and-assessment/code-of-practice/.

The 2011 Census quality assurance process

David Martin and Garnett Compton

4.1 Introduction

The United Nations (UN, 2009) sets out broad principles for the conduct of population censuses internationally and this includes the need for quality management systems. Broadly speaking, quality should be measured, the most important problems and their causes identified and corrective action implemented. This must all take place within reasonable limits to accuracy, timeliness and cost. The exact implementation of these quality assurance (QA) processes necessarily varies with the details of the censuses implemented in different countries, but the processes described in this chapter fit well within the UN framework.

Starting from 2001, the Office for National Statistics (ONS) has referred to the population statistics produced by the decennial census of population as 'census estimates' (Brown et al., 1999). This does not reflect a general decrease in the quality of those figures compared to previous censuses, but is, rather, an explicit acknowledgement that census statistics can only ever hope to be a very good estimate of the size and characteristics of the population. It is impossible to conduct a census which achieves a completely accurate response from every individual. Nevertheless, the census should achieve final estimates which are very close to the true values, measured by a range of different criteria. Indeed, it is this near-complete population coverage which makes the census unique as a data collection exercise and permits the production of a huge range of statistical information for small areas and sub-groups.

The UK approach to census quality is that assessment and correction should take place before results are published, a principle established in 2001 and termed the 'One Number Census', reflecting the intention that only one estimate be published for each census statistic. There is a continual cycle of cross-reference and learning between national statistical organisations, with particular attempts to learn both from previous census rounds and from contemporary trends and developments in other countries; hence many of the high-level principles outlined here will also be found reflected strongly in Statistics Canada (2002) or Eurostat (2011).

In the UK, the census is a devolved matter. The responsibility for conducting the censuses in the UK rests with: the ONS, for England and Wales; the National Records of Scotland (NRS), for Scotland; and the Northern Ireland Statistics and Research Agency (NISRA), in Northern

Ireland. Although the statistical organisations of the UK conducted simultaneous 2011 Censuses using very similar designs, there were differences in questions, outputs and important implementation details to meet local requirements and/or operational needs. To cite one example, administrative data were used directly in Northern Ireland to inform the census estimates whereas in England and Wales and Scotland they were used only as comparators. This chapter focuses on describing the QA processes applied by ONS to the census for England and Wales. Overviews of the broadly comparable QA processes for Scotland and Northern Ireland will be found in the respective census General Reports (NRS, 2015; NISRA, 2015) and more detailed QA strategy documents (NRS, 2011; NISRA, 2012) and important differences from ONS practice in England and Wales are highlighted here.

Quality assessment of the 2001 Census of England and Wales (ONS, 2005) had concluded that overall the census was a successful exercise. However, it was noted that there were a small number of local authorities in which the census population estimates had not been sufficiently adjusted to account for exceptional circumstances. The estimates were subsequently adjusted (15 out of 376 local authorities). To reflect these issues, adjustments were made to the mid-year estimates based on the 2011 Census, but not to the census database itself (ONS, 2004).

Therefore, a key strategic aim of the 2011 Census design, as outlined by Abbott and Compton (2014), was that it should aim to produce both the best possible national estimates and comparably high-quality population estimates across all areas. This was reflected both in an adaptive enumeration strategy, allowing greater resources to be targeted to areas which were expected to prove harder to enumerate, and a QA strategy which ensured that every local authority estimate was subject to the same assessment, with extra attention being paid to areas of most concern. The priority was to achieve the most accurate possible national and local population statistics, while minimising differences in quality between areas and population sub-groups, thus producing nationally consistent and comparable data.

The figures obtained by initial summation of information on valid returned 2011 Census questionnaires are termed the census counts. Census processing involved application of various adjustment methods to produce the census estimates, for example by taking account of under- and over-coverage. It is fully expected that census questionnaires will not be received from every member of the population and a carefully planned adjustment process was implemented. This was based on the use of a large Census Coverage Survey (CCS) to inform adjustment of the initial count for people and households who did not respond in order to produce final census estimates for publication. There are many complex factors leading to census non-response (see, for example, Carter, 2009), but, ultimately, users need to be confident in the quality of the published statistics and one of the aims of the QA process was to build confidence through a robust and transparent assessment (ONS, 2009). QA involves interrogation of this entire process, starting with operational intelligence and scrutiny of provisional figures, followed by active investigation of values and patterns which fall outside agreed tolerances and implementation of improvement processes prior to publication of the final estimates.

The key questions underlying this first level of census QA may be expressed in terms of the extent to which the census correctly recorded population and household totals and their basic demographic characteristics, across local authority areas. The accuracy of census estimates can be assessed in a variety of ways, including expert review of the census design and operation, which can provide broad assurance. A conundrum when attempting to quality assure a census estimate, either the population total or its characteristics, is that we ideally require another, independent measure of the same quantity which is of known very high accuracy. Of course, if we already had a full array of such alternative counts, there would be no need to conduct a census in the first place! Statistical assessment of the 2011 Census thus employed a range of alternative data sources, all of which can offer insights but none of which can be considered definitive.

Separately from assessing the level of census coverage described above, the accuracy with which respondents answered each question on the census questionnaire was also assessed. This assessment was undertaken by means of a separate Census Quality Survey (CQS), the findings of which were published as guidance to users rather than being directly incorporated in the census statistics. Additional QA work was undertaken on specific census operations, some of which are covered in Chapter 3.

The remainder of this chapter is arranged as follows. The central section describes the principal stages in the census QA process, beginning with the distinction between counts and estimates and followed by pre-adjustment cleaning and validation, the structure and role of the QA panels, comparator data sources, validation and national adjustment. We then present some examples of how this process worked for specific local authorities and point the reader to useful census quality resources available. The next section deals with the CQS. We conclude by summarising the overall quality of the 2011 Census and the ways in which confidence can be drawn from the QA processes employed.

4.2   Stages in the 2011 Census QA process

Counts and estimates

The household response rate measures the number of households from which a census return was received, divided by the estimate of the number of households containing residents. For England and Wales, the 2011 household response rate achieved was 95 per cent. For specific local authorities, this figure ranged from 82 per cent to over 99 per cent. The equivalent measure for individuals, excluding a measure of over-counting, was 94 per cent. More detailed analysis of these and other measures of coverage and response is provided in ONS (2012a). ONS was therefore faced with a set of census responses, from which initial counts could be derived, which were lower than what was eventually established to be the best estimate of the total population. A core part of the census processing was therefore the use of the CCS to estimate the structure of under- and over-coverage, which permitted assessment of the counts by demographic groups and geographical areas. Correction resulting from this process took the form of imputation of missing households and individuals by copying records with similar characteristics to those estimated to have been missed from elsewhere in the census database. Adjustment is made complex by variation in under- and over-coverage. At the heart of the census QA process were the steps necessary to transform these initial counts into the final estimates by demographic group and geographical area and the checking that these estimates were robust and plausible using all the comparable data sources and demographic methods available. An important consideration was that the same process should be applied to all 348 local authorities in England and Wales.

Pre-adjustment cleaning and validation

Census questionnaires were initially processed to create a census database comprising record-level data for persons and households. This involved the scanning of paper questionnaires and coding using optical character recognition software, and the combination of these records with those that had been returned online. This initial database inevitably contained a range of gaps, errors and inconsistencies – for example where respondents left questions blank, ticked more than one valid answer, gave logically inconsistent answers or the capture process created errors that looked similar. Many errors of this type are detectable by logical analysis of the database, which can then be amended, for example by the removal of false multiple responses, a process described in detail in ONS (2012b), and imputation of missing or incorrect values.

Clear requirements of this type of validation, edit and imputation are that it should work only to maintain census quality, that the number of changes to inconsistent data be minimised and that as far as possible it should be undertaken in a way that produces a complete and consistent database. ONS employed the Canadian Census Edit and Imputation System (CANCEIS) for 2011 Census processing, which had been developed by Statistics Canada for use in the Canadian census. More detail on the edit and imputation process is described in detail in ONS (2012c).

CCS and coverage assessment and adjustment stages

The principal QA tool of 2011 Census quality evaluation in England and Wales was a very large CCS (ONS, 2012d), which informed assessment and adjustment of initial counts to be applied to age/sex groups, household counts and sizes, ethnicity, students and armed forces populations at the level of the census estimation area. The CCS was a voluntary survey carried out independently of the census six weeks after census day, taking place over four weeks. The survey sampled 1.5 per cent of all postcodes in England and Wales and included nearly 340,000 households in 17,400 postcodes, stratified by a hard-to-count index, developed in advance by ONS to reflect expected difficulty of enumeration. A similar approach was taken in 2001, but the hard-to-count index was derived from the 1991 Census. For 2011, a modelling approach was used which included the use of up-to-date variables to define the hard-to-count index. The full CCS sample design is documented in Brown et al. (2011).

ONS (2012e) describes the coverage assessment and adjustment process. CCS responses were matched to the original census database to understand the extent to which different population groups had been captured. This work was undertaken for groups of neighbouring local authorities known as estimation areas to ensure sufficient sample sizes. Matched census and CCS data were used within a dual system estimator (DSE) to estimate the population in the areas sampled in the CCS (Abbott, 2009). The DSEs were used within a simple ratio estimator to derive population estimates for each estimation area. During census processing, various modifications were made to the DSE and ratio estimation process to ensure that the estimates were robust and to reduce variability (mostly resulting from small sample sizes) where appropriate. This included in some cases collapsing hard-to-count groups, collapsing age/sex groups and removing CCS sample postcodes with no data as described by ONS (2012f). Counts for communal establishments were assessed using the CCS for small establishments and administrative sources for large establishments (ONS, 2012g). Households and individuals estimated to have been missed were imputed onto the census database by replicating characteristics from enumerated individuals and households to match the characteristics of those missed. This included both the imputation of additional individuals into existing households and the imputation of additional households into empty census addresses which had not responded.

Over-coverage (ONS, 2012h) has not traditionally been seen as a significant challenge in UK censuses but increases in the complexity of household structures, living arrangements (e.g. 'living together apart' and children shared between households) and parallel internet and paper enumeration in 2011 increased the opportunities for the same individuals to be returned on more than one census questionnaire. Therefore, an exercise was undertaken to search for duplicated individuals within the census database and estimate over-coverage by age/sex groups which could be included in the adjustment process. The estimates resulting from the CCS-based assessment and adjustment processes formed the basis for the work of a series of QA review panels, described below, with some areas being referred back for further investigation and adjustment work to be undertaken.

Structure and purpose of QA review panels

The design of the census QA system, including coverage assessment and adjustment, was subject to independent review and published prior to the census (Plewis et al., 2011; ONS, 2011). The review concluded that the methods give confidence that the final census population estimates will be better than any other method and will be suitable for use in resource allocation and planning. As census processing and adjustment were completed for each estimation area and its constituent local authority district(s), an intensive QA process was applied to identify any estimates that might require further investigation and potential further adjustment. Emerging census estimates were compared with a range of alternative sources and focused particularly on any estimates that fell outside tolerance bounds. A series of QA panels was convened, ranging from an internal ONS steering group which screened every set of initial estimates to a High Level panel, which focused more on specific areas and issues referred up from the other panels and on broader methodological and national issues. Final agreement to publish the census estimates was considered by an Executive QA panel and signed off by the National Statistician only once all issues had been addressed and final recommendations made at each of the lower levels.

Initially, each set of census estimates was reviewed by an internal QA steering group, consisting of a methodologist from the coverage assessment and adjustment team, a census statistician responsible for the overall QA and a demographer from the ONS team responsible for population statistics. This group reviewed the estimates by sex and five-year age group and sex ratios against key comparators and tolerance bounds at the local authority level. The aim was to undertake a quick triage to ensure that the estimation process was generally sound before passing it on to the main QA panel for a more detailed review and assessment. The principal outcomes from this initial process were to reassess information from the CCS and/or collapse some age/sex and hard-to-count strata before referring estimates to the main QA panel.

The more detailed work was then undertaken by the census QA steering group, which reviewed the routinely produced evidence for all 348 local authorities. This initial review took the form of a series of core checks and included review of the coverage-adjusted estimates by sex and five-year age group against comparators and tolerance bands at local authority level; sex ratios; socio-demographic profiles of each area based on published data; operational intelligence from the census field process and diagnostics from the census coverage estimation process including confidence intervals around the census estimates. The specific comparator data sources used in this process are discussed in greater detail below. The principle of using tolerance bounds reflected the expectation that correct census estimates should be in broad agreement with, but not exactly the same as, comparator sources and that where disagreements exceeded preset levels of acceptability, this provided a steer on where to undertake further investigation. It was necessary to adopt such an approach due to the inevitability of different sources of error and variability within each of the comparator data sources. More detail on the setting of tolerance ranges is available from ONS (2011).

Results were then taken to the Main QA panel, which included both census staff and ONS and Welsh Government experts independent of the census process. This panel reconsidered the above data together with coverage-based estimates of short-term residents, students, ethnic groups, household number/size and armed forces, and paid additional attention to mortality and fertility rates implied by the census estimates. The Main QA panel had a wider range of outcomes which included making additional adjustments (such as correction of communal establishments found to be in the wrong geographic area), but also often sought to further explain apparent differences between the census estimates and comparator sources. These included, for example, identification of problems with rolled-forward population estimates from 2001; identification of incomplete GP patient registration at student ages and investigation of mismatches between council tax data and census estimates of households in relation to halls of residence and newly built properties. The Main panel could elect to escalate specific issues to the High Level panel. The Main panel eventually recommended that all 348 local authority estimates should be published.

The High Level panel included census/demographic experts and individuals independent of the census process. It also included academic expert membership, an expert former user and representatives from the Welsh Government and devolved administrations (NRS and NISRA). Scotland and Northern Ireland thus observed the ONS process, and subsequently applied generally similar arrangements for the QA of their own census estimates, as outlined in NRS (2011) and NISRA (2012). The High Level panel considered a wide range of issues referred up from the Main QA panel as well as examining estimates aggregated to regional and national levels, and reviewed national methodological issues and adjustments. Particular attention was paid to estimates for specific groups such as babies, students, armed forces and international migrants. The High Level panel confirmed confidence in the overall QA process, endorsed the approach adopted for national adjustment and adjustments for bias and over-count and eventually recommended to the Executive QA panel that the census estimates should be published.

The Executive QA panel was responsible for the final sign-off of the national and local census population estimates ahead of publication. The panel included the National Statistician, ONS Director General, ONS Executive Management and executive management representation from the Welsh Government. Before signing off the census population estimates, the panel considered the national estimates and the local authority estimates where inconsistencies with comparator data were greatest, and reviewed a range of other QA evidence.

Comparator data sources

As already noted, the challenge of assessing census estimates against alternative data sources is that none of the alternatives can be considered definitive. The ONS Beyond 2011 programme and subsequently the Census Transformation Programme have undertaken much relevant work on the extent to which it may be possible to replace future censuses with data from administrative sources, but the principal use of administrative datasets in 2011 was to provide external comparators for the census estimates. Table 4.1 from ONS (2012i) provides a summary of the alternative sources used in the QA checks, showing those aspects of the population for which each source has utility. A more comprehensive assessment of the alternative sources may be found in ONS (2012j).

An internal check against the rolled-forward annual population estimation system was possible by comparing 2011 Census estimates against the annual series of mid-year population estimates produced by ONS, based on the 2001 Census. This is therefore a modelled output which would itself be superseded and re-based on the 2011 Census population estimates, but it served to indicate the extent to which the emerging census results differed from what was expected.

These rolled-forward 2011 population estimates were the population estimates which were rolled forward from 2001 to mid-2010, including the latest improvements to migration statistics, and then extrapolated from June 2010 to March 2011. It was necessary to extrapolate to the census date because the components of population change used to create the annual estimates were not available at the time. Rolled-forward population estimates based on the actual (once available)

Table 4.1 Checks against comparator sources (all local authorities) core QA check (ONS, 2012i)

Core QA check	Comparator dataset
Age and sex (local authority and middle layer super output area (MSOA)/ lower super output area (LSOA))	• Birth registrations • GP Patient Register • Mid-year population estimates • School Census • Social security and revenue information
Household number and average size (local authority and MSOA/LSOA)	• Council tax • Census address register • GP Patient Register • Department for Communities and Local Government household projections
Ethnicity (local authority)	• Population estimates by ethnic group • Integrated household survey • School Census
Students (residential/communal) (local authority)	• Higher Education Statistics Agency (HESA) data • Further education data from Department for Business, Innovation and Skills
Armed forces (home/foreign) (local authority)	• Defence Analytical Services Agency (DASA) data • US Armed Forces
International migrants (local authority)	• GP Patient Register • ONS international migration estimates • Migrant Worker Scan
Short-term residents (local authority)	• ONS short-term migration estimates

components of change up to March 2011 are only marginally higher (52,000) at England and Wales level than those used in the QA process (ONS, 2012k).

The NHS Patient Register provides the most comprehensive single list of population, although it is known to overestimate the total population and to have differential coverage of population sub-groups. Council tax data were available for many local authorities and ONS work on the census address register allowed the creation of an alternative household count based on addresses, which could be compared with the number of households found by the census.

Further sources provide detailed alternative estimates of specific population sub-groups and were used to compare with the census just for those groups. HESA data provide details of higher education students. Birth registration data should provide a very close match to the number of under 1 year old children, while the annual School Census provides figures for children in school education. Benefits data from the Department for Work and Pensions Customer Information System (CIS) provided further estimates of the older population receiving state pensions and children for whom child benefit is being paid. In every one of these cases, it is important to understand that the administrative sources, even if maintained entirely correctly, would often differ from the true population totals, both nationally and locally, that the census is aiming to capture, for a variety of reasons. These include the inevitable time lags with which individuals register to use services (for example, delays in registering with a GP following changes of residential address) and important definitional differences between those eligible to register for specific services and the census definition of usually resident population. It is for these reasons that comparison of census estimates against alternative sources must be done with reference to tolerance bounds rather than simple comparison of the headline values.

Validation and national adjustment

A range of validation checks were applied to the census estimates themselves by calculating and then assessing the plausibility of fertility and mortality rates, and sex ratios. Taken together, these checks were particularly useful in understanding and identifying whether there were inconsistencies or anomalies in the estimates in a particular age and sex group. Comparing new census-derived fertility rates, for example, with the longer-term trends in fertility at the local level and national level gave added assurance that the estimates for women of child-bearing age were correct. Similarly, comparing the ratio of men to women by age provided a powerful indicator in the context of information around population change, particularly migration, by sex. This is because extreme ratios would have implied that implausible levels of migration had taken place.

Lastly, one of the lessons reinforced by the 2001 Census had been the need for a process and method to assess and adjust for any residual bias that could only really be detected at a national level after all of the estimation had been completed. This last check was called the national adjustment and is detailed in more information in ONS (2012l). In summary, the national (England and Wales) age/sex estimates and the associated sex ratios were examined, following the coverage assessment and adjustment phase, to determine whether there was evidence for a national adjustment. The evidence considered included: population change by age cohort from 1991 to 2011; an analysis of implied change from 2001 given the initial census population estimates with official migration figures for the period mid-2001 to mid-2010 and other indicators of immigration from administrative sources; the ONS Longitudinal Study (Lynch et al., 2015) and the Lifetime Labour Market Database (DWP, 2013).

This evidence was presented to the High Level panel and the Executive QA panel, who agreed that there was sufficient evidence to make a national adjustment, to be made through achieving a target sex ratio. The national adjustment would only be made to males, as females were assumed to be correct. The national adjustment would be an addition to the adjustments already applied and would be conservative to ensure that the population was not overestimated. Taking these criteria into account, the overall effect of the national adjustment was to add a further 303,400 to the male population, increasing the total population by 0.5 per cent. This total was cascaded down to the local authority level within the estimation process by applying adjustments to the ratio estimator. The level of adjustment was driven by the patterns of males missed in both the census and the CCS in each five-year age band.

4.3   Local authority examples

To help demonstrate some of the analysis and interpretation undertaken as part of the census QA process, a few examples are set out below. These and more examples can be found in Annex B of ONS (2012i). The graphical presentations used here are typical of the evidence considered by the QA panels in relation to each local authority. The graphs generally show separate lines derived from mid-year population estimates, GP Patient Register, census count and final census estimate.

For specific age groups, such as school-aged children, there are additional reference points from other sources that do not cover the full age range.

Sex ratio analysis (Example – Cornwall)

An assessment of sex ratios in comparator sources for Cornwall highlighted that the mid-year population estimates were inconsistent with other sources for 20–24 year olds, shown in Figure 4.1. The sex ratio of 110 (men per 100 women) in the mid-year population estimates is likely to be a function of the estimation of male and female migration over the previous ten years. The divergence between census estimates and the tolerance bounds was greater for men than for women. As international migration to and from Cornwall is relatively low, the issue is likely to be associated with the difficulty of accurately estimating male migration between Cornwall and the rest of the UK. Migration within the UK is estimated using GP Patient Register information on changes to the residential address postcode of people registered with a GP. The conclusion reached was that the tolerance range was unusually narrow at age 20–24 because the mid-year population estimates for men were too high.

Shape of bounds and consistency across ages (Example – Reading)

The range of values between upper and lower age/sex tolerance bounds reflected the level of consistency between the sources used as comparators. The tolerances had to be interpreted with caution as the width between the bounds was not consistent across the ages. Where the width between the bounds was narrow, comparisons were interpreted with caution.

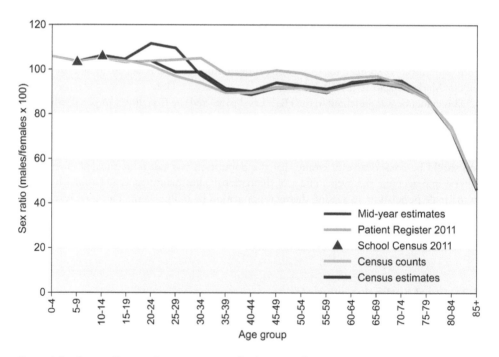

Figure 4.1 Cornwall sex ratio: census count/estimate and comparator datasets

Source: Adapted from ONS (2012i, Figure B1, p. 18).

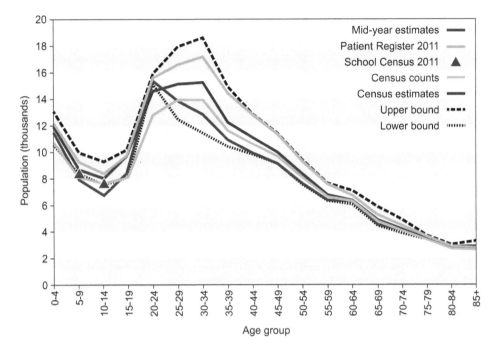

Figure 4.2   Inconsistent width in tolerance bounds (Reading)

Source: Adapted from ONS (2012i, Figure B5, p. 22).

The example of Reading, presented in Figure 4.2, shows very narrow bounds at ages 20–24 but very wide bounds at subsequent ages, particularly at 30–34. As suggested, this indicates that the narrow bounds (outside which the census estimates are falling) should be interpreted with caution. Assessing the census estimates against the tolerance bounds alone for Reading could also hide the very different age profile of the mid-year population estimates relative to the census count and the census estimate.

Inconsistencies with GP Patient Register at LSOA level (persons)

At LSOA level, census counts were generally highly correlated with the count of population on the GP Patient Register. In some local authorities, the GP Patient Register count was generally slightly higher but was consistent with evidence of list inflation. Investigation did highlight some larger differences at LSOA level which were attributed to student halls of residence having an incorrect location in the census data. These halls had been enumerated but had incorrectly been given the address of the university accommodation offices. The issue was generally associated with halls of residence having an incorrect LSOA but still being in the correct local authority. In a small number of cases there were halls of residence which had an incorrect local authority. These inconsistencies were systematically identified and corrected prior to the first release of census estimates.

The anonymised example shown in Figure 4.3 shows how the census estimates at LSOA level initially compared to the GP Patient Register. There are clearly inconsistencies either where the census estimate is higher or where the GP Patient Register is higher. Figure 4.3 also shows the same comparison after this has been corrected for. Remaining inconsistencies have been attributed to area-specific list inflation in the GP Patient Register.

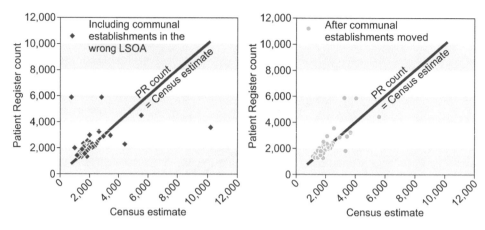

a. Before correcting hall of residence addresses   **b.** After correcting hall of residence addresses

Figure 4.3   Comparison of census estimates and GP patient registrations at LSOA level: a. before and b. after correcting student hall of residence addresses

Source: Adapted from ONS (2012i, Figure B6, p. 23).

Table 4.2 Useful census QA resources

ONS coverage assessment and adjustment process. Available at: www.ons.gov.uk/census/2011census/ howourcensusworks/howwetookthe2011census/howweprocessedtheinformation/coverageassessment-andadjustmentprocesses

ONS Data Quality Assurance. Available at: www.ons.gov.uk/census/2011census/howourcensusworks/ howwetookthe2011census/howweprocessedtheinformation/dataqualityassurance

ONS local authority QA – a variety of data packs including spreadsheets of 2011 census counts, estimates and comparator administrative sources at local authority level. Available from: UK Government Web Archive in The National Archives via http://www.ons.gov.uk/ons/guide-method/census/2011/census-data/2011-census-user-guide/quality-and-methods/local-authority-quality-assurance/index.html

Northern Ireland census evaluation and QA resources. Available at: https://www.nisra.gov.uk/statistics/ 2011-census/background/coverage-and-quality-surveys

Scotland's Census: QA resources. Available at: http://www.scotlandscensus.gov.uk/quality-assurance

Drawing on the processes described above, an extensive pack of QA information was published to help users understand the quality of the census population estimates and place the estimates in the context of other administrative data sources. This package of supporting information includes response rates by local authority by age and sex (Table 4.2); 95 per cent confidence intervals by local authority (ONS, 2012m); the size of the household bias adjustment, over-count and census estimate adjustments (ONS, 2012n); census estimates against other sources for each local authority, such as patient register, school census and child benefit, but also showing the tolerance bounds for each area (ONS, 2012o) and how the estimates were built from their count quantifying the effects of the various QA steps (ONS, 2012n). Similar resources were also made available by NRS for Scotland and NISRA for Northern Ireland. A range of relevant resources describing these processes in all parts of the UK can be found on their respective websites and are listed in Table 4.2.

4.4   Census Quality Survey: assessment of response accuracy

The Census Quality Survey (CQS), described in ONS (2014), plays a rather different role in the assessment of census quality from the CCS and processes described above. The focus here is on the accuracy with which respondents answer the various census questions and the objective is not to make alterations to the census data but rather to provide information for the guidance of census users. The CQS uses face-to-face interviews to re-ask respondents the census questions and then matches their answers back to their original census responses, allowing an assessment of the overall levels of agreement for different questions and thus an indication of the reliability of the different topics in the overall census data. The CQS was a much smaller survey than the CCS, from which it was independent, with results based on 5,172 households containing 12,103 individuals.

The CQS used a standard two-stage sample survey design, with initial selection of 288 post-code sectors, stratified on the hard-to-count index. 7,488 addresses from which census forms had been received were then sampled within these postcode sectors, stratified on response mode (paper or internet) and early or late census returns. The survey took place between May and August 2011, with the aim of interviewing all adults who had been a household member on census day. The overall CQS response rate was 75 per cent and 97.6 per cent of those interviewed were matched against a census record. Individual responses were weighted to reflect the inevitable fact that the households obtained from the stratified sampling do not comprise a set of individuals that is fully representative of the population. The principal output from the CQS was a series of agreement rates, based on the number of agreements between census values and CQS responses as a percentage of the total sample who responded to that question in the CQS. It is worth noting that this will suggest a level of agreement lower than that usually observed in the final census data, in which values such as age are often grouped.

There is a wide range of reasons why there may be differences between the census and CQS responses and these are quite informative with regard to the overall challenge of achieving a high-quality census result. The most obvious reason is that individuals gave a different answer in the census and CQS, perhaps because they made a mistake or, indeed, changed their mind as to the answer they wished to give; they may have interpreted the question differently on the questionnaire from when it was presented by the CQS interviewer. The original response may have been subject to edit or imputation in the census processing if a question was not answered in the questionnaire, or answered by a proxy respondent, in either case introducing answers different from those given by the individual themselves during the interview. There is also scope for errors to be introduced by scanning or character recognition, or by differences in the coding of answers. Finally, there may be entirely new errors introduced during the CQS, for example by the interviewer wrongly recording an answer.

Table 4.3 provides an overview of the CQS agreement rates for the individual questions. The general pattern is clearly apparent. Agreement rates are highest for sex (99.7 per cent) and country of birth (99.1 per cent) but much lower for year last worked (55 per cent) and national identity (60.4 per cent). In general, the highest rates are seen for those attributes which rarely, if ever, change, or which can unambiguously be asked and answered. More complex questions which require judgement or recall, which are liable to change or are harder to code generally achieve much lower rates.

In addition to these QA processes applied to the principal census estimates, ONS also undertook separate evaluation of many other aspects of the census process which are separately reported, such as the item edit and imputation process (ONS, 2012p) and the census address register (ONS, 2012q).

Table 4.3 Weighted agreement rates between 2011 Census and Census Quality Survey: Individuals England and Wales (ONS, 2014)

Question number	Census question	2011 CQS agreement rates (%)	2011 CQS Confidence interval width (+/– percentage points)
2	Sex	99.7	0.1
3	Date of birth	98.4	0.3
4	Marital and civil partnerships	98.1	0.3
5	Second address	97.1	0.4
7	School children/students	97.6	0.3
8	Term-time address	98.9	0.5
9	Country of birth	99.1	0.3
13	General health	68.2	1.2
14	Unpaid care	90.9	0.7
·15	National identity	60.4	1.4
16	Ethnic group (18 tick boxes)	94.7	0.8
18	Main language	96.3	0.7
20	Religion	90.4	0.9
21	Usual address one year ago	95.5	0.6
22	Passports	91.8	0.7
23	Limiting long-term illness	88.9	0.7
25	Highest qualification	67.6	1.0
26	Working status in previous week	91.2	0.6
27	Looking for work	96.2	0.6
28	Available for work	86.2	1.0
29	Waiting to start work	99.8	0.1
30	Reasons for not working	86.4	1.0
31	Ever worked	94.4	0.7
31	Year last worked	55.0	1.5
33	Self-employed or employee	94.7	0.5
34,35	Occupation code (Major group)	67.5	1.0
36	Supervisor	86.2	0.7
37,38	Industry code (Section)	74.2	0.9
40	Address of workplace (Postcode sector)	82.2	1.1
41	Travel to work	85.5	0.9
42	Hours worked	83.9	0.9

4.5 Overall quality statements

The success of any census hinges on producing relevant results to sufficient quality when they are required. Informing users about the quality of the data, and hence its limitations, is also crucial to aid user interpretation and understanding of the results. Various indicators of data quality were used to guide design and decision making about the 2011 Census. Key to the overall quality was the size of the 95 per cent confidence intervals and the overall level of response, and variation in response between local authorities. In these areas, the 2011 Census was a significant improvement on the 2001 Census.

The 95 per cent confidence interval achieved on the national population estimates of +/-0.15 per cent (+/- 83,000 people) is narrower than the equivalent confidence interval in 2001 of +/- 0.21 per cent (+/- 109,000 people), indicating more accurate population estimates. 97 per cent of local authorities had a 95 per cent confidence interval of +/- 3 per cent or better, compared with 94 per cent of local authorities in 2001. The overall response rate in 2011 was 93.9 per cent, slightly better than the 2001 overall response rate of 93.7 per cent prior to local authority adjustments. All local authorities had a response rate above 80 per cent and only 13 had a response rate below 90 per cent, compared with 2011 where 13 local authorities were below 80 per cent and 38 were below 90 per cent.

4.6 Conclusion

As described in this chapter, every QA process and panel considered an extensive range of evidence to assure the quality of the 2011 Census population estimates. The evidence assessed included operational intelligence and information provided by local authorities, the diagnostics from the coverage estimation process and data from comparator sources. Administrative data were used extensively for core checks on all 348 local authorities in England and Wales. Supplementary analysis focused on data discrepancies that were of particular concern.

Overall, the processes to quality assure the results were considered to be highly successful and enabled the key objectives to be met. Most importantly, the methods and data sources used were transparent and gave users confidence in the process and hence the census population estimates. This was a significant improvement on 2001, when the estimates for 15 local authorities were adjusted after the census results had been published. Modern censuses deliver an enormous range of data outputs and it is essential for users to understand the process by which the data they are using have been produced. In the case of the 2011 Census of England and Wales, great effort was put into achieving the best possible population estimates in a timely manner, taking account of household counts and sizes, ethnicity, students and armed forces populations. In assessing the quality of the census, users are also provided with information from the quality survey, which does not lead to further data adjustment.

The changing future nature of census will present changes to the principal quality challenges. Plans for 2021 (ONS, 2015) are likely to place greater emphasis on integration of, and calibration with, administrative sources throughout the census process but the greatly increased use of internet data collection will introduce new aspects such as the scope for mode bias between internet and paper responses. This is an issue which has already begun to be specifically addressed, for example, by the US Census Bureau in relation to the 2010 Census (US Census Bureau, 2012). A successful QA process for the 2021 Census will need to build on the successes and lessons from the 2011 Census while considering the new challenges and opportunities that will be available in 2021.

References

Abbott, O. (2009) 2011 UK Census coverage assessment and adjustment methodology. *Population Trends*, 137: 25–32.

Abbott, O. and Compton, G. (2014) Counting and estimating hard-to-survey populations in the 2011 Census. In Tourangeau, R., Edwards, B., Johnson, T.P., Wolter, K.M. and Bates, N.A. (eds.) *Hard-to-Survey Populations.* Cambridge University Press, Cambridge, pp. 58–81.

Brown, J., Abbott, O. and Smith, P.A. (2011) Design of the 2011 Census Coverage Surveys in England and Wales. *Journal of the Royal Statistical Society: Series A*, 174: 881–906.

Brown, J.J., Diamond, I., Chambers, R.L., Buckner, L.J. and Teague, A. D. (1999) A methodological strategy for a one-number census in the UK. *Journal of the Royal Statistical Society: Series A*, 162(2): 247–267.

Carter, M. (2009) *Explaining the census: Investigating reasons for non-response to the ABS census of population and housing.* Institute for Social Research, Swinburne University of Technology, Melbourne. Available at: http://www.sisr.net/documents/Census.pdf.

DWP (2013) Methodology Statement for Department for Work and Pensions Lifetime Labour Market Database Publications. DWP, London. Available at: https://www.gov.uk/government/uploads/system/uploads/attachment_data/file/243650/l2-methodology.pdf.

Eurostat (2011) European statistics code of practice for the national and community statistical authorities. Available at: http://ec.europa.eu/eurostat/documents/3859598/5921861/KS-32-11-955-EN.PDF/5fa1ebc6-90bb-43fa-888f-dde032471e15.

Lynch, K., Leib, S., Warren, J., Rogers, N. and Buxton, J. (2015) Longitudinal Study 2001–2011: Completeness of Census Linkage. *Office for National Statistics Series LS No. 11.* Available at: http://www.ons.gov.uk/ons/guide-method/user-guidance/longitudinal-study/what-s-new/ls11-for-publication.pdf .

NISRA (2012) Quality Assurance of the 2011 Census in Northern Ireland. Available at: https://www.nisra.gov.uk/sites/nisra.gov.uk/files/publications/2011-census-quality-assurance-strategy.pdf.

NISRA (2015) Northern Ireland Census 2011: General Report. Available at: https://www.nisra.gov.uk/publications/2011-census-general-report.

NRS (2011) Scotland's Census 2011 Data Quality Assurance Strategy. Available at: http://www.scotlandscensus.gov.uk/documents/methodology/cen-dat-qual-assur-2011.pdf.

NRS (2015) Scotland's Census 2011: General Report. Available at: http://www.scotlandscensus.gov.uk/documents/censusresults/Scotland's_Census_2011_General_Report.pdf.

ONS (2004) 2001 Census: Local Authority Population Studies: Full Report. Available at: http://webarchive.nationalarchives.gov.uk/20160105160709/http://www.ons.gov.uk/ons/guide-method/method-quality/specific/population-and-migration/pop-ests/local-authority-population-studies/2001-census---local-authority-population-studies--full-report.pdf.

ONS (2005) Census 2001: Quality Report for England and Wales. Available at: http://www.ons.gov.uk/ons/guide-method/census/census-2001/design-and-conduct/review-and-evaluation/evaluation-reports/quality-report/census-2001-quality-report.pdf.

ONS (2009) 2011 Census Data Quality Assurance Strategy. Available at: https://www.ons.gov.uk/file?uri=/census/2011census/howourcensusworks/howwetookthe2011census/howweprocessedtheinformation/dataqualityassurance/2011censusdataqualityassurancestrategy1_tcm77-189754.pdf.

ONS (2011) 2011 Census – Methodology for quality assuring the census population estimates. Available at: http://webarchive.nationalarchives.gov.uk/20160105160709/http://www.ons.gov.uk/ons/guide-method/census/2011/the-2011-census/processing-the-information/data-quality-assurance/2011-census---methodology-for-quality-assuring-the-census-population-estimates.pdf.

ONS (2012a) Response Rates in the 2011 Census. Available at: http://webarchive.nationalarchives.gov.uk/20160105160709/http://www.ons.gov.uk/ons/guide-method/census/2011/census-data/2011-census-data/2011-first-release/first-release--quality-assurance-and-methodology-papers/response-rates-in-the-2011-census.pdf.

ONS (2012b) 2011 Census Evaluation Report: 2011 Census Address Register. Available at: www.ons.gov.uk/ons/guide-method/census/2011/how-our-census-works/how-did-we-do-in-2011-/index.html.

ONS (2012c) 2011 Census Item Edit and Imputation Process. Available at: http://webarchive.nationalarchives.gov.uk/20160105160709/http://www.ons.gov.uk/ons/guide-method/census/2011/census-data/2011-census-user-guide/quality-and-methods/quality/quality-measures/response-and-imputation-rates/item-edit-and-imputation-process.pdf.

ONS (2012d) 2011 Census Coverage Survey Summary. Available at: http://webarchive.nationalarchives.gov.uk/20160105160709/http://ons.gov.uk/ons/guide-method/census/2011/census-data/2011-census-data/2011-first-release/first-release--quality-assurance-and-methodology-papers/census-coverage-survey-summary.pdf.

ONS (2012e) The 2011 Census Coverage Assessment and Adjustment Process. Available at: http://webarchive.nationalarchives.gov.uk/20160105160709/http://www.ons.gov.uk/ons/guide-method/census/2011/census-data/2011-census-data/2011-first-release/first-release--quality-assurance-and-methodology-papers/coverage-assessment-and-adjustment-process.pdf.

ONS (2012f) Tuning the Coverage Estimation Process. Available at: http://webarchive.nationalarchives.gov.uk/20160105160709/http://www.ons.gov.uk/ons/guide-method/census/2011/census-data/2011-census-data/2011-first-release/first-release--quality-assurance-and-methodology-papers/tuning-the-coverage-estimation-process.pdf.

ONS (2012g) Estimation and Adjustment for Communal Establishments. Available at: http://webarchive.nationalarchives.gov.uk/20160105160709/http://www.ons.gov.uk/ons/guide-method/census/2011/

census-data/2011-census-data/2011-first-release/first-release--quality-assurance-and-methodology-papers/coverage-within-communal-establishments.pdf.

ONS (2012h) Over-count Estimation and Adjustment. Available at: http://webarchive.nationalarchives.gov.uk/20160105160709/http://www.ons.gov.uk/ons/guide-method/census/2011/census-data/2011-census-data/2011-first-release/first-release--quality-assurance-and-methodology-papers/overcount-estimation-and-adjustment.pdf.

ONS (2012i) Quality Assurance of 2011 Census Population Estimates. 2011 Census: methods and quality report. Available at: http://webarchive.nationalarchives.gov.uk/20160105160709/http://www.ons.gov.uk/ons/guide-method/census/2011/census-data/2011-census-data/2011-first-release/first-release--quality-assurance-and-methodology-papers/quality-assurance-of-census-population-estimates.pdf.

ONS (2012j) Overview of Administrative Comparator Data Used in 2011 Census Quality Assurance. Available at: http://webarchive.nationalarchives.gov.uk/20160105160709/http://www.ons.gov.uk/ons/guide-method/census/2011/the-2011-census/processing-the-information/data-quality-assurance/overview-of-administrative-comparator-data-used-in-2011-census-quality-assurance.pdf.

ONS (2012k) Explaining the Difference between the 2011 Census Estimates and the Rolled-Forward Population Estimates. Available at: http://webarchive.nationalarchives.gov.uk/20160105160709/http://www.ons.gov.uk/ons/guide-method/method-quality/specific/population-and-migration/population-statistics-research-unit--psru-/difference-between-the-2011-census-estimates-and-the-rolled-forward-population-estimates.pdf.

ONS (2012l) Making a National Adjustment to the 2011 Census. Available at: http://webarchive.nationalarchives.gov.uk/20160105160709/http://www.ons.gov.uk/ons/guide-method/census/2011/census-data/2011-census-data/2011-first-release/first-release--quality-assurance-and-methodology-papers/making-a-national-adjustment-for-residual-biases.pdf.

ONS (2012m) 2011 Census Confidence Intervals. Available at: http://www.ons.gov.uk/ons/guide-method/census/2011/census-data/2011-census-user-guide/quality-and-methods/confidence-intervals/index.html.

ONS (2012n) 2011 Census Coverage Assessment and Adjustment Methods. Available at: http://www.ons.gov.uk/ons/guide-method/census/2011/census-data/2011-census-user-guide/quality-and-methods/coverage-assessment-and-adjustment-methods/index.html.

ONS (2012o) 2011 Census Local Authority Quality Assurance. Available at: http://www.ons.gov.uk/ons/guide-method/census/2011/census-data/2011-census-user-guide/quality-and-methods/local-authority-quality-assurance/index.html.

ONS (2012p) Item Edit and Imputation: Evaluation Report. Available at: http://webarchive.nationalarchives.gov.uk/20160105160709/http://www.ons.gov.uk/ons/guide-method/census/2011/how-our-census-works/how-did-we-do-in-2011-/evaluation---item-edit-and-imputation.pdf.

ONS (2012q) Data Capture, Coding and Cleaning for the 2011 Census. Available at: http://webarchive.nationalarchives.gov.uk/20160105160709/http://www.ons.gov.uk/ons/guide-method/census/2011/census-data/2011-census-user-guide/quality-and-methods/quality/quality-measures/data-capture--coding-and-cleaning/report--data-capture--coding-and-cleaning.pdf.

ONS (2014) 2011 Census Quality Survey. Available at: http://webarchive.nationalarchives.gov.uk/20160105160709/http://ons.gov.uk/ons/guide-method/census/2011/census-data/2011-census-user-guide/quality-and-methods/quality/quality-measures/assessing-accuracy-of-answers/2011-census-quality-survey-report.pdf.

ONS (2015) 2021 Census Design Document. Available at: http://www.ons.gov.uk/ons/guide-method/census/2021-census/progress-and-development/research-projects/2021-census-design-document.pdf.

Plewis, I., Simpson, L. and Williamson, P. (2011) Census 2011: Independent review of coverage assessment, adjustment and quality assurance. Available at: https://www.ons.gov.uk/file?uri=/census/2011census/howourcensusworks/howweplannedthe2011census/independentassessments/independentreviewofcoverageassessmentadjustmentandqualityassurance/2011censusreviewfinalrepor_tcm77-224561.pdf.

Statistics Canada (2002) Statistics Canada's Quality 2002 Assurance Framework. Available at: http://www.statcan.gc.ca/pub/12-586-x/12-586-x2002001-eng.pdf.

UN (2009) Principles and Recommendations for Population and Housing Censuses, Revision 2. *Statistical Papers, Series M, No. 67/Rev.2,* Department of Economic and Social Affairs, Statistics Division. United Nations. Available at: http://unstats.un.org/unsd/publication/SeriesM/Seriesm_67rev2e.pdf.

US Census Bureau (2012) 2010 Census Quality Survey. US Census Bureau. Available at: https://www.census.gov/2010census/pdf/2010_Census_Quality_Survey.pdf. The Routledge Handbook of Census Resources, Methods and Applications

Part III

Delivering different types of census data to users

UK census aggregate statistics
Characteristics and access

Justin Hayes, Rob Dymond-Green,
John Stillwell and Victoria Moody

5.1   Introduction

Census aggregate statistics (CAS) are essential and unique outputs from population censuses. They are extensive, varied and widely used data products released by the three UK national statistical agencies which provide information on the characteristics and distribution of the respective populations at a range of geographical levels. They are widely employed in local and national policy settings to inform decision making as well as in research settings where they are often the sole source of reliable information about small area populations. This chapter starts by describing the CAS and goes on to describe features of the different 2011 outputs. The ONS technical framework for dissemination of the CAS is outlined and selected online access interfaces are exemplified. Thereafter, attention is focused on providing a summary of the acquisition and processing of CAS data into a new multi-dimensional data framework with an online interface (*InFuse*) that allows users to build queries and extract data they require. Some conclusions are presented in the final section.

5.2   What are CAS and why are they important?

CAS are the principal and most widely used outputs from UK censuses. They have been produced by conducting census surveys of the entire UK population every ten years since 1801 and form the basis of much of the reporting that is required by law (Census Act 1920, s4; Census Act (Northern Ireland) 1969, s4). CAS from UK censuses are large datasets that describe the size, characteristics and geographical distribution of the UK population. They are typically counts of people or households possessing particular combinations of characteristics within defined geographical areas, which have traditionally been presented in tables. For example, the 2011 CAS tell us that in the local authority district of Birmingham, there were 9,421 individuals who were male, aged 16–24 and unemployed.

CAS are unique in several respects. They provide information about people or household characteristics for the whole population across the UK as compared with other non-census sources which are partial or relate to samples. They deliver information on individual characteristics or combinations of characteristics for small geographic areas such as output areas or wards.

Finally, the CAS 'estimates' are considered to be reliable in terms of quality and provide a high level of consistency in terms of spatial coverage. Accordingly, they are immensely important in facilitating analyses and comparisons of characteristics across the UK at a range of geographical scales from national to small local areas with populations of only a few hundred that are not possible with data from other sources. Moreover, comparison of CAS from successive censuses can identify patterns of historical change – important for understanding how communities are evolving and for forecasting future change to inform long-term resource allocation.

CAS are designed and produced to provide the agencies of central government with the information needed to make informed decisions on policy and expenditure based on reliable evidence, and so are very important because of the enormous impacts they have on many aspects of all of our lives. CAS also have many secondary uses. Beyond central government, there are many constituencies of users whose activities are heavily dependent on census data, both in isolation and to provide context to other statistics, including local government, national health agencies, water authorities, businesses, the academic sector and charities. They are used on their own, as context for other statistics, as well as providing a basis for annual statistics such as the mid-year population estimates which in turn are used to determine government expenditure across sectors and regions. They represent the definitive source of best-quality information and are available to anyone who wants to use them.

5.3   How are the CAS created and what do they look like?

In order to produce the CAS, censuses are conducted by the Census Offices to gather information about people and households at an individual or household level, as explained in Chapter 3 of this book. The information cannot be used in this raw individual form due to the identifiable personal information that it contains, so it is summarised in the CAS to anonymise the personal information provided by respondents and make the information more manageable and useful for most purposes. Individual-level data cannot be published without considerable protection due to personal confidentiality constraints, as laid down by Act of Parliament, and would not be useful for many purposes in this form due to the large volume and complexity of the individual returns. The aggregate statistics also contain counts of other objects, such as rooms, dwellings and families, together with a small number of rates (e.g. percentages or counts per thousand people).

The counts are produced by aggregating the information about individual units (persons, households) provided on the census questionnaire forms using households located by their addresses. The smallest fundamental building blocks for aggregation of unit data are the output areas (OAs), containing a minimum population of 40 households and 100 residents, which were first introduced in Scotland in the 1981 Census and elsewhere in the UK in 2001. In order to facilitate comparison between 2001 and 2011, the Census Offices have attempted to maintain stability in the definition of OAs and super output areas (SOAs) between 2001 and 2011, making modifications only where necessary due to significant population change, boundary changes at a higher level or a desire to improve social homogeneity. Merges were applied where an OA population was less than 100 people or 40 households, where a lower SOA less than 1,000 people or 400 households and where a middle SOA fell below 5,000 individuals or 2,000 households. This process resulted in a total number of 227,759 OAs in England, Wales and Scotland, and a further 4,537 small areas (SAs) in Northern Ireland. These geographies, together with the new 2011 geography of workplace zones, are described in more detail in the next chapter of this book, which also explains the higher-level geographies in each country for which the CAS are released. Table 5.1 illustrates some examples of variables taken from the 2011 CAS at different geographic levels.

Table 5.1 Selected examples of CAS from the 2011 Census

Statistic	Unit	Geographical area	Characteristics	Population base	Count
Count	Persons	Claremont Ward, Blackpool (E05001647)	In arts, entertainment and recreation or other service activity industry	All usual residents aged 16 to 74 in employment the week before the census	211
Count	Households	Brighton and Hove Local Authority District	Number of households containing one family headed by same-sex civil partners	All households	937
Count	Persons	Workplace Zone E33032181 (around the Barbican, City of London)	Employee working part-time	All usual residents aged 16 to 74 in employment in the area the week before the census	256

Table 5.1 shows us how various features combine to produce a single statistic which has one geography and a variable number of characteristics. So, in the third case, for example, we know that there were 256 usual residents aged between 16 and 74 working part-time in the workplace zone which contains the Barbican in London in the week before the 2011 Census. Traditionally, the statistics have been presented in tables which combine related characteristics so that, for example, full-time employment, part-time employment and self-employment can be viewed and considered together. Furthermore, combinations of characteristics are presented in all but the simplest area statistics, so that it is possible to look at characteristics in relation to groups defined by one or more additional characteristics.

Table 5.2 shows a table layout for ONS table LC4109EW as an example of the format of a CAS table. In this case, the table shows three characteristics: access to a car or van (which is a household characteristic that can be inherited by members of the household); age; and sex (which are both person-level characteristics). The table layout does not show actual data counts as such, but rather shows how the outputs are formatted. The table has an identification to indicate that it applies to a given database. The structure is different from a two-way table of frequencies in that totals are embedded into the table and often appear before the values for the sub-groups and sub-totals are also included within the table. Each cell contains an identifier for the statistic/count which would be contained within. The table layout usually specifies a general format which is constant for each geographical area for which it is produced. The geographical area will also have a unique identification number.

Greater value can be drawn from the figures once proportions or percentages are calculated to make comparisons between characteristics or areas. In order to do this, we need total counts for the population, although we might have some flexibility about how we define that population through our choice of denominator. So, drawing on Table 5.2, we might be interested in how many males aged 65+ in an area do not have access to a car. The variable of interest in LC4109EW is cell 0054. We would typically be interested in the proportion of male usual residents aged 65+ who do not have access to a car, which would be calculated by dividing the count obtained in cell 0054 by that obtained in cell 0053. However, it is plausible that we might alternatively wish to know the proportion of all usual residents in our area who are men aged 65+

Table 5.2 Table layout for table LC4109EW: car or van availability by sex by age

Table population: *All usual residents*

	All categories: car or van availability	No cars or vans in household	1 car or van in household	2 or more cars or vans in household
All categories: Sex				
All categories: Age	0001	0002	0003	0004
Age 0 to 15	0005	0006	0007	0008
Age 16 to 24	0009	0010	0011	0012
Age 25 to 34	0013	0014	0015	0016
Age 35 to 49	0017	0018	0019	0020
Age 50 to 64	0021	0022	0023	0024
Age 65 and over	0025	0026	0027	0028
Males				
All categories: Age	0029	0030	0031	0032
Age 0 to 15	0033	0034	0035	0036
Age 16 to 24	0037	0038	0039	0040
Age 25 to 34	0041	0042	0043	0044
Age 35 to 49	0045	0046	0047	0048
Age 50 to 64	0049	0050	0051	0052
Age 65 and over	0053	0054	0055	0056
Females				
All categories: Age	0057	0058	0059	0060
Age 0 to 15	0061	0062	0063	0064
Age 16 to 24	0065	0066	0067	0068
Age 25 to 34	0069	0070	0071	0072
Age 35 to 49	0073	0074	0075	0076
Age 50 to 64	0077	0078	0079	0080
Age 65 and over	0081	0082	0083	0084

Source: Table layout archived at The National Archives copyright@ons.gov.uk.

without access to a car, which would be calculated by dividing the count in cell 0054 by that in cell 0001. If we wished to compare proportions of men aged 65+ who do not have access to a car or van in two areas we would need to obtain variables LC4109EW0054 and LC4109EW0053 for both areas and perform the proportion calculation for each. This example reflects that, while the CAS can be conceptualised as tables, users often do not need to use whole tables; rather, their interest is in particular variables.

5.4 Different types of CAS

Over time, census outputs have been produced for an increasing range of smaller areas. At one time, data were released for local authority districts and small area data (sub-district) were produced by 'request only' in 1961. Aggregate outputs down to enumeration district were produced as core outputs from 1971 and output geographies were uncoupled from enumeration geographies in Scotland in 1981, followed by England and Wales, and Northern Ireland in 2001. In all cases, CAS outputs were specified in advance of the census as a series of cross-tabulations of variables and categories that were to be produced for geographies at different scales depending on disclosure considerations.

In the case of the 2011 Census, the ONS undertook a topic consultation in 2006 to solicit user needs which were to be taken into consideration along with a number of other factors such as data quality, public acceptability, respondent burden and operational concerns (ONS, 2006). This consultation informed the choice of questions that were asked in the Census Test in 2007 and the selection of final topics that were identified in the 2011 Census White Paper (Cabinet Office, 2008). National Records of Scotland (NRS) and the Northern Ireland Statistics and Research Agency (NISRA) conducted similar consultations and there is a degree of harmonisation between all three national statistical agencies in the questions asked and the variables included. However, variations do often occur in the variable classifications that are used for table outputs, and in the way in which they are presented and documented.

As indicated in the previous section, counts are reported on the basis of one or more variables such as sex, age and number of cars and vans, where each variable is represented using a number of categories. Sex is the simplest variable with three categories (persons, males, females) whereas there are over 100 single year of age categories. Variables may be drawn directly from responses to census questions or derived *post hoc*. Sex, for example, directly reflects responses to the question on the form, while age is derived from date of birth.

The counts in the CAS have traditionally been reported in census tables which may be either univariate, relating to one variable, or multivariate, where one variable is cross-classified by at least one other variable, such as the number of cars and vans by age and sex. In many cases, tables may contain different categorisations of the same variable; for example, tables are available for counts of individuals by single year of age, five-year age groups, ten-year age bands or a mixture of age ranges of different sizes.

The 2011 univariate Key and Quick Statistics were amongst the first families of statistics to be released in phases during 2013. The Key Statistics are a series of summary figures (tables commencing with KS) covering the full range of topics from the census and available at a range of geographies from national level down to output areas. They are often derived from more than one question on the census form. They provide data on the defining characteristics of the population grouped around the themes of 'who we are', 'how we live', and 'what we do' with information generally available as both numbers and percentages. The tables that comprise the Quick Statistics (QS) are also univariate and provide the most detailed information for single topics available from the census (i.e. they use the expanded versions of the variable classifications which have the most categories) and correspond with the 2001 univariate statistics.

There are two sets of multivariate statistics that form a substantial bulk of the 2011 Census CAS. The first of these are the Detailed Characteristics (DC) whose counts represent the cross-tabulation of two or more variables. Since the numbers in individual categories can become small, the Census Offices need to choose the categories carefully in order to maintain confidentiality and the lowest level of geography for the DC data tables is the middle layer super output area (MSOA) whose minimum size is 5,000 people or 2,000 households. However, data for a number of variables with more detailed classifications in the DC are only available down to local authority scale. The other set of multivariate counts are the Local Characteristics (LC), which provide the most detailed data possible for the smallest geographical areas (OAs). LC counts are not as detailed as their DC equivalents and are comparable with the Census Area Statistics derived from the 2001 Census.

One fundamental concept underpinning all these statistics is that of the 'population base'. As society becomes more complex and more mobile, there is increasing demand for population estimates that enumerate people at a geography other than their usual residence. There are questions about whether to include those people resident on a temporary basis, those visiting an area or those working in the area but living elsewhere, for example, as well as whether there should be separate counts of those living in private households and those living in communal establishments, as provided in the 2001 Census. In 2004, ONS conducted a user consultation (Baker, 2004) which identified the need for alternative population bases and the subsequent ONS working group supported the use of usual residents plus visitors as the main output base for the 2011 Census data, as reported by Smith and Jefferies (2006). In outputs, the most commonly used population base is that of 'usual resident', which excludes visitors, students living away in term time and short-term residents in the UK (ONS, 2014).[1]

The alternative population statistics are a further set of 2011 CAS tables that have been produced using a number of alternative populations, including second addresses, out-of-term populations, short-term residents, workplace and workday populations, and Armed Forces (AF). Some people divide their time between several addresses and therefore a second address is an address where someone stayed, for part of the week or year, which was not their usual residence. In some cases, this address may be the location of a second home but it might also be a vacation address or that of some form of accommodation that is used when working away from their place of usual residence. Students and boarding school children who reported a second address which was their home address are counted at that second address and this constitutes the out-of-term population. A resident defined as short-term in the 2011 Census is anyone born outside the UK who had stayed or intended to stay in the country for a period of three months or more but less than 12 months.

It is the information collected by the census on travel to work (and travel to study in Scotland) which enables statistics to be produced on workplace and daytime populations. Those who do not work are excluded. Alternatively, the workday or daytime population is an estimate of the population of an area during the working day which includes everybody who works or studies in the area, wherever they usually live, and all respondents who live in the area but do not work or study. People who work or study mainly at or from home, or who do not have a fixed place of work or study, are included in the area containing their home address. The daytime population will include shift and night workers such as hospital staff and security guards. One further population sub-category includes those usual residents who are serving as members of the Armed Forces (AF); this is only a relatively small population base which is likely to be geographically concentrated in specific local authorities but which is key to understanding service provision and demand. ONS has prepared 2011 Census tables that are equivalent to those produced in 2001 at ward level, specifically relating to the Armed Forces population (AF001 to AF004).

In addition to the standard CAS tables, the Census Offices also produce commissioned tables (CT, also known as ad hoc tables) in response to requests from users. The characteristics and content of the CT tables produced by ONS, whose contents have been constrained by statistical disclosure control, can be identified from the 2011 Census Index of Tables and Contents, available for download from the National Archives.[2] This Excel file provides details of all the tables produced by ONS (1,623 tables by August 2016) and a search tool which helps the user find the table(s) required and summarises the metadata, including the geographic breakdown, for the table selected. Table CT0179, for example, contains estimates that classify usual residents living in a household aged 16 and over by sex, by age and by car or van availability from national level to merged local authority level in England and Wales. Thus, table CT0179 has a different population base from table LC4109EW (shown in Figure 5.1) as well as providing estimates for those in single ages from 16 to 25 and from 60 to 74. The 'Table details' spreadsheet in the Index also indicates the availability of each table on the different ONS websites which will be summarised in the next section.

5.5   Realising the value of the CAS through user access

The benefits of the CAS are only realised when they are accessed, downloaded and used as evidence to support decision making or for research and analysis. In the past, this final stage of the process has often been overlooked and underfunded by the suppliers, which is one reason why the Economic and Social Research Council (ESRC) invested, in the 1990s and 2000s, in developing an infrastructure through the Census Programme to facilitate access to the different types of census data for use in academic research.

The ONS' Census Outputs Project for 2011 built on previous census data dissemination experience and the development of information and communication technologies (ICT), making use of the internet as the primary dissemination route. The proposed technical solution was to provide an Application Programming Interface (API) as the query management interface through which user agencies could access various forms of census (and other) data, in bulk format, where required. The open API was the proposed access arrangement for the ONS' own services such as *Neighbourhood Statistics* and *Nomis*, as well as UK Data Service-Census Support (UKDS-CS) and other user portals (ONS, 2015).

The *Neighbourhood Statistics*[3] (*NeSS*) service was launched in 2001 as part of the National Strategy for Neighbourhood Renewal (Cabinet Office, 2001) and enables direct online access to 2011 CAS within a neighbourhood (LSOA) or other selected area identified by a postcode which the user supplies. It also provides access to a number of other non-census open datasets and allows the user to specify a variable and establish how the chosen neighbourhood compares against its local authority and the nation as a whole. If, however, a user requires sets of data for more than one neighbourhood or area, then *Nomis*,[4] ONS' repository of UK labour market statistics, provides access to each of the various types of 2011 CAS identified in the previous section. The ability to find subsets of information from these very large sets of data is critical. As described previously, CAS outputs in 2011 were specified using table frameworks and one of the most valuable tools in *Nomis* is the Table Finder, which allows the user to identify the table required (from a list of hundreds) based on the topics of interest. Thus, if a user wishes to find data on car or van availability by sex and by age, as indicated in our earlier example (Table 5.1), the user selects the relevant keywords and picks the geographical scale to filter out the tables required (Figure 5.1). In this case, Table LC4109EW is identified in the second box on the list when the national level is chosen.

By hovering the cursor over the table name, two buttons will be revealed. The first is the Info button and simply provides metadata: details of the units, description of the data in the table, and details of the variables and their categories. The second is the Select button which enables the user to select an area and choose how the data should be displayed as a table on the screen. Figure 5.2 illustrates the output when age groups are positioned as rows and the numbers of cars or vans are selected as columns. This equates to the top panel in Table 5.1 but with the actual statistics rather than the codes. In this mode, it is not possible to show the data for males and females but this is easily achieved by querying the data to select for your chosen area(s) the sex categories wanted as well as the age and car or van categories, and downloading the data as .xlsx or xls files.

The note underneath the table shown in Figure 5.2 reminds the user that protection of personal information is a cornerstone of UK censuses. Statistical disclosure control (SDC) measures have been used with all the 2011 CAS data which involve targeted record swapping between different areas to protect information supplied by respondents to the census. Every household has a probability of being selected for record swapping and some records are created by imputation for coverage adjustment. The method has been designed to ensure there is sufficient doubt as to whether a value of 1 is a true value or one that has been created by imputation or swapping persons in or out of that cell. SDC is a useful reminder that the CAS are estimates and further details about confidentiality of all types of 2011 Census data are available from the ONS website.[5]

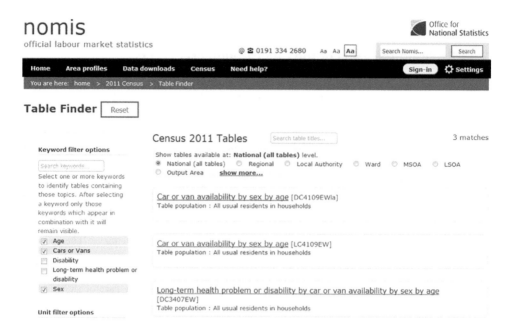

Figure 5.1   Table Finder in *Nomis*: example of finding Table LC4109EW

Source: https://www.nomisweb.co.uk/census/2011/data_finder.

LC4109EW - Car or van availability by sex by age

⬛ Download (.xlsx) - or older .xls format
Table population: All usual residents in households
Age by Cars or Vans

Units: Persons

Date	2011
Geography	England and Wales
Measures	value

	All categories: Car or van availability	No cars or vans in household	1 car or van in household	2 or more cars or vans in household
All categories: Age	55,071,113	10,671,316	21,482,846	22,916,951
Age 0 to 15	10,537,963	1,856,262	4,149,031	4,532,670
Age 16 to 24	6,253,980	1,425,049	1,917,038	2,911,893
Age 25 to 34	7,430,851	1,713,484	2,906,729	2,810,638
Age 35 to 49	11,857,419	1,754,420	4,367,859	5,735,140
Age 50 to 64	10,105,262	1,340,642	3,753,157	5,011,463
Age 65 and over	8,885,638	2,581,459	4,389,032	1,915,147

In order to protect against disclosure of personal information, records have been swapped between different geographic areas. Some counts will be affected, particularly small counts at the lowest geographies

Figure 5.2   Output from *Nomis*: number of cars and vans by age group, England and Wales

Source: https://www.nomisweb.co.uk/census/2011/LC4109EW/view/2092957703?rows=c_age&cols=c_carsno.

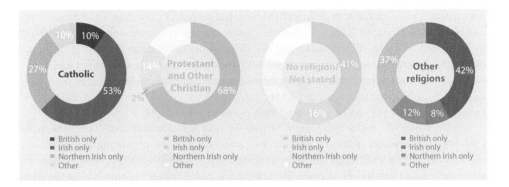

Figure 5.3   Part of infographic from *NINIS*: religion by national identity, Northern Ireland

Source: Neighbourhood Statistics (NISRA). Part of 2011 Census Analysis, Religion, released December 2014, http://www.ninis2.nisra.gov.uk/public/census2011analysis/religion/infographic.aspx.

Web interfaces have been developed by NRS and NISRA to provide user access to the 2011 CAS in Scotland and Northern Ireland respectively. Scotland's Census Tables Index performs the same function as *Nomis'* Table Finder and the data warehouse[6] provides options for downloading large volumes of Scotland's Census 2011 data. This is one of the components of NRS' *Census Data Explorer*,[7] through which it is possible to access: Area Profiles to compare the characteristics of different areas across Scotland; Maps and Charts to view selected results through map and graphic visualisations; and Standard Outputs to access selected KS, QS and DC tables. ONS also developed its own *Data Explorer*[8] for finding, viewing and manipulating data online, creating charts and maps and downloading data, as well as providing the capability for others to build applications by downloading data via the open API.

The *Northern Ireland Neighbourhood Information System* (*NINIS*), developed by NISRA, holds the 2011 Census data for Northern Ireland together with other non-census datasets, and allows the user to filter by the available geography and data subset. To view a map of an area, it is necessary to enter a postcode, street or area name in the search box and select the geography (e.g. ward) in the geography drop-down and then view the map by clicking on the relevant search results. Infographics have been used in the Census component of the *NINIS* interface to provide 'short stories'[9] based on certain characteristics such as religion (Figure 5.3), national identity and the elderly.

5.6   *InFuse*: an alternative approach to managing and disseminating the CAS

As we have outlined previously, CAS have traditionally been released as sets of tables containing various combinations of variables that are available for specific geographies. These table structures used by the Census Offices were replicated in *Casweb*, a system that was developed at Mimas (University of Manchester), with funding from the ESRC, to facilitate user access to all types of CAS from the 1991 and 2001 Censuses for England and Wales, Scotland and Northern Ireland, and from the 1971 and 1981 Censuses for England, Wales and Scotland (Stillwell et al., 2013). In this section, we explain the principles that underpin the development of a database system that facilitates user access to the CAS known as *InFuse* (Information Fusion) and illustrate how users interact with the online interface to build queries and download CAS data and metadata. The new system is based on the remodelling of the bulk data supplied by the three UK Census Offices in a multitude of separate datasets, each with separate descriptions constrained by a requirement to be printed on a two-dimensional page, to a multi-dimensional structure, with geography as one of the dimensions, searchable across the 2011 Census produced by each of the UK Census Offices, across and within the UK. *InFuse* offers users the opportunity to get more value out of the CAS by making the data easier to find and use, irrespective of the national boundaries of their collecting organisations and by extending their use to new audiences.

Ideally, a user of census data should be able to find (quickly and easily) and understand subsets of information from any part of the census dataset before making use of them in research or analysis. Although the CAS provide a wealth of information, their volume and structure can also present problems in finding specific pieces of information within the whole dataset (or discovering whether the information required exists or not within the dataset). Table-based outputs are prescriptive and reductive. They are based on a historic concept of CAS as a finite, fixed and final dataset that had to be published in physical volumes as the primary product and purpose of the census. They are supplied as thousands of separate datasets, each of which has been independently 'hand-crafted', leading to lots of inconsistencies in categories and labels. Each dataset comes with its own set of descriptions (in the form of table frameworks) that may, or may not, be consistent with descriptions for other datasets. If no variable identifiers are used, it becomes difficult to tell if category labels are consistent or not for the same variable in different tables. The need to preserve table frameworks results in heavyweight bespoke applications since the information needed to explore and subset each individual dataset needs to be coded into the application and this makes development more difficult. Moreover, while table frameworks provide a good way of presenting information on combinations of categories visually on a printed page, they do have a tendency of introducing inconsistencies and result in poor comparability within and between censuses and other datasets. Thus, CAS are still seen as a resource for experts due to the complexities of using them to develop applications.

Processing the CAS data from the Census Offices

One important shortcoming of the tabulation frameworks used by the Census Offices from a user perspective is that similar information is published across multiple tables, and in some cases, the data in one table is described differently from that contained in another. So the user first has to find the data required by initially looking through the table titles, opening a table with a title that sounds as though it contains the information required and then searching for the relevant cells in the table. Searching for the data counts required can be both time consuming and frustrating for a user, particularly if the data are described differently between tables. Thus, a major challenge was to develop a new system of finding data by making the metadata much more consistent such that statistics relating to a certain variable (e.g. age or ethnicity) are grouped together in topics and the categories or labels for these variables are harmonised.

A unique dataset has therefore been created by processing the 2011 Census data obtained from the three Census Offices in the UK so as to provide the user with a much clearer indication of how each variable (or topic) is used in the various univariate and multivariate tables that have been referred to previously and to select data in a more consistent manner. The (re)processing of metadata and data has been undertaken using structures derived from Statistical Data and Metadata eXchange (SDMX)[10] since this standard not only provides the structure needed to handle the concepts, topics and geographies associated with the data, but also has the ability for geographical areas to be assigned to multiple 'parent' geographies, a feature that avoids the duplication of

	A	B	C	D	E	F	G	H	I
1	Table **P01**								
2	**2011 Census: Usual resident population by single year of age and sex, England and Wales**								
3									
4	England and Wales								
5									
6	All usual residents [1]								Numbers
7									
8	Age [2]	Persons	Males	Females	Age [2]		Persons	Males	Females
9									
10	All ages	56,075,900	27,573,400	28,502,500					
11									
12	0 – 4	3,496,800	1,789,700	1,707,000	55 – 59		3,183,900	1,573,600	1,610,300
13	0	711,500	364,600	347,000	55		651,300	323,000	328,200
14	1	704,200	360,100	344,100	56		639,400	315,900	323,500
15	2	698,800	356,800	342,000	57		643,600	318,000	325,600
16	3	699,500	358,800	340,700	58		629,000	310,500	318,500
17	4	682,800	349,500	333,300	59		620,700	306,200	314,500
18									
19	5 – 9	3,135,700	1,604,900	1,530,800	60 – 64		3,377,200	1,658,000	1,719,200
20	5	666,200	340,300	325,900	60		624,700	307,000	317,700
21	6	633,800	324,000	309,800	61		644,900	316,400	328,500
22	7	627,400	321,000	306,400	62		661,300	325,000	336,300
23	8	604,300	310,300	294,100	63		720,200	353,600	366,600
24	9	604,000	309,300	294,700	64		726,000	356,000	370,000

Figure 5.4 Table P01 2011 Census: usual resident population by single year of age and sex, England and Wales

Source: ONS 2011 Census First Release.

geographical information that the alternative standard, Data Documentation Initiative (DDI),[11] would have required. Furthermore, SDMX is the standard used by Eurostat in attempting to create a hub of consistent 2011 Census data across Europe. Having decided to use SDMX, releases of 2011 Census data were downloaded when they became available, often in a variety of different formats and/or file types. The first sets of results were published by ONS using Excel tables created manually and designed to be aesthetically pleasing and easy to read. Figure 5.4 illustrates the contents of an extract of Table P01, a cross-tabulation of the population by age and sex for England and Wales.

In order to load this into the SDMX model, it is necessary to define the metadata (concepts, codelists and codes); in this case the concepts are the variables (age and sex), the units (persons) and the universe (England and Wales), although within our model the universe is broken up to enable cross-searching within the data structure (cube). The codelists are the lists of variable categories (e.g. single years of age) and the codes are the labels used for the categories (e.g. 0, 1, 2, 3, 4). To process Table P01, a consistent column name ID (a cell ID) is added to ensure consistency with other tables and a table prefix is used, which is the table ID, followed by 1 to n, where n is the number of columns in the table. This then creates an ID that will be unique within the system, and to which codelists and codes can be attached. It is apparent from the extract of P01 in Figure 5.4 that the columns which contain 'Males' are part of a Sex codelist, so this codelist can be created if it does not already exist in the library of codelists and 'Males' can be added to this codelist if it has not already been added. Likewise, we can assume that the rows beginning with 'All Ages' contain codes for an Age codelist. The title of the table suggests that the data do not refer to a special version of age; labels for Age could be used in more than one concept, for example age of household reference person or age at arrival into the UK. The Age column also has a footnote: (not shown in Figure 5.4) which helps inform which codelist it could be applied to. If the labels (codes) for each age listed are not present in the codelist, they can be added, but the codelist for Table P01 appears to be hierarchical as there is a label of 0–4 followed by rows labelled 0, 1, 2, 3 and 4. This relationship has to be defined because the name of the codelist listed in the table title suggests the table should only contain single ages. It becomes necessary to make a judgement between describing the data as listed versus the utility of having a single codelist for Age, the latter being chosen for use in *InFuse*.

Each CAS table produced by ONS has been processed in the same way. Table P05, for example, also contains data for Age by Sex but refers to counts for a finer geographical granularity (local authority). Here again the table title helps to indicate which codelists the table contains. As there is a single codelist for Age, the codes created in Table P01 can be applied. Thus, as Tables P01 and P05 are described using the same set of codelists and codes, the data they contain are grouped together and, in *InFuse*, a user will choose the topics of Age and Sex and see all the data in one place. As additional tables are processed in this way, the working model for the description of the data evolves. Consequently, assumptions made based on the data already processed may change and it then becomes necessary to work back through the data tables previously described to check that the proposed changes do not end up misrepresenting the data.

The working model was initially based on ONS releases 1 and 2 but changes in labelling were apparent in subsequent releases. For example, whereas age group labels $0 - 4, 5 - 9$ and so forth were used in early releases with univariate data, new age labels of 0 to 4, 5 to 9 and so forth were used in multivariate tables. Moreover, the topic names used in the table names did not remain consistent between early and subsequent releases. Given these changes, it became imperative to undertake rigorous quality assessment (cell by cell) to check the standardised description of each table had captured what the published table actually contained. Thus, in addition to creating a consistent set of metadata, the new system has avoided duplication of the same data held in different tables. The de-duplication process takes into account the geographic extent of the cell and its granularity, which then enables each instance of the data to be appended together, ensuring that coverage of the data is as full as possible.

Once the processing of CAS tables from ONS had been undertaken, it was then necessary to process the tables from NRS for Scotland and from NISRA for Northern Ireland. In the latter case, since the data had been described using the similar codelists and codes that the ONS had used, any differences which existed were relatively minor. The processing of Scottish CAS, however, was less straightforward, with a considerable amount of inconsistency between the way NRS described some of the concepts in comparison with ONS and NISRA. A further complication arose because NRS produced its CAS in (15) batches, which meant that when new variants of a concept appeared in later releases, all instances of that concept had to be reprocessed. Moreover, NRS changed the way its CAS data were produced, from straightforward csv files to providing table layouts that necessitated additional processing to extract the description text strings for each cell and then undertaking QA to verify that the information had been captured correctly. Table 5.3 shows that the number of cells for England and Wales and Northern Ireland was similar (at around 142,000) whereas there were approximately 54,000 cells in Scotland. However, the ONS

Table 5.3 Details of the changes made when processing CAS data for *InFuse*

Agency	Cells	Raw text descriptions	Additions (Universe and Units)	Matched raw text	Changed	% change
ONS	143,965	479,548	215,139	396,971	82,557	17.2
NISRA	142,372	438,471	248,933	155,052	283,419	64.6
NRS	54,300	152,965	123,539	30,967	121,998	79.8

CAS data only required changes to 17.2 per cent of the labels, whereas for the NISRA and NRS CAS, changes were made to 64.6 per cent and 79.8 per cent of cell labels respectively.

Once all the cells were described in a consistent way, it was then possible to match the metadata from NRS and NISRA into the existing set for ONS CAS, thus removing duplicates and creating a UK meta-dataset, with exceptions where concepts described were country specific. Information on the geographic coverage and granularity of each table was then combined with the list of datasets to determine where it was necessary to append data from different countries together to create composite data tables for the UK as a whole.

The final steps in the processing involved loading the metadata into the *InFuse* application and then checking the implications of some of the decisions made about concepts and codelists. For example, one issue that emerged was with the treatment of the economic activity variable, whose codelist contained around 200 codes and which therefore required some further simplification. The result of all this processing and QA has meant that a CAS dataset has been built for which concepts, codelists and codes are consistent, enabling end-users to find and download comparable data for geographies across the UK without hunting through a multitude of tables.

Using the InFuse interface

The *InFuse* interface is accessed through the UKDS-CS website and, once it is selected, users are presented with a screen that allows topics to be filtered according to user requirements and then geographical areas are picked. Figure 5.5 exemplifies the topic combinations when Age, Sex and Car or van availability are selected for persons. When the user identifies the combination required and clicks the Select button, the definitions of the topics (variables) concerned, the units and the population base are reported before the user selects the categories (labels) from within

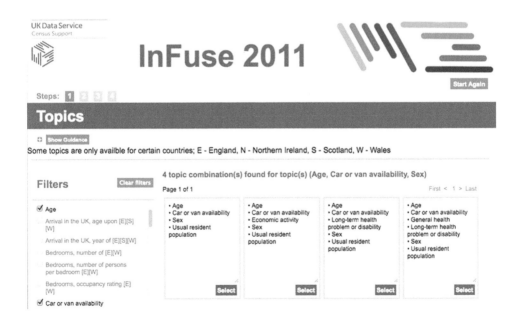

Figure 5.5 Topic selection in *InFuse*: age, car or van availability and sex for persons

Source: http://infuse2011.ukdataservice.ac.uk/InFuseWiz.aspx?cookie=openaccess.

topics. Once the category combinations have been selected, the user makes a selection of one or more geographical areas from multiple geographical levels if required. When the query is completed, a summary of the data to be downloaded is presented and clicking on the Download tab will result in two csv files being created, one containing the metadata describing the data and the other containing the data values themselves. A fuller description of the steps involved in building a query in *InFuse* can be found on the UKDS-CS website.[12]

The *InFuse* interface is therefore relatively lightweight and dynamically generated, with each screen dependent on information provided by an API in response to requests input on a previous screen. The *InFuse* API, developed in conjunction with the *InFuse* interface, provides controlled access to the descriptive information and data via a series of operations. This web-based graphical user interface (GUI) provides an excellent data discovery and extraction tool for novice to intermediate users. However, it is less well suited to meeting the more demanding needs of advanced users and the UKDS-CS has been working in collaboration with the School of Environmental Sciences at the University of Liverpool to develop a script-driven interface (*InFuseR*) to all the CAS data holdings in *InFuse* via the *Infuse* API (Moody, 2015).

5.7 Conclusions and recommendations

The CAS are the most fundamental outputs from the 2011 Census. This chapter has explained the basic layout of a CAS table, summarised the differences between the different types of CAS, and outlined some of the channels for users to access the CAS data. It has also provided some details of the way in which the data and metadata released by the Census Offices have been processed by the UKDS-CS team at Jisc into a new multi-dimensional database model with a user interface called *InFuse* that gives the user the opportunity to avoid searching through the multitude of CAS tables to find the data required but to see what data are available for topics and combinations of topics before building a query.

The value of the CAS far outweighs the costs of their production in providing essential information that cannot be obtained from other sources. The CAS that are provided via *InFuse* are used in research across many academic disciplines, usually in conjunction with other data sources. By providing – uniquely – searchable UK-wide data, *InFuse* supports the development of impact because it renders efficiencies in enabling UK CAS – seamlessly – to underpin or link with UK-level and constituent area-based research or policy development. In particular, *InFuse* offers the facility for baselining the characteristics and distribution of the respective populations at a range of geographical levels of the UK in relation to the impact of government policies on those populations, both comparably across small area geographies and between areas in the countries which make up the UK.

Research using the 2011 CAS citing *InFuse* has been published in a number of journals, contributing to knowledge about engaging new migrants in infectious disease screening; understanding the psychological distress of people with dementia and their caregivers; trends in patient profiles in a memory clinic over 20 years, and variations in car type, size, usage and emissions in the UK. Several of the case studies in the chapters in Part V of this book have used CAS from *InFuse*.

InFuse is used not only in the higher education sector but also in policy development in local and central government, and by civil society and commercial sector organisations. User analytics show that (amongst IP addresses which indicated organisational origin) 58 per cent of users come from UK further or higher education institutions and 10 per cent of users come from local and central government, commercial organisations, the NHS, charities and non-profit organisations, and non-UK universities. A range of organisations from these sectors have used *InFuse*. Census data downloaded from *InFuse* informed the Welsh Government with regard to policies to engage gypsy and traveller families in education, showing that over 60 per cent aged over 16 from these communities had no qualifications. Executive recruitment firm Sapphire Partners used census data from *InFuse* in a report on female representation on boards, revealing that 77 per cent of FTSE board members are men, and 70 per cent of new board appointments go to men. A study by the Marie Curie charity into the differing needs of Black, Asian and minority ethnic groups in Scotland for end-of-life care used *InFuse* to determine that the minority ethnic population in Scotland has doubled since 2001 from 100,000 to 200,000 – highlighting the need for greater and more appropriate provision. A Knowledge Transfer Partnership between homelessness charity Llamau, Cardiff University and other agencies used *InFuse* data to show that Welsh young homeless people participating in the study were over twice as likely to have left school with no qualifications compared to UK-wide figures for their age group and gender (Llamau, 2015).

The experience of processing the 2011 CAS for *InFuse* at Jisc has revealed, first, that the content of the CAS has been restricted and inflexible due to being pre-defined and constrained as a set of tables containing fixed combinations of variables and categories. Second, production of the pre-defined outputs has been slow due to the various components being individually and independently specified and produced. And, third, access to the CAS has been made difficult due to fragmentation of the data into thousands of separately, and often inconsistently, specified datasets without a global descriptive model required to quickly and efficiently query and operate across the entire set of CAS outputs.

Thus, opportunities for improvement in 2021 exist primarily in the areas of harmonisation by design of the census programmes of the three UK Census Offices in the planning stages, and in the dissemination modules and formats of the 2021 CAS outputs to obtain maximum value and accelerate the capacity for impact, once available. The focus should be on enabling the data; setting it up to be as widely used as possible. In particular, a set of agreed standardised variables and categories should be adopted across the three UK Census Offices using contemporary standards for the description of unit and aggregate data with variations only where additional to the standardised variables. The use of consistent structures will also improve comparability of the CAS outputs with data from other sources. The initial creation of a library of CAS outputs similar in content to those from 2011 will be essential for comparative analysis, but the library might then be extended by adding outputs from further dynamic queries of the 2021 unit data. The size of the CAS dataset could be managed by deleting data if not used for a certain period of time, and regenerating it if requested again.

Web delivery via open and standardised APIs in 2021 would provide comprehensive access to CAS for external developers, supported by documentation and application modules with the aim of making the development of applications to use the data as easy as possible and thereby encouraging external innovation. This might include the development of mechanisms to provide, on request, quick, comprehensive and automated web access via one or more APIs to metadata for (details of) any and all the aggregate data that it might be possible to produce from the full collection of unit results in order to allow developers and users to explore potential outputs, followed by the corresponding data if it is permissible within SDC constraints. There would need to be some moderation of requests in terms of data volumes and processing resource and time. The Census Offices should use these APIs to develop their own end-user interfaces as well as providing them to encourage embedding of CAS use in other applications and interfaces.

Other opportunities presented in support of enhancing timely access to CAS include the development of a UK geography framework (via API) to keep track of, manage and provide operations based upon the UK's complex and changing geographies. The use of descriptive models with structures that can be described by data description standards, for example DDI for unit data and SDMX for aggregate data, would enable the communication and transfer of descriptive information and data and more consistent, global descriptive structures. Description of the data down to unit level would facilitate the more effective exposure of CAS to the web of data. Opportunities also exist for automated SDC. These elements will enable faster and more comprehensive access to the 2021 CAS output.

As CAS provide the fundamental evidence base for policy and spending decisions in central and local government, the next generation development of *InFuse* will offer an open-source data portal to disseminate the UK Data Service socio-economic aggregate data via a single user platform which offers a model for data aggregation from the smallest UK geographies to, potentially, international geographies for data mapping, analysis and research (including flow data and boundary data with appropriate access controls) in an integrated data dissemination portal utilising integrated metadata and SDMX to facilitate resource discovery and the development of API-driven applications. The future development of *InFuse* will aim to enable researchers in all sectors to respond to local, national and international challenges using CAS as the contextual basis for analysis across disciplines and sectors.

Acknowledgements

The authors are grateful to Jo Wathan and Richard Wiseman for their comments and edits on a draft version of this chapter.

Notes

1 For information on population bases in Scotland see http://www.scotlandscensus.gov.uk/alternative-population.
2 http://webarchive.nationalarchives.gov.uk/20160105160709/http://www.ons.gov.uk/ons/about-ons/business-transparency/freedom-of-information/what-can-i-request/published-ad-hoc-data/census/index.html.
3 https://www.neighbourhood.statistics.gov.uk/dissemination/.
4 www.nomisweb.co.uk/.
5 www.ons.gov.uk/census/2011census/confidentiality.
6 www.scotlandscensus.gov.uk/ods-web/data-warehouse.html.
7 www.scotlandscensus.gov.uk/ods-web/home.html.
8 http://web.ons.gov.uk/ons/data/web/explorer.
9 www.ninis2.nisra.gov.uk/public/census2011analysis/index.aspx.
10 https://sdmx.org/.
11 www.ddialliance.org/.
12 http://infuse.ukdataservice.ac.uk/help/guide.html.

References

Baker, L. (2004) Covering all bases: Early thoughts for population bases for the 2011 Census. *Population Trends*, 116: 6–10.
Cabinet Office (2001) A New Commitment to Neighbourhood Renewal: National Strategy Action Plan. Report by the Social Exclusion Unit, Cabinet Office, London.
Cabinet Office (2008) Helping to Shape Tomorrow: The 2011 Census of Population and Housing in England and Wales. Cmnd 7513. The Stationery Office, Norwich.
Llamau (2015) *Study of the Experiences of Young Homeless People (SEYHoPe): Key Findings & Implications.* Llamau, Cardiff. Available at: www.llamau.org.uk/creo_files/default/seyhope_report_final_v3.pdf.
Moody, V. (2015) Innovation Fund project: InfuseR, providing scripted access to tabular census data. Available at: http://blog.ukdataservice.ac.uk/esrc-innovation-fund-project-infuser-providing-scripted-access-to-tabular-census-data/

ONS (2006) The 2011 Census: Assessment of initial user requirements on content for England and Wales. *ONS Information Paper.* Available at: http://webarchive.nationalarchives.gov.uk/20160105160709/ http://www.ons.gov.uk/ons/about-ons/consultations/closed-consultations/2011-census---responses/ index.html.

ONS (2014) 2011 Census Glossary of Terms. Available at: http://webarchive.nationalarchives.gov. uk/20160105160709/http://www.ons.gov.uk/ons/guide-method/census/2011/census-data/2011-census-user-guide/glossary/index.html.

ONS (2015) 2011 Census General Report for England and Wales. Available at: www.ons.gov.uk/ census/2011census/howourcensusworks/howdidwedoin2011/2011censusgeneralreport.

Smith, C.W. and Jefferies, J. (2006) Population bases and statistical provision: Towards a more flexible future? *Population Trends*, 124: 18–25.

Stillwell, J., Hayes, J., Dymond-Green, R., Reid, J. Duke-Williams, O., Dennett, A. and Wathan, J. (2013) Access to UK census data for spatial analysis: Towards an integrated Census Support service. In Geertman, S., Toppen, F. and Stillwell, J., (eds.) *Planning Support Systems for Sustainable Urban Development*. Springer, Heidelberg, pp. 329–348.

6

Geographic boundary data and their online access

James Reid, James Crone and Justin Hayes

6.1   Introduction

The UK decennial census of population is unique in terms of its geographical coverage and its depth, not least because of the geographical detail (small area) at which it captures primary socio-economic and demographic data. Census geography products − typically expressed as digitised boundary data files for use in Geographical Information Systems (GIS) − support the census statistical information in published reference tables (the 'aggregate' data) and associated datasets (typically 'lookup tables' which describe the relation of one geographical area to another and cross-walk area codes so that it is possible to traverse the census geography hierarchy), but contain no information about the population themselves. They are thus the areal 'buckets' into which the aggregate count information can be poured. This chapter outlines the main census geography products available from each of the national Census Offices in the UK and overviews a range of tools that are available to access and use the data.

6.2   The UK geography hierarchy and its fundamental building blocks

The range of census geography products available reflects the geographical complexity for which the census aggregate information is produced, and the relationships between them. The 2011 Census geographical areas form a complex hierarchy, summarised in Figure 6.1 (see also Table 6.1).

Output areas (OAs) are created for census data, specifically for the output of census estimates. The OA (or small area (SA) in Northern Ireland) is the lowest geographical level at which census estimates are provided and typically contain counts for around 300 persons in England and Wales, 114 in Scotland and 400 in Northern Ireland. OAs were introduced in Scotland at the 1981 Census and subsequently in all the countries of the UK at the 2001 Census. The 2011 Census OAs/SAs form the '2011 Statistical Building Blocks', which reflects the UK Government Statistical Service's (GSS) Geography Policy (GSS, 2015) and enshrines several principles relating to the creation, management and use of different UK geography types.

Proponents in favour of retaining the UK census in its current form regard this so-called 'small area' information as an unparalleled source for base-lining population characteristics and for ensuring information richness at a scale that provides intra-area and inter-area comparison

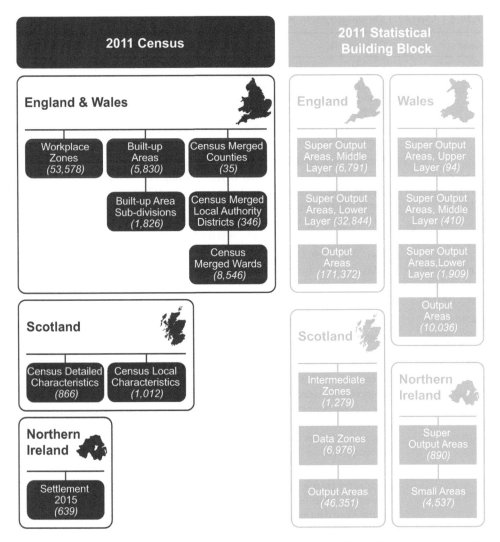

Figure 6.1 The hierarchy of '2011 Statistical Building Blocks' and derivative census geographical areas in the UK

Source: Adapted from the ONS Hierarchical Representation of UK Statistical Geographies (April 2016).

at a sufficiently localised scale to enable policy and planning interventions to be meaningfully expressed 'on the ground'. Local councils, for example, use the census information at small area level, *inter alia*, to forecast school enrolment populations, estimate demand for local health provision and inform infrastructure provision for housing demand.

The UK census of population is the only reliable national survey that offers such a fine grained and data rich description at small area level – the mosaic of which covers the entire physical UK land mass. These small areas comprise the individual pieces of the jigsaw puzzle that together provide the spatial view of contemporary society and allow for macro aspects of household and population changes to be discerned, for example the ageing population structure or the rise of single person households. In many respects, the census is thus a 'bottom–up' exercise that has its physical expression as the ground truth at small area level.

Table 6.1 Main 2011 Census geography products from ONS for England and Wales

Geography type	Commentary
Output areas (OA)	Maintaining stability as far as possible was key for the 2011 Census although changes due to, for example, local population change or administrative boundary changes to which OAs align inevitably mean that 2011 OAs differ slightly from the 2001 versions. A useful summary of the changes is provided by the ONS publication 'Changes to Output Areas and Super Output Areas in England and Wales, 2001 to 2011' (Tait, 2012).
	The total number of 2011 OAs is 171,372 for England and 10,036 for Wales.
Super output areas (lower layer) (LSOA)	Super output areas (SOAs) were designed to improve the reporting of small area statistics and are built up from groups of OAs and were originally planned to come in three different aggregation levels – Lower, Middle and Upper, although the latter has only been used in Wales in 2011 (see below). LSOAs in 2011 have a population of 1,000–3,000 people and 400–1,200 households.
	There are 32,844 LSOAs in England and 1,909 in Wales.
Super output areas (middle layer) (MSOA)	MSOAs in 2011 have a population between 5,000 and 15,000 and a household count between 2,000 and 6,000.
	There are 6,791 MSOAs in England and 410 in Wales.
Super output areas (upper layer) (USOA)	Upper SOAs were created for Wales only, in 2011.
	There are 94 USOAs.
Census merged ward (CMW)	Electoral wards/divisions are the key building blocks of administrative geography in England and Wales. They are the spatial units used to elect local government councillors in metropolitan and non-metropolitan districts, unitary authorities and the London boroughs in England; and unitary authorities in Wales. Electoral wards are found in most of England but in Wales, the Isle of Wight and some of the new unitary authorities the equivalent areas are legally termed 'electoral divisions', although they are frequently also referred to as wards.
	They are used to constitute many other geographies – CMWs are a frozen geography created specifically for 2011 Census Detailed Characteristics (DC) tables. DC tables have a higher minimum population threshold than other tables as the more detailed information carries an increased risk of identifying persons or households using the lower population threshold (100 persons) applied to other tables. If the census estimate for a ward falls below 1,000 persons or 400 households, the ward is merged with a neighbouring ward, or wards, until the aggregated census estimate for the merged wards is above both the minimum person (1,000) and household (400) thresholds.
	Merged wards include 2011 wards whose estimates are above the higher threshold, without the need to merge with another 2011 ward, as well as merged wards. This gives complete coverage of England and Wales. These merged wards have the same boundaries as their equivalent 2011 ward, but are given a different entity code to the entity code of the 2011 ward.
	There are 8,546 CMWs in 2011.
Census merged local authority districts (CMLAD)	Local authority districts (LADs) is a generic term to describe the 'district' level of local government in England and Wales. It includes non-metropolitan districts, metropolitan districts, unitary authorities and London boroughs in England; and Welsh unitary authorities. The areas are made up of best-fit amalgams of CMWs.
	There are 346 CMLADs in 2011.

Geography type	Commentary
Census merged counties (CMC)	Amalgams of CMLADs. There are 35 CMCs in 2011.
Built-up areas (BUA) and built-up area sub-divisions (BUASD)	BUAs and their BUASDs are a new geography, created as part of the 2011 Census outputs. These data provide information on the villages, towns and cities where people live, and allow comparisons between people living in built-up areas and those living elsewhere. Census data for these areas (previously called urban areas) have been produced every ten years since 1981. A new methodology to capture the areas was used in the 2011 version, but it still follows the rules used in previous versions so that results are broadly comparable. As previously, the definition follows a 'bricks and mortar' approach, with BUAs defined as land with a minimum area of 20 hectares (200,000 square metres), whilst settlements within 200 metres of each other are linked. Note that there are areas included in the boundary data not in the census tables. These BUAs and BUASDs have been identified as areas that have not been allocated a population. In most cases, this is because they do not have any residential buildings – for example, industrial estates, airports, theme parks, etc. There are 337 BUAs where population has not been allocated (305 in England and 32 in Wales) and 133 BUASDs where population has not been allocated (123 in England and ten in Wales).
Workplace zones (WZ)	WZs are a new output geography for England and Wales and were produced using workplace data from the 2011 Census. WZs are designed to supplement the OA, LSOA and MSOA geographies. OAs were originally created for the analysis of population statistics using residential population and household data. As a result, they are of limited use for workplace statistics as there is no consistency in the number of workers or businesses contained within an OA. OAs are designed to contain consistent numbers of persons, based on where they live; WZs are designed to contain consistent numbers of workers, based on where people work. This means that WZs are more suitable for disseminating workplace-based statistics and outputs. WZs have been created by splitting and merging the 2011 OAs to produce a workplace geography that contains consistent numbers of workers. The WZs align to the existing OA hierarchy. They have been constrained to MSOA boundaries to provide consistency between the OA and WZ geographies, and to allow comparison of the 2001 and 2011 Census workplace outputs at the MSOA level.

As Figure 6.1 implies, the UK census, whilst occurring simultaneously in all parts of the UK, is conducted separately by three different agencies. The responsible body in England and Wales is the Office for National Statistics (ONS), in Scotland, the National Records of Scotland (NRS) and in Northern Ireland, the Northern Ireland Statistics and Research Agency (NISRA). Each responsible agency defines the census geographies used for its country and the methodology and thresholds it uses to derive them. However, each Census Office attempts to capture population characteristics at a geographical level, which ensures the anonymity of individual census respondents balanced against the utility of having the highest resolution of disaggregation of respondent counts for areas as is practicable.

The methodology for the construction of these small areas has varied over time and is exhaustively described elsewhere (Cockings et al., 2011). However, in the last few censuses, a fundamental objective has been to try and maximise consistency both across space and over time – the OA (and its equivalents) now provides the basic geographical unit for describing areal characteristics and is the 'building block' against which all higher geography types are derived. OAs also provide the atomic unit against which bespoke custom geographies may be built. The rationale for this is to provide a convenient mechanism by which new geographies can be constructed by aggregation, whilst at the same time allowing for flexibility in comparing areas and their change over time; one of the outstanding features of UK geography (not just census geography) is its mutability over time, which complicates temporal analyses and is frustrated by the well-known issues associated with the so-called Modifiable Areal Unit Problem (MAUP) (Openshaw, 1984).

A range of techniques have been proposed to try and counter the comparability problems that changing geographies present to analysts, from grid count based solutions (Martin et al., 2011) to dasymetric mapping, a technique which depicts quantitative areal data using boundaries that divide the area into zones of relative homogeneity with the purpose of better portraying the population distribution (see, for example, Langford, 2007). In any case, cross-comparison between censuses remains an ongoing topic of research and any inter-censal analysis must account for the fact that UK census geography and the definition of 'small areas' has varied over time and represents as much a convenience mechanism for conducting the census as for definitively expressing geographical variation across space in a consistent and time comparable way. Census users of small area data must therefore temper their conclusions by recognising that there is always a degree of 'apples and oranges' in comparability, no matter how sophisticated the approaches adopted to redress the small area problem.

6.3 Boundary data and the census

As noted above, the primary geography outputs from the census are expressed as digitised boundary data files. Digitised boundary datasets, sometimes referred to as 'DBDs' or 'boundary data', are a digitised representation of the underlying geography of the census. They are often used within GIS or Computer Aided Design (CAD) systems. Census-related geography outputs first became available to researchers and teachers in the UK Higher Education sector in the late 1980s and, by the time of the 1991 Census, had become a more mainstream census output that was deemed of sufficient utility to that sector to merit investment by the Economic and Social Research Council (ESRC). That investment led to the creation of a range of bespoke online tools for census delivery to a principally academic audience and was reinforced by continued investment at the time of the 2011 Census.

However, the 2011 Census saw two independent developments which have impacted upon current delivery of the census from the time of the initial ESRC investments. The first of these was a general need to rationalise the ESRC data infrastructure portfolio. This included the census and consequently the UK Data Service-Census Support (UKDS-CS) was born under the auspices of the larger data delivery and support platform, the UK Data Service. UKDS-CS provides census-specific tools and support and coalesces the previously extant tools providing access to the full range of census products. It also covers the historic data portfolio of census products ranging from the 1971 through to the 2011 Censuses. No other UK resource provides online access to this time series.

The second trend which has a bearing is the move towards the release of 'open data' by government agencies. The political imperative for transparency and the reuse of data captured at public expense has taken on a momentum that has snowballed into other sectors. Undoubtedly, the advantages of access to open data for researchers and the general public is in principle laudable, although this belies the fact that the delivery of the same itself incurs a cost – typically borne by the data disseminator. In the case of UKDS-CS, a broad move to publishing data under an Open Government Licence (OGL), with a concomitant removal of access impediments such as mandatory registration and login, has led to an upsurge in shipped data volumes (conservatively in the order of several hundred per cent). The long-term implications of this in terms of wider impact, sustainability of the required infrastructure and the cost of continued investment at the time of writing remain an open question. Whatever the broader context, it remains that most of the census digital boundary data and associated products, such as lookup tables, are now classed as open data under liberal licence and are disseminated as such by UKDS-CS – very few products are restricted under 'Special Licence' to academics only and the services are free to use for whatever purpose by anyone, not just academics (subject only to citation clauses under OGL).

Thus, the digitised co-ordinates (points, lines, areas) which make up these census geographies are freely available as DBDs. These are the boundaries which form the areal representation against which various census statistics, such as the counts of households or the ratio of males to females (again, freely available – see Chapter 5), can be associated and subsequently visualised and analysed. Census area statistics contain a pointer (generally a code such as 'E09000022' which represents the 2011 code, in this case for the London Borough of Lambeth) to the geographical census areas to which they relate and by linking census area statistics with the corresponding digitised boundary datasets for a specific census year, the census attributes can be readily visualised as a map (see Chapter 12).

Mapping census datasets allows for an exploration of the characteristics of census datasets geographically and may provide additional demographic, socio-economic and cultural insights into the census data, particularly when used in conjunction with other non-census data. A GIS allows for further spatial analyses of the census data, and their combination with other non-census geographically referenced datasets, such as crime or health data, can lead to new insights and spawn new research questions. Illustratively, the digitised boundary datasets can be used for:

- map production for research articles;
- data synthesis and development of residential neighbourhoods;
- geostatistical analysis of demographic or employment change;
- small area analysis and deprivation studies;
- health care research – incidence mapping and analysis; and
- historical demographic research.

Tables 6.1 to 6.3 elaborate upon the summary information provided in Figure 6.1 and give a flavour of how rich UK census geography digitised boundary outputs are for England and

Table 6.2 Main 2011 Census geography products from NRS for Scotland

Geography type	Commentary
Output areas (OA)	OAs form the main building bricks for the Scottish census areas. Most OAs are created by aggregating a small number of neighbouring postcodes, although some postcodes are large enough to become a single-postcode OA. All higher geographies (e.g. Health Board areas) are built up from these OAs. Any area for which census output is produced is the aggregation of OAs. OAs aggregate exactly to Council Areas but not necessarily to any other higher geography. The aggregations of OAs for these other higher geographies are termed 'best-fit', as the boundaries of the aggregations approximate the true boundaries of the geography. This is because the boundaries of the individual postcodes do not follow existing administrative and political boundaries.
	There are 46,351 Census 2011 OAs.

(Continued)

Table 6.2 (Continued)

Geography type	Commentary
Data zones (DZ)	Broadly similar to LSOAs. DZs are aggregations of OAs and in 2011 had a population range of between 500 and 1,000 people.
	There are 6,976 DZs in 2011.
Intermediate zones (IZ)	Broadly similar to MSOAs. IZs are a statistical geography that sit between DZs and local authority districts, created for use with the Scottish Neighbourhood Statistics (SNS) programme and the wider Scottish public sector. IZs are used for the dissemination of statistics that are not suitable for release at the DZ level because of the sensitive nature of the statistics, or for reasons of reliability. IZs were designed to meet constraints on population thresholds (2,500–6,000 household residents), to nest within local authorities, and to be built up from aggregates of DZs. IZs also represent a relatively stable geography that can be used to analyse change over time, with changes only occurring after a census.
	Following the update to IZs using 2011 Census data, there are now 1,279 covering the whole of Scotland.
Census local characteristic postcode sectors (CLCPS)	CLCPSs represent a hierarchical level above the individual postcode level. There are 1,131 postcode sectors in Scotland, but no census data are actually produced for true postcode sectors. In order to reach the required threshold to publish census statistics, some postcode sectors are grouped together.
	This results in 1,012 CLCPSs.
Census detailed characteristic postcode sectors (CDCPS)	CDCPSs are the second type of a postcode sector geography used in census output. CDCPSs further group together Scotland's 1,131 true postcode sectors to reach the required thresholds for publishing more detailed census statistics.
	This results in 866 CDCPSs.

Table 6.3 Main 2011 Census geography products from NISRA for Northern Ireland

Geography type	Commentary
Small areas (SA)	Broadly comparable to OAs. The 2011 SAs were introduced after the 2011 Census by amalgamating 2001 Census OAs which were built from clusters of adjacent postcodes. SAs nest within the 890 SOAs and the 582 Electoral Wards in Northern Ireland but not the new Local Government Districts, which are best-fit only.
	There are 4,537 SAs and they are designed specifically for statistical purposes. They are the lowest geographical areas the NI 2011 Census results are released for. The average SA represents 400 people and 155 households, although they range in size from 59 households and 98 people to 988 households and 3,072 people.
Super output areas (SOA)	Due to changes since 2001 to local government structure in NI, the existing SOA statistical building blocks needed to be modified slightly and a 'Modified Super Output Area' geography was created as the basis for aggregations of the 2011 SAs.
	There are 890 modified SOAs.
Settlement development limits (SDL)	Headcount and household estimates are published for 2011 settlements which are derived from individual georeferenced census returns and available as SDL boundaries. They are available for settlements exceeding the thresholds of 20 or more households and 50 or more usual residents.

Wales, Scotland and Northern Ireland respectively. Note that these represent just a subset of the data available from UKDS-CS but describe the most frequently used 2011 Census geography products.

6.4 Challenges and issues

UK census geographies provide a range of issues and challenges to prospective users. First, the fact that there are three separate, independent but co-operating census agencies adds to the complexities – each has its own principal census website, its own dissemination mechanisms, its own publishing timescales and, more importantly, its own individualistic approach to data capture and release. The UKDS-CS attempts to ameliorate this by providing a single point of entry to all official census geography outputs and commits significant resource to managing and ingesting data from the various agencies, quality assuring and processing these data.

An illustration of value added by quality assuring is the issue of 'sliver' polygons – often arte-facts introduced at the time the boundaries are created due to the semi-automatic approaches used to generate OAs. Slivers are very small polygons enclosed within the larger parent polygon which represents the OA itself. Often not apparent except at very large scale, these small residual artefacts can become problematic as OAs are aggregated to produce higher geographies and they can become visually intrusive when used to produce maps. They also produce incorrect polygon counts for area types as the slivers are regarded as legitimate areal features in most GIS. UKDS-CS has developed an ingest work-flow which automatically tests for slivers and removes them so that the derivation of the higher geography types does not suffer from excessive polygon counts due to their inclusion.

Additionally, the data are 'value added' by providing functionally rich online tools as well as the provision of additional 'harmonised' data products that provide UK-wide census products for key census geographies. UKDS-CS, via its primary website, allows for the discovery of data holdings and then provides pointers to the tools that can be used to extract and format the data as the user requires. The specific tools for the extraction of the census geography boundary data and related files are discussed in Section 6.5, but it is worth noting that the discovery of the census boundary data collection (and indeed other census products) is mediated via the creation of collection-level metadata which can be searched via faceted browsing on either the UKDS-CS website or directly from the UK Data Service parent site.

The 'Discover' application is the principal data finding aid provided by the UK Data Service and is a flexible search tool for filtering search results and, as the name suggests, discovering more detail on any specific data resource that the UK Data Service delivers. This includes a census filter to allow casual users an intuitive entry point to locate census outputs in the first instance. Frequent users more familiar with the range of census products and services available will likely short-cut direct to the various census tools themselves and use the bespoke filtering and search tools provided to refine their data requests. Additional dataset-level metadata, as supplied by the original data owner and which include details of product lineage and provenance (useful for determining fitness of purpose to a user's task), are also maintained internally by UKDS-CS for the purposes of dealing with support requests and are available to users upon request. As an aside, the process of gaining clearance to release the pre-2011 Census products under OGL was ham-pered by not having sufficient corporate memory about the creation and lineage of individual datasets – something that metadata at time of creation would have assisted with.

Discovery is, of course, only the first step towards use. Users face further challenges of format usability (can they deal with the format the data are delivered in?); data volumes (digital boundary data are often 'over-digitised', meaning they contain too much extraneous detail, which has the effect of bloating file sizes and can retard download times in restricted bandwidth situations); downstream tooling (does the user have the required software to use the data?); and, even with adequate metadata (can the user actually meaningfully interpret and use the data as supplied?), UKDS-CS offers a range of approaches for dealing with these issues for the digital boundary data resources.

Format conversion

The DBDs are provided in a range of popular GIS-ready data formats as well as 'attribute-only' comma separated value (csv) format to ensure that most proprietary software packages can cope with the data once downloaded. Log analysis from operational use of the census boundary services highlights the fact that the vast majority of the format requests favour the ESRI shapefile and *MapInfo* format varieties and these are fully supported by the digitised boundary data delivery tools.

Default data compression

All data extractions are, by default, zipped and compressed for delivery in order to reduce file-size and bandwidth consumption. The option to avoid compression is offered but is rarely used.

Provision of 'simplified' datasets

A number of the more popular and complex datasets are offered in multiple versions – 'generalised' variants which simplify the co-ordinate information in the boundary file and produce smaller files (typically one-tenth the file size), with less complex boundary shapes. These are necessarily 'lossy' and more suited for thematic mapping purposes, whereas the full-resolution boundary datasets are better suited for analytical purposes (Figure 6.2a). Additionally, there can also be variation in the areal coverage of the boundaries. Boundary sets can be prepared to 'extent of the realm' or 'clipped to the coastline'. 'Extent of the realm' boundary sets typically extend to UK mean low water height, although they can extend to islands off the coast; for example, Avonmouth ward in the City of Bristol extends to the islands of Flat Holm and Steep Holm in the Bristol Channel. 'Clipped to the coastline' boundary sets, derived from the 'extent of the realm' boundaries, show boundaries to mean high water. Usually prepared for visualisation of data, such boundaries more closely represent map users' expectations of how a coastal boundary should look. Whereas 'extent of the realm' boundaries adjacent to an inlet or estuary may join at a point midway across the water, 'clipped to the coastline, boundaries permit the more precise identification of the waterside (Figure 6.2b). UKDS-CS offer permutations of these variants (generalised/ full resolution, clipped/full extent) to allow users to choose which best suits their specific needs.

Online and offline support

A significant and salient aspect of UKDS-CS is that it provides both online help and offline one-to-one support for users. Queries, if not able to be resolved by consultation with the extensive online support material, can be submitted via an online query form. Queries received in this fashion are routed to UKDS-CS data experts conversant with the specific data and queries are triaged and responded to in as timely a fashion as resources permit. Most users receive a satisfactory resolution and/or additional guidance within a few hours of their initial query. Two online eLearning modules are also available for interested users to undertake self-paced learning and cover: 'Understanding the geography of the UK' and 'Geographic analysis and visualisation of census data'.

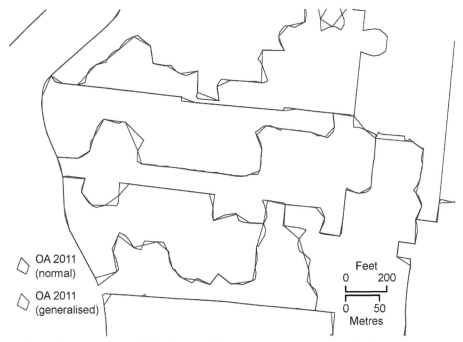

a. Normal and generalised OAs for part of London showing the lossy effects of generalisation

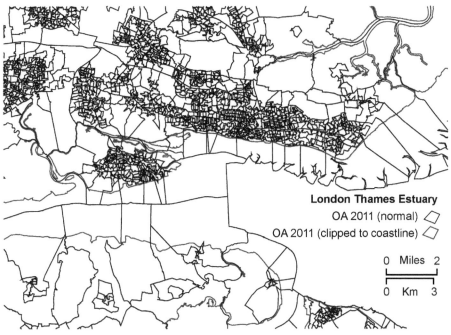

b. Normal and clipped to coastline OA variants around the Thames Estuary, London

Figure 6.2 Example boundaries at: a. different levels of generalisation and b. the 'extent of the realm'

6.5   UKDS-CS geography tool suite

In order for UKDS-CS users to better make the most of census geography outputs, a range of online tools have been developed and are currently supported under funding from the ESRC. In most cases, these tools have extensive online help which guides a novice user through their use and, over time, the applications have been refined to assist users to get to the data they want as quickly as possible. As the tools and their forerunners have supported census users for over two decades, inevitably they have undergone several iterations of implementation across different hardware and software platforms. The move in 2012 to an integrated UKDS-CS service has provided the opportunity to rationalise and upgrade various components of the architecture which are summarised below.

Until 2012, the majority of the UK academic census needs were fulfilled by a range of different specialised university-based 'delivery units', loosely badged under an ESRC 'Census Programme' banner. Under an open call, issued by the ESRC in 2011, the old Census Programme census delivery units collaborated to run an integrated service under the auspices of the new UK Data Service, which was commissioned by ESRC in parallel. UKDS-CS is a value-added component of the UK Data Service and operates within the broader context and operational environment set by the parent. A first immediate priority for the new UKDS-CS service was to acquire a consistent look and feel brand style from the UK Data Service. Following the commissioning of a UK Data Service style guide dictating basic colours, logo design, element placement rules and font styles for both digital and hard-copy resources, UKDS-CS adapted its various existing services to the new house style as well as establishing a new census website (census.ukdataservice.ac.uk), to act as the front door for resource discovery and for primary information pertaining to all census resources.

At the same time, the new open data publishing ethos was becoming established and the UK Data Service took a lead in pushing for an 'open by default' basic principle towards data publishing – notwithstanding the need for continuing to support operational modes protecting safeguarded and confidential data where issues of privacy and licensing are paramount. Many of the existing census resources had previously been subject to what were known as 'Special Conditions' with access restricted to registered users only, a prerequisite of registration being that a user held a valid UK Access Management Federation for Education and Research (UKAMF) credential. The UKAMF provides a single solution to accessing online resources and services for education and research and had been used by the pre-UKDS-CS delivery units. However, with the general move towards open data publishing by UK government departments, and the various census agencies themselves rapidly adopting the permissive OGL, it became apparent that the inherited authentication and authorisation model needed to be adapted to allow for a broader constituency of user beyond the traditional confines of academia. Consequently, the current operational model provides access to all open data resources without registration and without login but also retains a login-based authentication service for registered users where census-related data does not fall under an OGL regime and still inherits 'Special Conditions'. The result of this is that for the majority of users (academic and non-academic alike), services are open to the web and do not require registration/login. For a small proportion of data, principally historical datasets sourced from the Great Britain Historical GIS project and the pre-open era postcode directories where intellectual property rights remain in force, users are still required to register and agree to the terms ('Special Conditions') defined for that dataset – mostly this translates as a user must be in possession of UKAMF credentials and use the data only for non-commercial purposes.

UKDS-CS also inherited a number of other operational criteria as a result of being a value added component of the UK Data Service – not least that all census tools and products should migrate from their existing hardware platforms for delivery by the ESRC-funded technical infrastructure comprising the UK Data Service delivery capability. As a consequence, the range of re-branded, open census tools have also undergone technical migration. A description of the available census geography tools is provided below and, where appropriate, comment on the impact of the infrastructure redeployment is noted, especially where it has led to functional changes or performance improvements to the services offered.

The delivery of the 2011 (and indeed the 1971–2001) census geography products by UKDS-CS is currently managed through the suite of online applications described below.

Easy Download (ED)

This is a simple web application which provides a tab-based interface to access and directly download pre-bundled national coverage datasets reflecting the most popular census geography boundary data (Figure 6.3). It is organised by country and year and also provides a separate tab for 'lookup tables' – files which provide various areal designation codes mapped to other codes for

Figure 6.3   The *EasyDownload* interface

Source: https://borders.ukdataservice.ac.uk/easy_download.html.

CS = UKDS-CS

'higher' geography types, for example a census OA may have a lookup table providing the mapping between the OA code and the county code/name. The lookup tables are generally provided by the various census agencies as ancillary census products for the convenience of end-users. Drilling down by country and year yields a series of links to the actual boundary data provided in a range of formats – csv for just the attribute/field information (no complex geometry information other than a simple point geometry), Keyhole Markup Language (KML) for use in *Google Earth*; *MapInfo* TAB format for use in *MapInfo* and ESRI shapefile format for *ArcGIS* users and general GIS usage. UKDS-CS has created separate amalgams of the various national datasets to provide consistent UK coverage for various geographical scales that can be used with the aggregate statistics provided by *InFuse* and these are available separately under the '*InFuse* United Kingdom' tab (see Chapter 5).

Each download link points to a pre-bundled zip file containing the actual census boundary data and a 'Terms and Conditions of Use' file. During the re-architecture of the extant application suite, all pre-bundled datasets were rebuilt to update the Terms and Conditions to reflect licensing changes as a result of OGL adoption as well as a rebase of the range of formats provided – seldom used or deprecated data formats from the old applications were dropped and greater consistency in dataset and attribute naming was enforced with a view to easier ongoing maintenance and legibility for end-users.

Boundary Data Selector (BDS)

Unlike *ED*, which is just a simple front end to a webserver delivering static pre-built datasets, *BDS* allows for greater flexibility in terms of the data selected and in delivery options. *ED* provides the census boundaries as full coverage national extent packages, whereas *BDS* allows users to filter and subset datasets. *BDS* again uses a tabbed interface with the primary 'What and Where?' selection tab (Figure 6.4), which allows users to filter all available boundary datasets by country (England, Wales, Scotland, Northern Ireland) or aggregation (Great Britain, United Kingdom) and also includes coverage for a number of Irish datasets.

The 'Geography' filter allows users to determine the geography type they are interested in and uses the standard geography categorisation as used by ONS. Finally, filtering by year allows quick determination of the datasets of interest that meet the selection criterion and, once applied using the 'Find' button, all available boundaries are shown in the 'Boundaries' box. Selecting an item from the list provided allows further refinement and the 'List Areas' button, where applicable, can be applied to show all the sub-geography areas for that dataset. Where a dataset has a multi-level hierarchy, further drill down may be possible by repeated use of the 'Expand Selection' button, which will drill into the highlighted selected geography sub-areas. For example, choosing 'English Census Merged Wards 2011', clicking Find and choosing 'Leeds' in the 'Areas of Interest:' box allows the user to drill down further via 'Expand Selection' which results in the ability to single out 'Headingley' as the geographical area for which Census Merged Wards 2011 are required.

The Map tab on the interface provides a simple preview of the data selected based on the geographical extents of the selection chosen and the Format tab allows the user to redefine the delivery format (ESRI Shape (default), KML, *MapInfo* or csv) and compression option (Zip or None). Finally, clicking the Extract Boundary Data button sends the request to the back-end database.

As some data extractions can be quite large and compute-intensive, a number of 'tricks' are utilised to minimise resource consumption. All *BDS* queries are checked against recent queries run by other users (as a concurrent application, several dozens of users may be accessing the same datasets at any one time) and if an identical query has a confirmed and still viable result (the results of data extraction requests are maintained for a finite time to allow users to bookmark their data and return to it later), the previously extracted data are presented to the user rather than running

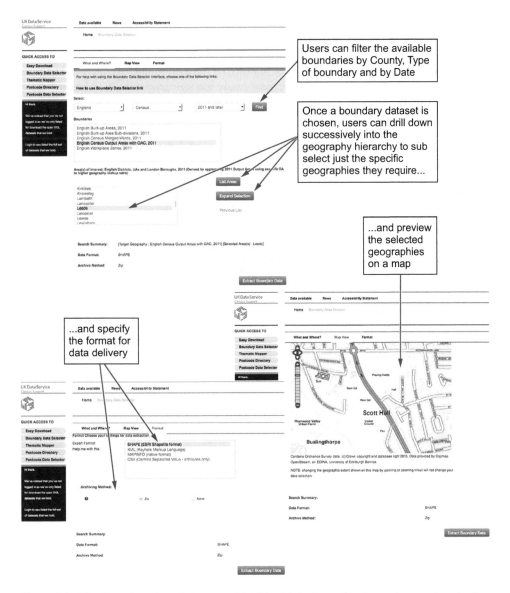

Figure 6.4 The *Boundary Data Selector* multi-tabbed interface allowing refinement and selection of geographical subsets of datasets

Source: Adapted from https://borders.ukdataservice.ac.uk/bds.html.

a fresh extraction. Moreover, data extraction requests that equate to full national coverage for a dataset are automatically redirected to the *ED* versions which have already been extracted. In this fashion, repeat requests for the same data from different users do not choke the system.

BDS uses the open source *Postgresql* database with the *Postgis* spatial extension to manage the selection and extraction of the census boundary data. As part of the migration from the old application architecture to the new UKDS-CS version, *BDS* was completely overhauled to use only open source tools. Whilst *Postgresql/Postgis* had been a staple feature of the pre-existing application, a range of other components have been deprecated in the migration, including the proprietary Safe Software *Feature Manipulation Engine* (*FME*) which was used exclusively to handle the spatial format conversion tasks. As a result of the move towards a more public service, one that serves more than just an academic audience and no longer for the most part requires registration or login, the use of the existing commercial software would have breached the (very generous) Education Licence under which the software was formerly used – by delivering data to potentially commercial users, a commercial licence would have been required. Consequently, an open source alternative was sought and the underlying application logic changed whilst ensuring that the functionality users were familiar with was retained. *BDS* data extractions now use the open source vector manipulation OGR library and utilities to handle all boundary data format conversions and, as a happy by-product, *BDS* is now based on a fully open source software stack.

In addition to the digital boundary data available through the applications above, UKDS-CS also provides access to a range of other key products that are often used in conjunction with census data (similar to the various census lookup tables) but provide a valuable research resource in their own right. Key amongst these are the postcode directories, which are special purpose, quarterly updated lookup tables providing a lookup between every UK postcode and a range of higher geography codes.

The UK postcode system was devised by the Royal Mail to enable efficient mail delivery to all UK addresses. Initially introduced in London in 1857, the system as we now know it became operational for most of the UK in the late 1970s. There are approximately 1.8 million unit postcodes in use and each postcode covers an average of about 15 properties, although in reality it can contain anywhere between 1 and 100 properties. Postcodes are not static. New properties are constantly being built and old ones demolished, and the Royal Mail sometimes has to re-code within existing areas to maintain or improve distribution efficiency. These 'introductions' and 'terminations' are recorded in updates to the Postcode Directory. Unlike the boundary data, these postcode directories are simply large tabular files (around 2.3 million rows by around 30–40 columns) that only contain a point co-ordinate (a British National Grid easting/northing pair) that provides the spatial location of the individual postcode centroid.

The UK postcode is often used in research as a location proxy and is extremely common in social surveys (indeed, it is critical in the creation of the core census output products and in planning and delivering the UK census). The ability to allocate postcodes to their associated higher geography is often critical for the production of count summaries from surveys – for example, how many respondents live in a particular health authority based on knowledge of their home postcode. Whilst the various postcode directories available via UKDS-CS have been known under various names at various times (e.g. the All Field Postcode Directory (AFPD) or the National Statistics Postcode Directory (NSPD)), they are essentially the same. Variants of the core NSPD became available from ONS in 2012 as open data products licensed under OGL by removing some of the fields maintained by Royal Mail which required payment of royalties. Use of the earlier postcode directories from UKDS-CS requires registration and acceptance of 'Special Conditions' as noted above. Current versions of the directories, now referred to as the ONS Postcode Directory (ONSPD) Open Edition, and the National Statistics Postcode Look-up (NSPL) are available without registration to all users. The directories differ slightly in the method used to assign geographies to individual postcodes, which has an impact on their accuracy for certain applications; in most cases, users of the postcode directories should use the ONSPD and the NSPL should be used by producers of National Statistics.

Postcode Download (PD)

Similar to *ED*, *PD* provides full product 'grab-and-go' packaged versions of the various postcode directories. Open users will only see directories dating back to May 2011, but registered users agreeing to the relevant 'Special Conditions' (see above) can access directories dating back as far as 1980. Each data bundle provides the full directory as supplied by ONS along with any user guides and ancillary code lookup tables.

Postcode Data Selector (PDS)

Analogous to *BDS*, *PDS* allows for greater user flexibility in what and how data are extracted. Figure 6.5 provides an overview of the main interface for *PDS*. Under the 'Where' heading,

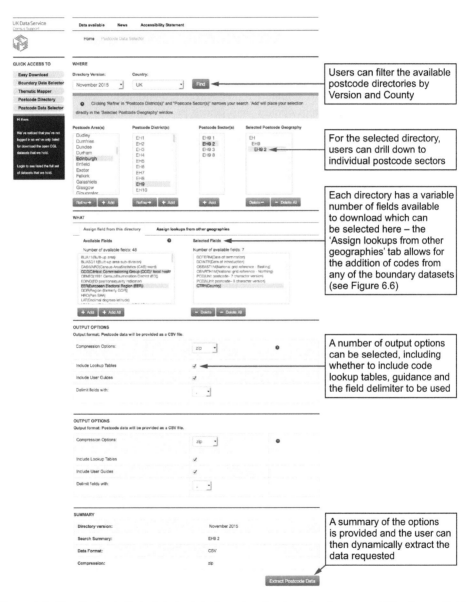

Figure 6.5 The *Postcode Data Selector* interface showing how to choose and filter the available postcode directory data

Source: Adapted from https://borders.ukdataservice.ac.uk/pds.html.

users can decide which version of the directory they are interested in and also filter by country of interest. As with *PD*, only directories appropriate to the type of user will be displayed – open users will only see current to May 2011 versions.

UK postcodes are hierarchical, the top level being the postcode area, for example 'GU' (for Guildford) or 'B' (for Birmingham). The next level is the postcode district, for example 'GU1' or 'B22'. This is also commonly known as the 'outcode' and generally provides the primary bulk sorting office routing information for the Royal Mail. Next comes the sector, for example 'GU16 7', and, finally, the unit, for example 'GU16 7HZ'. The combination of sector and unit (the '7HZ' bit) is often called the 'incode'. This structure is used in the *PDS* to allow users to filter and 'mix and match' the postcodes of interest to them. The display box on the left hand of the interface under the 'What' heading and on the 'Assign field from this directory' tab shows all the available fields (columns) that the selected postcode directory contains whilst the right-hand box shows the selection that the user has added. By default, six mandatory fields are provided to ensure that each extraction request contains a minimum number of valid and useful fields. Below the field choice boxes, a number of output options are provided to allow users to define what character delimiter they prefer for the output csv file; for inclusion of the directory-specific user guides and documentation, including code lookup tables; and the compression options. A summary footer displays the current search summary and options chosen. The Extract Postcode Data button commences the data extraction and allows users to bookmark and return at a later time if the extraction is long running. Ultimately, a PostcodesData.zip file will be generated, which contains: the extracted data, a Terms and Conditions zip archive containing the UK Data Service terms and conditions and the original official data licences as supplied by the relevant statistical agency, and a plain text README summary of what was extracted.

A notable functional addition to this application is available under the 'Assign lookups from other geographies' tab on the main interface. Users can assign, to any postcode selection, further additional codes derived from any of the available *BDS* datasets allowing for the creation of custom, bespoke lookup tables. For example, after selecting postcodes for areas in Edinburgh, a user might be interested in knowing which Police Force Area these current postcodes would have been assigned to in 1991. By using the 'Assign lookups from other geographies' tab and using the filters (identical to those in *BDS*), the user can add the codes for Police Force Areas from the boundary file 'Scottish Police Force Areas, 1991' to their data extract, thus creating an on-the-fly bespoke lookup table appended to their postcode directory data selection (Figure 6.6). This feature uses a point-in-polygon operation within the *Postgis* spatial database to assign the attribute codes from the polygon boundary data to the point data (the postcodes).

GeoConvert

GeoConvert is an easy-to-use web interface with unique functions that enable users to find relationships between areas of common UK geographies (e.g. postcodes, output areas, wards), and to convert data between them. It can also provide a range of commonly used data (e.g. indices of deprivation) for some areas, primarily postcodes.

A large number of different geographies (e.g. wards, districts, output areas, postcodes) are used in the publication of data from different producers, and with different topics and uses across the UK. Some geographies form sets with neat hierarchies in which smaller areas from geographies at lower levels nest neatly within larger areas from geographies at higher levels, but many do not. Most of the geographies change over time as areas change their names or extents, or are terminated, split or amalgamated to form new areas. This can make it difficult to compare or combine data published for areas belonging to different geographies, or from the same geography at

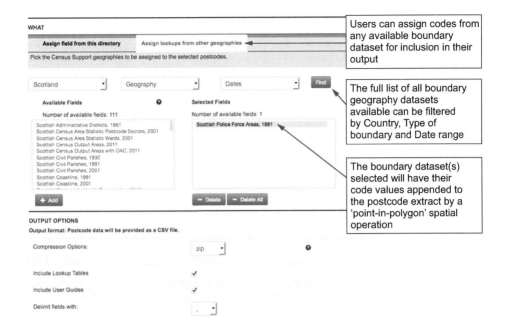

Figure 6.6 Dynamic assignment in *PDS* using a point-in-polygon operation to attach fields from the geography boundary datasets to the postcode data extract

Source: Adapted from https://borders.ukdataservice.ac.uk/pds.html.

different times. *GeoConvert* simplifies and automates matching and conversion processes using the best information and methods available to identify and record relationships between individual areas within and between geographies.

Behind *GeoConvert*'s functions is a data model that records information about areas belonging to a wide range of UK geographies, as well as relationships that have been identified between areas. In the current version of *GeoConvert*, these area relationships have been derived using information from the three UK census agencies about the populations of the sets of special enumeration postcodes that were used as the fundamental building blocks of areas used for outputs from the UK 2011 Census, together with their relationships with larger geographical areas. This information allows the populations of all the areas related to postcodes to be disaggregated to constituent postcodes, which can then be reconstituted into different areas to generate area relationship coefficients between pairs of areas based on the spatial distribution of population recorded in the 2011 Census. *GeoConvert* enables users to supply sets of identifiers for areas of one geography, and have them matched to identifiers for areas of another. It also enables data supplied for one area to be reapportioned to different areas based on the area relationship coefficients between them.

The current area relationship coefficients are based on the distribution of 2011 Census postcode populations. Conversions using these coefficients are most appropriate for characteristics that would be expected to be directly related to overall population distribution (numbers of males and females, for example). Conversions for characteristics that are not directly related to population distribution (numbers of dairy cattle, for example) would not be appropriate. There is potential for further development of *GeoConvert* to area relationship coefficients based on a range of different kinds of relationships (proportional areal overlap, for example) that would extend the range of matching and conversion operations that could be carried out appropriately using *GeoConvert*.

6.6 Alternative access to census boundary data

Whilst UKDS-CS offers an open and free to use comprehensive range of tools for accessing census geography products (current and historic), it is one of several notable UK applications providing data; a few of the more noteworthy are outlined below. The information is provided for reference and the convenience of prospective census geography users – it is not an endorsement of any particular application by UKDS-CS. Readers/users requiring any assistance with UK census geography are advised to contact UKDS-CS in the first instance.

ONS geoportal[1]

The ONS is the UK Government's main survey organisation and its main producer of official statistics. It is also the central co-ordinating agency for the wider GSS. ONS Geography is an internal department of the ONS and provides a geographic referencing framework to enable the collection, production and presentation of accurate and consistent statistics. ONS Geography aims to:

- provide a single definitive source of geographic information (GI) and metadata;
- provide advice and best practice guidance on using GI data, and GIS tools and software to reference, analyse and present statistics; and
- lead on initiatives across the GSS, central government and beyond, to harmonise the use of GI in the production of statistics.

ONS Geography disseminates its products principally via the national geoportal and these products include the 2011 Census geography boundaries. The *geoportal*, similar to *ED* and *BDS*, provides access to ready-packaged and bespoke extraction facilities to enable users to choose the data they need for the areas they require, and similarly provides access to lookup tables and postcode directories. Unlike UKDS-CS, however, it does not supply pre-2011 census geography products but does provide a broader thematic range of products than just solely census-related products. ONS Geography supports an Application Programming Interface (API) onto a range of their geographic products. API access to UKDS-CS resources is similarly available on request. Note, however, that ONS' geographical remit covers England and Wales and, whilst acting as lead agency on behalf of Scotland and Northern Ireland for some issues, remains primarily a dissemination agency for English and Welsh geography products.

NRS Census Data Explorer[2]

NRS is a non-ministerial agency of Scottish Government (the UK-devolved administration for Scotland) and is responsible for civil registration, the census in Scotland, demography and statistics, family history and the national archives and historical records. The census has its own website,[3] which provides a variety of online tools via its *Census Data Explorer* (*CDE*) for accessing and viewing Scottish census outputs. *CDE* provides quick access to Area Profiles (topic summaries by area) via a map interface and is available for both 2001 and 2011 Censuses. Additional 'Maps and Charts' and 'Standard Outputs' sections of the website provide further interactive tools for exploring and downloading the census outputs.

NRS' *Data Warehouse* provides two options for downloading large volumes of Scotland's 2011 data as: (i) 'Standard data files', which offer the standard output tables in csv format with textual descriptions of each geography; and (ii) 'Bulk data files', which provide tables in csv format with geographic areas identified by a unique 'S number' (e.g. 'S12000033', which relates to Aberdeen

City). A 2011 Census index provides a lookup for users to access and cross-match textual descriptions of each geography against the unique S numbers. Again, given the geographical remit of NRS, products only cover the national extent of Scotland.

Northern Ireland Neighbourhood Information Service (NINIS)[4]

NISRA is the principal source of official statistics and social research on Northern Ireland and has responsibility for conducting and reporting the results of the 2011 Census. Census results are available online via the *NINIS*,[5] which offers online views, a variety of bulk data download options, and tabular and map views of the results. In addition to the conventional geographic output geographies, NISRA also releases a grid square product which provides basic 2011 Census statistics for a combination of 1 kilometre and 100 metre grid squares in Northern Ireland.

6.7 Conclusions

The UK census of population, and in particular the small area geographies and statistical products it provides, are a rich resource for researchers and analysts as well as a challenge, given the complexity, variability and sheer wealth of the data offered. The UKDS-CS offers a range of tools and help to assist potential users, irrespective of skill abilities, to better access and use census products. Whilst not unique in this regard, UKDS-CS provides additional facilities and a single point of entry to UK-wide (as distinct from nation-only access) current and historical census data. This chapter has provided an overview of the main tools and features that UKDS-CS offers as well as pointers to the relevant national census agency offerings and noted along the way some of the unique aspects of UK census geography and the issues and challenges that arise in their use.

Notes

1 http://geoportal.statistics.gov.uk/.
2 http://nationalrecordsofscotland.gov.uk.
3 www.scotlandscensus.gov.uk.
4 www.nisra.gov.uk.
5 www.ninis2.nisra.gov.uk.

References

Cockings, S., Harfoot, A., Martin, D. and Hornby, D. (2011) Maintaining existing zoning systems using automated zone-design techniques: Methods for creating the 2011 Census output geographies for England and Wales. *Environment and Planning A*, 43: 2399–2418.

GSS (2015) GSS Geography Policy. Available at: https://gss.civilservice.gov.uk/wp-content/uploads/2012/12/GSS-Geography-Policy-is-now-available.pdf.

Langford, M. (2007) Rapid facilitation of dasymetric-based population interpolation by means of raster pixel maps. *Computers, Environment and Urban Systems*, 31: 19–32.

Martin, D., Lloyd, C. and Shuttleworth, I. (2011) Evaluation of gridded population models using 2001 Northern Ireland Census data. *Environment and Planning A*, 43(8): 1965–1980.

Openshaw, S. (1984) The modifiable areal unit problem. *Concepts and Techniques in Modern Geography No. 38.* Geo Books, Norwich.

Tait, A. (ed.) (2012) Changes to Output Areas and Super Output Areas in England and Wales, 2001 to 2011. ONS. Available at: http://www.ons.gov.uk/ons/guide-method/geography/products/census/report--changes-to-output-areas-and-super-output-areas-in-england-and-wales--2001-to-2011.pdf.

Census interaction data and their means of access

Oliver Duke-Williams, Vassilis Routsis and John Stillwell

7.1 Introduction

Flow data – often referred to as 'origin–destination data' or 'interaction data' – are a specialised form of aggregate data that involve movements from one geographical location to another. In a population census context, as we shall see in due course, the datasets are about flows of individuals or households but, more generally, flow data might also include the movement of physical goods such as manufactured commodities or of intangible assets such as flows of capital between banks, or of telecommunications between countries.

Recent population censuses in the UK have captured the movement of people or groups of people in households between places of usual residence in the 12 months before each census, together with the movement of people between where they live and where they work. These flow or interaction datasets, referred to collectively as the Origin–Destination Statistics (ODS) by the Office for National Statistics (ONS) in the context of the 2011 Census, also contain counts of students and those with second residences as well as migrants and commuters to work or study. This chapter outlines the different types of flow data that are available from the 2011 Census and explains how their characteristics have changed since previous censuses. Changes in the process of statistical disclosure control mean that whilst there is no small cell adjustment of the 2011 flow data as occurred in 2001, a three-tier hierarchy of access conditions – open, safeguarded and secure – has been introduced. The chapter will also exemplify how users can access flow data at different levels using the *WICID* software interface that is part of the UK Data Service. Chapters 27 and 28 of this book provide examples of how flow data are being used in research to understand the changing patterns of internal migration and commuting to work between 2001 and 2011.

7.2 Census flow data: concepts and definitions

Census flow data are generated from the responses to certain questions on the census form. In recent decades, two sets of questions have been used to generate flow data relating to migration and to commuting, but new questions in the 2011 Census have permitted the expansion to include new flows. Results from the 'flow' questions on the census form can be seen in many census data outputs: there are 'origin' (outflow) and 'destination' (inflow) fields to varying levels of geography in cross-sectional and longitudinal microdata sets, and there are a number of aggregate observations in the small area data. This chapter, however, focuses only on those origin–destination outputs published specifically as part of the flow data products that are large and complex (relative to other census datasets) and less widely used.

Flow datasets from the 1966 (sample), 1971 and 1981 Censuses were produced on demand at cost to the customer (Denham and Rhind, 1983, p. 67); more latterly they have been produced as part of planned outputs, and distributed by ONS. Rees et al. (2002), writing before the 2001 flow data were released, described the migration outputs produced as part of the 1991 Census and compared them to expected outputs from the 2001 Census; a similar comparison of 1991 and expected 2001 journey to work outputs is reported in Cole et al. (2002).

In order to fully understand census migration flow data, it is useful to set out some concepts and definitions. The first of these is that of 'usual residence', the address at which a person lives all or most of the time. For most people, this idea is wholly unambiguous, but there are many individuals for whom the concept is more blurred – for example, students who may be considered to have both a parental domicile or 'home' address as well as a 'term-time' address, seasonal workers who may spend large parts of the year at an address other than their 'home', and persons with second residences who may regularly spend intervals of time at different 'homes'. The children of divorced parents are another specific population sub-group whose place of usual residence may be divided between two or more locations.

Migration data – as with any other data relating to some change of status – can be viewed at varying levels of granularity. We may view a migration, a permanent change of usual address, to be a discrete event, and try to capture and record all such events over a time period; these are known as 'moves or event data'. Alternatively, we may look at the net effect of an individual's change in location over a time period in order to ask whether a change has occurred at some stage during that period; these counts are referred to as 'migrant or transition data'. Census migration data in the UK adopt the latter approach, and look at the net effect of migration over a one-year period. The two approaches give different counts of migration (Rees, 1977). Census migration data compare a person's address at census date with that 12 months previously and, where there is a difference, a single act of migration is recorded. In practice, a person may have changed address more than once during the transition period; each of these intermediate moves would be recorded by event data, but would not be recognised in transition data.

Further differences arise due to the demographic accounting framework imposed. For individuals to be identified as migrants in the census data, it is necessary for them to be alive at both the beginning and end of the transition period. Persons who have left the country in the transition period will not be recorded (and will thus not be identified as migrants), nor will persons who have changed residence during the transition period but who have died before census day. Likewise, infants born during the period will not be included as migrants, an issue that has caused ONS to generate estimates of migrants aged under one year of age in the last two censuses from data on births and migration rates of women of child-bearing age. Assuming that each migration event can be captured, event data have a number of advantages over transition data, not least in identifying the true propensity to migrate, and can be used to estimate transition counts. Event data are typically captured through population registers or administrative sources (such as the National Health Service Central Register) and cannot be recorded by an instrument such as a census as easily as transition data. The incidence of these two types of migration data, together with lifetime migration, in countries around the world is documented in Bell et al. (2014).

Journey to work data are in many ways simpler than migration data as they do not measure change over a transition period. However, they also have their own ambiguities, relating to both place of work and mode of travel to work. For many people, their place of work is fixed, and can thus easily be described. However, for many other people this is not so easy, as their workplace may vary from day to day, in either a predictable or an unpredictable fashion. The pattern of journey to work may also vary from day to day, both for people with and without fixed workplaces: a journey pattern may include intermediate locations such as a school or caring commitment, or may involve retail destinations as part of the trip; separate journeys to work and returning home may take different routes and may involve different types of transport. Again, a census form necessarily has to try and frame these ideas within a limited amount of space, typically with only a single question, and cannot collect the same level of trip event data as would be possible in a specialised travel survey. Problems arise for the census agencies over how to deal with homeworkers *vis-à-vis* commuters in the broadest sense and, more specifically, the distinction between those who work at home and those who work from home when reporting mode of transport.

7.3   The 2011 Census flow data questions

Migration questions

A number of questions relating to migration were included on the 2011 Census form. The same main question wording and the same response categories were used in England and Wales (Q21, Form H1), in Scotland (Q10, Form H0), and in Northern Ireland (Q13, Form H4). The response categories were: 'The address on the front of the questionnaire'; 'Student term time/ boarding school address in the UK'; 'Another address in the UK'; and 'Outside the UK'. For the second and third options, space was given to write in a UK address; for the last option, space was given to write in a country name. Those persons who indicated via this question that 12 months prior to the census they were not living at the address on the front of the census form (that is, the person's usual residence at the time of the census) were identified as migrants. This definition thus included all persons who changed usual address, both within the UK and those who had entered the UK.

Very similar questions about migration (with minor variation in the question wording, and in the tick box options) have been used in all recent UK censuses. In the 1971 Census, two questions were included, one asking about usual address one year previously, and the second asking about usual address five years previously. A number of other questions on the 2011 Census form related specifically to international immigrants, including questions identifying how long someone had been in the UK, and the length of intended stay.

A sequence of questions were asked in 2011 that are particularly relevant to the generation of census flow data. Question 9 on the form used in England and Wales asked for each individual's country of birth, with tick box options for the most common responses and an additional write-in option. Persons who were not born in the UK were then asked in the next question to give the month and year of most recent arrival in the UK, with a note indicating that short visits outside the UK should be ignored. Finally, those persons indicating that they had been born outside the UK *and* who had most recently arrived within the last year were asked to state the length of intended stay in the UK, with three options: less than six months, six to 12 months, and 12 months or more. Equivalent questions were asked in Northern Ireland, although in Scotland only the month and year of arrival were requested. Where data were collected about both time of arrival and length of intended stay, it is possible to use the definition adopted by the United Nations (1998) that a long-term migrant is one who has moved to another country for at least 12 months, and to restrict tables of results to long-term migrants only. This definition relates to movement across international boundaries; it does not require 12-month residency at any particular residential address within the country.

The question about a respondent's usual address one year prior to the census is often referred to as 'the migration question' as it is the key identifier of a 'migrant'. Similar questions are used in many censuses around the world, with some variation in the length of time over which migration is observed (Bell et al., 2014). The responses to the question allow the generation of a flow matrix from origin (location of usual address one year prior to the census) to destination (location of usual address at the time of the census). Given that locations are fundamentally captured at the address level (assuming that addresses given are correct and complete), they can be aggregated to any level of geography. In the case of migration data, the most fine-grained level of aggregation is the output area (OA) level. This is often a far more detailed level than required for analysis, but gives the option of arbitrary re-aggregation into any other geography that can be represented as an amalgamation of these units. Spatial aggregation is frequently required for data to meet confidentiality constraints and is the reason why ONS releases only aggregate flows at this geographical scale.

Student questions

As described above, one of the response options on the migration question was to identify a named UK address as being a student term-time or boarding school address. This tick box option had not been included in prior UK censuses (someone with this status would have simply recorded their address as being 'another address in the UK') but is of a significance belied by its small footprint on the physical page. By filtering on this option, it has been possible for the national statistical agencies to generate a subset of migration statistics relating to students, specifically to those persons who had a student address (or boarding school address) one year prior to the census. Of these, some will still have been students at the time of the census, whereas others will have completed their studies. The response to this question thus gives us some idea of how students and recent former students migrate. This is of interest as students are both highly mobile and often poorly captured in data. The 1991 Census recorded students at their parental domicile, for example, and a special table (Table 100) was released for student flows, whereas term-time address was used in 2001 (Duke-Williams, 2009).

In an ideal world, the question about address one year ago might be supplemented with others asking about other statuses one year ago, for example employment status, occupation and housing tenure, all of which would help analysts to understand the context of the migration transition as well as the spatial aspects. However, such questions would, of course, burden the respondent and significantly increase the length of the form, and this is deemed impractical by the census agencies. However, the 'student' tick box does allow analysts to gain a richer idea of the context of migration, although may, of course, prompt further questions when interpreting the results. Data disaggregated by student address status one year ago have only been available from the 2011 Census.

Journey to work and place of study questions

Another question that captures address data is that relating to the respondent's place of work. Unlike the migration question, however, the wording of the commuting question in 2011 varied in different parts of the UK. In England and Wales, Q40 asked: "In your main job, what is the address of your workplace?", with space to write in an address. Tick box categories allowed people to indicate that they worked mainly at or from home, on an offshore installation, or that they had no fixed place of work. The responses to this question allow an origin–destination matrix to be constructed of 'home' to 'workplace' locations.

In Scotland (Q11) and in Northern Ireland (Q43), the census form asked a different question: "What address do you travel to for your main job or course of study (including school)?" This captures data on the journey to work in the same way that the question used in England and Wales does, but additionally captures data on the journey to a place of study for students and schoolchildren. The tick box options had slightly different wording as required to refer to both groups, but there was also an additional tick box option: 'Not currently working or studying'. This fundamentally changes the audience for the question. Whereas the question in England and Wales only addresses those self-employed or in employment,[1] the question used elsewhere in the UK can be answered by all respondents. A result of this is that care must be taken if comparing results in England and Wales with those in the rest of the UK to ensure that a like-with-like comparison is being made.

For many people in Scotland and Northern Ireland, the distinction between place of work and place of study is not problematic in that people have one of these but not both, and the form design is unambiguous. However, for anyone who both works and studies, a problem arises in that only one address can be given. The form wording directs people to use the place at which they spend most time, but this is subjective, and the balance may well vary over the course of the year. In flow datasets produced from the 2001 Census, the additional availability of data about journey to a place of study led to the creation of different data products for Scotland and for the rest of the UK. For residences in Scotland, journeys to work and place of study were reported in the 2001 Special Travel Statistics (STS), whereas for residences in England, Wales and Northern Ireland, results were reported in the 2001 Special Workplace Statistics (SWS). The 2001 STS could be broadly thought of as a superset – in terms of table structure and content – of the 2001 SWS, but direct comparison between the two sources was awkward for users.

As with the migration data, since these locations are gathered at the address level, they can be freely aggregated at the data processing stage and the finest level at which the 'home' location is observed is the OA level. At the workplace (or destination) end, however, OAs are not the ideal spatial units to use for these data. Some workplaces are physically large (for example, an airport) but do not have permanent residents, and thus cannot easily be accommodated in a residential-focused geography. Some parts of the country – for example, the City of London – have large numbers of workers, but very few residents. In England and Wales, an alternative geography was developed (Martin et al., 2013) of workplace zones (WPZs or WZs also often used in documentation) (ONS, 2014). WPZs are designed such that they can be smaller than OAs in areas to which a large number of workers travel (thus allowing the City of London to be split up into small areas), but larger than OAs in residential areas where there are fewer employers (reducing problems of flows of very few people). Consequently, there are 53,578 WPZs in England and Wales, as compared to 181,408 OAs, with WPZs nesting hierarchically into Middle Super Output Areas (MSOAs). The journey to work data are thus somewhat more spatially complex than the migration data, in that their base origin and destination geographies are not the same and the matrices of flows from OAs to WPZs are asymmetric.

Whilst various census questions relate to employment, for example occupation undertaken, responsibilities and hours worked, one particular question refers directly to the interaction between home and work, namely a question on the method of transport used to travel to work. This question has been included (with evolving response categories) on all censuses from 1966 onwards and is of major importance for use in transport planning. Chapter 28 in this book examines how modal split has changed between 2001 and 2011 at local authority district scale in England and Wales.

Questions on the workplace have been asked in all recent UK censuses; the overloading of the question to allow place of study to be recorded was used in Scotland for the first time in 2001, and in both Scotland and Northern Ireland in 2011. Like any question that has a written answer (as opposed to a tick box selection), details on occupation and addresses are difficult and expensive to code. Workplace data were coded for a 10 per cent sample of census forms in 1966, 1971 and 1981. The 1961 Census used a short form/long form approach, with 10 per cent of households receiving a long form (with workplace and occupation questions only appearing on the long form), and all other households receiving a short form; this is the only time that the UK has used this two form approach (ONS, n.d.a).

Second residence questions

The 2011 Census in England and Wales included two related 'flow data' questions that have not previously been asked in UK censuses. The first of these (Q5) asked: "Do you stay at another address for more than 30 days a year?", with space to write in either a UK address or the name of a country outside the UK. A follow-up question (Q6) then asked about the nature of that address, with options of: 'Armed forces base address'; 'Another address when working away from home'; 'Student's home address'; 'Student's term time address'; 'Another parent or guardian's address'; 'Holiday home'; and 'Other'. Similar questions were trialled in tests in Scotland prior to the 2011 Census (NRS, 2007) but were not used in the full census. The test census form used in Scotland included similar identification of reason for having a second residence, but added two further questions regarding the number of days per week and weeks per year that the second residence was used.

As the response categories suggest, the nature of a 'second residence' can be diverse and covers a broad range of the mobility spectrum (Bell and Ward, 2000), from annual/seasonal movements in the case of a holiday home to weekly (or more frequent) moves between two locations in the case of commuting/work-based residences, and of children alternating time between separated parents. The processing of the data gave rise to a number of different matrices which comprise the Second Residence Statistics (SRS): from location of usual residence to location of second residence; from location of second residence to location of workplace (for persons who had a second address for work purposes); and a combined category of either primary or secondary residence to workplace location. There is little precedent for a wide-ranging question on second residences in the census, although some aspects are picked up in censuses elsewhere. For example, the 2010 Census of Switzerland[2] included a question for employed persons: "From which address do you normally leave for work?", with provision for the respondent to indicate whether it was the same address as used on the front of the census form, or a different address, to be written in the space provided, with a parallel question for school children and students. No previous UK census has included a question of this kind.

7.4 Origin–destination statistics in 2011

A large number of tables of flow data have been produced as part of the outputs of the 2011 Census. They can be primarily grouped into four families of tables: the Special Migration Statistics (2011 SMS); the Special Workplace Statistics (2011 SWS); the Special Residence Statistics (2011 SRS); and the Special Student Statistics (2011 SSS). Within these families, groupings can also be considered in terms of the structure of the data tables, the spatial level and the security or access level. This set of groupings is summarised in Figure 7.1.

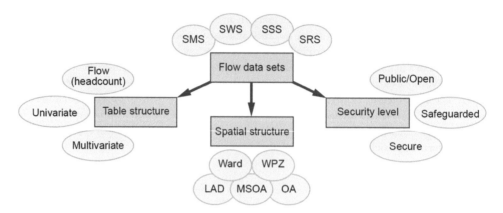

Figure 7.1   2011 Census flow data outputs

Table structure

The structure of the data tables describes the amount of attribute detail used to tabulate results. There are three levels: at the most aggregate, flows are reported in terms of total numbers of persons only and are referred to as 'flow' or 'headcount' tables. There is scope for ambiguity with this terminology: in this chapter the term 'flow data tables' refers generically to all data tables from flow data outputs, regardless of structure, whilst the term 'flow tables' refers to a specific subset of flow data tables, namely those that report total counts only. 'Univariate' tables report a given flow disaggregated by categories of a single variable (such as age), whereas 'multivariate' tables report a given flow by cross-tabulated categories of two or more variables (such as age and sex). The terms 'univariate' and 'multivariate' are used slightly loosely in this respect, in that the data are already disaggregated by two further characteristics, an origin location and a destination location.

Spatial structure

The spatial level or geography used to report flow data also varies between tables of output. This is represented in part in Figure 7.1, although the actual set of geographies used varies within and between countries of the UK. At the finest scale, OAs are used across the whole of the UK to tabulate flow data. WPZs are the finest geography used to tabulate workplace destinations in England and Wales but were not constructed in Scotland or Northern Ireland at the same time as those for England and Wales.

Both these base polygons can be aggregated into larger units, although they have separate pathways. OAs can be aggregated into Lower Layer Super Output Areas (LSOAs), which in turn can be aggregated into MSOAs. WPZs can be aggregated into MSOAs but they do not nest within LSOAs. MSOAs are thus the lowest level geography at which migration and commuting data can be easily compared with an entirely consistent geography, for the majority of journey to work observations. Many tables are made available at the local authority district (LAD) level, a composite of several different administrative units across parts of the UK: London boroughs (plus the City of London), metropolitan districts, non-metropolitan districts, unitary authorities (plus the Isles of Scilly), council areas and district council areas, as explained in Chapter 6. For the 2011 flow data, it is common for the City of London to be reported as an aggregate unit together with the City of Westminster, and for the Isles of Scilly to be similarly aggregated with the Cornwall Unitary Authority, due to small residential population sizes in both cases.

The 2011 flow data offer an improvement over the 2001 outputs in the handling of overseas origins. Whereas migration data in 1991 included country of origin (major origin countries, and regionally grouped smaller countries), in 2001 this level of reporting was removed, with only 'total from overseas' counts being included. The 2011 migration data are generally divided into 'UK' and 'non-UK' variants, with the non-UK versions tabulating international flows. For the non-UK tables, some have a broad geography, in which 59 countries or groups of countries are recognised, whilst some are 'detailed' and use a standard coding in which up to 184 countries or groups of countries can be identified. As with other levels of spatial detail, there is a trade-off between detail and ease of access to the data.

Geographic scope

As well as the reporting geographies used to disaggregate output tables, it is useful to consider the geographic scope of the different outputs. This is more complex with the 2011 outputs than with outputs from earlier censuses. Some outputs can only be generated for some countries, due to differences in questions asked. Thus, flow data tables relating to second residences are only available for people with (primary) residences in England and Wales, as the relevant questions were not asked of respondents living elsewhere in the UK. A question about Welsh language skills was asked in Wales, and a flow dataset disaggregated by Welsh language ability is available for residences in Wales only. Similar flow tables about language ability in Scotland were published as part of the outputs from earlier censuses, but have not yet been published as part of the 2011 outputs.

Regardless of the main scope of any table, the outputs usually contain cross-border flows. Thus, tables published for residences in Scotland might potentially include flows to destinations in England (or Wales or Northern Ireland), and so on. Cross-border flows are more problematic in the 2011 outputs than was the case with outputs from earlier censuses as they sometimes feature asymmetric geographies: thus, flows within a country may be tabulated at ward-to-ward level, but cross-border flows will be tabulated at a ward-to-district level. Some tables have been specified at a UK level, whereas others are only published for England and Wales, Scotland or Northern Ireland.

Summary of tables

Table 7.1 provides a summary of the numbers of flow data tables produced at different levels of attribute detail, by country and by 'family' of outputs. The total of 223 tables represents a

Table 7.1 Summary of origin–destination tables from the 2011 Census

	SMS			SWS		SRS	SSS		Total
Attribute detail	UK	EW	W	UK	EW	EW	UK	SC	
Flow count	5	–	–	4	6	24	3	–	42
Univariate	22	13	2	43	38	34	9	3	164
Multivariate	12	2	–	–	2	–	1	–	17
Total	39	15	2	47	46	58	13	3	223

Note: These statistics refer only to the 'open' and 'safeguarded' tables explained in Section 7.5.

considerable expansion on the number of tables published as part of the 2001 Census (across three spatial levels) which comprised a total of 16 migration tables, 14 journey to work tables and 14 'travel' tables (Stillwell and Duke-Williams, 2007).

Whereas the geographies used in the 2001 and earlier outputs were relatively straightforward, those used in the 2011 outputs are more diverse. Not only is a wider range of geographies used, but the names and scope of these geographies varies across component parts of the UK, resulting in a set of output tables which in some cases only apply to part of the UK. Table 7.2 summarises the range of geographies used in the flow data tables and the total number of zones in each country.

Table 7.2 Summary of the most common 2011 Census geographies used in flow data tables

Name	Breakdown	Remarks
United Kingdom Output Areas	Total: 232,296 England: 171,372 Wales: 10,036 Scotland: 46,351 Northern Ireland: 4,537*	* In the vast majority of cases, ONS has not used OAs for Northern Ireland in the official 2011 Census data dissemination. In most of the 2011 Census flow datasets containing lower level geographies, Super Output Areas (SOAs) were used instead.
United Kingdom Workplace Zones	Total: 100,819 England: 50,868 Wales: 2,710 Scotland: 46,351* Northern Ireland: 890*	* As of January 2016 no WPZs had been released for Scotland and Northern Ireland. As part of this geography dataset, Scotland OAs and Northern Ireland SOAs were added.
United Kingdom Lower Layer Super Output Areas (LSOA)	Total: 42,143 England: 32,844 Wales: 1,909 Scotland: 6,500* Northern Ireland: 890*	* Northern Ireland SOAs and Scotland Data Zones (DZs) have been used in this geography dataset.
United Kingdom Middle Super Output Areas (MSOA)	Total: 9,326 England: 6,791 Wales: 410 Scotland: 1,235* Northern Ireland: 890*	*Only England and Wales MSOAs have been released. As part of this geography dataset, Scotland Intermediate Zones (IZs) and Northern Ireland SOAs were added.
United Kingdom Wards	Total: 9,505 England: 7,689 Wales: 881 Scotland: 353 Northern Ireland: 582	
United Kingdom Merged Local Authorities	Total: 404 England: 324 Wales: 22 Scotland: 32 Northern Ireland: 26	
United Kingdom Regions	Total: 12 England: 9 Wales: 1* Scotland: 1* Northern Ireland: 1*	* Only England regions have been released. In *WICID*, Wales, Scotland and Northern Ireland are represented by their country geographical code.

7.5 Security and access control

Perhaps the most significant development in the 2011 origin–destination data has been the introduction of a revised approach to data security. Various methods of preserving confidentiality have been used in past censuses; whilst some similar approaches were retained in the 2011 outputs, much of the emphasis has been placed on controlling availability of the data. In previous censuses, a range of statistical disclosure control approaches were used to 'protect' the contents of tables of results (Duke-Williams and Stillwell, 2007). Outputs from the 2001 Census were (for residences in England and Wales, and in Northern Ireland) protected using an approach called 'small cell adjustment method' (SCAM) that probabilistically altered values of 1 and 2 to 0 or 3.

SCAM proved problematic for flow data in particular, as these data are characterised by having a very large proportion of the non-zero flows being small values. This method was dropped for 2011 Census data; instead, the liability was moved down from ONS to census users who are now responsible for the protection of low flow counts in their research outputs.

ONS has used a method called record swapping to help protect the data. Record swapping is a pre-tabulation method applied to all the 2011 Census data (including the microdata) and involves swapping information between a pair of households within a proximate area (ONS, 2012). This method is performed on all records so all counts have the same potential to contain swapped records. Record swapping ensures that any 'attacker' making a claim to have identified an individual within the census outputs could be met with an argument that it is possible that the 'person' in question may in fact be contained within a swapped record: thus, there is always uncertainty that any given data point is accurate. An advantage of record swapping over post-tabulation forms of disclosure control is that there is consistency (in terms of values aggregating to the same total) within and between published data tables, a feature that is not possible with approaches such as SCAM which undermine the concept of a 'one-number' census.

The record swapping approach to disclosure control is not sufficient to fully protect all of the data; there remain a large number of small values within the outputs, some of which are 'genuine' non-swapped values, and thus may provide a risk of disclosure. The proportion of records that were swapped is unknown, and thus the level of uncertainty in the data cannot be quantified. Further measures have been put into place to protect the small values that frequently occur in flow data, namely controlling who can access different sets of outputs.

The 2011 Census origin–destination data outputs have been divided into three groups (ONS, n.d.b) that fit with an existing ONS taxonomy of access control:

- *Public* data, also referred to by users as 'open data' and as 'Open Government Licence (OGL) data', are openly available to all users, from two main sources as outlined in Section 7.3;
- *Safeguarded* data are not considered to be personal (in the context of a legal definition of 'personal data'), but there may be a risk of disclosure if they are linked to other data (including *a priori* knowledge); and
- *Secure* (or 'controlled') data may be identifiable, and thus are potentially disclosive. In the case of origin–destination data, secure data have a high level of attribute and/or spatial detail, giving the risk that someone may claim to be able to identify individuals (subject to record swapping).

The number of tables in each access category is indicated in Table 7.3. Certain access conditions have been put in place for safeguarded and secure data. Safeguarded data are available to users who are identified as being from the public sector, subject to publication restrictions. Users are advised by ONS – via an agreed text displayed on the website providing access to the data – that

Table 7.3 2011 Census flow tables in each access category

	Open	Safeguarded	Secure	Total
SMS	5	51	176	232
SWS	9	84	193	286
SSS	1	15	33	49
SRS	6	52	117	175
Total	21	202	519	742

Note: 'Open' and 'safeguarded' tables are held in *WICID*.

small values (that is, values below 3) in outputs must be protected. ONS suggests that data *could* be rounded, but note that this is only one possible method. In practice, responsible users will need to take care that any measures applied cannot be reversed (for example, by reference to unmodified sub-totals). Alternative approaches are up to the user, but might include cell suppression or aggregation. Controls on publication of small numbers also apply to non-tabular outputs, such as graphs or maps. A consequence of the restriction on publication of small numbers is that the data cannot be re-distributed in their original form. Researchers outside the public sector who wish to use the safeguarded data are obliged to visit the ONS Virtual Microdata Lab (VML), a safe-setting controlled access facility where they can also access the secure tables.

The 'secure' datasets are available for use under controlled conditions by researchers who have Approved Researcher accreditation, and who have had a specific project approved. Data are currently available in comma separated value (csv) and (in some cases) *SPSS* formats. As with other controlled-access datasets, outputs produced by researchers must satisfy disclosure risk assessments before they can be published or otherwise disseminated by researchers. Researchers are required to book a terminal in advance, and then travel to one of the ONS offices at which the VML can be accessed.

7.6   Obtaining flow data

As outlined in the previous section, there are different routes to using the census origin–destination data that are dependent on the security or access rating of each table. The public flow data are available through *Nomis*[3] (Townsend et al., 1987) and through *WICID*[4] (Stillwell and Duke-Williams, 2003). *WICID* (Web-based Interface to Census Interaction Data) has been developed and extended through a series of Economic and Social Research Council (ESRC)-funded projects, and is part of the UK Data Service-Census Support. The open nature of these data mean that it is likely that redistribution will also occur through other routes as well over time.

WICID also provides access to the safeguarded flow data and is the primary access route for these data for users in the public sector. As well as providing an interface to plan and download data extracts, the system also hosts download files in both csv format and in a format suitable for use in conjunction with *SASPAC* (Rhind, 1984), a system which is widely adopted in local authorities. The secure flow data can only be accessed via the ONS VML and a user must have Approved Researcher status. The VML is also the access route for non-public-sector users of safeguarded flow data.

This section focuses on *WICID*, which is available via the UK Data Service website, and describes how it can be used to create a data extract. Users are stepped through the process of constructing a query in two main phases, after which the design and structure of the download

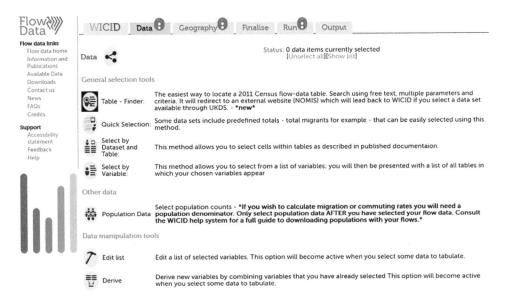

Figure 7.2 The data selection tools interface in *WICID*

Source: https://wicid.ukdataservice.ac.uk/cider/wicid/query/select/data/data.php.

can be selected. The two main phases are data selection and geography selection. We explain each of these in turn.

In the first step, the user is required to select some data of interest. A number of approaches (selection tools) are available (Figure 7.2). *Quick Selection* allows simple access to a number of pre-defined totals (typically, total persons in a flow) for a variety of tables. The data are structured as consisting of tables (grouped into sets), each made up of a number of variables. The related metadata permits two selection methods: *Select by Variable* and *Select by Dataset and Table*. The *Select by Variable* tool lists all identified thematic variables. Thus, the term 'age' is listed as a single variable, rather than the large number of variants of age reflecting different age groups used in different tables. Having selected a variable, the user is iteratively shown a list of matching tables, together with a modified list of variables, limited to those that are used in conjunction with the currently selected variable(s). The *Select by Dataset and Table* tool allows browsing through all tables. In both tools, after selecting a table, the user is shown a representation of the table, from which all or some of the table cells can be selected and added to the query.

In contrast with the outputs from 1991 and 2001, data in a much larger number of tables were disseminated as part of the 2011 Census Flow Data release by ONS, often with somewhat generic titles. This has made browsing by dataset and table more impractical. Working together with ONS and *Nomis*, a new 2011 Census 'table-finder' tool was developed for this purpose. *WICID* provides easy access to this tool by integrating control from and to it. As part of this implementation, a prominent button that redirects to the *Table-Finder* has been placed in the main *WICID* data selection page (Figure 7.2).

The *Table-Finder* tool provides interactive selection of 2011 Census tables by allowing users to search using free text and multiple parameters and criteria. All 2011 Census open and safe-guarded flow data tables contain links back to *WICID* when selected. Users can also choose to visit *Nomis* webpages instead (open data only) or contact ONS to request permission to access the dataset in their VML. It may be worth noting that the *Table-Finder* only lists tables with their natively supported geographies; therefore, tables of flows aggregated to higher level geographies which are supported by *WICID* will not be displayed.

The *WICID* link redirects back to the *WICID* system at the table information webpage; this page provides detailed information on the census table that has been selected such as totals, supported geographies (both native and via aggregation), variables and citation information. It also contains a button to select the table via the *WICID* queryable tool and, wherever available, buttons to download the table in csv or *SASPAC* format via the downloads page. Access control remains the same; users will need to log in via Federated Access (using their institutional user id and password) in order to be able to query safeguarded data, whilst a simple anonymous guest-login is enough to acquire access to the open datasets.

Having selected data to tabulate, the user is then directed to select geographical areas of interest as part of the second phase of query building. As with the data phase, there are a number of tools in the geography phase to assist the user. The *Quick Selection* tool allows the user to add all areas in a chosen geography (e.g. all wards, all districts, etc.) to their query. The *List Selection* tool allows the user to select a geography, and then a list is shown of all components of that geography, from which the user can select one or more areas. This is practical for geographies with a relatively small number of areas, but looking through a list is impractical for geographies with thousands or tens of thousands of areas. In such cases, the *Type-in* box selection method is more helpful – users can enter an area name (or wildcarded partial name) or, if known, alphanumeric code, and will be shown a list of all matches, from which they can select one or more areas. The *Map Selection* tool allows the user to select areas using the new *WICID* interactive map through a modern interface. The selected areas are seamlessly adjusted with other possible selected areas using any of the rest of the geography tools available in *WICID*.

A related tool is *Postcode Selection*, which allows the user to type in a full or partial postcode, and generate a list of all areas that intersect with that postcode. For all tools, the user is initially shown a list of geographies in order to choose one from which to select areas. This list is modified based on the currently selected data; thus, the user can only select areas from geographies that are compatible with their previous data selection. As well as familiar geographies, there are also aspatial 'geographies' which group together various totals and special categories for some datasets and represent these as 'areas'. Thus, for example, a user might select all flows from a set of regular areas to the aspatial destination 'workplace unknown'.

For origin–destination data, it is necessary to select two sets of areas. In many cases, the user will want to carry out a symmetric selection – that is, the set of origin and destination areas will be identical. A short cut tool, *Copy Selection*, enables this to be done easily, setting the destinations to be the same as the currently selected origins, or *vice versa*. Once the Geography part of the query has been completed, the query can be run to generate the flows. Figure 7.3 shows a *WICID* query that has been constructed in order to extract and download total migration flows within and between 12 regions of the UK from SMS Table MM01CUK_all, which contains data within and between LADs.

WICID includes a large number of lookup tables converting from one geography to another, which permit on-the-fly aggregation of base areas. Thus, whilst any particular dataset has been formally published at one spatial scale, *WICID* can seamlessly present it as also being available for aggregates of that base geography, as indicated in the example shown in Figure 7.3. Within the *List Selection* mode, it is also possible to exploit these lookup tables in another way. After initially selecting a geography for which results are to be tabulated, the user has the option of selecting an additional aggregate geography to help select areas. For example, a user might select wards as the initial geography, but then rather than typing in ward names or selecting from a list of wards,

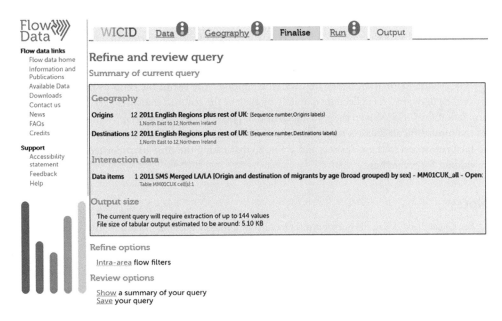

Figure 7.3   *WICID* query for extraction of regional migration flows from the 2011 SMS

Source: https://wicid.ukdataservice.ac.uk/cider/wicid/query/refine.php?mode=show.

would select districts as the selection geography. In this scenario, the user would be shown a list of districts and, on choosing one, all wards within that district would be added to the query.

7.7   Conclusions

The 2011 origin–destination outputs offer a number of advantages over the outputs of previous censuses. In particular, they include entirely new datasets (SSS and SRS) which will allow researchers to explore mobility patterns in new ways; there is a larger number of output tables than has previously been the case, especially at a spatially detailed level; and the SCAM disclosure control approach used with the 2001 outputs, which was particularly problematic in the case of flow data, has not been repeated.

These advantages have come with certain costs. Most significantly, the trade-off for more detailed data and less heavy-handed disclosure control has been the introduction of access controls. For public sector users, access is not especially onerous: the data are available for download and for online use. The restrictions on publication may prove somewhat trickier, given that spatially detailed flow data are characterised by a large number of cells with small values, including large numbers of cells with the value 1 or 2. Dealing with this will require a two-pronged attack for educators and trainers: first, it will be important to make sure that users of the data are aware of the requirements; and, second, advice will need to be given over how users might modify the data appropriately.

A second problem is posed by the large number of tables: these are considerably harder to navigate than has been the case in the past. The situation is complicated by the fact that some of these tables are overlapping in the way that they provide data for different parts of the UK, with variant scales for cross-border flows. It will be helpful to derive portmanteau datasets that group together data from different source tables to a UK level to aid the user experience. Part of the problem with table navigation seems to arise from the simple state of being flow data tables: the table titles used by ONS attempt to incorporate labelling of the spatial structure and base population of the table as well as the actual content (that is, the cross-classifying variables in the table), and become long, unwieldy and hard to differentiate.

The origin–destination tables are invariably amongst the last of the outputs from any census since they draw on data collected separately by each of the national statistical agencies and require conflation (by ONS). Given their characteristic of possessing a dual geography, they are also some of the most difficult datasets to process, disseminate and analyse. The *WICID* interface attempts to facilitate access to these datasets and we hope that taking the time to understand them will be rewarding for researchers.

Notes

1 For England and Wales, the question was preceded by a routing question that asked "If you had a job last week", with those persons who did have a job then directed to the 'place of work' question.
2 http://unstats.un.org/unsd/demographic/sources/census/quest/CHE2010en.pdf.
3 www.nomisweb.co.uk/census/2011.
4 https://wicid.ukdataservice.ac.uk/.

References

Bell, M. and Ward, G. (2000) Comparing temporary mobility with permanent migration. *Tourism Geographies*, 2(1): 87–107.

Bell, M., Charles-Edwards, E., Kupiszewska, D., Kupiszewski, M., Stillwell, J. and Zhu, Y. (2014) Internal migration data around the world: Assessing contemporary practice. *Population, Space and Place*, 21(1): 1–17.

Cole, K., Frost, M. and Thomas, F. (2002) Workplace data from the census. In Rees, P., Martin, D. and Williamson, P. (eds.) *The Census Data System.* John Wiley, Chichester, pp. 269–280.

Denham, C. and Rhind, D. (1983) The 1981 Census and its results. In Rhind, D. (ed.) *A Census User's Handbook.* Methuen, London, pp. 265–286.

Duke-Williams, O. (2009) The geographies of student migration in the UK. *Environment and Planning A*, 41(8): 1826–1848.

Duke-Williams, O. and Stillwell, J. (2007) Investigating the potential effects of small cell adjustment on interaction data from the 2001 Census. *Environment and Planning A*, 39(5): 1079–1100.

Martin, D., Cockings, S. and Harfoot, A. (2013) Development of a geographical framework for census workplace data. *Journal of the Royal Statistical Society: Series A (Statistics in Society)*, 176(2): 585–602.

NRS (2007) 2006 Census Test Form. Available at: www.scotlandscensus.gov.uk/documents/preparation/2006-census-test-form1.pdf.

ONS (2012) 2011 Census: Methods and quality report confidentiality protection provided by statistical disclosure control. Available at: www.ons.gov.uk/ons/guide-method/census/2011/census-data/2011-census-user-guide/quality-and-methods/methods/statistical-disclosure-control-methods/confidentiality-protection-provided-by-statistical-disclosure-control.pdf.

ONS (2014) Workplace Zones: A new geography for workplace statistics. Available at: http://webarchive.nationalarchives.gov.uk/20160105160709/https://geoportal.statistics.gov.uk/Docs/An_overview_of_workplace_zones_for_workplace_statistics_V1.01.zip.

ONS (n.d.a) Census 1911–2011. Available at: www.ons.gov.uk/ons/guide-method/census/2011/how-our-census-works/about-censuses/census-history/census-1911-2011/index.html.

ONS (n.d.b) 2011 Census Origin–Destination Data User Guide. Available at: http://www.ons.gov.uk/ons/guide-method/census/2011/census-data/2011-census-prospectus/release-plans-for-2011-census-statistics/subsequent-releases-of-specialist-products/flow-data/origin-destination-data--user-guide.pdf.

Rees, P. (1977) The measurement of migration, from census data and other sources. *Environment and Planning A*, 9: 247–272.

Rees, P.H., Thomas, F. and Duke-Williams, O.W. (2002) Migration data from the census. In Rees, P., Martin, D. and Williamson, P. (eds.) *The Census Data System.* Wiley, Chichester, pp. 245–267.

Rhind, D. (1984) The SASPAC story. *BURISA*, 60: 8–10.

Stillwell, J. and Duke-Williams, O. (2003) A new web-based interface to British census of population origin–destination statistics. *Environment and Planning A*, 35(1): 113–132.

Stillwell, J. and Duke-Williams, O. (2007) Understanding the 2001 UK census migration and commuting data: The effect of small cell adjustment and problems of comparison with 1991. *Journal of the Royal Statistical Society: Series A (Statistics in Society)*, 170(2): 425–445.

Townsend, A.R., Blakemore, M.J. and Nelson, R. (1987) The NOMIS data base: Availability and uses for geographers. *Area*, 19: 43–50.

United Nations (1998) Recommendations on Statistics of International Migration, Revision 1. *Statistical Papers, Series M, No. 58, Rev. 1*, Department of Economic and Social Affairs, Statistical Division, United Nations, New York.

8

Census microdata
Cross-sectional samples

Jo Wathan, Paul Waruszynski and Orlaith Fraser

8.1 Introduction

Census microdata are unaggregated outputs containing samples of individual records to facilitate flexibility in their reuse. They look like the sort of data collected if you were to conduct your own survey as they contain a wide range of varying characteristics for each person. The name 'census microdata' has, in the 2011 Census round, become synonymous with *cross-sectional* microdata that are drawn from one point in time (i.e. a single census). This is in contrast to longitudinal microdata which have different features and which are described in the next chapter of this volume.

Each census microdata product is characteristically available as a large flat file in which individual records (or 'cases') are recorded in rows and characteristics which vary from person to person (known as 'variables') are in columns. This is demonstrated in Figure 8.1, which shows a corner of the 2011 England and Wales teaching file as seen when opened in the *Stata* statistics package. One case (4) and one variable (econac1t) have been highlighted. The actual dataset contains over half a million rows and 18 columns. Files usually contain all of the census topics, but the total amount of information in each file is limited to the level appropriate for the setting in which it is made available (as we will see later in the chapter).

The chapter gives an overview of census microdata, and the 2011 UK microdata files in particular. We start by providing some historical and international context in Sections 8.2 and 8.3. In Sections 8.4 and 8.5, we summarise the files that have been made available and their characteristics, including two of their key features: that microdata are available only for samples and that population bases are flexible. Work that has been done to ensure that the confidentiality of respondents is protected is covered in Section 8.6. Access routes to the data are described and the research utility of the data is illustrated, drawing on examples from the literature in Sections 8.7 and 8.8. Finally, we reflect on the future of microdata before concluding.

8.2 The development of census microdata in the UK

Historically, census outputs were always in the form of census counts; over time these became more extensive, but they were not flexible. When users wanted information which was not already in a published tabulation, the only recourse was to have a table produced to order, a

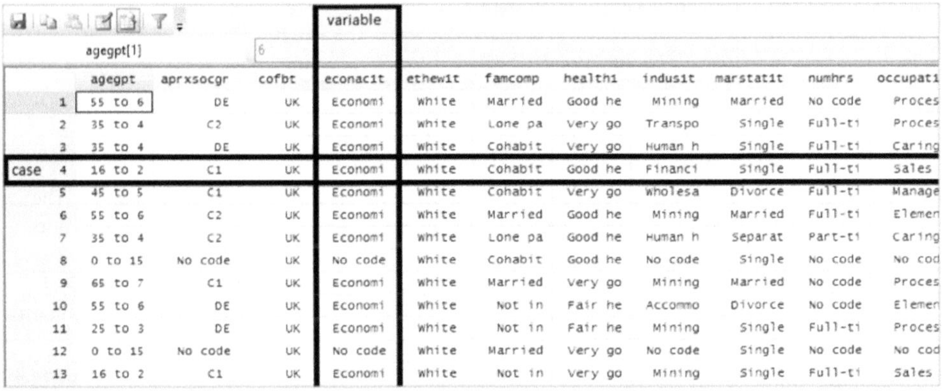

	agegpt	aprxsocgr	cofbt	econacit	ethew1t	famcomp	health1	indus1t	marstat1t	numhrs	occupati
1	55 to 6	DE	UK	Econom1	white	Married	Good he	Mining	Married	No code	Proces
2	35 to 4	C2	UK	Econom1	white	Lone pa	Very go	Transpo	Single	Full-ti	Proces
3	35 to 4	DE	UK	Econom1	white	Cohabit	Very go	Human h	Single	Full-ti	Caring
case 4	16 to 2	C1	UK	Econom1	white	Cohabit	Good he	Financi	Single	Full-ti	Sales
5	45 to 5	C1	UK	Econom1	white	Cohabit	Very go	wholesa	Divorce	Full-ti	Manage
6	55 to 6	C2	UK	Econom1	white	Married	Good he	Mining	Married	Full-ti	Elemen
7	35 to 4	C2	UK	Econom1	white	Lone pa	Good he	Human h	Separat	Part-ti	Caring
8	0 to 15	No code	UK	No code	white	Cohabit	Good he	No code	Single	No code	No cod
9	65 to 7	C1	UK	Econom1	white	Married	Very go	Mining	Married	No code	Proces
10	55 to 6	DE	UK	Econom1	white	Not in	Fair he	Accommo	Divorce	No code	Elemen
11	25 to 3	DE	UK	Econom1	white	Not in	Fair he	Mining	Single	Full-ti	Proces
12	0 to 15	No code	UK	No code	white	Married	Very go	No code	Single	No code	No cod
13	16 to 2	C1	UK	Econom1	white	Not in	Very go	Mining	Single	Full-ti	Sales

Figure 8.1   The 2011 Census microdata teaching file for England and Wales, showing the first 11 variables and 13 cases

Source: ONS. 2011 Census, UKDS-CS (2015) Census 2011 Microdata Teaching File for England and Wales. [data collection]. UK Data Service. SN: 7613, http://dx.doi.org/10.5255/UKDA-SN-7613-1. Contains public sector information licensed under the Open Government Licence v3.0.

so-called commissioned table. By restricting outputs to counts, the majority of the information gathered by the censuses was locked away, both figuratively and literally.

The first ever census microdata sample (the Public Use Microdata Sample (PUMS)) was drawn from the 1960 Census in the United States (Dale et al., 2000). At this time, computing power was increasing, enabling both computer-based data production and analysis. Additionally, the potential audience for detailed data increased as higher education grew and produced a new cadre of social researchers who sought more sophisticated data products. In the UK, planning for longitudinal microdata files started in the 1960s, resulting in the Longitudinal Study for England and Wales drawn from 1971 Census data for secure and controlled use.

By 1981, two different reports had recommended the addition of cross-sectional anonymised samples from UK censuses which could be more easily accessed (Norris, 1983). With growing demand and a continued need to adhere to legally binding promises of confidentiality, the 1991 Census White Paper (Her Majesty's Government, 1988) recorded the heavily qualified view that "[census abstracts] . . . in the form of samples of anonymous records . . . would also be considered, subject always to the overriding need to ensure the confidentiality of individual data" (reported in Marsh et al., 1991, p. 20).

In response, Marsh et al. (1991) formally assessed the low risk of releasing two non-overlapping samples from the 1991 Census: one was a proposed sample of individuals with large local authorities identified, the other a sample of private households (including the individual members) with regional geography. It was in response to this evidence, alongside the accompanying demonstration of user need, that samples from the 1991 Censuses were produced. The first census microdata files known as 'Samples of Anonymised Records' (SARs) were produced by the then Office of Population Censuses and Surveys (OPCS) and purchased by the Economic and Social Research Council (ESRC) for reuse.

8.3   Census microdata in the international context

It is now common rather than exceptional for census offices to create microdata for reuse. This is most clearly manifest in the breadth of the International Public Use Microdata Sample (IPUMS)

Table 8.1 Selected international census microdata

Country	Census microdata arrangements
United States	PUMS are now based on the American Community Survey since this rolling sample replaced the census long form in the 2010 Census round. See www.census.gov/main/www/pums.html.
Canada	Public-use microdata files (PUMFs) are available from the census until 2006. The Canadian long form was famously discontinued for the 2011 Census round and replaced by a 1% sample survey called the National Household Survey. See www12.statcan.gc.ca/nhs-enm/2011/dp-pd/pumf-fmgd/index-eng.cfm. At the time of writing, however, the new Canadian Government had announced the return of the long form in time for the 2016 Census round.
United Kingdom	Files are available for UK countries between 1991 and 2011 based on a traditional census design. In 2011 they were produced separately for each census office and are available either through that Census Office or through the UK Data Service.
Germany	Traditional population censuses were replaced in 1991 by a 'micro census' which produced a 1% sample from a long-running annual, now continuous, survey. In 2011, however, a new registered census was undertaken and microdata products are being produced. At the time of writing, two files – one a 10% sample of households, the other microdata on housing – were available in a safe setting. See www.forschungsdatenzentrum.de/bestand/zensus_2011/index.asp.
Spain	Census microdata are available from a traditional census for 1991 to 2011. http://www.ine.es/en/censos2011_datos/cen11_datos_microdatos_en.htm.
France	The French census was reformulated to constitute a rolling survey which is designed to generate national coverage over a period of five years. Microdata are available, covering a wide range of topics and with large samples. See www.insee.fr/fr/themes/detail.asp?ref_id=fd-rp2012&page=fichiers_detail/RP2012/doc/documentation.htm.
Japan	A change to the Japanese statistics Act in 2009 has enabled the production of microdata from official data for the first time for 2000 and 2005 for approved use. Work is progressing on data for microdata for the 2010 Census and to enable the production of data for smaller geographies. General information on data for secondary analysis can be found at: www.soumu.go.jp/english/dgpp_ss/seido/2jiriyou.htm.
International	The IPUMS files are a collection of microdata samples from 1960 onwards from over 80 countries which are harmonised to enable comparability. See https://international.ipums.org/international/.
Europe	The Integrated Census European Microdata files are similar to those in the IPUMS but for Europe alone.

collection (Ruggles et al., 2003), which now contains microdata from over 80 countries. However, the nature of the microdata necessarily reflects an increasingly diverse range of census collection methods, with 'census' microdata often also including data generated from surveys or administrative data that now form part of modern census-taking across the globe, as demonstrated in Table 8.1.

8.4   2011 Census microdata files

The three UK Census Offices have produced and released a variety of microdata samples from the 2011 Census round in order to satisfy different user needs and requirements. The microdata samples have been harmonised across the UK, wherever possible, but releases remain specific to

Table 8.2 2011 Census microdata samples for England and Wales, Northern Ireland and Scotland

Territory/ Country	Access type		
	Open	Safeguarded	Secure
England and Wales	**Teaching file** Released January 2014 1% sample individual records 18 variables	**Region sample** Released December 2014 5% sample individual records 122 variables	**Individual sample** Released November 2014 10% sample individual records 258 variables
		Grouped LA sample Released March 2015 5% sample individual records 120 variables	**Household sample** Released November 2014 10% sample household records 245 variables
Scotland	**Teaching file** Released June 2014 1% sample individual records 17 variables	**Region sample** Released November 2015 5% sample individual records 103 variables	**Individual sample** Released December 2015 10% sample individual records 135 variables
		Grouped LA sample Released November 2015 5% sample individual records 75 variables	**Household sample** Released December 2015 10% sample household records 131 variables
Northern Ireland	**Teaching file** Released January 2014 1% sample individual records 18 variables	**Northern Ireland sample** Released November 2015 5% sample individual records 122 variables	**Individual sample** Released March 2015 10% sample individual records 220 variables
		Grouped local authority district (LAD) sample Released November 2015 5% sample individual records 120 variables	**Household sample** Released March 2015 10% sample household records 200 variables

each of the three territories (i.e. England and Wales, Scotland and Northern Ireland), taking into account separate questions, requirements and confidentiality considerations. The different types of samples have been designed to maximise the amount of information that can be made available for each of the three types of access routes (Table 8.2).

Open access teaching files

The open access teaching files are designed as 'taster' files to provide insight into the more detailed microdata products available and to encourage wider use of census data, including in teaching. Each contains a 1 per cent sample of individual records and fewer than 20 variables; geography is at national level outside England and region level (e.g. South East, South West, etc.) within England.

Regional and grouped local authority safeguarded files

Users requiring easy-to-access microdata products with considerably more research utility may find the safeguarded individual census microdata files of use. Each Census Office has produced two of these files, each of which contains a 5 per cent sample of person records: (i) a file with grouped local authority geography; and (ii) a file with regional geography in England or country in Wales, Scotland and Northern Ireland.

The grouped local authority file identifies local authorities which have populations of 120,000 usual residents or more. Smaller local authorities with populations below this threshold are grouped together where necessary. In this sample, a small number of local authorities have been split where authorities that were present in 2001 had merged by 2011; for example, Cornwall is now a single unitary authority but the former local authorities of Caradon and North Cornwall were merged to achieve the population threshold. These files have between 75 variables (Scotland file) and 120 variables (other UK files).

The regional files have less geographic detail but more socio-economic content. In most of the UK, the regional files have two additional variables as compared with the comparable grouped local authority file. In Scotland, there are as many as 28 more variables than in its grouped local authority equivalent. However, many variables in the regional file contain more detailed categories. For example, in the England and Wales safeguarded files, there is more information on ethnicity in the regional file (18 categories) than in the grouped local authority file (13 categories). The former has greater detail about mixed ethnic groups in particular.

Individual and household secure files

The secure samples offer the largest sample sizes, scope to work across households, the most socio-economic detail and the smallest geography. They therefore offer a very high degree of research utility. The Office for National Statistics (ONS) (for England and Wales), Northern Ireland Statistics and Research Agency (NISRA) (for Northern Ireland) and National Records of Scotland (NRS) (for Scotland) have each produced two secure files: an individual file and a household file. Every secure file contains a 10 per cent sample of records. The variables in each can be considerably more detailed than those in the safeguarded files. For example, in the England and Wales files, the country of birth variable in the safeguarded regional file has 26 categories, compared with 274 categories in the secure individual file.

Individual secure file

This 10 per cent file contains unlinked individual records. Accordingly, and in common with the safeguarded and open files, this file does not permit additional household variables to be created. It does, however, include those individuals who live outside private households in communal establishments. Because some variables are specific to this group, this tends to be the largest file produced by each Census Office. The individual secure file in England and Wales contains 258 variables compared to 220 in Northern Ireland.

Household secure file

The 10 per cent household file is a sample of individuals drawn from the records of those living in private households. Each household has a unique household ID number, allowing users to link household members to each other. This has the advantage of enabling researchers to undertake a range of analyses that would not be possible using an individual file, such as analysing at the household level, using household context or linking household members on the basis of their relationships.

8.5   The production of the 2011 Census microdata

Census data are no longer 'sold' in the way that they were in 1991. Rather, datasets are produced where it is deemed to be part of the Census Office's core work. A necessary first step in the production of outputs is to establish a business case. Keith Dugmore was therefore commissioned by the ESRC-funded SARs support team at the University of Manchester in 2009 to interview SARs users to assess the impact arising from their analyses. He demonstrated that research using 1991 and 2001 SARs had accounted for over £1.5 million worth of ESRC funding and had been associated with record inward investment in Northern Ireland from the US of $150million (Dugmore, 2009).

The samples

The 2011 Census Microdata (as they are now called) contain samples which are used to generate estimates of population characteristics. They are therefore created to be representative of the population. They are designed to be large enough to identify minority groups whilst small enough to protect the identity of respondents. Households to be included in the household samples were flagged for selection first, to ensure whole households were selected. These were then ineligible for selection of the individual samples to ensure that there was no overlap between the household and individual samples. The individual samples in the safeguarded and open files were drawn from the larger secure sample, and contain fewer cases as well as less detail.

Sampling error

All census outputs are subject to errors. Census results vary from 'true' values because of imperfections in the data collection process and the intrinsic variability of population characteristics from day to day. However, unlike census products which are based on the whole population, microdata are also subject to sampling error, arising because the sample is one of many possible samples which could have been drawn. Estimates produced from one sample therefore differ from estimates which could have been produced from other possible samples of the same size drawn in the same way from the same population. Therefore, users who require accurate population totals should refer to published census outputs wherever possible and not microdata samples; the latter are designed for multivariate analyses or specialist statistics not available from summary tables and are less precise.

The sampling errors are usually measured through standard errors of sample estimates. The standard error measures how much the estimate will vary across possible samples. If the samples had been drawn from a simple random sample without any complications to the sample design this would be straightforward to calculate. For example, the standard error of a percentage can be calculated using the following:

$$se(p) = \sqrt{\frac{p \star (100 - p)}{n}} \tag{8.1}$$

where $se(p)$ is the standard error of a percentage p and n is the number of cases in the sample used to produce the percentage calculation. This value is used in calculating the precision of estimates. We would expect the true population value to fall within the range approximately twice this value either side of the estimate 95 times out of 100. However, it is worth bearing in mind that standard errors may be affected by sample design, as we shall see.

For general analyses of the whole sample, users may find that the large size of the census microdata samples mean that the chances of drawing invalid inferences from the data are small. Users should take care when considering weak relationships as they may appear statistically significant when they are substantively insignificant or artefactual.

Stratification

Stratification improves the precision of the sample by ensuring that the sample covers a full range of values for the specified characteristics. This is done by splitting the population into non-overlapping groups and sampling from each of them. All 2011 Census microdata samples were stratified geographically by census output area within each local authority. This method ensures good representation of data, as the sample members are evenly spread between local authorities and within each authority. This method is consistent with the user requirement for a multipurpose product that can be used for a wide variety of analyses. It also controls against extreme sample selection, ensuring, for instance, that an entire output area is not selected at random.

The exact method used in the ONS samples was as follows:

- Each case was allocated a random number.
- The data were reordered by local authority, output area and then random number; in other words, at this stage all cases in the first output area in the first local authority would be at the top.
- Random start points were chosen and cases were then selected systematically.

This sampling method has some particular features. First, sampling below local authority may be said to be implicit. That is, that although the data were not split into separate strata at the output area level, the method had the effect of improving the quality of the sample coverage at this lower level. Second, because a systematic random approach is undertaken, the cases are selected with a probability proportionate to the size of the local authorities. This means that the approach does not introduce any bias across local authorities. Third, the method avoids potential pitfalls of sampling systematically by geography as the cases are randomised within output area. The overall effect of stratification is to improve the precision of estimates as compared to a simple random sample.

Clustering

Clustering occurs when sampling selects groups of cases rather than individual cases. This is not problematic with the individual data, but for the secure household sample the households constitute naturally occurring clusters. Clustering will reduce the precision of estimates if the data are used to undertake individual-level analyses for multiple household members. The effects can be quite marked for some characteristics; in 1991, for example, the effects were largest for ethnic group, with the standard error of estimates of Pakistani, Bangladeshi and Indian groups being 2 or higher. This means that without adjustment, standard error calculations which do not account for the sample design will underestimate the true value by a factor of two or more (Census Microdata Unit, 1994). For this reason, the individual file is preferable for individual-level analyses.

Population bases

The main population base for published statistical tables from the 2011 Census is the usual resident population as at census day, 27 March 2011. The 2011 Census microdata samples, however,

Table 8.3 Population bases for each of the 2011 Census microdata samples

Country	Microdata files	Population bases available
England and Wales	Teaching file, safeguarded individual sample (region), safeguarded individual sample (grouped local authority), secure household sample, secure individual sample	Students at their non-term-time address, short-term residents, usual residents
Scotland	Teaching file	Usual residents
Scotland	Safeguarded individual sample (grouped local authority), safeguarded individual sample (region)	Usual residents, students at their non-term-time address
Northern Ireland	Teaching file	Usual residents
Northern Ireland	Safeguarded individual sample (grouped local government district (LGD)), safeguarded individual sample (region), secure household sample, secure individual sample	Usual residents, students at their non-term-time address, short-term residents

include data from the total UK population, which includes usual residents, short–term residents (in England and Wales and Northern Ireland only) and students living away from home during term time. This provides extra flexibility for researchers but all users need to be aware of this feature to ensure that their analysis correctly reflects the intended population.

There are differences in population definitions between countries (Table 8.3). For the purposes of the 2011 Census in England and Wales, and in Northern Ireland, individuals were counted as usual residents if they had lived, or planned on living, in the UK for a period of 12 months or more, or had a permanent UK address and were outside the UK and intended to be outside the UK for a period of less than 12 months. Those who had stayed, or planned on staying, in the UK for more than three months but less than 12 months were counted as short-term residents. Scotland's 2011 Census aimed to enumerate all people who were usually resident in Scotland for six months or more. Students and schoolchildren in full-time education studying away from the family home were counted as usually resident at their term-time address. Basic demographic information only (name, sex, age, marital status and relationship) was collected at their non-term-time ('home' or vacation) address. This introduces a potential to double count students when analysing these basic characteristics using files which contain data collected from both term-time and vacation addresses.

Selecting a population base

A population base variable is available where appropriate so that users can define the population base for the analysis. In England and Wales, and Northern Ireland files, this is called popbasesec. So, for example, in order to select only usual residents one would select only those cases that take the value 1 for popbasesec. In Scotland, the equivalent variable is called termind in order to reflect the difference in the variable in Scotland compared with the rest of the UK. Residents in private households and in communal establishments (hotels, prisons, hospitals, student halls of residence, etc.) are included in all individual-level microdata samples. Household-level samples contain only individuals living in private households.

Adjusting the census estimates

Unlike most surveys, the census microdata are drawn from a database which includes both enumerated responses and estimated responses where data are believed to have been missed by the enumeration process. For users, this simplifies the process of using the data, but the method used to produce the full database requires considerable effort on the part of the Census Offices.

Underenumeration

The 2011 Census is the most complete available source of information on the population. However, despite efforts to reach everyone and obtain the most accurate information possible, no census is perfect and some people are inevitably missed. A Census Coverage Survey (CCS) was undertaken in order to improve the accuracy of census results by estimating the number and characteristics of people missed by the census (NRS, 2012; ONS, 2013). From this, it was possible to impute responses which were missed by census enumeration. Further information about this process is given in Chapters 3 and 4 of this book.

As the microdata samples are taken from the complete census database, all microdata products include imputed persons. This means that there is no need for users to use additional methods to deal with non-response; for example, weights are not required. The process does, however, introduce some potential sampling error as the process is based on the results of a large sample survey. Confidence intervals have been published to assess the accuracy of census estimates (NISRA, 2012; NRS, 2013; ONS, 2012a).

Item non-response

Where respondents failed to answer a question, provided inconsistent information or gave invalid responses, the 2011 Census Item Edit and Imputation Process (ONS, 2012b) was implemented to correct these inconsistencies and estimate missing data whilst preserving the relationships between census characteristics. After item editing and imputation, all of the returned questionnaire records were complete and consistent. Item non-response, editing and imputation rates are available for the 2011 Census population and housing estimates for England and Wales (ONS, 2012c).

Improvements compared with 2001 samples

The microdata products resulting from the 2011 Census show a number of improvements over the 2001 microdata samples. In particular, the overall sample sizes are increased, with secure files now containing a tenth of all records. There are also two published files with 5 per cent samples. The grouped local authority microdata file is a response to user demand for files in which a large number of local authorities can be identified, but which have more variable detail than the Small Area Microdata File of 2001.

Public use microdata have also been created for the first time for 2011 Census data for England and Wales, Northern Ireland and Scotland, enabling users to access data without restriction or registration. Public use files offer a significant advantage in terms of expanding the range of users able to access the data and increasing equality of access (UNECE, 2007).

8.6   How is confidentiality protected in 2011 Census microdata?

It is widely accepted by all three Census Offices of the UK that protecting the confidentiality of personal information is of paramount importance. This is the highest priority and a key challenge when it comes to making microdata available; so much so that protecting confidentiality permeates all aspects of data design and access. This section summarises how the privacy of census respondents has been protected in respect of the microdata samples and the balances which are struck between the content of samples and their accessibility.

The confidentiality imperative

The confidentiality of census data is protected by law, in accordance with the legislation applicable to each territory. In each territory, breaches are punishable by law and can result in up to two years, imprisonment:

- In England and Wales, these data are protected by the Census Act 1920, the Statistics and Registration Service Act 2007 and the Data Protection Act 1988.
- In Northern Ireland, the confidentiality of personal census information is protected in accordance with the Census Act (Northern Ireland) 1969, as amended by the Census (Confidentiality) (Northern Ireland) Order 1991.
- In Scotland, the confidentiality of personal census information is protected in accordance with the Census Act 1920, as amended by the Census (Confidentiality) Act 1991, and the Census (Scotland) Regulations 2010.

This legal obligation is joined by ethical and practical considerations. Census respondents are promised, by the National Statistician, that their responses are confidential. Such promises are considered essential to gain the trust necessary for respondents to share their personal information. If respondents' responses were to be disclosed, this would be a breach of both trust and ethics. The Census Offices are very aware that failure, or perceived failure, to protect the promised confidentiality of census responses could lead to mistrust in statistics, as well as potentially greater resistance in participating in future surveys undertaken by the offices.

We may summarise the approaches involved as those that: (i) make the data safer; (ii) seek to ensure users' safety; and (iii) ensure the safety of settings and outputs. We now consider each of these in more detail.

Making the data safer

Microdata are drawn from the census output database. Accordingly, some of the protections applicable to all census outputs are discussed here as well as those additionally applied to microdata. There are two strategies which are employed across all census outputs to protect the confidentiality of respondents: record swapping and limitation on the total amount of information in outputs (ONS, 2012d).

Targeted record swapping is designed to introduce some uncertainty into the overall data. The method is applied to the census database before the outputs, including the microdata samples, are generated. This technique assesses every individual and household for how unique or rare the record is. A sample of records is taken (weighted by their uniqueness to include a higher proportion of unusual cases) and swapped with other similar records in a different nearby geographical area. This technique ensures that each record has a non-zero probability of being swapped. Accordingly, it creates a level of doubt about whether individual records are in the correct area.

The second strategy involves the restriction of attribute detail, particularly at lower levels of geography. For tabular output, this means that tables do not contain a large number of attributes, and that tables with more geographical detail tend to have less information about socio-economic characteristics. This strategy is also applied to the microdata; however, as we will see later, it is combined with strong access control strategies to ensure that very detailed data are still possible. It is worth noting that the edit and imputation process introduces additional uncertainty as records may contain information that has been imputed. In order to benefit from this uncertainty as to whether any value is 'true' (i.e. not imputed) it is not possible to identify imputed values in the 2011 Census microdata.

The *safe data* approach to protecting confidentiality is most visible in the open access teaching files. These public files are easily accessible online and are therefore subject to no access controls. Without access controls, safeguards to ensure that responses remain confidential have to be built into the data themselves. Two approaches are used: restricting the total amount of information in any files, and sampling. The most obvious step in anonymising data is to ensure that direct identifiers are removed. This involves the removal not only of names and addresses but also date of birth and detailed geography. In addition to removing direct identifiers, the amount of detail about any individual is restricted, so that those characteristics cannot be used to indirectly identify the individual. This affects the number of variables and the amount of detail within those variables. For this reason, geography, for example, is very limited in the teaching files.

Safeguarded data have some access controls (as described below) but also need to be adequately protected to prevent casual data disclosure, for example, if they are considered alongside other published sources. In order to balance user demand for socio-economic detail *and* sub-regional geographic detail, it was necessary to produce two separate files. One file had good socio-economic detail and regional geography; the other file contains grouped local authority geography, but some socio-economic detail was subject to additional restriction.

The open teaching files have a small sample size of 1 per cent, meaning that only one in 100 people will actually be in the sample. Consequently, there is a very good chance that unusual cases in the files may actually be one of many with the same characteristics in the population. This sample size, when combined with the targeted record swapping mentioned earlier, introduces *sufficient uncertainty* as to whether any case is either as unusual as it appears or located where reported. To establish 'sufficient' uncertainty, methodologists calculated the number of records that were unique in the population as a proportion of those that were unique within the sample. Further detail about the proportions that were unique, or the proportions that were swapped is confidential.

In summary, microdata without access controls have more socio-economic detail than tabular outputs. However, they are protected by loss of geographical detail and a small sampling fraction.

Safe users

Because of their highly restricted content and small sampling fraction the teaching files might be best considered as 'taster' files, which do not have the research utility usually associated with census microdata. Accordingly, additional controls are required in order to produce files with a higher utility. Confidentiality protection is ensured by regulating who can use the data and how they can be used. These principles apply quite differently to safeguarded and secure data.

Safeguarded data are more detailed than open data and they have larger sample sizes. Before being granted access to the data, users of the safeguarded samples have to register with the UK Data Service and agree to a series of terms and conditions regulating the use of the data. These terms and conditions impose restrictions and limitations on users. Users, for example, must not attempt to or claim to have identified an individual in the data. Users must not share the data with others who have not registered to use the data themselves and must keep passwords secret.

The secure samples rely mainly on access controls to ensure that data can be used without posing a confidentiality risk. The secure files have more detail than can be published; accordingly, they are subject to legal protections under the Statistics and Registration Services Act. This requires all users to have attained Approved Researcher status for their research project before they can have access to the data. An Approved Researcher is someone who has demonstrated both that they have the knowledge and experience to handle potentially disclosive personal information and that they are accessing the data for a suitable research project.

Safe settings and safe outputs

Finally, census microdata are protected by the environment they are held in and how access to that environment is managed. This occurs especially with the secure microdata, which are only available to access in the Virtual Microdata Laboratory (VML) or the NISRA Secure Environment for Northern Ireland. These environments have stringent physical and computer security to ensure that only persons who meet the criteria and have been approved can physically access the data. Outputs can only be removed from the safe setting after the Census Offices have performed checks on them and they are satisfied with the risk of disclosure.

8.7   Accessing UK census microdata

The way in which users access data is clearly dependent on the amount of detail in the file in question. In this section, we give practical guidance on how users can access each type of data.

Open samples

The teaching file for England and Wales can be downloaded from the ONS website. For Scotland, the sample is available from the NRS website, and the sample for Northern Ireland is available from the NISRA website. All three teaching files are available in csv format and all come with a user guide (ONS, 2014a; NISRA, 2015a; NRS, 2015). The user guides provide supplementary information about microdata, the purpose of the teaching files, what the samples contain, as well as how the samples were drawn. A quality document, showing how distributions for key variables compare between the teaching file and the population is available for the England and Wales sample. Also available from the ONS website is a teaching aid, providing examples of possibilities for data exploration using the England and Wales teaching file (ONS, 2014b). These data are available under an Open Government Licence (OGL) and can be reused as desired.

Safeguarded samples

Safeguarded microdata files are available through the UK Data Service[1] under the auspices of concordats with each Census Office. The data are available to use only by those who have registered first to use the service and have ordered the data for a particular project. Registration is usually a short online process which includes agreeing to an End User Licence (EUL) as described above. Breaches will lead to immediate termination of access and may result in legal action. The process of registration is most straightforward for users who belong to research institutions that are members of the UK Access Management Federation[2] but other users are welcome to obtain a dedicated login as part of their registration process, although applications outside the UK or to sell data are referred to the Census Offices.

Users can find the microdata in the census subsite[3] or by searching for 'census microdata' (for 2011 products) or 'SARs' for earlier products in data search tool Discover.

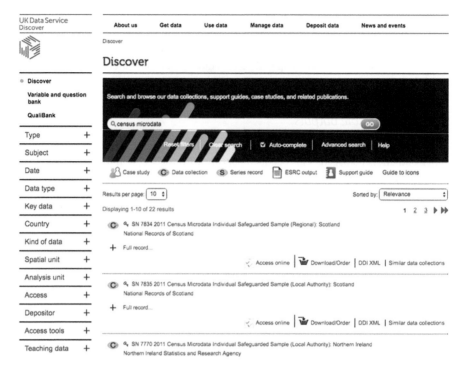

Figure 8.2   Results for a Discover search on 'census microdata'

Source: https://discover.ukdataservice.ac.uk/?q=census+microdata.

The first link in the list of results shown in the screenshot in Figure 8.2 is for the '2011 Census Microdata Individual Safeguarded Sample (Regional): Scotland'. By scrolling down and clicking on the resulting link in the results, users will find that each file has a catalogue entry which contains key facts about the dataset and links to any documentation as well as a link (at the top of each page) to access the data.

Additionally, many census microdata datasets are available to browse in *Nesstar* (Ryssevik and Musgrave, 2001). This facility allows unregistered users to view univariate frequencies, whilst registered users are able to interrogate the data to produce two- or three-way tables and graphs online, or to download the data or subsets in a wider variety of formats, including, for example, *SAS*. Figure 8.3 shows the *Nesstar* summary of the numbers of 'Workers in generation one of family' from the individual safeguarded (local authority) sample for England and Wales.

The Census Support service offers specialist support for users of these data that can be contacted through the UK Data Service's Helpdesk. Staff also provide training, including webinars, as part of the wider Census Support training programme.

Secure samples

The secure samples, those with the greatest level of detail, are only available from the VML for Great Britain and the NISRA Secure Environment for Northern Ireland. These facilities are environments enabling researchers secure access to data for research on approved projects. Researchers can find codebooks and user guides on the ONS and NISRA websites.

Figure 8.3 Exploring the 'Workers in generation one of family' variable in *Nesstar*

Source: http://nesstar.ukdataservice.ac.uk/webview/index.jsp?v=2&mode=documentation&submode=abstract&stud
y=http://nesstar.ukdataservice.ac.uk:80/obj/fStudy/7605&top=yes.

The VML is available at various sites across Great Britain, and approval, as well as Approved Researcher accreditation, is required before access can be granted. More information on Approved Researcher status and on gaining accreditation is available from the ONS website (ONS, 2015). Once projects have been approved, specific training will be provided to researchers on using the VML. The VML desktops contain a number of different statistical and office software such as *SPSS*, *Stata* and *Office* for researchers to conduct their analysis. Once research has been completed, users place outputs and material that they wish to export from the VML in a folder to be checked, which are then released to the user if deemed non-disclosive. Researchers wanting more information or to use data held in the VML should contact the following email address: vml.service.desk@ons.gsi.gov.uk.

A very similar arrangement is in place in Northern Ireland. Access to the NISRA Secure Environment is managed by the NISRA Research Support Unit, which can be contacted to discuss any proposals to use the data. As with the VML, an application must be completed to become an approved researcher. Additionally, users are subject to a security check. Access is available at NISRA offices in Belfast and outputs will be checked prior to release and publication. Further information is on the NISRA website (NISRA, 2015b).

8.8   The research value of census microdata

As with all census outputs, use of the data extends across sectors. However, compared with other census outputs, microdata of all types have historically been disproportionately adopted by serious research users with the analytical skills and statistical tools to most fully exploit the data. Census microdata are one of the few data types which provide sub-regional geography with sufficiently large samples to permit useful analysis. The large sample size is itself a critical strength permitting users to identify groups which would normally be too small to identify in a general-purpose sample. Finally, the microdata are drawn from the same data source as the aggregate statistics, providing scope for linkages. The following discussion summarises some of the ways in which the data have been used in academic settings and by commercial and policy settings, drawing heavily on Dugmore's 2009 business case (Dugmore, 2009). Further examples of the use of 2011 Census microdata for substantive and methodological research are contained in Chapters 19, 24 and 28 in Part V of the book.

Understanding ethnicity

The creation of census microdata in the UK with their unusually large sample sizes coincided with the introduction of questions on ethnicity in the UK censuses for the first time in 1991. As individual ethnic groups have typically constituted only small proportions of the entire population, large sampling fractions have been necessary to capture large enough sample sizes of these minorities within a general population. Census microdata are therefore unique in having a general purpose sample that can also be used to explore ethnic differences.

Finney has made a major contribution to understanding the relationship between migration and ethnic segregation using microdata to unpick the factors which influence migration. She has explored propensity to move in response to events over the life course (Finney, 2011). Clark and Drinkwater (2009) were able to compare the employment rates for ethnic groups in 1991 and 2001, noting that not all groups had improved their employment rates equally. This was something that was brought into international context by Li (2010), who compared ethnic minority labour dynamics between the UK and USA, noting that, with only one exception, ethnic minorities in both states experienced disadvantages. Clark and Drinkwater noted in particular the impact of religion and living in an area of deprivation as being influential. More generally, the addition of a religion variable has allowed greater scope for research into religious minorities and the inter-relationship of ethnicity and religion (e.g. Johnston et al., 2010).

Understanding health and caring

Only recently have censuses in Scotland and Northern Ireland contained questions about specific conditions. In earlier censuses and in England and Wales now, health questions have been limited to questions on longstanding limiting illness and, more recently, general health. Nevertheless, these topics are partial indicators of mortality and have been used to good effect. Popham (2006), for example, explored the effect of being Scottish on stated health. Using modelling techniques he was able to distinguish between the impact of being born Scottish, living in Scotland and socio-economic characteristics on health. Bambra and Popham (2010) explored regional worklessness levels and their relationship to health, concluding that economic characteristics of an area might have been underestimated in previous research. More recently, Graham (2015) has used 2011 Census microdata to understand the health of the Jewish population.

The addition of caring questions on the census has generated considerable interest. Norman and Purdam (2013) have explored the geography of caring and how this is impacted by the co-residence of a household member with long-term limiting illness. A very different type of caring was explored by Nandy and Selwyn (2013). They were able to use the secure version of the 2001 data to take advantage of the deprivation indicators on the data to demonstrate that kinship carers of children (i.e. carers of children who are related to them, but not parents) are more likely to be in deprived areas.

Methodological work

Over time two main features of the census microdata files have been exploited for methods development. First, the microdata contain large samples with a good range of key socio-economic indicators. Second, the microdata are drawn from the general census output allowing linkages to be made between these data and other census outputs. Accordingly, they have been used most widely to explore statistical approaches to bridging the gaps between outputs and as a basis for synthetic data. For example, the SARs have been used in work on microsimulation, including the MoSeS project (Birkin et al., 2009). In a different setting, the 2011 teaching microdata have been used to generate longitudinal study teaching data, drawing on cross-sectional data and applying transitions drawn from longitudinal data to generate synthesised longitudinal data (Dennett, 2014).

Commercial and policy use

Dugmore (2009) reported that private sector survey designers routinely used census microdata. Samplers used the data to inform decisions on how best to design the sample and operationalise it in the field. Many major studies were designed in this way, including studies which were undertaken to monitor government policy and administration such as digital switchover, the experience of people with disabilities in the workplace and understanding the major groups of users of health care.

Carers UK funded extensive work, undertaken by Yeandle et al. (2007) with census microdata in the period 2006–07 using a mixture of local surveys and census microdata. In some areas, surveys were supplemented with results from microdata for hard-to-reach minority communities. In other areas, analysis of the Small Area Microdata (SAM) was done in lieu of a survey. The resulting reports informed new advice which was disseminated to policy makers. The most economically impressive use Dugmore exposed was by the Office for the Deputy First Minister in Northern Ireland. DTZ had undertaken research on behalf of the Office to explore changing inequalities of key groups with respect to employment and found that, whilst differences between religious groups were narrowing, inequalities remained for those with disabilities and between genders.

8.9   Where next for microdata?

At the time of writing, it is clear that census-taking is in transition and the future of microdata is intimately connected to these changes.

2021 and beyond

Census microdata provide invaluable datasets for many researchers and organisations, offering a level of flexibility and the potential for much more complex analyses than can be provided through standard census outputs. The Census Offices recognise the value of the microdata products and are committed to continuing to provide data consistent with users' needs. The 2021 Census will be an online census with an increased use of administrative data to enhance population statistics (ONS, 2014c). The detail of the content of the 2021 Census and the format of its outputs have not yet been decided. The expectation is that there will be a range of microdata products for the 2021 Census that will build upon those available for 2011.

With a current recommendation that 2021 will be the last traditional census, any decision on the provision of any outputs, including microdata, will be dependent on the evaluation of the 2021 Census amongst other considerations. No further information is available at this time. However, as we saw earlier, there are international precedents for creating microdata under alternative data collection regimes.

Extending backwards

Whilst we move towards the next census, there is also work underway to extend the census microdata collection backwards to incorporate censuses from another three decades. The Extending and Enhancing Historical Microdata project was funded by the ESRC to generate sample datasets from existing electronic files. The files have been restored to a usable state for preservation, and data from fully coded 10 per cent files are being repurposed as census microdata files for reuse.

Whilst the files have some known issues arising as a result of their long-term storage, these will nevertheless extend the collection backwards, adding data from 1961, 1971 and 1981. At present we anticipate that files will have a very similar structure to those from 2011, with three levels of access. For each year there will be: a small open file with 1 per cent samples of individuals; two accessible files available under licence, one with a 5 per cent sample of individuals and one with a 0.95 per cent sample of households; and a 9 per cent sample of households in a secure file which contains more detail.

8.10 Conclusions

Census microdata are produced in order to maximise the utility of the data whilst protecting the confidentiality of respondents to whom legally enforceable promises of privacy were made. Census microdata are unique sources which have considerable research potential due to their flexibility, source and size. The 2011 outputs have a number of improvements compared with 2001 SAR data. Furthermore, the 2011 Census microdata are the third instalment of a data series, providing potential for considering longer-term change than before. This historical scope will be strengthened by the addition of historical samples dating back to 1961.

Looking ahead, there are unanswered questions about the still-evolving proposals for census transformation which mean that our crystal balls are still misty. There is much to be learned from international agencies who have experienced similar transformations, but who still seek to make powerful flexible microdata available for research.

Notes

1 www.ukdataservice.ac.uk.
2 www.ukfederation.org.uk/.
3 https://census.ukdataservice.ac.uk/

References

Bambra, C. and Popham, F. (2010) Worklessness and regional differences in the social gradient in general health: Evidence from the 2001 English census. *Health and Place*, 16(5): 1014–1021.

Birkin, M., Turner, A., Wu, B. and Xu, J. (2009) MoSeS: A grid-enabled spatial decision support system. *Social Science Computer Review*, 27(4): 493–508.

Census Microdata Unit (1994) *User Guide to the SARs*. Second edition. CMU, Manchester.

Clark, K. and Drinkwater, S. (2009) Dynamics and diversity: Ethnic employment differences in England and Wales, 1991–2001. In Constant, A.F., Tatsiramos, K. and Zimmermann, K.F. (eds.) *Ethnicity and Labor Market Outcomes*. Research in Labor Economics Volume 29, Emerald Group Publishing Limited, Bingley, pp. 299–333. Available at: http://dx.doi.org/10.1108/S0147-9121(2009)0000029014.

Dale, A., Fieldhouse, E. and Holdsworth, C. (2000) *Analysing Census Microdata*. Arnold, London.

Dennett, A. (2014) Synthetic Data for the UK Longitudinal Studies – SYLLS. Available at: http://calls.ac.uk/2014/01/10/synthetic-data-for-the-uk-longitudinal-studies-sylls/.

Dugmore, K. (2009) A Business Case for Microdata from the 2011 Census. Available at: https://census. ukdataservice.ac.uk/media/347240/SARs2011businesscase2009.pdf.

Finney, N. (2011). Understanding ethnic differences in the migration of young adults within Britain from a lifecourse perspective. *Transactions of the Institute of British Geographers*, 36(3): 455–470.

Graham, D. (2015) Health and Disability in Britain's Jewish Population. Institute of Jewish Policy Research. Available at: www.jpr.org.uk/documents/Health_and_disability_in_Britains_Jewish_Population.pdf.

Her Majesty's Government (1988) *1991 Census of Population*. Cmnd 430. HMSO, London.

Johnston, R., Sirkeci, I., Khattab, N. and Modood, T. (2010) Ethno-religious categories and measuring occupational attainment in relation to education in England and Wales: A multilevel analysis. *Environment and Planning A*, 42(3): 578–591.

Li, Y. (2010) The labour market situation of minority ethnic groups in Britain and the US – an analysis of employment status and class position (1990/1–2000/1). *ISC Working paper 2010–01*. Institute of Social Change, University of Manchester, Manchester.

Marsh, C., Skinner, C., Arber, S., Penhale, B., Openshaw, S., Hobcraft, T., Lievesley, D. and Walford, N. (1991) The case for samples of anonymized records from the 1991 Census. *Journal of the Royal Statistical Society: Series A*, 154(2): 305–340.

Nandy, S. and Selwyn, J. (2013) Kinship care and poverty: Using census data to examine the extent and nature of kinship care in the UK. *British Journal of Social Work*, 43(8): 1649–1666.

NISRA (2012) Population and Housing Estimates for Northern Ireland: Quality Assurance Report. Available at: http://webarchive.nationalarchives.gov.uk/20160105160709/http://www.nisra.gov.uk/archive/census/2011/results/population-estimates/quality-assurance-report-16-07-2012.pdf.

NISRA (2015a) Northern Ireland 2011 Census microdata teaching file. File to download and user guide available at: https://www.nisra.gov.uk/statistics/2011-census/results/specialist-products/microdata-teaching-file.

NISRA (2015b) Northern Ireland 2011 Census microdata secure samples. Variable lists and user guide available at: https://www.nisra.gov.uk/statistics/2011-census/results/specialist-products/secure-microdata.

Norman, P. and Purdam, K. (2013) Unpaid caring within and outside the carer's home in England and Wales. *Population, Space and Place*, 19(1): 15–31.

Norris, P. (1983) Microdata from the British census. In Rhind, D. (ed.) *A Census User's Handbook*. Methuen, London, pp. 301–319.

NRS (2012) Estimation and Adjustment Strategy. Available at: www.scotlandscensus.gov.uk/documents/methodology/census-est-adjust-strategy-nov2012.pdf.

NRS (2013) Confidence intervals of population estimates, by age group and gender. Available at: www.scotlandscensus.gov.uk/documents/censusresults/release1b/rel1bconfidenceintervals.pdf.

NRS (2015) Scotland 2011 Census microdata teaching file. File available to download and user guide available at: http://scotlandscensus.gov.uk/microdata.

ONS (2012a) 2011 Census: Confidence intervals. Available at: www.ons.gov.uk/ons/guide-method/census/2011/census-data/2011-census-data/2011-first-release/first-release--quality-assurance-and-methodology-papers/census-confidence-intervals.xls.

ONS (2012b) 2011 Census Item and Edit Imputation Process. Available at: www.ons.gov.uk/ons/guide-method/census/2011/census-data/2011-census-user-guide/quality-and-methods/quality/quality-measures/response-and-imputation-rates/item-edit-and-imputation-process.pdf.

ONS (2012c) 2011 Census: Census item non-response, item editing and item imputation rates. Available at: www.ons.gov.uk/ons/guide-method/census/2011/census-data/2011-census-user-guide/quality-and-methods/quality/quality-measures/response-and-imputation-rates/item-non-response--editing-and-imputation-rates.xls.

ONS (2012d) 2011 Census: Assessing our methods to protect your confidentiality. Available at: www.ons.gov.uk/ons/guide-method/census/2011/confidentiality/assessing-our-measures-to-protect-your-confidentiality/index.html.

ONS (2013) Quality and methodology information for the 2011 Census statistics for England and Wales: March 2011. Available at: www.ons.gov.uk/ons/guide-method/method-quality/quality/quality-information/population/population-and-household-estimates.doc.

ONS (2014a) England and Wales 2011 Census microdata teaching file. Available at: www.ons.gov.uk/ons/guide-method/census/2011/census-data/census-microdata/microdata-teaching-file/index.html.

ONS (2014b) England and Wales 2011 census microdata teaching file – Possibilities for exploring the data. Available at: www.ons.gov.uk/ons/guide-method/census/2011/census-data/census-microdata/microdata-teaching-file/possibilities-for-exploring-the-data/index.html.

ONS (2014c) The Census and Future Provision of Population Statistics in England and Wales: Recommendation from the National Statistician and Chief Executive of the UK Statistics Authority. Available at: www.ons.gov.uk/ons/about-ons/who-ons-are/programmes-and-projects/beyond-2011/beyond-2011-report-on-autumn-2013-consultation--and-recommendations/national-statisticians-recommendation.pdf.

ONS (2015) Approved Researcher Accreditation. Available at: www.ons.gov.uk/ons/about-ons/business-transparency/freedom-of-information/what-can-i-request/approved-researcher-accreditation.html.

Popham, F. (2006) Is there a 'Scottish effect' for self reports of health? Individual level analysis of the 2001 UK census. *BMC Public Health*, 6: 191.

Ruggles, S., King, M.L., Levison, D., McCaa, R. and Sobek, M. (2003) IPUMS International. *Historical Methods: A Journal of Quantitative and Interdisciplinary History*, 36(2): 60–65.

Ryssevik, J. and Musgrave, S. (2001) The social science dream machine: Resource discovery, analysis and delivery on the web. *Social Science Computer Review*, 19(2): 163–174.

UNECE (2007) Managing statistical confidentiality and microdata access – Principles and guidelines of good practice. United Nations. Available at: www.unece.org/fileadmin/DAM/stats/publications/Managing.statistical.confidentiality.and.microdata.access.pdf.

Yeandle, S., Bennett, C., Buckner, L., Fry, G. and Price, C. (2007) Managing Caring and Employment. *Report No. 2, Carers, Employment and Services (CES) Report Series*, University of Leeds, Leeds. Available at: www.sociology.leeds.ac.uk/assets/files/Circle/carers-uk-report-2.pdf.

9

Longitudinal studies in the UK

Chris Dibben, Ian Shuttleworth, Oliver Duke-Williams and Nicola Shelton

9.1 Introduction

The longitudinal studies (hereafter collectively referred to as the 'LSs') are the most complex of all the outputs from the UK censuses. Whilst introduced at separate times, they are now considered to form a distinct family of data resources. There are three separate studies: the ONS Longitudinal Study (ONS LS), the Scottish Longitudinal Study (SLS) and the Northern Ireland Longitudinal Study (NILS). The ONS LS, which covers England and Wales, was the first of these studies to launch and does not refer to its spatial coverage in its name; the later SLS and NILS have more clarity in their titles. In order to reduce ambiguity and to aid consistency with the other two studies, the ONS LS is sometimes referred to by users as 'the England and Wales LS', but this informal label will not be used in this chapter.

The three LSs have a largely similar design with individuals linked between censuses, although the linking approach in Northern Ireland is different from that used in England and Wales, and in Scotland. The availability of data from different time points allows a number of different types of analysis to be conducted, for example prospective analysis of census characteristics, prospective analysis of event-level data (from linked administrative data) and retrospective and prospective analysis between census and event data, as well as international comparisons with equivalent data elsewhere.

The three studies vary considerably in other aspects, such as the sample size, number of census years linked and amount of other data for individuals that is also linked. Data for different censuses and for different administrative items are held in separate tables. When a user data request is being prepared, records are joined across multiple source tables as required, and a single set of output data records is produced for the user to analyse.

The LSs provide sets of research data that allow researchers to explore a diverse range of demographic and social issues, using a variety of analytical methods. The longest-running study is the ONS LS, and the CeLSIUS website[1] identifies, at the time of writing, over 800 research outputs by users of the data, including journal papers, research reports and conference presentations. Research has covered many areas including, for example, associations between unemployment and mortality; variations in social mobility; lone parenthood; household structures; trends in fertility and labour market behaviour. Extensive research has taken place on links between inequality and health, with work using the ONS LS informing the Black Report (Black et al., 1980), the Acheson Report (Acheson, 1998) and the Marmot Review (Marmot et al., 2010). The ONS LS was also a key dataset used in calculations of trends in life expectancy used in the Turner Report (Turner et al., 2005) reviewing pension savings in the UK and advising on related policy.

9.2 Definitions

Longitudinal studies are surveys that involve repeated observation of individuals over a period of time; in the case of human studies the period of time may extend to many years. A number of possible designs exist for such studies, members might be selected through having a common characteristic such as a birth date (or year), or through experiencing a common event (e.g. the cohort of people who left school in a particular year or the cohort of people who have had a particular medical intervention). In the context of UK social studies, there are a number of well-known non-census studies, including birth cohort studies (from the 1946 National Survey of Health & Development (Wadsworth et al., 2006)), the Millennium Cohort Study (Plewis, 2004) and, most recently, studies such as Understanding Society (Buck and McFall, 2011) based on a sample of households.

Census LSs are specifically built around samples drawn from the census, but they also contain additional linked data including life events, and health-related and other administrative data. The LSs could therefore be seen as 'just' a series of cross-sectional observations that couple detailed demographic data with life events. However, such a view of the LSs would be to miss their most important characteristic: the fact that individuals can be observed at multiple time points across the life course allows the researcher to identify associations between past experience (housing, education, employment, etc.) and later life outcomes. This permits richer analysis of cause of death, for example, than would be possible solely using the individual-level data recorded on the death registration. Furthermore, it should be noted that to view LS data solely as cross-sectional is also to miss the point that the Samples of Anonymised Records (SARs), as discussed in Chapter 8, may well be a better resource for cross-sectional analysis of individual-level data.

9.3 The studies

The three LSs share broadly similar content. They contain data on sample members, who are selected on the basis of their birth date. These data are captured from administrative sources and from census returns. Given the context of this book, more focus is placed in this chapter on the census data aspects of the LSs, but it is important to recall that much of their strength lies in the administrative data that are linked to the census records. In terms of census data, the three studies share common benefits: that data are captured for sample members and also for 'non-members'.

'Non-members' are other residents in members' households, as identified in each census. Census data are retained for the non-members in a similar fashion to the data for members – that is, with similar coding used for variables. As a sample member ages, so the nature of the associated non-members is likely to change: for a sample member who is a child, the non-members usually consist of their parent/s and possible siblings; by the time that same sample member has become an adult, the associated non-members are more likely to include a partner and the sample member's own children. Whilst LS members are explicitly linked between censuses, this is not the case for non-members; thus, for a hypothetical adult sample member, it is not possible to definitively state that the spouse observed as a non-member in the 2011 Census is in fact the same person as the spouse observed as a non-member in the 2001 Census. Of course, it is possible to draw inferences from other available data about whether or not this is the case; for example, it can be identified whether a spousal LS non-member in one census has the same date of birth as in a previous census. The 2001 and 2011 Censuses contained detailed 'relationship matrix' questions showing the relationships between all household members. Given that sample members are selected on birth date, it is feasible that a single household will contain more than one sample member. The likelihood that this will occur varies across the three studies as they all have different sampling fractions.

Unlike birth cohort and other panel studies, people cannot opt in or opt out of the LSs; all persons who are born on one of the LS birth dates are included in the sample automatically. There are a number of ways that people may 'enter' or 'exit' the sample; these are enabled through both census records and administrative data. Migrants entering the UK from overseas will be recognised as sample members based on birth date when they enter into the National Health Service (NHS) administrative system (with variations in practice in the different member countries of the UK). Similarly, babies born on a qualifying date effectively become sample members at birth (once the birth is administratively recorded) – they do not have to 'wait' until the next census to be identified as sample members. People may 'leave' the sample through either death or embarkation (emigration). Whilst death is administratively well documented for almost all people, embarkation is more problematic; it is possible for people to leave the UK without notifying the NHS or other administrative data sources. When people do 'exit' from the study, their records are retained so that they can still be used for analysis and, for those who have emigrated, for future continued usage should that person re-enter the country.

When a new wave of census data is added to an LS, it is necessary for tracing to take place; this is an administrative process by which census records for an individual are linked to national health records, which then permits linkage to earlier census data for the same person. Not all persons who have qualifying dates of birth can be traced, and it is obvious that no tracing and linking processes will be perfect. Ambiguity can arise from errors in form completion, and it is also possible for multiple census records to exist for a given person (e.g. students recorded at both a term-time address and at a parental address, and in the most recent census, persons with both an internet-collected record and a paper-collected record). Where multiple records exist, one preferred record has to be identified.

It is useful to recall that the three LSs are all implemented as independent studies; even though there are common sample-membership birth dates, a person moving from (say) Northern Ireland to England would be seen as leaving the NILS, and as a new entrant to the ONS LS. The LS in the destination country would not 'inherit' their earlier census records from their origin country.

Data from the LSs have the potential to be highly disclosive: even for a single census, the combination of responses to census questions (including place of usual residence to an aggregate level) may be unique, but when responses are linked across multiple censuses, the probability of uniqueness rises as the number of attribute fields in each record grows rapidly. The birth dates which are used to draw the samples are not disclosed, which offers a defence against attempts to identify individuals. It is important that identification is not made, because any such identification would disclose one of the sample birth dates, thus revealing that all persons with that birth date were sample members. Thus considerable emphasis is placed both on security of access to the data and also on responsible use of the data by those who are working on approved projects.

ONS Longitudinal Study England and Wales

The original sample for the ONS LS was drawn in 1974 (Office of Population Censuses and Surveys (OPCS), 1973) from individuals recorded in the 1971 Census. Two concerns were identified, justifying a decision to commence a longitudinal study (Hattersley and Creeser, 1995):

first, that more information on fertility and birth spacing was required (the 1971 Census had included additional questions on fertility); and, second, that occupational data as recorded on death registrations were not ideal for determination of occupational mortality rates, as changes in occupation over a person's life were not recorded.

Hattersley and Creeser (1995) identified a number of methodological developments that permitted a linkage design to be established with a sample drawn from the 1971 Census. First, the 1971 Census included a question asking for respondents' dates of birth, rather than age. This was the first time (excluding the 1966 Sample Census) that full date of birth had been gathered. Similarly, birth and death registrations had included date of birth from 1969 onwards, permitting potential linkage on a date of birth basis. Finally, general advances in information technology in the 1960s had made such a linkage study feasible.

The sample was drawn by selecting four birth dates, giving a sampling fraction of 4/365, or 1.1 per cent of the population of England and Wales. As with all of the studies, these birth dates are not disclosed. The 1971 sample consisted of around 500,000 people, with a similar number of persons (allowing for overall population growth) being sampled at each subsequent census. Sample members are included in all censuses for which they are present and enumerated. More than 200,000 people have been enumerated in five successive censuses (from 1971 to 2011) (Lynch et al., 2015).

In the transition between any two consecutive censuses, some sample members will be lost to the sample either through death or emigration, whilst others will be added to the sample, through birth or immigration. Thus, any child born with an LS sample birth date will automatically become a sample member; similarly, someone entering the country (once they enter into the NHS registration system) with an LS sample birth date will become a sample member. Successful linking clearly depends on the individual being included in the census data capture, and therefore people may effectively enter or leave the record set through enumeration or failure to be enumerated in the census. Blackwell et al. (2003) reported tracing rates from 1971 through to 2001 for the ONS LS. These varied from 98.4 per cent in 1991 to 99.3 per cent in 2001; the tracing rate in 2011 was 98.8 per cent (Lynch et al., 2015)

As well as census data, the ONS LS contains linked data on birth and death registrations of sample members, on live births to sample mothers (and, for some time points, on fatherhood), on immigration and emigration (as observed via NHS registration), on cancer registration and on widow(er)hood (death of a sample member's spouse).

Scottish Longitudinal Study

Although a sample similar to the ONS LS was extracted from the Scottish 1971 Census data, the 1 per cent sample (around 50,000 people) was argued to be too small to allow research on many of the epidemiological and socio-demographic questions of importance to Scotland. The original Scottish study was, therefore, discontinued in 1981 and, unfortunately, the original data from the 1970s erased. Given that Scotland had, compared to England, relatively few longitudinal databases, combined with a growing recognition that a set of fairly unique demographic and health issues were facing policy makers in the country in the 2000s (e.g. mortality rates higher, fertility rates lower, a population ageing faster and more people living in deprived circumstances than in England and Wales), the idea of a Scottish LS was revisited.

Various factors made the construction of a longitudinal study more feasible in the 2000s. The growing awareness of the value of longitudinal data in answering a range of complex research questions amongst a number of academic and government researchers led to valuable support for the funding requests. Improvements in computing power, data linkage techniques and the quality of electronically held administrative datasets since the 1970s also meant that embarking on such a linkage study was more technically feasible. On the other hand, attempting to locate and transcribe information from some of the historic census and life events records raised a series of challenges. A group of academics requested funding from the then Scottish Higher Education Funding Council (SHEFC), now the Scottish Funding Council (SFC), to establish the Longitudinal Studies Centre – Scotland (LSCS), which is responsible for the establishment, maintenance and support of the SLS. Because of the problems associated with the 1 per cent sample size, identified when the Scottish component of the LS was abandoned, this funding allowed for a 2 per cent nationally representative sample, based on eight birth dates (four of these matched those used in the ONS LS to allow future comparative studies). Further funding to establish the study was then secured from the Scottish Chief Scientist's Office (CSO), which allowed the sample to be extended to 5.5 per cent, based on 20 birth dates. Funding from the Scottish Executive (now Scottish Government) and, more recently, from the Economic and Social Research Council (ESRC), has since enabled the establishment of the SLS support team that provides tailored, free support to academic researchers wishing to use the dataset.

The SLS is similar to the ONS LS but routinely links not only to life events data but also to secondary health care and, more recently, to education census and outcomes data (since 2007) and prescribing data (since 2009). Life events data collected for the SLS members include: births of new SLS members into the study (those born with one of the 20 birth dates), births, stillbirths and infant mortality occurring to sample members (where the mother and/or the father is the SLS member), widow(er)hoods (where the SLS member is the surviving spouse), deaths, cancer registrations, hospital records, marriages (where the bride and/or groom is the sample member), divorces (where the husband and/or wife is the sample member; note that the information on divorces will become available shortly), emigrations out of Scotland and re-entries after earlier emigrations. These events have been added for the period 1991–2013. It is also planned to include fertility events between 1974 (when the information on life events was first collected electronically) and 1991, allowing the construction of complete fertility histories for some women in the study.

The health data are provided by the Information and Services Division (ISD) of the Scottish NHS. Unlike the vital events data, which are linked into and held on the SLS database along with the census data, due to the dynamic nature of these health data they are linked on a project-by-project basis. As with the ONS LS, these include cancer registrations which occur to sample members. Unlike the ONS LS, though, it has also been possible to link hospital episode information, allowing studies of a wide range of morbidity outcomes. Recently, use has also been made of the link within the maternity record between mother and child to produce a new cohort, child of the SLS members (COTS) who can be followed up in the health care data.

Until now the only data on educational experience and attainment of SLS members have been the 1991 and 2001 Census data on educational attainment. In 1991, only tertiary qualifications were noted. In 2001, people reported which qualifications they had attained, ranging from O levels to degree level, but with no details or indication of when they were attained. To augment this, data were obtained from ScotXed, the agency within the Scottish Government responsible for collecting and co-ordinating data from schools. These consisted of the following: School Census data for every pupil in local authority (LA) funded schools, data on attendance, lateness and exclusion from school for the same pupils, and attainment data originally collected by the Scottish Qualifications Authority (SQA) giving details of the results for all SQA-accredited qualifications for candidates in the school years 2007–08 to 2010–11. The School Census data were obtained for censuses conducted in the September of 2007–10. Attendance and lateness data were collected for these same school years, although the collection was done at the start of the following year. The SQA data were for qualifications examined in the equivalent school years.

Recently, the SLS has been starting to look backwards in time. For a cohort of study members born in 1936, two additional sets of records have been collected and linked: a cognitive ability test they sat in 1947 (aged 11), and the 1939 National Register for them and their family aged 3. Together with the data collected as part of the main study, this has delivered a 1936 cohort with early life conditions, cognitive ability, school outcomes data, middle life occupational information and detailed information after the age of 55. It is hoped to extend this work as increasing amounts of historic administrative data are made machine readable.

Northern Ireland Longitudinal Study

The NILS was established in 2006. It arose out of discussions between the academic community in Northern Ireland and the Northern Ireland Statistics and Research Agency (NISRA), prompted by the existence of LSs in other constituent countries of the UK but not in Northern Ireland. It differed from both the ONS LS and the SLS in that its structure was based on health card registrations rather than the census. Crucially, this data spine provides a link to other health and social care data and regular six-monthly address updates with the possibility also of linking census data for NILS members and members of their households. Equally important is the sample fraction of the NILS. The sample uses 104 birth dates including those used in the SLS (which itself includes those birth dates used in the ONS LS). At about 28.5 per cent, this fraction is the largest of all the UK LSs and so, although the absolute number of NILS members at each census is about 500,000 (a similar number to the ONS LS), it is possible to do finely grained analyses down to the level of super output areas (SOAs) as small numbers in small areas do not raise disclosure problems.

The NILS started by linking just the 2001 Census but its census data holdings rapidly expanded. The 2011 Census link was completed by 2013 but this was swiftly followed, with support from the ESRC, by retrospective links to the 1981 and 1991 Censuses in 2014 and 2015 respectively, and since a 2021 Census will be taken in Northern Ireland, it is expected that by the middle of the next decade there will be five censuses linked to the NILS which cover 40 years of rapid social, economic and political change. There is, however, more to the NILS than this. Information from the Valuation and Land Agency (VLA) has routinely been linked to the NILS from its inception. This provides data on rateable value and other housing characteristics. Via the health and social care spine, there are also routine linkages which provide data on births and deaths of NILS members and also births to NILS members. There is the potential also for data linkages to be made on request to explore marriages and widowhoods. The NILS data framework also supports the Northern Ireland Mortality Study (NIMS). This is a way to access data on 100 per cent of deaths in Northern Ireland. There is a 1991 NIMS linked to the 1991 Census, and also a 2001 and a 2011 NIMS. Given that this covers all deaths, it is possible to deal with detailed causes of mortality once sufficient deaths have built up as time elapses from the base census. Finally, there is the possibility of linking data not routinely available but held by the health and social care system via the mechanism of Distinct Linkage Projects (DLPs) which provide an ethical and tested way to expand the data available to researchers. Examples of these include the use of prescription and cancer screening data.

Looking forward, the prospect of linking the 2021 Census to the NILS has already been mentioned, and this will extend the already rich research potential of the NILS within its current institutional setting. However, the advent of the Administrative Data Research Network (ADRN) across the UK, and the regional Administrative Data Research Centre for Northern

Ireland (ADRC-NI), increases the probability of linking the NILS to other administrative datasets from education, justice and social welfare, and thus potentially takes the NILS into new territory. These datasets are of research interest but they very often lack covariates, so there is a limit to what can be done using them. The potential to link them to the NILS will offer the chance to do analyses with temporal depth and to consider how an individual's current personal and household circumstances (e.g. whether they are on jobless benefits or not) can be understood in a life course framework by relating the present to their situation in 2011, 2001, 1991 and 1981. This will add value to the NILS and to the data to which it might linked.

9.4 Data use arrangements

Owing to their disclosive nature, the LSs have much stricter access arrangements than most other census-related datasets. The arrangements for each of the three studies are similar but not identical, but they share strict concerns about the risks of breach of confidentiality. The path to preventing any such breach is to adopt a number of inter-related strategies. The data can therefore only be used by approved researchers working on approved projects, and working under specific access conditions.

The LSs, and their associated support units, have been running for a considerable number of years. Consequently, access arrangements have adapted over time as permitted by developments in technology and changes in user expectations, but nevertheless have continued to be guided and (relative to other data resources) limited by the overriding concerns of data security. A common aspect of all three support units is that the support that is given to researchers working in projects is free at the point of use.

CALLS Hub

The Census and Administrative Data Longitudinal Studies (CALLS) Hub was commissioned by the ESRC alongside the re-commissioning of the three research support units for an initial five-year period from 2012 to 2017. Its stated role is to co-ordinate, harmonise and promote the work of the three LS research support units (CeLSIUS, SLS-DSU and NILS-RSU, described below), with the intention of providing a streamlined experience for users. One of the key purposes of the Hub is to act as an initial point for researchers who are contemplating using one or more of the studies. The Hub is a collaboration between the University of St Andrews, University of Edinburgh and University College London, though the management group also includes the directors of CeLSIUS, SLS-DSU and NILS-RSU. The Hub acts to combine information about the studies and to provide resources, including copies of all relevant census forms, and an integrated data dictionary.

CeLSIUS

The Centre for Longitudinal Study Information and User Support (CeLSIUS) provides support for UK academic, statutory and voluntary sector users of ONS LS. Additional users are supported directly by ONS. CeLSIUS is an ESRC-funded research support unit. It was based at the London School of Hygiene and Tropical Medicine under the directorship of Professor Emily Grundy during the period 2001–12, and moved location to University College London (UCL) when re-commissioned for the period 2012–17, under the directorship of Dr Nicola Shelton. Prior to the establishment of CeLSIUS, support for the ONS LS was provided through the Social Statistics Research Unit at City University from 1982 and from 1998 at the Centre for Longitudinal Studies, Institute of Education.

In order to use the ONS LS, researchers must follow a number of stages. A Research Proposal Form must be submitted, which details the purpose of the intended research, together with an LS Supplementary Form which identifies the specific data items and population which are required for the research to be carried out. The research proposal must name all researchers who will be involved in a project, including those who will not necessarily directly carry out analysis (e.g. a PhD supervisor). All named researchers must hold ONS Researcher Accreditation, which involves meeting certain criteria and then making an Accredited Researcher Application. As part of gaining accreditation it also necessary to complete certain training; upon completion of this training, researchers are asked to sign an Accredited Researcher Declaration. ONS Researcher Accreditation lasts for a period of five years.

Each project will be assigned to a Project Officer who will assist with the application and will support the user during the analysis. In practice, researchers are encouraged to contact a support officer prior to submission of the research proposal who will discuss the project and likely variables required.

Once accreditation has been gained and a research proposal approved, a project-specific data extract is prepared, and there are then two ways in which researchers can use the data. Direct access to the data extract is possible by using a terminal in a secure setting. A session must be booked in advance, and most ONS LS researchers use terminals at the ONS London offices in Pimlico. No data or notes may be taken out of the secure environment: results can be subsequently sent to users if disclosure control criteria (such as minimum cell counts) are satisfied. Alternatively, users may remotely submit a script for use with one of the supported software packages; scripts are run by a support officer, who will then return results, again, subject to them meeting the disclosure control criteria.

SLS-DSU

Use of the SLS is supported by a unit from the University of Edinburgh based in offices within the Scottish National Statistical Agency, National Records of Scotland (NRS) under the directorship of Professor Chris Dibben. The unit was originally set up by Professor Paul Boyle, then at the University of St Andrews and based there until 2014, when it moved, with Dibben, to Edinburgh.

Using the SLS requires some preparatory steps before a researcher can access data. Prior to contacting the support unit it is recommended that a researcher attends an SLS training session, reads through the information on the unit's website, in particular the data dictionary, and have developed a set of research questions. Because SLS data are quite different from other types of social science and health data, it is always helpful to have an early conversation with a support officer, who will have had many years of experience using LS data to further scope what may be possible. Researchers wishing to use NHS data in their analyses are required to complete an approved Safe Researcher training course and all researchers need to acquire Approved Researcher status (assessed on application by the unit). All research needs to be feasible and robust and therefore requires an application to the SLS Research Board, who assess whether it should be supported and may provide some advice on how it could be improved. Final approval is granted after both SLS Research Board and all appropriate ethical board approval is gained.

Because of the sensitive nature of the data, direct access to the SLS is only possible on non-networked computers in a safe setting in Edinburgh, though the support team are able to run syntax provided remotely. The safe setting computers have standard statistical software such as *SPSS, SAS, R* and *Stata*. After running analyses (or having them run remotely), output files must be cleared by the SLS team before they can be released. The process for clearing final outputs protects the SLS by reducing the risk of disclosure, ensuring that the study and data are properly described and that the data have been used appropriately.

NILS-RSU

The arrangements for accessing the NILS for research purposes have much in common with those for the other UK LSs in their generic features. The NILS has been in operation for ten years and the application process from beginning an application to getting data can be speedy and completed within four months. The process starts with researcher validation – an evaluation of whether the person is a 'proper person' to conduct research – and this is assessed by means of a statement of research experience, membership of learned bodies and relevant publications. The researcher must also be a 'safe researcher'. This means that they have undertaken training on data security and NILS procedures.

Once these hurdles are overcome, research ideas can be submitted to the NILS Research Support Unit (NILS-RSU) where guidance is offered on the feasibility of these, the range of available data and the application process, including advice on completing the application form, although researchers can make considerable solo progress using online resources such as the NILS data dictionary and metadata. As might be expected there are standard items that are requested, such as project title, abstract and the intellectual context for the work, but there are some features that are not seen in the other LSs since relevance to health and social research must be demonstrated. There is also a requirement to show plans for dissemination, especially with regard to policy relevance.

The NILS-RSU website provides example forms to guide researchers. The NILS does not provide all the variables held in the database to researchers but only those that are requested. The application should provide a rationale for the variables to be chosen, especially when dealing with sensitive information such as religion, which is deemed in the NILS data dictionary to be restricted. The applications are assessed by the Research Approvals Group (RAG), which meets every two months and includes representatives from academia, NISRA, the Public Health Agency and the Social Care Business Services Organisation. It considers applications using 18 criteria but with two (a longitudinal element and relevance to health and social care research) being essential. The RAG may simply approve the application or it may return it to the researcher with requests for clarification or suggestions for improvement.

Once the project has been approved the requested data are extracted by the NILS-RSU staff. Users receive large text tables which they must import into their chosen statistical software (*SPSS* and *Stata* are supported) where the data can be labelled and prepared as they wish. Typically, several different data tables are received (e.g. 2001 Census individual data, 2001 Census household data, 2011 Census individual data and 2011 Census household data) and these are linked by the researcher using the index fields provided by NISRA to create the analytical database. Once this is done, the researcher is free to work on his or her project. Normally this is done in the NILS-RSU (there is no facility to work remotely), although it is possible to submit by email *SPSS*, *Stata* and *R* code which can then be run by NILS-RSU staff.

Outputs can be either intermediate or final. Intermediate outputs may be shared within the project team amongst those who are signed up as researchers on that specific project. Final outputs can be released beyond the project team. No counts of less than ten are permitted either for intermediate or final outputs and in this the NILS differs from the ONS LS. This restriction is policed by the NILS-RSU which also checks for factual errors in the ways that the NILS has been cited or described. At least one output has to be longitudinal and at least one output also has to have relevance to health and social care. Researchers are also strongly encouraged to pursue policy relevance and dissemination. The NILS-RSU keeps a record of intermediate and final outputs and they (or the links to access them) are made available on the NILS-RSU website. Projects are not kept live indefinitely. From the start, they have a fixed end date and this can be extended up to three times at the discretion of the RSU, but if the researcher wants to extend it then further, approval must be sought from the RAG.

9.5   Toward UK level LS datasets

An obvious question from an outside perspective is why there is not a UK-level LS dataset. The simple answer is that whilst the three studies are conducted within parts of the UK, they remain legally separate and cannot be easily commingled. A practical constraint lies in the fact that all three studies are accessible only via secure arrangements; in order to create a common data file (ignoring differences in sample size and content), it would be necessary for at least two of the studies to permit export of their data to the third study (or for all three to be centralised at a fourth location). This would not be consistent with the secure storage and access conditions of the data.

There are, however, a number of different levels of integration that can be considered; advances have been made in a number of ways, and it is possible to speculate about the potential for further integrative work. For example, a relatively simple way of aiding understanding of population dynamics in each study would be to allow the partial tracing of members in administrative data in other parts of the UK. Whenever a new census is linked, it will be the case that there are a number of members who had been previously observed in one or more earlier censuses, but were not captured in the census being linked, and for whom there is no administrative record of death or embarkation. One plausible explanation is that the member has moved to another part of the UK. Thus, it would be useful for each agency (ONS, NRS, NISRA) to be able to transmit a set of minimal identifiers (such as an NHS number) to the other two agencies, who could then respond for each person identified indicating whether there are any administrative data to suggest that that person was resident in their respective territories at the time of the census.

We describe below three developments that are intended both to encourage wider use of the LSs and to make cross-LS analysis easier. The CALLS data dictionary is a catalogue of variable information which pools metadata across all three studies; e-DataSHIELD is an analysis technique that allows statistical operations to be carried out on multiple data sources without requiring those sources to be located in the same place; and two types of synthetic data have been developed.

Combined metadata: the CALLS data dictionary

The data producers for each of the three studies also maintain an ecosystem of supporting materials for their own study, and included amongst these are detailed data dictionaries which list each field in the data tables, and give information about coding. The data dictionaries are invaluable resources for carrying out analysis. As part of the CALLS Hub development plans, a combined data dictionary was developed which provided a single metadata repository providing information about all three studies. Whilst the separate dictionaries have all been developed to serve a similar purpose, they have different metadata structures and different implementations. Thus, creating a single integrated dictionary is not simply a case of merging together the separate dictionaries.

The integrated dictionary therefore provides both a uniform way of querying the metadata and a consistently formatted set of results. More importantly, it is designed to enable cross-study

Table 9.1 Similarity scores in the CALLS integrated data dictionary

Score	Meaning
0	No match found in other LSs, but some guidance notes given
1	Question wording similar, but categories incompatible
2	Question wording similar, categories may be aggregated to a common basis
3	Question wording similar, only minor differences in categories
4	Question wording similar, categories identical/near identical
5	Question wording identical, but categories incompatible
6	Question wording identical, categories may be aggregated to a common basis
7	Question wording identical, only minor differences in categories
8	Question wording identical, categories identical/near identical

work, by allowing a user (or potential user) to determine whether a given variable exists in multiple studies. Whilst the query entry box encourages simple terms to be entered, it also supports wildcard characters and a number of Boolean modifiers, allowing advanced users to construct more complex queries. Advanced queries can include or exclude particular search terms, and can also allow the user to supply alternate search terms, should thematically similar variables be known to have different names in different studies.

Key to the development of the integrated dictionary has been the production of similarity scores for pairs of variables. For relevant variables, the search results will include a 'Similar' column, which gives guidance as to whether related variables are similar or not. Table 9.1 shows the set of possible scores that are reported. For each pair of variables (the variable currently being reported, and a potential equivalent), scores may vary from 1 (wording is similar, but question responses are not compatible) to 8 (question wording and responses are identical or near identical). It should be noted that the scoring is necessarily a broad brush approach – it is still contingent on the researcher to look closely at the variables involved (and to seek assistance from the research support teams if needed), but the intention with the similarity scoring is that it is possible to do an initial assessment of whether combined analysis is feasible or not.

Following on from the cross-study comparison, the similarity scoring can also be applied to cross-census consideration within the same study. Again, this allows an initial exploration to be done prior to potential research, to determine whether an apparently similar variable in different censuses (within the same country) can in fact be validly compared in analysis.

The similarity scores were developed in order to support users with one typical question ('Are these variables the same?') that might be asked when considering applying to use the data. A second common question is to ask whether there are sufficient numbers of people with a given characteristic to make analysis feasible, especially when those people are to be further disaggregated by other characteristics. A second development of the integrated dictionary has been to capture and store frequency information for certain variables. An initial group of core variables has had such information added, with plans to extend the frequency data over time. Figure 9.1 shows part of the output of the data dictionary for a sample variable (ECOP1 – Economic activity in the NILS 2011 members table). The image shows the variable values (as included in data for analysis), the associated text labels and, finally, an observed frequency in the sample data. Here, the labels have been re-ordered in frequency rank. Frequency observations are reported as being within a given range when small values might otherwise breach publication thresholds.

Coding labels / Frequencies	VALUE / LABEL	▲ FREQUENCY
	02 / Economically active (excluding full-time students): In employment: Employee, full-time	124205
	XX / No code required	103781
	15 / Economically inactive: Retired	76255
	01 / Economically active (excluding full-time students): In employment: Employee, part-time	47035
	18 / Economically inactive: Long-term sick or disabled	26731
	16 / Economically inactive: Student	19290
	07 / Economically active (excluding full-time students): Unemployed: Seeking work and available to start in 2 weeks, or waiting to start a job already obtained	16821
	06 / Economically active (excluding full-time students): In employment: Self-employed without employees, full-time	16472

Figure 9.1 Partial screenshot of CALLS data dictionary

Source: Northern Ireland Longitudinal Study, illustrated in CALLS Hub Data Dictionary, http://calls.ac.uk/variables/.

e-DataSHIELD

Being able to compare the different parts of the UK, or simply to increase the sample size available to a researcher, makes the combination of LSs an attractive option. Comparison of results between the different studies may be carried out by running separate analyses in the relevant safe haven and comparing the published reports (e.g. Popham and Boyle, 2011). This approach has several disadvantages. One can never be sure that the datasets and variables, which are nominally the same, are really comparable. An analysis that adjusts for covariates in each individual agency will not be identical to what one would obtain if the raw data were pooled. Tests for study-by-covariate interactions are not readily carried out from published reports. A similar situation has arisen in the analysis of genomic data, where a pooled analysis of small individual studies is required for adequate inference, but the individual centres do not wish to share their data.

The DataSHIELD system[2] was developed in response to this and implements a joint analysis by linking the computer in each centre to an analysis computer (AC). The AC holds no raw data, but receives summary statistics from each of the individual studies, combines them, and passes the combined summaries back to the individual centres. This allows joint analyses such as generalised linear models (GLMs) to be fitted by iterating this exchange of summary statistics. The interface between the AC and the other centres prevents any raw data being exchanged. Because security concerns would not allow LS centre computers to be linked in this way, an adapted procedure of exchanging summaries between agencies by email has been adopted by the LSs. Routines in *R* have been developed to allow such analyses to be carried out via the e-DataSHIELD protocol.

Synthetic data

A further recent innovation has been the development of two types of synthetic data. First, a synthetic 'spine' dataset has been developed based on data and observations that are open licensed and thus easily disseminated. Second, a set of tools has been produced which can derive a synthetic set of output data from a sensitive (and non-shareable) set of input records. These approaches are designed to encourage new users and to make analysis more practical for existing users. In the longer term, it is plausible that these techniques could be used to produce datasets which could be held in the same location, thus enabling easier UK-level analysis.

The synthetic LS 'spine' dataset (Dennett et al., 2015) includes transitions of key demographic variables included in the national LSs. It was created using the 2011 England and Wales teaching SAR dataset, available from ONS, and a series of 2011 back to 2001 transitional probabilities taken from the England and Wales LS. A new LS-like dataset with plausible distributions was developed by first estimating the numbers of individuals in particular age groups undergoing each longitudinal state transition and then allocating transitions to the appropriate number of individuals in the SAR microdata set. Transitions applied include general health, marital status, religion and approximate social grade. In addition, live births to females were estimated and added, and likelihoods of death over the ten-year period were modelled. The initial synthetic dataset was produced for England and Wales, and was published together with the *R* source code for all algorithms applied. The same approach has been used to develop a similar dataset for Scotland, with a Northern Ireland version also in preparation.

The 'synthpop' project was led by SLS-DSU (Nowok et al., 2015) and produced an *R* package to generate bespoke synthetic datasets for individual research projects. The data are protected by removing variables and replacing them with synthetic versions. Replacements for categorical or continuous variables are generated by drawing from conditional distributions fitted to the original data using parametric or classification and regression tree models. Users can request synthetic versions of the data they request from the LSs for use outside the secure microdata laboratories, subject to confirmation by the data holders. These synthetic versions will allow for simple tasks such as the refining of analysis scripts to be carried out more easily and we are confident that the synthetic data will be good enough to produce analysis results very close to those that would be carried out on the real data. After developing analyses on the synthetic data, users will have the option of having them repeated on the actual LS datasets; indeed, users are advised that they should not claim statistical validity in their results unless and until they have repeated their analysis on the true LS data.

9.6   Conclusion

The chapter has described three closely related sets of data: the ONS LS, the SLS and the NILS. These are all complex resources, and thus the differences between them are at times subtle but significant. Together, they form a very rich family of research data. Longitudinal datasets have also been developed in other countries in the world, and if the UK is to remain confident that its own studies are of a gold standard, we must continue to maintain and improve them. With complex data, it is fair to say that analysis will always require researchers to have a background of suitable and specific training. The aspects of the studies which can be improved lie in ease of access and usability, and these are areas where the UK community cannot rest on its laurels.

We still see no real feasibility of a UK LS in the short term, although we remain hopeful that a legal route to permit access to multiple data sources will one day be found. In the absence of the easy ability to work on multiple LSs directly, a more hopeful avenue may be via synthetic data. If synthetic data derived from 'real' data are allowed out of safe settings (whether that be to the researcher's desktop or to a virtual secure environment), it will become easier to carry out multi-country analysis.

When considering the future, we can also look towards future datasets. The next census will take place in 2021, after which the LSs will be extended with additional census data. This will give a total time span, especially for the ONS LS, that remains internationally impressive: six censuses spanning 50 years. It is interesting to observe that in 1971 the median population age (UK,

Chris Dibben, Ian Shuttleworth, Oliver Duke-Williams and Nicola Shelton

rather than England and Wales) was 34.1 (Smith et al., 2005). Moving these people forward 50 years gives an age of 84.1, which is greater than current (national) life expectancy (and quite a bit higher than life expectancy in 1971) – so, the older half of the 1971 sample have passed their life expectancy. We currently have no knowledge of the final set of questions that will be in the 2021 Census round, but note that there are some variables that were new in 2011, and we will have our first chance in the 2021 sample to see whether and how these have changed (assuming that they are asked again). At the same time, we are hopeful that the emergence of the ADRN might facilitate wider linkage of administrative data with the LSs, further enriching the research potential they offer. Just as there are three separate LSs, with distinct characteristics, so there are four administrative data research centres (there are separate centres in England and in Wales) and thus progress is likely to be at different speeds in different contexts.

Notes

1 www.ucl.ac.uk/celsius/.
2 www.datashield.org.

References

Acheson, D. (1998) *Independent Inquiry into Inequalities in Health Report.* The Stationery Office, London.
Black, D., Morris, J., Smith, C. and Townsend, P. (1980) *Inequalities in Health: Report of a Research Working Group.* Department of Health and Social Security, London.
Blackwell, L., Lynch, K., Smith, J. and Goldblatt, P. (2003) *Longitudinal Study 1971–2001: Completeness of Census Linkage.* Office for National Statistics, London.
Buck, N. and McFall, S. (2011) Understanding society: Design overview. *Longitudinal and Life Course Studies,* 3(1): 5–17.
Dennett, A., Norman, P., Shelton, N. and Stuchbury, R. (2015) A synthetic longitudinal study for the United Kingdom. *CASA Working Paper Series 201.* Centre for Advanced Spatial Analysis, University College London, London.
Hattersley, L. and Creeser, R. (1995) *Longitudinal Study 1971–1991: History, Organisation and Quality of Data.* The Stationery Office, London.
Lynch, K., Lieb, S., Warren, J., Rogers, R. and Buxton, J. (2015), *Longitudinal Study 2001–2011: Completeness of Census Linkage, Series LS No. 7.* Office for National Statistics, Titchfield.
Marmot, M., Allen, J., Goldblatt, P., Boyce, T., McNeish, D., Grady, M. and Geddes, I. (2010) *Fair Society, Healthy Lives The Marmot Review Strategic Review of Health Inequalities in England Post-2010.* The Marmot Review, London.
Nowok, B., Raab, G.M. and Dibben, C. (2015) synthpop: Bespoke Creation of Synthetic Data in R. University of Edinburgh. Available at: https://cran.r-project.org/web/packages/synthpop/vignettes/synthpop.pdf.
OPCS (1973) *Cohort Studies: New Developments.* Studies on Medical and Population Subjects No. 25. The Stationery Office, London.
Plewis, I. (2004) *Millennium Cohort Study First Survey: Technical Report on Sampling.* Third edition. Centre for Longitudinal Studies, Institute of Education, University of London, London.
Popham, F. and Boyle, P. (2011) Is there a 'Scottish effect' for mortality? Prospective observational study of census linkage studies. *Journal of Public Health,* 33(3): 453–458.
Smith, C., Tomassini, C., Smallwood, S. and Hawkins, M. (2005) The changing age structure of the UK population. In Chappell, R. (ed.) *Focus on People and Migration.* Palgrave Macmillan, Basingstoke.
Turner, A., Drake, J., Hills, J. and Turner Commission (2005) *A New Pension Settlement for the Twenty-First Century: The Second Report of the Pensions Commission.* The Stationery Office, London.
Wadsworth, M., Kuh, D., Richards, M. and Hardy, R. (2006) Cohort profile: The 1946 national birth cohort (MRC National Survey of Health and Development). *International Journal of Epidemiology,* 35(1): 49–54.

Part IV
Visualising 2011 Census data

Using graphics to drive user engagement

Experiences from the 2011 Census

Alan Smith and Robert Fry

10.1 Introduction

Data visualisations of the 2011 Census were a key component in the Office for National Statistics' plans to drive user engagement and awareness of key messages in the data. Interactive graphics were a feature of most census releases and were designed to attract a broad audience, from analysts through to citizen users. Social media and online syndication were used to bring the content to the widest possible audience. Within the broad genre of data visualisation, a variety of products were published, ranging from narrative through to exploratory interfaces. A final experiment saw the use of gamification techniques to make census data more personal, while challenging perceptions of official figures. This chapter illustrates some of the visualisations of data from the first and second releases, together with the 'How well do you know your area' quiz.

10.2 Census first release

The first 2011 Census results were released on Monday 16 July 2012. These included the headline national figure for England and Wales (with breakdown by individual year of age/sex) together with sub-national usually resident population estimates by five-year age/sex bands. The first release of census data is always national news. From a presentation point of view, this offers both opportunities and challenges; wider public interest in the data is high but demand for fresh insight is not helped by much of the richer data being part of later releases.

The Office for National Statistics (ONS) Data Visualisation Centre prepared two interactive visualisations based on these data, with the intention of maximising outreach by: (i) providing users with an immersive, engaging environment for exploring first release data and stories; (ii) syndicating the content with media agencies; and (iii) encouraging and observing the use of social media to share and promote the content. The published graphics were picked to occupy different positions on the visualisation spectrum – one was primarily a data explorer interface; the other largely a 'storytelling' visualisation, featuring a significant narrative element.

The first interactive visualisation was a dual population pyramid, which allows a user to compare the size and structure of any two areas in England and Wales down to local authority level. Animation was employed to allow the user to overlay any two areas for visual comparison, while the display itself can be switched from viewing the size (population estimates) to the structure (percentage of

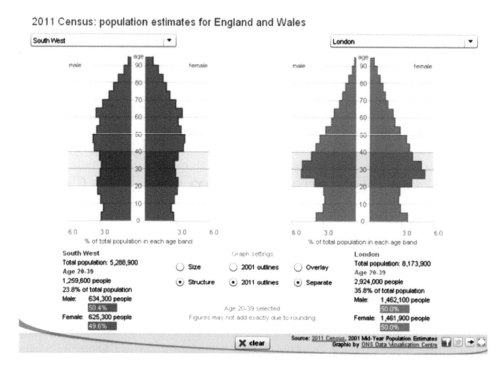

Figure 10.1 Example of the population pyramid visualisation

Source: UK Government Web Archive in The National Archives.

population in each age band), facilitating hierarchical as well as magnitude comparisons. Figure 10.1 enables comparison of the age/sex structure of the populations of the South West and London.

Population estimates for 2001 are also available for visual comparison, so that users are able to explore change over time as well as make geographical comparisons. Finally, users can export images from the application and share their own discoveries via Twitter; tweets are automatically generated by the application using the geographical areas currently displayed in the tool as Twitter hashtags.

The second visualisation was an interactive, animated population pyramid looking at national data for England and Wales over the 100 years from 1911 to 2011 (1911 is the earliest year for which individual age/sex band data are available). As well as an interactive population pyramid, this interface features extensive narrative that details the demographic changes over a century. 'Stories' are linked to the data, with hyperlinks. For example, clicking on the link 'post-World War II baby boom' in the narrative highlights the appropriate part of the graph and displays the relevant data. The narrative does not relate solely to the census data being displayed – it also references data from other sources (for example, total fertility rate) to try and explain the patterns in the population structure. The text for the 'stories' is also offered in the application as an MP3 audio stream, synchronised to user interaction; if a user navigates to a particular time period, the audio stream follows. This functionality is switched off by default to ensure compliance with web accessibility standards – screen readers used by some users would otherwise be in conflict with the audio. In addition to 'stories', the animation contains key events of wide public interest from each census year supported by a photo, together with a link to further information about that year on Wikipedia. Inspired by the charts of William Playfair, the graphic also includes references to the reigning monarch and prime minister, providing further historical context.

Both of these products were published on the ONS website as part of the first release content. They were also offered to the media for direct syndication – and were taken and published by the BBC, *The Guardian* and *The Daily Telegraph*. Traffic for the interactives on the ONS website was modest (<10,000 views in the first week) but it was interesting to note that the exploratory graphic was the more popular of the two. The syndication metrics, however, boosted consumption of the content into hundreds of thousands of page views – with the narrative graphic explaining population change over time being the more popular of the graphics.

Of course, quantitative metrics are useful but cannot give direct measures of how much users appreciated or engaged with the content. For this kind of qualitative feedback, the content was monitored on Twitter by following comments directed towards @statisticsONS (the official ONS Twitter account at the time) and tweets including hashtag variants of #census. Most tweets were very positive indicating that users were engaged with the content and keen to share – for example, there were repeated uses of words such as 'fun', 'fascinating', 'cool' and 'interesting'.

The appeal to users of the '100 years animation' – the interactive graphic with narrative – was neatly summarised later in the week by a blog post on coffeespoons.me, a blog dedicated to 'examining how we measure our lives', noting "what makes this different [from standard population pyramids] is the context that the ONS has added that the strict data pulls lack". The post goes on to emphasise that the added context of the narrative helps users to understand 'why' in addition to 'what'.

10.3   Census second release

The second main tranche of census data was released on 11 December 2012, and consisted of the main topic-level data such as data on health, housing, ethnicity and religion. Data were released at various geographic levels, initially down to lower super output areas (LSOAs). Delivery of the data ran very close to the set publication date, offering little opportunity for any deep analysis into the data, and so our solution was tactical in providing data explorer interfaces to allow users to explore the data and patterns within the data. A ten-year gap between censuses can mean huge changes in the characteristics of the population. Some of these characteristics are not routinely collected through other official statistics and so being able to look at these data at such a low geographic granularity is a rare opportunity.

On the day of the data publication, ONS released a suite of interactive maps looking at some of the key topic areas. The aims of these maps were: (i) to illustrate the broad national picture of changes; and (ii) to allow users to explore changes and differences relevant to them at their neighbourhood level of LSOAs. The maps were underpinned by the *Google Fusion* data tables service which offered several advantages, including free hosting, a mapping interface that many users will have been familiar with and a mapping Application Programming Interface (API) with a rich feature set, including address and postcode lookup.

Where possible we provided a dual-panel mapping interface – a map of 2001 set alongside a map of 2011. This was not always possible because of changes in definitions and so a comparison would not have been possible or appropriate and there were further challenges in that the LSOA boundaries had changed to reflect on-the-ground changes in neighbourhoods, such as large housing developments. In these areas, a like-for-like comparison was not possible, but we plotted the data on the respective boundaries.

As the saying goes, "a picture is worth a thousand words", and the maps revealed striking wholesale changes in some aspects of society – for example, the increase in the percentage of people stating they have no religion or the increase in those privately renting versus those owning their own house with a mortgage. Changes in the percentage of those in Private rented: Private

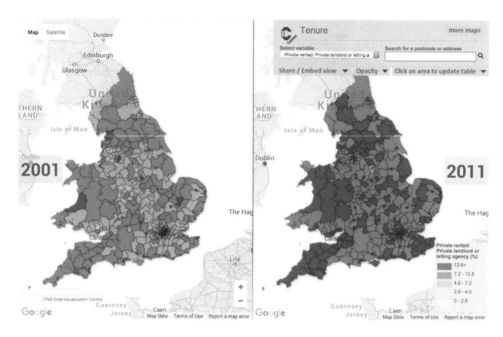

Figure 10.2   Interactive census mapping, tenure in 2001 and 2011

Source: 2011 Census, KS402 Tenure, www.nomisweb.co.uk/census/2011/ks402ew.

landlord or letting agency can be identified from the 2001 and 2011 maps shown in Figure 10.2 for local authority areas across England and Wales.

Another important requirement of the interface was to allow users to explore their neighbourhood and surrounding areas. By entering a postcode, the user was taken to a different zoom level, switching from the local authority view to LSOA level. This allowed a more personalised look at the data. Like the interactive pyramids, users were able to share dynamic links to specific views within the data. We saw plenty of examples of this in action on Twitter, with users sharing links and discussing the changes they saw with friends, family and colleagues. It was encouraging to see the use of this feature to discuss the effects of policy on different areas; for example, discussion around the introduction of bike lanes and their impact on the number of people cycling to work (Figure 10.3).

Like the pyramids, the maps were offered to the media for syndication, with one difference: that the maps remained hosted on the ONS site, but could be embedded via an iframe. There was widespread take-up of the maps by various media outlets – *The Guardian, The Telegraph*, the *Daily Mail* – as well as a whole host of local media outlets using the maps to paint their local pictures. The media used the maps to support their own narrative, each with their own slant on the findings presented.

This high level of syndication led to an impressive number of views in the days following the release, with the individual maps collectively achieving well over 1 million views. Since then, the maps have remained an important resource for users. To this day, well over three years since first published, the maps still attract a lot of use. In particular, some of the maps where the theme remains topical and is somewhat unique to the census, such as religion or ethnicity, still attract close to 10,000 page views per month.

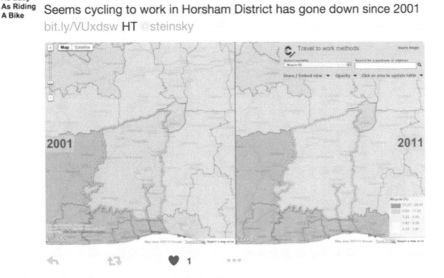

As Easy
As Riding
A Bike

Mark Treasure @AsEasyAsRiding · 13 Feb 2013

Seems cycling to work in Horsham District has gone down since 2001
bit.ly/VUxdsw **HT** @steinsky

Figure 10.3 Twitter discussion based around the interactive maps

10.4 Know your area quiz

A final experiment with 2011 Census outputs was the personalised interactive quiz 'How Well Do You Know Your Area?'. The concept for this product was influenced by many convergent factors including: (i) the emergence of gamification as a technique for generating richer interactions with users, particularly through survey design; (ii) the trend of what has been described as 'participative data visualisation', as exemplified through interactive graphics from the *New York Times* and the BBC;[1] and (iii) evidence, from surveys like the Ipsos MORI 'Perils of Perception' series, that the general public have poor intuitive feel for official data. For example, respondents were asked to guess the proportion of people in the country who were Muslim – the mean answer was 24 out of every 100, whereas the official 2011 Census figure is around five per 100. The psychologist, Daniel Kahneman (2011) describes how "we can be blind to the obvious and we are also blind to our blindness" (p. 26), citing media reporting (of exceptional events) and individual ad hoc experiences as powerful methods of distorting population heuristics. We were keen to explore if these misconceptions might present at the local level using census data.

To develop the quiz, we used the open source visualisation library *d3js* and the Googlemaps API for the presentation layer, and the Neighbourhood Statistics Data Exchange (NDE) API to provide the underlying geography lookup and census data. The quiz asked users for a postcode (England and Wales only) which was mapped to a ward and a customised quiz for the area prepared. We chose seven questions from 2011 Census Key and Quick Statistics that could be answered via a standardised range of 0–100 (per cent answers and median age). Each question was posed alongside an interactive slider with graphical feedback by way of icons and bars. Once the user pressed 'submit', we made modest use of animations and transitions to reveal the census figure – users who are a long way off with their guesses have to suffer the indignity of

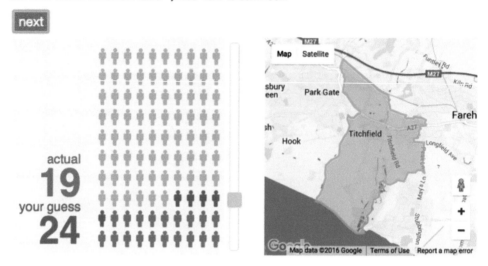

Source: 2011 Census Key and Quick Statistics (rounded to nearest whole number)
Created by ONS Digital, powered by NeSS Data Exchange and ONS Open Geography Portal
Find out more about plans for the future of the census and population statistics.

Figure 10.4   How well do you know your area?

Source: 2011 Census, KS402 Tenure, https://www.nomisweb.co.uk/census/2011/key_statistics.

a longer animation to reveal the difference. At the end of the seven questions, a custom overall score (0–100) is prepared based on how close to the actual census figures the users' guesses were (Figure 10.4).

Once published, the impact of the quiz far surpassed our own expectations of what was essentially an experimental product. Despite census data being over three years old by the time of release, *The Guardian* syndicated the product prominently on its homepage, as did numerous local media sites (the local focus no doubt helping). Collectively, this helped generate around 250,000 views of the app in the first week of release. Sharing results on social media generated plenty of online debate about the census in a way that traditional products would not have done. It was clear to see that many people were prepared to have their preconceptions of their local area shattered by what the quiz revealed to them.

Reviewing the quiz as an experiment, we were pleased with what this technique was able to add to census dissemination. The format was subsequently mimicked by other media agencies and the ONS has used similar techniques again on non-census related outputs. However, because we were surprised by the extent of the success, there are certain elements that we would change if we were to run such an exercise again. Perhaps the most interesting of these would have been to capture the individual answers to questions – over 250,000 responses would have made for a large survey in its own right and the opportunity to use that kind of data to produce perception maps of the country, for example, is a very interesting proposition.

10.5 Conclusions and recommendations for future activities

Overall, the ONS data visualisations of 2011 Census data proved to be very popular and achieved the aim of driving user engagement and awareness of the key messages within the data. Key to the success was the creation of high-quality and varied content. Content was delivered to an ever-increasing user expectation of quality, set largely by an increasingly active data journalism community. Delivering to this quality meant that the data were readily picked up and reused by the media through syndication, allowing the content to reach far and wide, beyond the traditional borders (and users) of the ONS website.

Standards in the field of data visualisation are increasing at a fast pace to meet increasingly sophisticated demands. For example, now a user would expect to be able to view content on a variety of devices, including mobile phones and tablets. At the time of publication this was not such a strong requirement for our users but it would be impossible to produce content like this now without mobile interaction and display being a key design objective.

Our 2011 content was, at times, limited by technological limitations imposed on us by the web infrastructure of ONS. In particular, we were restricted in our ability to use the visualisations as part of a wider narrative; instead, each visualisation stood alone on a separate web page. Over time these limitations have been overcome – the launch of the new ONS website and sister site Visual ONS[2] exposes data visualisation techniques similar to those mentioned in this chapter, but with additional narrative and analysis enhancing the user experience.

Notes

1 *New York Times* dialogue quiz 'How Y'all, Youse and You Guys Talk' and the *BBC*'s Olympic Body Athlete Match.
2 http://visual.ons.gov.uk.

Reference

Kahneman, D. (2011) *Thinking, Fast and Slow*. Farrar, Straus and Giroux, New York. Available at: https://vk.com/doc23267904_175119602.

11

Contrasting approaches to engaging census data users

Jim Ridgway, James Nicholson, Sinclair Sutherland and Spencer Hedger

11.1   Introduction

Census data are 'open data', and can be downloaded by anyone with access to the internet. However, there are two important barriers to use. Multi-dimensional UK data are distributed across the websites of different Census Offices – the Northern Ireland Statistics and Research Agency (NISRA), National Records of Scotland (NRS) and the Office for National Statistics (ONS) – and many users will not have the skills to synthesise data from multiple sites. UK-wide datasets follow on much later than the original releases, provide only univariate data, and are not disaggregated by parliamentary constituency. A deeper problem relates to strategies for data dissemination; official agencies traditionally have seen their primary role as creating data and making them available to an audience familiar with spreadsheets and Application Programming Interfaces (API), and have not set out to engage with a more general audience. The ONS strategy for the 2011 Census has been to increase user engagement (especially with non-traditional users) with census data, and to present key messages from the data in public forums.

Here, we describe a collaborative venture between Durham University and the House of Commons Library (the Library), where an attempt was made to facilitate the work of this key information provider. The Library is the first port of call for politicians who need information; it deals with around 60,000 enquiries each year. A key justification for a national census is that the data can be used to guide decision making – deciding on national policy in ignorance of the gross and fine details of the population being served would be foolhardy, and so politicians need access to census data (preferably in an accessible form). Voting patterns are a central concern of politicians and their aides. The Library has extensive evidence on such matters. For the ONS, however, disaggregating data by political party could be seen as a political act that would be contrary to their constitution.

The main focus of this chapter is to describe the *Constituency Kit* – an accessible collection of resources built around census data that resulted from collaboration between Durham University and the Library. The work takes the ONS dissemination strategy to places ONS itself cannot go. Here, we describe our approach to user-oriented design, the development process and use by users, then draw some conclusions about the potential for collaboration between national statistics offices and other agencies about strategies for disseminating information. First, though, we describe an earlier attempt to present census data in a way that facilitates user interaction and engagement, on a purpose-built website.

11.2   Interactive displays: the mousetrap model

The SMART Centre[1] has developed a variety of interfaces to present open data in ways that facilitate exploration. A key feature of data relevant to social issues is that it is multivariate – a number of variables are relevant to any social phenomenon, and relationships between variables are rarely linear. Educational attainment at age 16 years, for example, varies according to sex, ethnicity and eligibility for free school meals with some interesting interactions; for example, white boys eligible for free school meals have lower attainment on average than black boys receiving free school meals, despite the higher average attainment of white pupils (Ridgway et al., 2013). Our first attempt at making census data more accessible was to present it in multivariate displays. These included line graphs, population pyramids and multivariate interactive displays. Figure 11.1 provides an example of an interactive display showing long-term health problem or disability by age, sex and religion in Northern Ireland.

Users can explore phenomena by dragging variable names into different places, and moving sliders. Such displays have considerable virtues. They can present multivariate data in ways that are accessible to statistically naïve users, and allow users to explore the data and draw their own

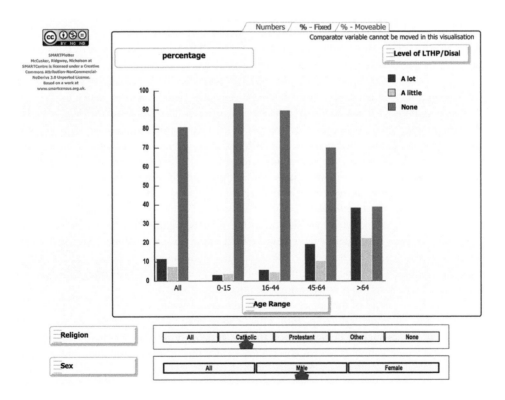

Figure 11.1   Long-term health problem or disability by age, sex and religion in Northern Ireland

Source: smartcensus website, www.smartcensus.org.uk/index.php?option=com_content&view=article&id=116&Itemid=630.

conclusions. On the basis of trials in school classrooms, we have evidence that naïve users (aged 14 years) can draw sensible conclusions from multivariate data, and can 'invent' concepts such as 'interaction' that are deemed too difficult for students studying A-level statistics (Ridgway and Nicholson, 2010).

This approach to user engagement can be characterised as the 'mousetrap' model. "Build a better mousetrap, and the world will beat a path to your door", to paraphrase a remark by Ralph Waldo Emerson. What Emerson was unaware of was that more than 4,400 patents have been issued for mousetraps in the USA since 1838 (Kassinger, 2002). Our website was underwhelmed with hits. The obvious lesson is that the web is not a level playing field. Users access their favourite websites; they have particular data needs, not a general desire to browse resources; and major providers have ways to attract traffic to their sites. Consequently, there is rather little interest in custom-built websites that provide census data – even if they are displayed in novel and interesting ways.

11.3   The *Constituency Kit*

As a result of this rather bruising experience, we set out to establish a partnership with a group whose *raison d'être* was to provide information to influential groups, who have very high credibility as a source of information, and who communicate directly with decision makers. The collaboration between Durham University and the Library adopted a very different approach to the mousetrap model. Rather than embedding a specific dataset in a variety of general-purpose data visualisations, we set out to create data visualisations designed to serve particular user needs – assembling a rich data source that included census data. The context was the 2015 UK General Election. People involved in elections identify seats that were marginal in earlier elections, seek out demographic information about constituencies, and devise local and national election strategies based on this information. Census data are an essential element in these activities.

The *Constituency Kit* was designed to synthesise a great deal of data from a variety of sources about every constituency in a form that was easy to use by both sophisticated users (Library staff) and relatively unsophisticated users such as MPs and their aides, journalists and citizens (where we assume no sophistication). The *Constituency Kit* comprises three components: the *Constituency Explorer*; a *Quiz* that runs on mobile phones; and a document for every constituency that can be downloaded in pdf format from the *Constituency Explorer* (over 12,000 pages in total).

One version of the *Constituency Explorer* is shown in Figure 11.2. Each circle represents a constituency, and constituencies are shown within regions. Figure 11.2 shows the percentage of residents in each constituency who describe their ethnic group as White (selected from the pull-down menu). London has been chosen, and the colours show the political party to which each MP belongs. The vertical black bars show the mean value for each region. It can be seen that London stands out from other regions, demographically.

The *Quiz* is a variant of the one developed by the Data Visualisation Centre in ONS – a version is shown in Figure 10.4 (see p. 166). Users choose their constituency, and are asked seven questions on topics such as demographics, and the proportion of the electorate who voted in the 2010 General Election. It returns correct answers for the chosen constituency and the UK.

User-oriented design

We are grateful to Alan Smith (then at ONS Digital) and his team for an inspirational graphical design that has largely been incorporated into the *Constituency Explorer*, and for the underlying source code (which we have modified extensively). We are also grateful to Alan's team for the

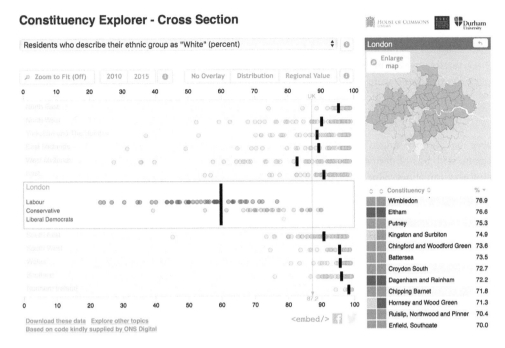

Figure 11.2 The *Constituency Explorer*

Source: Constituency Explorer website, www.constituencyexplorer.org.uk/explore/ethnic_group.

design and code of the *Quiz* (which we modified by changing geographical boundaries and using data calls via the *Nomis* API[2]). We began by setting out some criteria for the final design. These were modified during the development process, based on insights from the development team and user groups. A summary set is mapped out here. The evolution of the design, and the shifts in design criteria, along with the rationale for changes is set out in Sutherland et al. (2015). Design criteria included the following:

- *Data issues*: UK-wide data should be used, wherever available; the *Kit* should include a 'one stop shop' for constituency data, and should facilitate access to as much data as possible that can be disaggregated at constituency level; data must be downloadable, along with metadata; and displays should be themed (e.g. demographic data should be grouped together in the same display) and it should be easy to navigate between displays.
- *User criteria*: components of the *Constituency Kit* should run on most mobile and desktop devices; each display should be attractive, and should engage users; the display should be easy to use; an (optional) tutorial was needed for the *Constituency Explorer*; there needed to be multiple ways to access, sort and display information; multiple versions of the *Constituency Explorer* were needed for different uses – such as a general-purpose version that provided detailed information about every constituency, and an 'election special' that showed transitions between political parties between the 2010 and 2015 elections; the *Constituency Explorer* and the *Quiz* needed to link to social media; and resources needed to be 'embeddable' – readily incorporated into the website of anyone who wanted to use them.
- *Statistical criteria*: on-screen explanations of statistical concepts used in the display (box plots, mean, etc.) were needed, and metadata issues had to be addressed. Metadata had to be readily

171

available on screen in a simplified form; full descriptions of metadata had to be accessible from the display; and metadata had to be an integral part of any data download.

- *Cognitive criteria*: rescaling would be available, but only ever as a result of active user choice, variable by variable; and the dangers of 'infographics' – pretty tricks irrelevant to understanding the data – were to be avoided.
- *Political criteria*: regions were to be ordered in terms of their NUTS (Nomenclature of Territorial Units for Statistics) code; and the exact colours used by political parties were to be used.

For the *Quiz*, we judged Alan Smith's quiz to be almost entirely fit for purpose. It is an interactive game with lots of desirable features. The only substantive change was to provide the user with information about values for the UK, as well as for their chosen constituency (this modification was later incorporated into Alan's version).

11.4 The development process

The development was based on two data visualisations created by Alan Smith (Ridgway and Smith, 2013; Chapter 10 of this book): one showed data on educational attainment disaggregated by local authority within regions, and the other was a *Quiz* written to run on smartphones. For the *Quiz*, new boundaries and maps were used, along with new questions, but in other respects it remained largely unaltered. The relevant code was modified extensively to create the *Constituency Explorer*. Technical aspects of the development are discussed elsewhere (Sutherland et al., 2015); here we discuss aspects related to user engagement.

The most important group of users is the Library. We had regular team meetings to discuss all aspects of the design. The Library members were instrumental in decisions to incorporate UK-wide data wherever possible; political sensitivities were high on their agenda; they wanted to provide links to extensive Library resources directly from the *Constituency Explorer*; they were keen to have a version that showed election results from both the 2010 and 2015 general elections. An extension suggested by the Library was to present a table with values on every variable available in the visualisation for a particular constituency, its region, and the UK as a whole.

We used Trello[3] as our major communication channel. Trello offers, *inter alia*, a shared space to support the working of spatially disparate teams. We used it to document versions of the *Constituency Explorer* as they were created, and the reasons underpinning the change – the *Constituency Explorer* is now in its fifteenth iteration.

A second key user group is politicians and their aides. We ran two focus groups at different stages in the development project, attended by MP aides. Aides brought along their own devices, so we were able to test the *Constituency Explorer* and the *Quiz* on a variety of platforms and operating systems. There were no crashes, just minor functionality issues with some smartphones. As well as creating enthusiasm for the *Constituency Explorer* – "I love it – I've shown it to my minister" – important design ideas came from the groups. One was a request for a facility to 'pin' constituencies. The 'pin' function allows a user to identify a constituency (or constituencies) via a pin on the display, and this pin continues to identify the constituencies of interest, as different variables are selected, and the display changes. A further feature is that when a user who has pinned constituencies swaps from one set of variables to another, all the pins are carried forward to the new display. Users also wanted more and different data about the results of the 2010 General Election, and that tweets should link to a specific display, not to the starting position of the data visualisation.

11.5   Characteristics of the *Constituency Explorer*

The *Constituency Explorer* (Figure 11.2) runs on a wide variety of platforms, including mobile devices, and has the following features: over 150 variables can be displayed; UK-wide data are presented where available; census variable descriptors are adjusted 'intelligently' for Scotland and Wales; counts, currency and percentages can be displayed; users can search by constituency name or can locate a constituency via the map, or can click on a circle in the display to identify a constituency; the display facilitates secondary sorting (e.g. by party, then by percentage unemployed); users can access metadata on screen, at different grain sizes; there are links to detailed data about each constituency provided in pdf format; users can download data (which always includes the associated metadata), can link to social media, and can embed the display in their own website.

In terms of functionality, users can compare constituencies on a wide range of variables (e.g. to explore demographics), and can identify outliers; safe and marginal seats can be identified by plotting the majority of votes, and can be pinned to explore underlying variables such as demographics; the display maps the distribution of political parties spatially; regions can be compared directly via means and box-plots; and one version shows the changes in political party, dynamically, between the 2010 and 2015 elections, as a function of any of the 150+ variables.

Characteristics of the Quiz

The *Quiz* (see Figure 10.4, p. 166) has the following features: it runs on mobile devices and desktops; it presents seven questions about constituencies (e.g. "For every 100 registered voters how many actually voted in the 2010 General Election?"); and it provides answers for the constituency and the UK, and scores each performance. The *Quiz* allows users to search by postcode or constituency name. When users have finished the *Quiz*, there are links to both a relevant *Constituency Explorer* display, and to social media, so that users can share their scores.

Use by users

The *Constituency Kit* was designed primarily for use by the Library, and the staff have found it to be very useful in their work – they received a number of requests from newly elected MPs who wanted information about their own constituencies. However, the Library is not simply a reference resource for Parliamentarians; it is well connected in networks concerned with all aspects of political life. The Library engages with academics, journalists, pressure groups, charities and non-governmental agencies as part of its responsibility to promote the work of Parliament.

The Library prepared press releases and blogs about the *Constituency Kit*. One aspect of promoting the *Constituency Explorer* was a seminar for a rather heterogeneous group that included journalists, academics, Fullfact (the leading UK organisation for checking claims by politicians, and in the media), and the Independent Parliamentary Support Agency (IPSA). A result has been that Fullfact uses the *Constituency Explorer*, and there is a link to it on the Fullfact website. At the time of writing, we are discussing a version of the *Constituency Explorer* to display data on MPs' expenses with IPSA.

Patterns of use

In each of March, April and May 2015 (the period around the General Election), there were over 4,000 hits. To October 2015, there were 15,500 users in 19,500 sessions with an average session duration of 3.5 minutes. This excludes any access from Durham University. We did not want to include access by the development team – but probably have excluded a number of genuine users (see below). These numbers are not large in absolute terms, but are pleasing given that we chose to disseminate information to selected audiences.

In addition, there has been a large number of references to the *Constituency Explorer* on a variety of websites, including some devoted to political science, social work, business, the labour market and local authorities, as well as sites that discuss data visualisation. It is being used increasingly to support teaching (in political science, contemporary history, geography and mathematics). The *Explorer* has also been referred to in media such as the BBC and *The Guardian*. It was an important element in Durham University's initiative to engage a broad audience with academic work around the General Election, which won an award (2015) from the Chartered Institute of Public Relations for 'Best Use of Media Relations'.

11.6   Current developments

The Library continues to use the *Constituency Kit*. It now has the source code, and the facility to upload new datasets. Data about constituencies will continue to be an important source of questions asked of the Library, and it plans to continue to use the *Constituency Explorer* for internal use as a data repository, to present a wide variety of data relevant to constituencies, and as a data source that others can be referred to. It has been updated and used for the 2017 General Election.

11.7   Lessons for data gatherers and others

Gathering census data carefully is an expensive process. In a climate of austerity, hard questions will be asked about the benefits of this census-taking. If very little use is made of the information, few people will resist the idea of adopting cheaper (and perhaps less reliable) methods. An important defence is to show the benefits that have arisen from the use of census data. This will be easier to do if key potential users are actively engaged in the process of reconfiguring census data into a form that makes it easy for them and their client groups to use. This might appear to be a resource-intensive exercise; however, compared with the costs of the original data collection, these additional costs are modest. The development of the *Constituency Kit* shows that custom visualisations can be designed in collaboration with key users to present census data as part of a larger data resource. Embedding UK-wide census data along with other data of central interest to a specific user group in a tool that is regularly used is likely to increase the perceived value of the census data, and to justify the original costs of data collection. ONS open sourcing the code for data visualisations made the work reported here possible.

Acknowledgements

This work was supported by the House of Commons Library Innovation fund, and by the Economic and Social Research Council via grant ES/M500586/1.

Notes

1   www.dur.ac.uk/smart.centre/.
2   www.nomisweb.co.uk/api.
3   https://trello.com.

References

Kassinger, R. (2002) *Build a Better Mousetrap.* John Wiley & Sons, New Jersey.

Ridgway, J. and Nicholson, J. (2010) Pupils reasoning with information and misinformation. In *Proceedings of the International Conference on Teaching Statistics (ICOTS8): Data and Context: Towards an Evidence-based Society.* Available at: http://iase-web.org/documents/papers/icots8/ICOTS8_9A3_RIDGWAY.pdf.

Ridgway, J. and Smith, A. (2013) Open data, official statistics and statistics education – threat and opportunities for collaboration. In *Proceedings of the First Joint International Association for Statistical Education and the International Association for Official Statistics: Statistics Education for Progress, Macau.* Available at: http://iase-web.org/documents/papers/sat2013/IASE_IAOS_2013_Paper_K3_Ridgway_Smith.pdf.

Ridgway, J., Nicholson, J. and McCusker, S. (2013) Open data and the semantic web require a rethink on statistics teaching. *Technology Innovations in Statistics Education,* 7(2): 12.

Sutherland, S., Hedger, S., Ireland, M. and Ridgway, J. (2015) Designing interactive displays to promote effective use of evidence. In *Proceedings of the International Association for Statistical Education Conference: Advances in Statistical Education.* Available at: http://iase-web.org/documents/papers/sat2015/IASE2015%20 Satellite%2037_SUTHERLAND.pdf.

The *Thematic Mapper*

James Reid and James Crone

12.1 Introduction

The *Thematic Mapper* (*TM*) is an online tool for the creation of thematic maps, one of several productivity tools offered by the UK Data Service-Census Support and introduced in Chapter 6 of this book. Thematic maps are map visualisations used to present attributes or statistics about geographical areas (such as counties, statistical wards or census output areas), the spatial patterns of those attributes, and relationships between them. For example, a thematic map might represent the base population of census, electoral, health, postal or other administrative geographical areas. We focus in Section 12.4 on a census-based case study as an exemplar – although the general principles and the applicability of *TM* hold true for the wider non-census use cases too. In the next two sections we provide some rationale for this tool, step through the work-flow tasks needed to construct a basic choropleth map using the wizard, and discuss some of the technological issues involved.

12.2 Rationale and overview

Over a number of decades, servicing the demands of the UK academic sector for access to and delivery of the range of geography products from the UK national population censuses, a recurring and frequent use case has emerged. Aggregate statistics in raw and tabular form, whilst informative in themselves, can only impart so much and data visualisations have much to offer in summarising and presenting census data such that they are more readily consumed by students, 'lay people' and policy makers. A common and popular form of (geo)visualisation is through the creation of maps. Choropleth maps, in particular, have a long pedigree (Friendly, 2009) and whilst they do pose certain interpretative issues, they arguably remain the most widespread means of map-based communication for statistical data. Consistently, from survey responses from users of the various UK census services, maps are seen as providing compelling and informative means for conveying census information and this is evidenced by that fact that all of the UK Census Offices now support some form of basic map realisations of their census data. This is hardly surprising and merely underlines the more general observation made by Dykes et al. (2005) that "the increasing importance and use of spatial information and the map metaphor establishes geovisualisation an essential element of 21st century information use" (p. 4).

However, the popularity and demand for maps belies the fact that that consumer demand originates in an audience with a broad range of spatial literacy and data manipulation abilities. The diversity of the demand base and the implicit information skill required for appropriate use is succinctly summarised in an Economic and Social Research Council (ESRC) report entitled *Review of Spatial Resource Needs and Resources* (Owen et al., 2009). Again, this throws into relief the fact that *TM* provides an intuitive interactive online tool which can be used to quickly create customised thematic maps.

TM provides a directed work-flow specifically for the creation of choropleth maps (a map which shows area differences by using shading or colours), and provides a 'wizard'-like approach to guide users through the various steps of map creation. In summary, *TM* offers the following: user-specified creation of choropleth maps from any csv file containing supported geographic identifiers; real-time data upload, editing and styling; wizard-based data upload; an extensive range of polygon geometries provided by the UK Data Service-Census Support geographies database; modern, responsive application with contextual help; colour palettes based on *ColorBrewer*;[1] contemporary Ordnance Survey (OS) base maps; the facility to export maps as png image and pdf documents; and the opportunity to export source map data as a shapefile with style information presented in an Open Geospatial Consortium (OGC) Styled Layer Descriptor (SLD) file for interoperability and reuse in a desktop Geographical Information System (GIS). The individual steps in this wizard process are given in Table 12.1 along with notes summarising the main aspects at each stage.

Table 12.1 Wizard steps in creating a basic choropleth map in *Thematic Mapper*

Step	Work-flow task	Notes
1	Upload a csv dataset	• Allows users to upload a local csv file containing their own data to be mapped. • The uploaded csv file should have a header row as the first row in the file giving the column name and each column name in the header row must be no more than ten characters in length (a limitation imposed by the final shapefile output format). • Any records with missing values in the field chosen to create the map from will not be included in the output choropleth map.
2	View and make edits to uploaded data	• Uploaded data is shown in a paged tabular view for review and editing. • Any violations of fieldname lengths and missing data are also flagged. • Double-clicking on the data cell values allows for inline edits.
3a	Identify output geography used in uploaded data	• The uploaded csv file should have had at least one column holding geographic identifiers e.g. census output area codes, local authority codes, etc. • This stage allows users to identify which of the geographies available for mapping matches the data they have uploaded, e.g. if the uploaded csv file contains geographic identifiers for lower super output areas (LSOAs), this can be filtered for and selected as the target geography for mapping. • The full *TM* help pages provide for each available geography a description of what each geography is together with the permissible code ranges of its geographic identifiers.
3b	Choose the column with the values that you wish to map	• A list of fields from the uploaded csv file is presented from which the user can choose the variable they wish to map – only valid columns containing numerical data (positive/negative integer and decimals) are allowed.

(Continued)

Table 12.1 (Continued)

Step	Work-flow task	Notes
3c	Specification of geographical id column	• A drop-down list of fields from the uploaded csv is presented for the user to choose which field represents the geographic identifier. This will be used to join the data to be mapped against the target geography selected at step 3a. • At this point the data to be mapped will be joined to the target geography and become ready for map customisation.
4	Customise generated map	• Once the data selection and joining wizard process is completed the user is re-presented with a basic choropleth map which can then be customised further to refine the data classification scheme employed, colour selection, map annotations and print options. • These options are explored more fully in Section 12.4.

12.3   Underlying technologies: challenges and issues

TM pre-dates the emerging wave of online mapping tools (such as *ArcGIS Online* and *Google Fusion*) and, whilst in some respects it is functionally less rich than these offerings, it remains relevant due to the range of contemporary and historic UK census-related boundary information against which users can choose to map their data. *TM* uses an open source technology stack comprising the following core components: *Mapserver, Geotools, Geoserver, Postgis* and *OpenLayers* as well as a number of open source *Java* libraries for csv and image manipulation. These components are orchestrated using bespoke code as part of the overall work-flow.

One of the significant challenges, given the openness (free access) of the service, is mitigating the potential for malicious use. All file uploads in the initial mapping work-flow are quarantined and virus/malware checked before being passed to the next stage of the work-flow, which undertakes initial analysis of the file contents and performs basic integrity and validity checking, for example presence of a file header, field lengths and missing values inspection. If these pass basic heuristics, the data are presented back to the user via the interface for any final editing and, once the user is happy to proceed, the next stage is to relate the tabular data to the geographic (boundary) data and to download the relevant spatial data and join the two data sources using the parameters supplied by the user at steps 3a–3c in Table 12.1. The data are then grouped and a *Mapserver* mapping file is automatically created for presentation of the results via the main *OpenLayers* map client interface at step 4.

12.4   Case study example and walk-through

To further illustrate the functionality provided by *TM*, we present a simple case study and walk-through, demonstrating the creation of a map of the percentage of the Scottish population describing their health as 'very bad' in the 2011 Census. The raw data for this case study were sourced from the UK Data Service aggregate statistics application, *Infuse* (see Chapter 5 for details), and, for convenience, the data are provided online[2] so readers may download and follow the walk-through to gain a better understanding of the application and its potential. As variations in health and life-expectancy indicators across Scotland are a popular topical focus of public interest post-census, this case study provides a simple but instructive application of choropleth mapping.

The target geography in this instance is local authority level, of which there are 32 (council areas) in Scotland and these will provide the geography identifier for relating the tabular data to the digital boundary data. Having extracted the data from *Infuse*, the first task before uploading to *TM* is to ensure that the csv file is suitably prepared. This means ensuring that the csv file has a header row as the first row in the file giving the column name and that each column name in the header row is no longer than ten characters in length (a limitation imposed by the final shapefile output format). It is also useful at this point to check the data for missing values or to otherwise manipulate the values as required for mapping, for example if rates are required for mapping purposes then these could be computed pre-upload. Once the final csv is prepared it can be uploaded via the *TM* wizard and the choropleth generation may commence by clicking Make a Map on the menu bar of the *TM* home page. Note that it is often useful to upload a csv with multiple variables for mapping, as the entire mapping process can be repeated to produce multiple maps from a single csv upload. Once the uploaded file undergoes some preliminary checking, the user is presented with a tabular editing view that reports any issues related to fieldnames and content and allows for final review and editing. Once confirmed, the next stage is to identify the relevant boundary dataset that relates to the geographic identifiers in the uploaded csv file. *TM* provides a drop-down selection of the available geography types and a simple means to filter the selection, for example by country and/or by boundary type.

The next stage of the process asks the user to identify the variable to be mapped. This is provided by a drop-down which reflects the column headings in the uploaded csv file. Users select the variable to be mapped from the drop-down and confirm by clicking the Apply button, which reveals another selection drop-down that identifies which of the columns from the uploaded csv holds the geographical identifiers – in this example, the column that holds the local authority codes. *TM* provides online help pages which detail for each available boundary dataset a description of what each geography is, together with the permissible code ranges for its geographic identifiers. At this point, the user can choose to review or change the selections for geography boundary datasets, variable to be mapped or the column containing the geographic identifiers – the application automatically refreshes and updates as any of these variables is changed.

Once satisfied that the correct combination of geography boundary dataset, the field to be mapped and the field containing geographic identifiers has been identified, clicking the Apply button at wizard step 3c initiates the application in performing a join between the datasets and automatically creates a basic map of the variable the user selected for mapping.

The user is now able to customise the basic map in a variety of ways:

- *Classification*: refers to how the values of the mapped variable are categorised and can be done by either using manual user-supplied break ranges or via an automatic approach which allows for varying types of classification such as natural breaks (using the Jenks method – an optimisation method to classify the data into groups, so that the variance within each group is as small as possible, whilst the variance between the groups is as high as possible), equal interval, quantile or unique interval.
- *Colour selection*: provides a range of sequential or divergent colour palettes that can be applied and may be filtered for their relevance to several criteria: colour blindness, photocopy safe and print friendly.
- *Legend*: dynamically updates based on colour palette and classification options applied by users.

- *Annotation*: allows users to supply basic map composition elements such as a title, name and description of attribute being mapped. These appear in the final output map.
- *Output*: allows for the composed map (including annotation, if any) to be output as a simple png (image) or in pdf format. The download option provides a shapefile format of the boundary data joined with the user-uploaded tabular data and provides an SLD file of the colour and classification customisations. These allow for easy import into a desktop GIS system such as *QGIS*. Figure 12.1 illustrates the distribution of the percentage of the Scottish population describing their health as 'very bad' in 2011. It shows a classic east/west split, with Glasgow in particular illustrating high levels of self-reported very bad health and relatively modest levels in the Highlands.

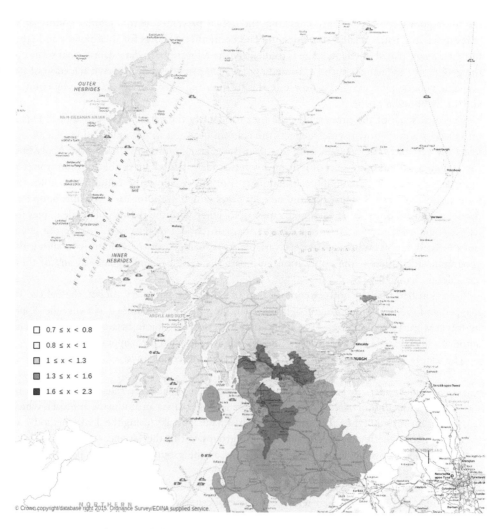

Figure 12.1   Population with 'very bad' health in Scotland by local authority area, 2011

12.5   Conclusion

Using a simple example, we have described an online tool which forms part of the broader suite of tools for UK census users provided by the UK Data Service and which allows users to produce simple choropleth maps using their own data. Whilst aimed at census data users, the tool can be used with non-census data to provide flexibly customisable online choropleth map visualisations.

Notes

1 http://colorbrewer2.org/.
2 www.dropbox.com/sh/tnqbd69iis6byqk/AACvCpLZpyHROSFAmTGRA4Vwa?dl=0.

References

Dykes, J., MacEachren, A.M. and Kraak, M-J. (eds.) (2005) *Exploring Geovisualization*. Elsevier, Amsterdam.

Friendly, M. (2009) Milestones in the history of thematic cartography, statistical graphics, and data visualization. Available at: www.math.yorku.ca/SCS/Gallery/milestone/milestone.pdf.

Owen, D.W., Green, A.E. and Elias, P. (2009) Review of spatial resource needs and resources. *Working Paper No. 88*, German Council for Social and Economic Data (RatSWD), Berlin. Available at: www.ratswd. de/download/RatSWD_WP_2009/RatSWD_WP_88.pdf.

13

An automated open atlas for the 2011 Census

Alex Singleton

13.1 Introduction

This chapter describes the background, motivations and methods used to create the 2011 Census open atlases, each of which maps a comprehensive selection of census attributes at the output area (OA) scale for each local authority in England and Wales. The maps were produced and available to download from February 2013, with a revision produced in February 2014 which improved the code, enabling some additional map adornments and making the pdf file navigable through a table of contents. The entirety of the code, software used and output atlas are freely available within the public domain under an open licence. The output open atlases are now available to download from the Economic and Social Research Council (ESRC) Consumer Data Research Centre (CDRC) data portal.[1]

13.2 Background

Prior to the 2001 Census, data outputs were only available through a limited number of suppliers, and although free at the point of use within the academic sector, data were bound by restrictions that limited public dissemination or the creation of derivative products without financial implications (Openshaw, 1995). However, since the 2001 Census, such data have been made more openly available, and are now disseminated with an Open Government Licence (OGL).[2] In parallel, key census geographic boundary data are also now supplied free from similar restrictive licensing arrangements as part of wider Open Data initiatives (Shadbolt et al., 2012) by the Ordnance Survey (OS) and Office for National Statistics (ONS). As illustrated throughout this book, the availability of such data enables a plethora of public-facing applications to be created.

Historically, the census has been our richest source of spatial data about the characteristics of populations and the places in which they live. For many end-users of the census, creating maps for their local area is a common initial task after the data are released, and is repeated for multiple variables and nationally between various geographic contexts. Such mapping typically comprises fairly basic cartographic conventions, with attributes mapped as appropriately styled choropleth layers, alongside map adornments including legends, scale bars or contextual labels. For multiple variables and local areal extents, these mapping tasks can be very repetitive and time consuming, which was a primary motivation for creating an automated method of production. Taking all of the maps which were produced by this atlas project as an example, and the assumption that it would take five minutes to create a map, it would take 467 days to replicate the same output, with the maps produced continuously without breaks! In the very unlikely event that you could recruit a team of people for this monotonous task, it would likely be very expensive, and fortunately there are now alternatives.

13.3 Automated maps

There are numerous approaches that can be adopted to automate map production. One possibility is to create a web interface such as that presented on the *DataShine* website[3] which is discussed in Chapter 15. This application consolidates census data which sit within a database as attributes, and these are rendered into mapped output on the fly as cartographically rich maps based on user-specified queries. This is an excellent solution, where bespoke queries comprising variable geographic extents and contextual details are required. However, this atlas project took an alternative approach. Rather than mapping a single variable with richer cartography, this project aimed to map every variable for some fixed geographic extents, providing the outputs as fixed image pdf documents. This aimed to supplement the resource-intensive and repetitive map production work that is common in many facets of the public sector. However, creation of such outputs is not well suited to traditional GIS tools, which typically require more manual intervention. Although, through programming languages such as *Python*, GIS software such as *ArcGIS* or *QGIS* can also be automated, these were not used for this project, as this approach would have added additional complexity and the software were proprietary.

A tool which has gained traction in recent years is the open source statistical programming language called *R*.[4] Base *R* functionality enables data import, manipulation, model creation and visualisation; however, through an extension framework, it has also been provided with a wide variety of additional functionality, for example, the import and manipulation of spatial data. Furthermore, *R* is able to interact with the terminal/command line, which also enables additional functionality by calling external libraries.

This chapter does not present the code used to produce the maps and compile these outputs into the census atlases, as these are presented elsewhere.[5] However, the process is described in simple terms to illustrate the principles of how this might be replicated for different contexts.

The first stage was to download all of the 2011 Census Key Statistics bulk data from *Nomis*[6] at the OA level. These data have coverage for England and Wales and include multiple substantive topics that span a range of tables. Each of these tables was read into *R* from the csv files through an automated process. In addition to counts, the tabular outputs contained pre-calculated percentage scores for each of the attributes, so separate calculations were not required. In addition to the raw data, metadata giving the table name and attribute descriptions were also read from the downloaded files and saved. The automation of these processes was possible given the standard format used for both the data and metadata.

A similar process automated the download of shapefile boundaries from the ONS including OAs, which were used to create the choropleth layer, and wards to provide contextual labels. An additional lookup from the ONS was downloaded that provides a link between OA codes and those used for local authority district areas. The process for creating a single map will now be described using pseudo code:

1   Extract a census data variable to map for a given local authority district.
2   Extract the OA boundaries for a given local authority district:
    •   join the census data to be mapped.

3   Plot a choropleth map:
   - select a colour palette using *ColorBrewer*;
   - calculate the break points in the attribute being mapped using natural breaks (Jenks method); and
   - apply colour palette to break points and plot the results.
4   Plot the ward boundaries.
5   Plot the ward labels in a way that avoids overlap.
6   Find the extent of the mapped area x, y co-ordinates:
   - using these attributes, plot a scale bar in the bottom left.

This map was exported as a vector pdf from *R* and then an external library, *PDFCROP*,[7] was used to remove white space around the map. A legend for the map was generated separately within *R*, and again exported as a vector pdf and white space cropped. Thus, for each attribute mapped, two images were created, the first being the plot and the second the legend. The file names of these two images were also stored. This basic process was repeated within two nested loops:

1   Select a local authority district from a list of all local authority districts.
2   Select an attribute to map from a list of all census attributes:
   - create the map and legend outputs (as detailed above); and
   - map the next attribute and continue until there are no further attributes to map.

At the end of these two loops, maps and legends for every Key Statistics variable had been created for the selected local authority and sat within a local directory as a series of pdf images. An example output is shown in Figure 13.1.

13.4   Creating an atlas

A process was then required to combine all of the separate images together into a single pdf. The document preparation language *LaTex*[8] was identified as a solution to this problem as it enables very sophisticated and automated layout, labelling and navigation functions. In this instance, *LaTex* code was written by *R* to a text file, and then using further calls to the terminal was rendered into a pdf using the *Pandoc* library.[9]

The anatomy of the *LaTex* document for each local authority atlas comprised a series of components. The first was a document header that set out the title and author of the atlas, the page dimensions and those additional features required such as a list of figures which is automatically populated by the captions used for the figures included in the document (Figure 13.2).

The second component of the *LaTex* document was created using a loop that creates the necessary code to lay out each of the maps, associated legend and various blocks of text on a single page. This comprised a series of components which are shown in Figure 13.3.

Page layout code was required for each map that had been created for the local authority, and as such required a further loop to output a *LaTex* code block containing the layout options, including positioning of the images on the page. Additional *R* code extracted details from the metadata associated with the mapped attribute to provide a figure title and a copyright statement, which were both added to the *LaTex* code block. These code blocks were created sequentially for all maps.

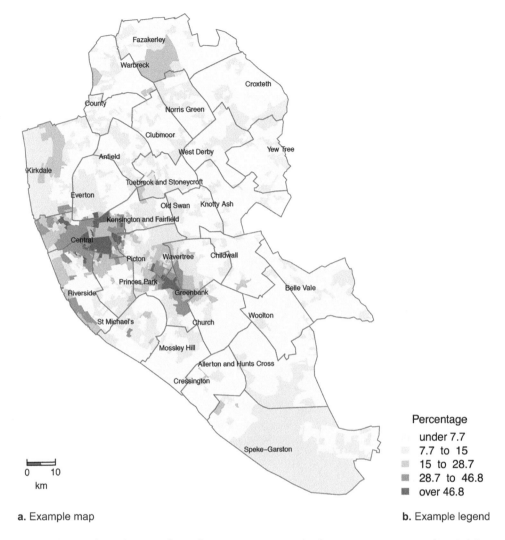

Percentage

under 7.7
7.7 to 15
15 to 28.7
28.7 to 46.8
over 46.8

a. Example map   **b.** Example legend

Figure 13.1   Selected output from the map creation code showing age structure (20–24) for those within the Liverpool Local Authority District

Source: http://www.alex-singleton.com/Open-Atlas/.

2011 Census Open Atlas (Liverpool / E08000012)

Alex D Singleton

January 24, 2014

List of Figures

Figure 13.2   Example header rendered from *LaTex*

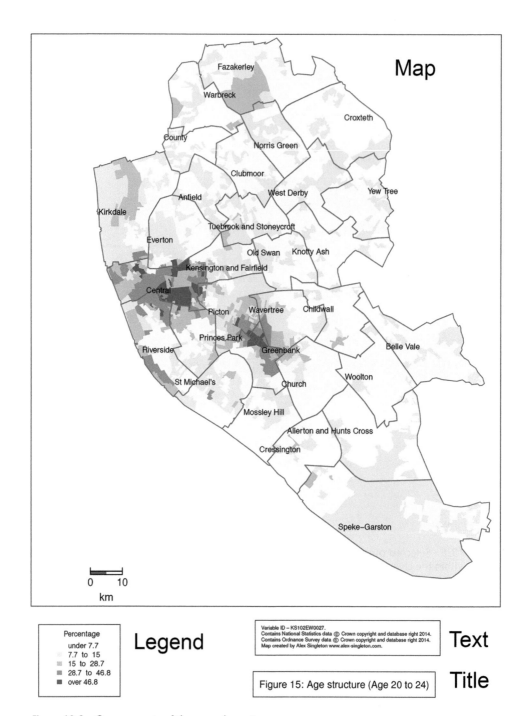

Map

Fazakerley
Warbreck
Croxteth
County
Norris Green
Clubmoor
West Derby
Yew Tree
Anfield
Kirkdale
Tuebrook and Stoneycroft
Everton
Old Swan   Knotty Ash
Kensington and Fairfield
Central
Picton   Wavertree   Childwall
Princes Park
Riverside   Greenbank
Belle Vale
St Michael's
Church
Woolton
Mossley Hill
Allerton and Hunts Cross
Cressington
Speke–Garston

0   10
km

Legend

Percentage
under 7.7
7.7 to 15
15 to 28.7
28.7 to 46.8
over 46.8

Variable ID – KS102EW0027.
Contains National Statistics data © Crown copyright and database right 2014.
Contains Ordnance Survey data © Crown copyright and database right 2014.
Map created by Alex Singleton www.alex-singleton.com.

Text

Figure 15: Age structure (Age 20 to 24)

Title

Figure 13.3   Components of the page layout

Source: Variable ID – KS102EW0027. Contains National Statistics data © Crown copyright and database right 2014. Contains Ordnance Survey data © Crown copyright and database right 2014. Map created by the author, www. alexsingleton.com.

The final stage in creating the local authority atlas was to write the header code, the map layout code blocks and a line of code which sends the *LaTex* document to a text file. This was then run through *Pandoc* to render a pdf output for the main body (i.e. internal pages) of the atlas. The final stage was then for *R* to make a further terminal call to the *PDFTK* library,[10] which enabled a cover to be joined onto the atlas body. Once the local authority district atlas had been created, all of the temporary images were removed, and then the process of image generation would begin again for the next atlas being created. An example of the final output can be seen in Figure 13.4.

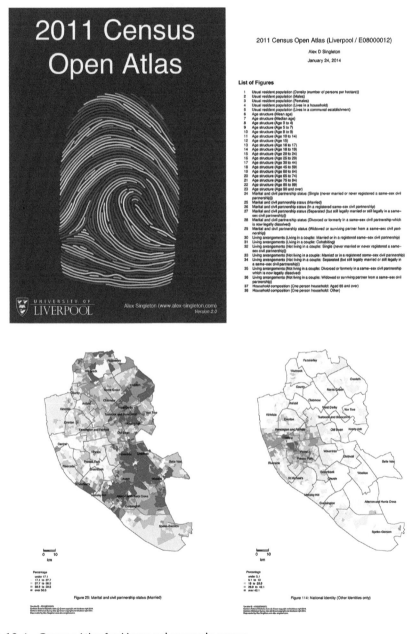

Figure 13.4   Census Atlas for Liverpool example pages

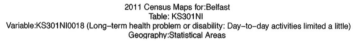

2011 Census Maps for:Belfast
Table: KS301NI
Variable:KS301NI0018 (Long–term health problem or disability: Day–to–day activities limited a little)
Geography:Statistical Areas

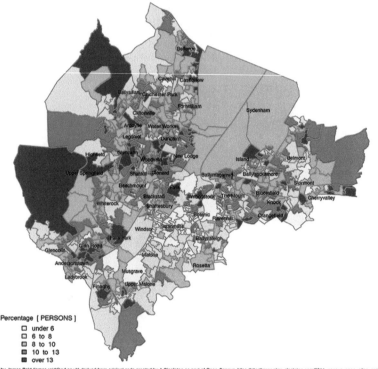

Percentage [PERSONS]
☐ under 6
☐ 6 to 8
☐ 8 to 10
▣ 10 to 13
▣ over 13

Map created by James Reid (james.reid@ed.ac.uk) derived from original code created by A.Singleton as part of Open Census Atlas (http://www.alex–singleton.com/2011–census–open–atlas–project/)

Source: NISRA : Website: www.nisra.gov.sit. NISRA Digital boundaries are supplied under the Open Government Licence.
This work is licensed under a Creative Commons Attribution 3.0 Unported License.

a. Northern Ireland Census Atlas

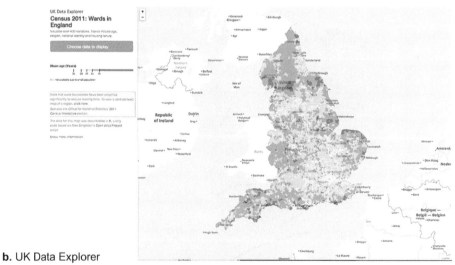

b. UK Data Explorer

Figure 13.5 Independent outputs using the atlas creation code

13.5 Dissemination: open code and open maps

An advantage of the open approach to creating the atlases was that, in addition to the outputs being fully within the public domain, it was also possible to place the code used for their construction into a code repository (https://github.com/alexsingleton/Open-Atlas). This enables other users to repeat and adapt the analysis for their own purposes. This is additionally made possible by all the software tools used for this project also being open and available within the public domain. Although there may be other examples of where the code has been used, two examples are highlighted in Figure 13.5.

The first example was created by James Reid (http://edina.ac.uk) and produced similar outputs to the England and Wales atlas presented here, but for the Northern Ireland extent. The second example was created by James Trimble (http://ukdataexplorer.com) and utilised a portion of the *R* code that provided the automated download and processing of the raw census data. Such examples illustrate the benefit of sharing the code openly.

The census atlas project was well received publicly, initially picking up coverage in *The Guardian* newspaper.[11] By November 2015,[12] there have been around 600 atlas downloads covering 161 local authorities.

13.6 Conclusion

This mapping project was the first of a number of similar projects which have now mapped a census in Japan and attributes for two substantive areas including transport[13] and internet usage behaviour.[14] The technique is very replicable for different contexts and given the usage statistics is clearly a popular output with end-users. In future, it would be useful to add additional contextual details to the maps presented, although this would require a balance between additional vector information and ensuring a reasonable file size for the downloads.

Notes

1 https://data.cdrc.ac.uk/product/cdrc-2011-census-open-atlas.
2 www.nationalarchives.gov.uk/doc/open-government-licence/version/3/.
3 http://datashine.org.uk/.
4 www.r-project.org/.
5 https://github.com/alexsingleton/Open-Atlas.
6 www.nomisweb.co.uk/census/2011/bulk/r2_2.
7 http://sourceforge.net/projects/pdfcrop/files/.
8 www.latex-project.org/.
9 http://pandoc.org/.
10 www.pdflabs.com/tools/pdftk-the-pdf-toolkit/.
11 www.theguardian.com/news/datablog/2013/feb/08/census-data-mapped.
12 http://www.alex-singleton.com/census_atlas_japan/.
13 www.alex-singleton.com/Transport-Map-Book/.
14 https://data.cdrc.ac.uk/product/cdrc-internet-user-mapbook.

References

Openshaw, S. (1995) *Census User's Handbook.* John Wiley & Sons, Chichester.
Shadbolt, N., O'Hara, K., Berners-Lee, T., Gibbins, N., Glaser, H., Hall, W. and Schraefel, M.C. (2012) Linked open government data: Lessons from data.gov.uk. *IEEE Intelligent Systems*, 27(3): 16–24.

14

Ethnic identity and inequalities

Local authority summaries

Ludi Simpson

14.1 Introduction

This chapter describes the *Ethnicity Profilers* that put 2011 Census information about local ethnic identity and inequalities clearly and straightforwardly on the screen of researchers and practitioners, intentionally repeating the most efficient and telling national analyses learned from previous decades. By describing some of the strategies we used, the chapter also hopes to stimulate others to disseminate other standard analyses in accessible spreadsheet or database applications for localities.

14.2 Background

Ethnic Identity and Inequalities in Britain (Jivraj and Simpson, 2015) reviews the highlights of what the 1991, 2001 and 2011 Censuses reveal about the dynamics, growth, separation and diffusion of ethnically defined populations, the changing nature of those definitions officially and for individuals, the different meaning of Britishness for each ethnic group, and the trajectory of inequalities over three censuses in housing, health, education, employment and deprivation. The book was written to give more permanent and considered form to the analyses (CoDE, 2015a) that had been published as rapid response briefings to the 2011 Census results as they emerged over 15 months from December 2012 to early 2014. Those rapid responses gained a lot of media and professional publicity by focusing on plain English explanations of tabular and graphic analyses that were mostly able to be rehearsed before the release of census results by thorough study of the planned outputs and of successful analyses of previous censuses. Their success has led to the continuing issue of local briefings.

The *Ethnicity Profilers* (CoDE, 2015b) showcased here have been designed to give analyses for all local authority districts in England and Wales, plus all 353 multi-member wards in Scotland, which was impossible in either the rapid response briefings or the book, both of which were limited by space to maps, analyses of types of locality and illustrative local examples.

The aim of the entire body of work was to highlight major characteristics and comparisons of policy interest in ways that were easily accessible to non-technical audiences. In the case of the *Profilers*, Microsoft Excel was the chosen platform, undertaken with standard Excel functions and without *VBA* programming. Undoubtedly a similar impact could be gained in other ways, for example with a database application on the internet. Excel was the instrument sufficiently known to the author. It has the advantage of being very common among the intended audiences, who therefore could have access to the data as well as the automated analyses in a format easily transferred to their own reports.

Automation of the *Profilers* by the author involved (i) offering selection of a district or districts and, on occasion, an ethnic, religion or national identity category or categories, using drop-down lists enabled with an *Active-X* combi-box from Excel's standard Developer menu; (ii) the preparation of a data sheet on which all the necessary data for the selected district-ethnicity combination appear in a single row; (iii) the extraction of data for the selected district-ethnicity combination using Excel's vlookup function; (iv) the construction of tables and graphics from the data; and (v) the construction of textual summaries from text strings which change depending on the data for the selected district.

14.3 Available analyses

The analyses, displays and text in the summaries are a subset of those possible. The aim has been to provide easy access to tried and tested analyses, transparency about methods to give confidence to those reproducing them and the ability to explore the data beyond the provided analyses. The examples on the printed page cannot properly illustrate the dynamic tool that is available for download, but intend to show what can be achieved with three of the eight *Profilers* that are available at the time of writing. The full list, available online,[1] includes:

- Censuses 1991, 2001 and 2011: diversity, ethnicity, religion, country of birth, national identity, language and more, for local areas in England and Wales.
- Census 2011: the age structure of ethnicity, religion and national identity, for local areas in England and Wales.
- Census 2011: ethnic composition of England's most deprived neighbourhoods.
- Census 2011: the age structure of ethnicity, religion and national identity, for local areas in Scotland.
- Censuses 2001 and 2011: diversity, ethnicity, religion, country of birth, national identity, language and more, for local areas in Scotland.
- Census 2001 and 2011: inequality indicators for local areas in England and Wales.
- Census 2001 and 2011: ethnic mixing and migration for local areas in England and Wales.
- Census 2001 and 2011: ethnic mixing and migration for local areas in Scotland.

The *Ethnic Inequality Profiler* illustrated in Figure 14.1 shows how a district's social indicators compare with England and Wales in 2011, for each of six broad ethnic groups, and how the inequalities have changed since 2001. Ethnic groups are only shown where the population for the indicator is at least 100. As reproduced in Figure 14.1, the *Profiler* produces text to summarise the results for one chosen ethnic group, in this case 'Minority' as a whole. But much more can be gleaned from the charts than is in the text summary. For example, Figure 14.1 shows that the Black group fared relatively well on education and health in Barking and Dagenham, though it did not do so in England and Wales. The percentage of young adults with no qualification reduced for each ethnic group between 2001 and 2011 but was highest for White British in both years in Barking and Dagenham. In contrast, for adults aged 25 and older, unemployment was less for White British than for Black adults. The *Profiler*, of course, raises questions. Will the

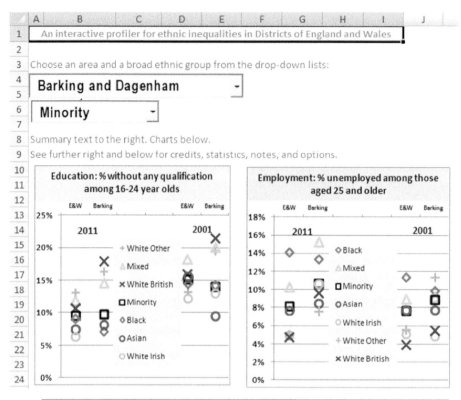

Barking and Dagenham's Minority population was 93,962 (50.5% of its total) in 2011.
Its Index of Multiple Inequality ranked 298 for Minority populations
    among the 348 districts of England & Wales (1=worst). It ranked 169 in 2001.
In 2011, the Minority population in Barking and Dagenham had:
    9.9% of 16-24 year olds with no qualifications, (White British 17.9%).
    10.7% unemployed among adults aged 25 and older, (White British 9.7%).
    15.4% long term limiting illness rate, (White British 23.3%).
    31.2% of households overcrowded, (White British 12.1%).

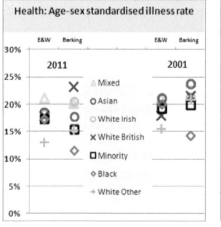

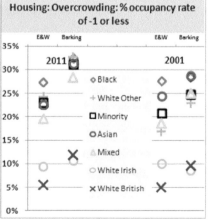

Figure 14.1   *Ethnic Inequality Profiler*, areas in England and Wales, 2001 and 2011 Censuses

Source: http://www.ethnicity.ac.uk/research/data-sources/area-profilers/.

relatively high proportion of young Black people in Barking and Dagenham with at least some qualification translate into less unemployment at older ages in future years?

The age structure *Profiler* provides pyramids, tables and further charts for comparing one or two districts, and within those districts either all people or one of the categories of religion, ethnic group or national identity. There is a similar *Profiler* for Scotland, its council areas and multi-member wards. In the illustration in Figure 14.2, the young age structure of Hindus in Bristol, with many young adults and relatively few elderly, is contrasted with the old age structure of those recording themselves with Irish national identity. Each group's children under 16 and people aged 65 or older are shown in the tables and other charts of this *Profiler* (only part is illustrated in Figure 14.2). Bristol's Muslim population has 39 per cent aged under 16 compared to its Hindu population with 16 per cent aged under 16. Can the reader suggest which ethnic group population has 45 per cent aged below 16? Look it up if you want to check your hunch.

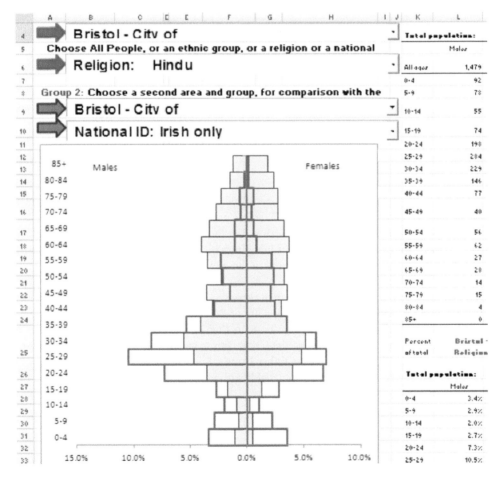

Figure 14.2 The age structure of ethnicity, religion and national identity, for local areas in England and Wales, 2011

Source: http://www.ethnicity.ac.uk/research/data-sources/area-profilers/.

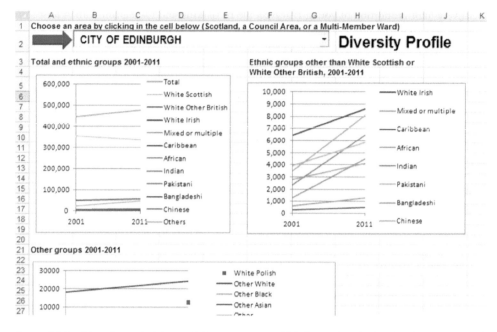

Figure 14.3 Diversity, ethnicity, religion, country of birth, national identity, language and more, for local areas in Scotland, 2001 and 2011

Source: http://www.ethnicity.ac.uk/research/data-sources/area-profilers/.

The diversity *Profiler* showcases the range of information from the 2011 Census related to ethnicity, and compares the number in each ethnic group with the previous census. Figure 14.3 illustrates the Scotland version, which contains data for the 353 electoral wards of Scotland and the 32 council areas of which the City of Edinburgh is one. The equivalent *Profiler* for England and Wales contains data for the 348 districts in those two countries.

Scotland's categories are not all the same as those used in England and Wales. White Scottish is shown separately and is the largest group in every ward and council area. 'White Other British' is the largest minority ethnic group in the City of Edinburgh as in most areas. Three-quarters of Scotland's 'White Other British' were born in England. Another category that appears in Scotland's census but not in the rest of the UK is Polish, which, in 2011, was the largest 'ethnic group' after White Scottish and White Other British. This *Profiler* highlights other statistics for each local area. What proportion of households with more than one person has more than one ethnicity recorded?

14.4 Conclusion

The *Profilers* were successful in reaching policy organisations and local government research, as the 2011 Census results emerged. Together with written briefings, they influenced the tone of media reports. Since ethnicity and related information are rarely updated between censuses, they will retain an active role at least until 2024.

Acknowledgements

The *Profilers* were funded by Joseph Rowntree Foundation, and hosted and supported by CoDE (Centre of Dynamics of Ethnicity), University of Manchester.

Note

1 http://tinyurl.com/ethnicityprofilers.

References

CoDE (2015a) *Dynamics of Diversity, Census Briefings.* Centre on the Dynamics of Ethnicity, University of Manchester, Manchester. Available at: http://www.ethnicity.ac.uk/research/outputs/briefings/dynamics-of-diversity/.
CoDE (2015b) *Ethnicity Profilers.* Centre on the Dynamics of Ethnicity, University of Manchester, Manchester. Available at: http://tinyurl.com/ethnicityprofilers.
Jivraj, S. and Simpson, L. (2015) *Ethnic Identity and Inequalities in Britain: The Dynamics of Diversity.* Policy Press, Bristol.

Mapping travel-to-work flows

Oliver O'Brien and James Cheshire

15.1 Introduction

The 2011 Census captured and reported the flows of people by transport mode from home to work in unprecedented detail. The national-level, small-area datasets are extremely large and complex, rendering them inaccessible to many users without the requisite skills and computing resources to handle them. In order to facilitate access to the freely available flow data detailed below, the *DataShine Commute* websites were created.[1] These form part of a broader suite of *DataShine* web mapping applications funded by the Economic and Social Research Council (ESRC) between 2013 and 2016.[2]

15.2 *DataShine* websites

DataShine Commute uses the same statistical geographies for both home and work areas, namely the middle level super output areas (MSOAs) for England and Wales. Due to different timescales for the release of the Scottish census data, a second website entitled *DataShine Scotland Commute*[3] was created at a later date using the MSOA equivalent – intermediate geographies (IGs), sometimes referred to as intermediate zones. Commutes across the boundary between England and Wales and Scotland are not included in the two small-area versions, but are included in a third, inter-regional version of the website, *DataShine Region Commute*,[4] which also includes flows to/from and within Northern Ireland. This latter version groups flows by local authority – so it does not show fine-grained flows but allows local authorities to see the commute traffic entering and leaving them. We took the decision to create three separate websites with different granularities for the various origin–destination data which were released by the national statistics agencies. Table 15.1 outlines the differences between the versions, which are hereafter collectively referred to as the *DataShine Commute* websites.

Table 15.2 shows the form of the flow and statistical area datasets. MSOAs typically have a residential population of 7,500, while IGs typically have a population of approximately 3,500. It should be noted that, as these are defined by lower and upper thresholds on their residential population, the workplace population of some MSOAs is extremely low (perhaps under 500), typically representing local shops and services serving the local population only, while for others

Table 15.1 Summary of the different *DataShine Commute* websites

Website	Flow data provider	Provider's table reference	Extent	Geographical granularity	No of centroids*	Largest single inter-area unidirectional flow (all modes)
DataShine Commute	ONS	WU03EW	England and Wales	MSOA	7,207	1,906
DataShine Scotland Commute	NRS	WU03BSC _IZ2011	Scotland	IG	1,286	1,339
DataShine Region Commute	ONS	WU03UK	All UK	Local authority (LA)	408	46,072**

* Includes four non-area/extra-area flows: mainly work at/from home, offshore installation, no fixed place and outside UK. Also, for *DataShine Commute* and *DataShine Scotland Commute*, this includes the super-aggregated flows to/from the other countries of the UK. None of these flows are mapped, but they do appear in the table of results on the websites.

** This is the commute from Wandsworth (in south-west London) to Westminster and the City of London (which are combined in the statistics), which is almost double the extra-home flow within Wandsworth itself.

Table 15.2 Sample records from the flow (a.) and statistical area (b.) datasets for *DataShine Region Commute*

a.

Origin Local authority reference	Destination Local authority reference	All-flows	Work-at-home	Metro	Train	Bus	Taxi	Motorbike	Car-driving	Car-passenger	Bicycle	Foot	Other
E41000067	E41000065	1,337	0	1	14	38	1	4	1,136	85	13	40	5
E41000067	E41000067	38,368	0	10	63	3,172	99	247	23,151	2,688	1,278	7,590	70
E41000067	E41000069	2,401	0	1	42	264	1	14	1,776	165	17	114	7
E41000067	OD0000001	5,605	5,605	0	0	0	0	0	0	0	0	0	0
E41000067	OD0000003	3,112	0	4	30	94	67	16	2377	239	37	178	70
E41000067	S12000006	767	0	0	26	36	0	2	634	44	7	13	5

b.

Local authority reference	Name	Longitude (EPSG:3857)	Latitude (EPSG:3857)
OD0000001	Mainly work at/from home	N/A	N/A
OD0000002	Offshore installation	N/A	N/A
OD0000003	No fixed place	N/A	N/A
OD0000004	Outside UK	N/A	N/A
95AA	Antrim	−696258.618	7303114.710
S12000005	Clackmannanshire	−416578.354	7588437.458
E41000067	Carlisle	−312180.349	7356277.248
W40000022	Merthyr Tydfil	−374153.136	6753420.442

it is far higher than the residential numbers, such as the MSOA which largely contains the City of London (with likely over 300,000 workers).

Workplace-area-based flow statistics, based on workplace zone (WZ) geographies, have become available subsequent to *DataShine Commute* being created.

15.3 Description of the *DataShine Commute* websites

The *DataShine Commute* websites use population-weighted centroids for the MSOAs/IGs and simple geographical centroids for the LA boundaries. Each centroid is represented by a clickable circle, and all flows in/out of the corresponding area are shown as straight lines between the selected centroid and the linked origin/destination centroids. The thickness of each line is directly proportional to the number of people who commute between the areas. The lines are partially translucent, to show the underlying geography and also indicate where there are multiple overlapping lines, and they are coloured red for home-outwards flows and blue for work–inwards flows. A filter allows a single transport mode to be selected with the line thicknesses adjusted to reflect the mode-specific counts. *DataShine Scotland Commute*'s statistics vary slightly, as they include journeys to study as well as work, and the mode breakdown is simplified.

Figure 15.1 shows the user interface of *DataShine Commute*, the initial website. All three websites share a common codebase, and are *JavaScript*-based, with the *OpenLayers 3* and *JQuery JavaScript* APIs used.

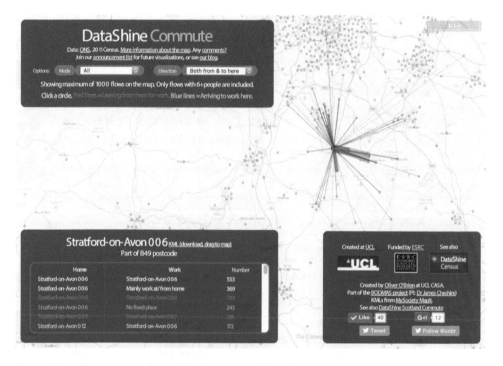

Figure 15.1 The user interface of the initial *DataShine Commute* website, with a sample MSOA centroid selected, and flows in/out of this centroid shown

15.4   Design and cartographical considerations and justifications

Flows smaller than a threshold (six journeys for the larger-scale maps and ten for the regional map) are excluded. This is to avoid overwhelming the map with lines, to speed up display of the map on the browser (as the lines are vectors rendered locally by the browser) and to stop the results table from being unnecessarily large. Note that even for low-work-population areas, there are typically a great many such small flows, and while they are individually insignificant, summing them up can represent a considerable flow (typically up to a third of the flows that are shown); this should be borne in mind if carrying out quantitative analysis based on the maps. As for major employment hotspots, there can be a great number of distinct flows to/from such a selected area individually exceeding these limits; an overall maximum of showing the largest 1,000 flow lines/table rows also applies, for the same reasons.

The flow data are overlaid on a raster background map. The websites use the regular Open-StreetMap 'Mapnik' render that appears by default at openstreetmap.org, with a translucent filter to lighten and partially fade it – so that, while the necessary detail is retained, it becomes more like a background map.

MSOA reference IDs and names are not in themselves normally useful for someone to identify with an area – the name variant simply being the name of the parent local authority followed by an alphanumeric code – so a technique is employed whereby postcode outcodes (the first part of the postcode), which have a sizeable population within the MSOA, are included. For example, MSOA E02006918 is named 'Hackney 028' but to add further context, it is additionally described in *DataShine Commute* as being 'Part of E5 postcode'. The E5 postcode will be familiar to those living in the locality as being the postcode for Clapton. Intermediate geographies, on the other hand, are supplied with familiar names (for most, if not all, of the Scottish local authorities).

15.5   Limitations and improvements

The *DataShine Commute* websites have been well received with over 10,000 visitors in the first year of launch to the initial (England/Wales) website. Particular views of the maps have been incorporated into consultancy reports, and there have been positive social media and blog comments.

A common user feedback request, following the launch of the initial website, was to aggregate the small-area flow statistics into a single measure showing flows in/out of local authority areas; this would allow local authority transport planners and analysts to visualise their cross-border flows. This feedback led to the development of the *DataShine Region Commute* website, which presents the data in this super-aggregated form.

Mapping flows in this way reveals some limitations of the underlying dataset; for example, unfeasible flows, such as long-distance flows between major cities where the primary method of travel is walking or cycling. The data are aggregated from census form returns so in some cases the spurious results may be due to employees specifying the head office address of their work-place, rather than the local office at which they normally work. The spurious flows are minimised on the websites by the small-flow cut-off numbers, as such errors would normally represent a small population. However, they can be seen in some cases, such as a Dorking (south of London) to Manchester flow, a distance of well over 124 miles (200 kilometres), which the results suggest some people carry out by walking. A Dorking satellite office of a Manchester corporation is one possible explanation.

We were keen to create a mapping tool that was straightforward to use, was ready to release in a matter of days after the data became available and maximised the resources available.

Given these constraints, we acknowledge a number of limitations. First, by employing straight lines, which do not move, fade or otherwise change when interacted with by the user, some results can be obscured if they happen to be in the path of another flow from a close by area or one on or nearly on the same axis as the first. The translucency attempts to mitigate this but it can be difficult to distinguish between overlapping lines. Multiple lines can also cause a local area to become almost completely opaque, obscuring the contextual geographical detail (e.g. town name) beneath. Using different scales for the vector lines at different zoom levels alleviates this, allowing different links to be visible on a more separated basis as the interested viewer zooms in.

The use of population-weighted centroids also implies most travel starts/ends near the centroid. Areas which contain at least two population groupings (e.g. villages), separated from each other, will appear to have the flows only going to one, not the other. This may be exaggerated by area boundaries that do not generally appear on the map (except local authority boundaries, which appear on the underlying raster map at certain zoom levels and may be obscured by the flow lines). The exclusion of such boundaries is partially to simplify the display of the map (as administrative geographies are typically not well-recognised by viewers) and partially to reduce load times, as vector geometries across wide areas can be complex and bandwidth consuming to load (except where heavily simplified).

15.6   Usage examples

Flows into the City of London, by mode of transport

By selecting the MSOA containing the City of London, a major workplace area at the heart of London with only a small residential population, and then filtering by mode type, it is possible to gain a picture of how City workers travel to their offices each day. Figure 15.2 shows the results for each of the available transport type groupings, at the same scale. The overall boundary of (Greater) London is also included on the graphic, to illustrate that most of the observed flows are coming from outside London altogether.

The mapped results show clear patterns. Commuter rail dominates, followed by driving. Car passenger commutes are negligible. The biggest single flow in by train is not from another area of London, but from part of Brentwood in Essex. Taxi flows into the City mainly come from very affluent neighbourhoods in the west of Zone 1 (Mayfair). Cyclists come from all directions, but particularly from the north/north-east. Motorbikes and mopeds, however, mainly come from the south-west (Fulham). The Tube flow is from north London mainly, but that is because that is where the Tubes are. Finally, the bus/coach graphic shows both good use throughout inner-city London (Zones 1–3) but also special commuter coaches that serve the Medway towns in Kent, as well as in Harlow and Oxford. 'Other' shows a strong flow from the east – likely commuters getting into work by using the Thames Clipper riverboat services from Greenwich and the Isle of Dogs. The detail of the population and flows from the rest of London is largely obscured by the mass of lines caused by the large and concentrated population across London, which allows the flows from outside of London itself to be more clearly seen (O'Brien, 2015).

The commuter belt around London, Edinburgh and other cities

A characteristic flow pattern is highly visible in *DataShine Commute*, between largely residential areas (outer suburbs and satellite towns) surrounding the large, traditionally structured cities,

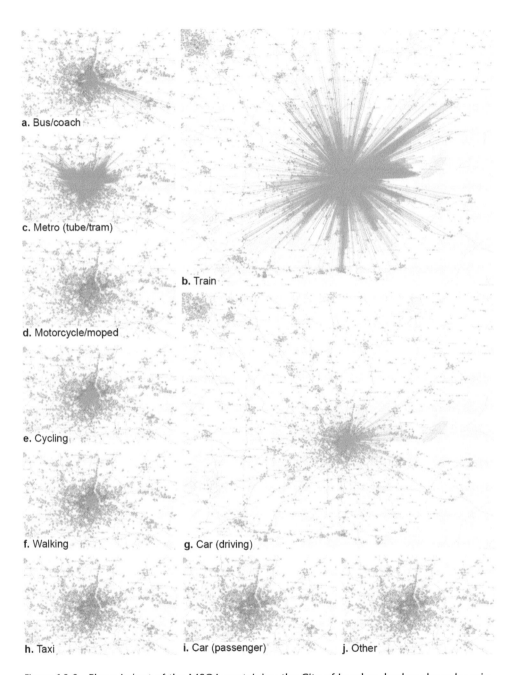

a. Bus/coach

c. Metro (tube/tram)

b. Train

d. Motorcycle/moped

e. Cycling

f. Walking

g. Car (driving)

h. Taxi

i. Car (passenger)

j. Other

Figure 15.2 Flows in/out of the MSOA containing the City of London, broken down by principal mode of travel

Source: Data from the ONS 2011 Census Special Workplace Statistics.

and the central business areas typically in the centre of each city. When such a residential area is selected, a strong, radial and unidirectional flow out of the area, into the centre, is seen, with only a small number of flows outwards to other areas. Conversely, people commuting into the selected area typically come from all directions.

Trans-national commuter flows in the UK

DataShine Region Commute allows transnational travel-to-work flows, such as from Gretna in Scotland to Carlisle in England, to be seen, including less obvious smaller ones such as between Great Britain and Northern Ireland. Some long-distance flows, such as from the centre of Edinburgh to the centre of London by rail, can also be seen.

15.7 Conclusions

Visualising origin–destination datasets at high resolution is necessarily complex, due to the size of the data and its multi-dimensional nature. This chapter has detailed a straightforward way of mapping such a dataset, on travel-to-work flows, and presented some examples of insights which can be gained from its use, such as the diversity and spatial extent of various transport types into central London, the commuter belts around England and Wales's cities, and cross-border travel to work.

More sophisticated visualisations of origin–destination datasets, such as arbitrarily aggregated flows and different techniques for minimising obscuring details and highlighting interesting results, will further help with gaining insight into the movements of people.

Acknowledgements

Graphics in this chapter contain OpenStreetMap data which is under an Open Database Licence (ODbL) and OpenStreetMap imagery which is under a Creative Commons Attribution (CC-By) licence. The data/imagery are © OpenStreetMap contributors. *DataShine Commute* is part of the *DataShine* project which is an output of the BODMAS (Big, Open Data: Mining & Synthesis) research grant, awarded by ESRC (grant reference ES/K009176/1).

Notes

1 http://commute.datashine.org.uk.
2 ESRC project reference: ES/K009176/1.
3 http://scotlandcommute.datashine.org.uk.
4 http://regioncommute.datashine.org.uk.

Reference

O'Brien, O. (2015) The City of London Commute. Available at: http://oobrien.com/2015/07/the-city-of-london-commute/.

16

Circular migration plots

Nikola Sander, John Stillwell and Nik Lomax

16.1 Introduction

The visual representation of migration flows aids the identification of patterns and relationships that are often difficult to deduce without visual displays. From the familiar migration flowmap to interactive visualisations of migrant routes across the Mediterranean,[1] visual representations help us better understand the global, regional and local flows of people. Traditionally, migration flowmaps were static graphics included in printed volumes. With the advances of internet technology, the number of interactive and dynamic visualisations of migration is expanding rapidly (see Dennett, 2015, for a concise summary). However, creating effective visual displays of migration flows is not trivial. Representing the multi-dimensional information contained in a matrix of origin–destination flows in a two-dimensional graphic requires custom software tools and some coding skills.

No matter how sophisticated the tools and coding skills are (e.g. *R*, *ArcGIS* or *Tableau*), creating an effective visualisation hinges upon an iterative design process that ensures the readability of the graphic. The key element of the design process is the encoding of information. In a successful design process, the encoding takes into account human cognition and visual perception, allowing the reader to easily decode (i.e. read) the graphic. In conventional maps, migration flows are typically encoded using the visual attributes of line width to indicate volume, spatial positioning to indicate origin and destination, and colour or arrows to indicate direction. In the mid-1840s, Charles Joseph Minard used these encodings to produce flowmaps of migration flows (Minard, 1869), but to date the key limitations of flowmaps have yet to be overcome.

Perhaps the most fundamental problem with flowmaps is that the lines representing flows are geo-located on a map, meaning that the line connects the true geographic locations of origin and destination regions. The geo-location leads to the occlusion of flows, which obliterates the structure of the migration system the map is supposed to reveal. In this chapter, we demonstrate an alternative way of visualising migration flows that avoids the problems of geo-locational encoding by moving from a true geographic to a circular layout (Abel and Sander, 2014). We summarise the graphical elements of the circular migration plot, provide some hands-on information on which software to use for creating the plots and discuss the key design challenges. Our aim here is to bridge the software requirements and coding elements with the design aspects of visualising migration flows. We conclude by providing some ideas for applications of the circular plot beyond migration flow data.

16.2   Introducing circular migration plots

The UK census is one of many sources of migration flow data, and the number of available migration datasets is growing rapidly thanks to projects like IPUMS (Integrated Public Microdata Series) and the emergence of new, alternative data sources. As the (geographical) level of detail at which migration flows are available increases, understanding the spatial structure of migration flows is becoming more and more challenging. Visual representations can effectively complement conventional statistical summary indicators and modelling approaches for analysing spatial structures and trends over time. But their use has thus far been hindered by the problems discussed earlier with the encoding of migration flows in conventional maps. Visualisations of migration flows have to depict a large amount of information. The most important items are the locations of origin and destination, the volume of movement, and the direction of the flow. In addition, visualisations should highlight the relationships between individual place-to-place flows, the main corridors of movement, and the key senders and receivers of migration.

The circular migration plot is an alternative visual representation of flow data that moves beyond the true geographic layout. In a nutshell, circular migration plots integrate flowmaps and circular network diagrams to depict the flow of people. As in the conventional map, flows are encoded using the visual attributes of line width to indicate volume and colour to indicate direction. But rather than encoding the spatial positioning as the true geographic position on a map, we adopt the circular layout commonly used to depict network relationships to indicate the approximate relational distance between origin and destination region. The key advantage of the circular migration plot over conventional flowmaps is that the plot more effectively depicts differences in the volume of flows and counterflows across the entire migration system.

The representation of relational distance depends of course on the underlying geography. For example, arranging the counties bordering Greater London in a circular layout is much more straightforward than imposing a circular positioning on the UK's local authority districts. How the zones of origin and destination are arranged in the circle is just one of a series of decisions that collectively make up the design process involved in creating circular plots.

With a basic level of coding skills, it is relatively easy to use *R* or *Circos* to create a basic circular plot. But, as with other network diagrams, the key problem with using software default settings, and thereby skipping the design process, is that the plot looks like a hairball and is difficult to read. The key to an effective, easy-to-read circular visualisation is the omission of small flows to avoid the plot becoming too dense, and the arrangement of regions in an order that limits the number of flow crossings. Due to the complexity of migration data and the fact that the spatial structure of each system of flows is different, there is no one-size-fits-all solution to designing effective circular plots.

16.3   Design aspects of circular plots

Four key design elements need to be considered when creating a well-organised and readable circular migration plot: filtering, ordering, sorting and colouring. Let us illustrate this using an example. The circular plot shown in Figure 16.1 depicts the flows between the 12 Nomenclature of Territorial Units for Statistics (NUTS) 1 regions of the UK. The plot was created by filtering out all flows containing less than 9,300 migrants (i.e. selecting the largest 40 flows), by ordering the regions in a way that approximates geographic location (i.e. northern regions at 12 o'clock,

Figure 16.1 Migration flows in excess of 9,300 people between NUTS 1 regions of the UK for 2010–11

Source: Office for National Statistics, 2011 Census: Special Migration Statistics (United Kingdom) [computer file]. UK Data Service-Census Support. Downloaded from: https://wicid.ukdataservice.ac.uk.

southern regions at 6 o'clock), by sorting the inflow and the outflows of each region based on their volume from largest to smallest, and by choosing a colour scheme that makes it easy to distinguish between individual regions.

The decisions we make during the iterative design process aim to highlight the largest flows, the level of population redistribution and the degree of connectivity. Figure 16.1 reveals that in the UK the largest movements occur between South East, London and East of England. The overall spatial pattern of migration flows between NUTS 1 regions in the UK is one of flows and counterflows between neighbouring regions, with London and South East being the only regions that exchange migrants with most other regions.

One indicator of migration impact is migration effectiveness, the degree of asymmetry in the network of flows between regions of a country (Stillwell et al., 2000). The lower the effectiveness of migration in redistributing population and the more equal the sizes of the flows in the migration system are, the more crucial becomes the design decision about the threshold

205

below which flows are filtered out. A too low threshold results in a plot that resembles a hairball, whereas a too high threshold results in too many sizeable flows being omitted. The decision about which threshold to use becomes easier if migration effectiveness is relatively high. This is typically the case with international migration and with internal movements from rural to urban areas in less developed countries (Rees et al., 2016). A viable solution for visualising internal migration in more developed countries is to avoid the hairball phenomenon by showing net flows rather than gross flows, thereby halving the number of flows to be displayed.

Another important design decision is whether to show within-region flows. Given that the majority of movements occur over short distances within the same region, including intra-regional flows in the circular plot reduces most inter-regional flows to rather thin lines. In Figure 16.1, we wanted to show inter-regional migration in the UK and hence decided to omit intra-region flows. Figure 16.2 highlights the prevalence of short-distance movements,

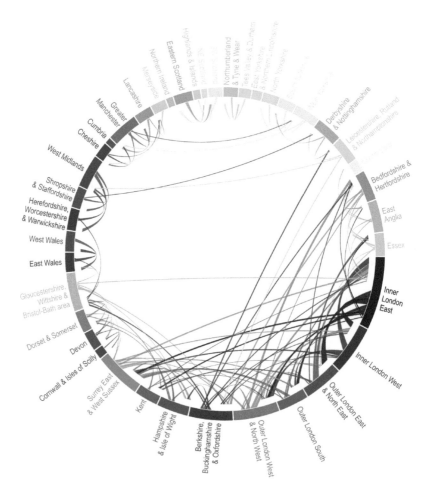

Figure 16.2 Migration flows in excess of 3,200 people between NUTS 2 regions of the UK for 2010–11

Source: Office for National Statistics, 2011 Census: Special Migration Statistics (United Kingdom) [computer file]. UK Data Service-Census Support. Downloaded from: https://wicid.ukdataservice.ac.uk.

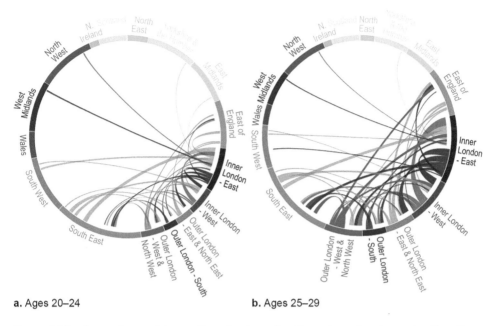

a. Ages 20–24   **b.** Ages 25–29

Figure 16.3 Screenshots of interactive circular migration plots showing migration flows between London (at NUTS 2 level) and the rest of the UK (at NUTS 1 level) for selected age groups

Source: Office for National Statistics, 2011 Census: Special Migration Statistics (United Kingdom) [computer file]. UK Data Service-Census Support. Downloaded from: https://wicid.ukdataservice.ac.uk.

especially in the northern and eastern parts of the UK. Except for movements into and out of London that go through the centre of the circle, the flows between 40 NUTS 2 regions are mostly aligned along the perimeter of the circle. The degree of clustering of flows along the perimeter is inversely related to the average distance moved within a system of flows, but it is directly affected by the positioning of the regions around the circle. This suggests that every effort should be made to position neighbouring regions close to each other in the circular plot.

Printed graphics like those shown in Figures 16.1 and 16.2 are limited in the number of regions that can be arranged around the circle; experience has shown that this number should not exceed 50 regions. Moreover, changes in the spatial patterns of migration over time or across age groups can only be portrayed through a series of (smaller) multiples. Interactive versions of the circular migration plot offer more flexibility since they allow for smooth transitions between time periods (or age groups), and between two nested geographies (e.g. region and district level). User interaction is further enhanced through tooltips indicating the numbers of migrants for a given flow when hovering over the plot with the mouse. The screenshots in Figure 16.3 show migration flows between London (at NUTS 2 level) and the rest of the country (at NUTS 1 level) for two selected age groups. The interactive circular migration plots were created using a *D3.js* library originally developed for the project The Global Flow of People.[2] Figure 16.3 reveals the mobility patterns of 20–24 year olds, who exhibit high migration intensities but relatively low effectiveness, and of 25–29 year olds, where a large proportion of movement occurs within London. Consideration of the patterns and drivers of mobility at different age groups is the focus of Chapter 27 of this book, while the interactive online visualisation of UK 2011 Census data for NUTS 1 and 2 regions by different age groups is available online.[3]

16.4   Creating circular migration plots

The circular plot can be used to visualise any migration dataset that can be arranged in a matrix format with rows corresponding to origins and columns to destinations. The matrix has to be symmetric with an equal number of origins and destinations. Within-region flows may be included in the visualisation or set to zero in the input data file.

Circular plots can be created using three alternative open-source software packages: *Circos*, originally developed for the graphical representation of genomic data (Krzywinski et al., 2009), *R* via the *circlize* package,[4] and the *JavaScript Library D3* (Bostock et al., 2011). The decision about which software to choose depends on the user's preferences. *Circos* offers the greatest flexibility in design and functionality regarding circular plots, while *R* may be more appealing for those who are already familiar with it. The *JavaScript Library D3* has become the state-of-the-art tool for interactive data visualisation. A custom circular plot library is available online.[5] Sander et al. (2014) provide a detailed manual for creating circular plots.

16.5   Conclusions

The circular plot was originally developed to visualise the global system of migration flows in a single graphic that could depict the relative size and direction of flows around the world.[6] The application of the plot has since been extended to other types of human mobility, including internal migration (see, for example, Charles-Edwards et al., 2015; Stillwell et al., 2015), but the visualisation can also be used with other types of flow data, such as commuting flows from the census or non-census data such as trade in endangered species and journal citations. We believe that the circular migration plot opens up new avenues for understanding migration at different spatial scales, for telling the stories hidden behind large datasets, and for more effectively communicating science to stakeholders and the wider public.

Notes

1   See International Organisation for Migration interactive website at: http://migration.iom.int/europe/.
2   www.global-migration.info.
3   www.nikolasander.com/uk2011census.
4   Available at: https://cran.r-project.org/web/packages/circlize/index.html.
5   Available at: https://github.com/null2/circular-migration-plot.
6   See www.global-migration.info.

References

Abel, G.J. and Sander, N. (2014) Quantifying global international migration flows. *Science*, 343(6178): 1520–1522.

Bostock, M.V., Ogievetsky, V. and Heer, J. (2011) D3 data-driven documents. *IEEE Transactions on Visualization and Computer Graphics*, 17(12): 2301–2309.

Charles-Edwards, E., Wilson, T. and Sander, N. (2015) Visualizing Australian internal and international migration flows. *Regional Studies, Regional Science*, 2(1): 432–434.

Dennett, A. (2015) Visualising migration: Online tools for taking us beyond the static map. *Migration Studies*, 3(1): 143–152.

Krzywinski, M., Schein, J., Birol, I., Connors, J., Gascoyne, R., Horsman, D., Jones, S.J. and Marra M.A. (2009) Circos: An information aesthetic for comparative genomics. *Genome Research*, 19(9): 1639–1645.

Minard, C.J. (1869) Carte figurative des pertes successives en hommes de l'Armée Française dans la campagne de Russie 1812–1813, Paris. Available at: http://en.wikipedia.org/wiki/File:Minard.png.

Rees, P., Bell, M., Kupiszewski, M., Kupiszewska, D., Ueffing, P., Bernard, A., Charles-Edwards, E. and Stillwell, J. (2016) The impact of internal migration on population redistribution: An international comparison. *Population, Space and Place*, published online, DOI: 10.1002/psp.2036.

Sander, N., Abel, G.J., Bauer, R. and Schmidt, J. (2014) Visualising migration flow data with circular plots. *Working Paper No. 2/2014*, Vienna Institute of Demography, Vienna.

Stillwell, J., Bell, M., Blake, M., Duke-Williams, O. and Rees, P.H. (2000) A comparison of net migration flows and migration effectiveness in Australia and Britain, 1976–96. Part 1, Total migration patterns. *Journal of Population Research*, 17(1): 17–41.

Stillwell, J., Lomax, N. and Sander, N. (2015) Monitoring and visualising sub-national migration trends in the United Kingdom. In Geertman, S., Ferreira, J., Godspeed, R. and Stillwell, J. (eds.) *Planning Support Systems and Smart Cities*. Springer, Dordrecht, pp. 427–445.

Part V

Using 2011 Census data for research

17

Creating a new open geodemographic classification of the UK using 2011 Census data

Chris Gale, Alex Singleton and Paul Longley

17.1 Introduction

Geodemographics formalises the relationship between people and place and is "the analysis of people by where they live" (Sleight, 2004, p. 16). Social similarity, independent of locational proximity, and the spatial autocorrelation of like-minded individuals are the core components on which geodemographics is based. Instead of dealing with individuals, geodemographics acts to simplify complex population characteristics into distinct groups whilst maintaining a spatial dimension. A geodemographic classification is therefore a mechanism for distilling large multi-variate datasets into succinct descriptions of the population, thereby providing summaries of the social, economic, demographic and built characteristics of small geographical areas (Adnan et al., 2010). Within the context of a large multivariate dataset like the census, this mechanism provides a way to delineate population variations that occur across the country.

The UK has a history of producing geodemographic classifications based on census outputs. The earliest published work focused on Liverpool, but was later expanded to give national coverage (Webber and Craig, 1976, 1978; Webber, 1977). Similar classifications were also created for the 1981 (Charlton et al., 1985) and 1991 Censuses (Blake and Openshaw, 1994, 1995). The 2001 Output Area Classification (2001 OAC) (Vickers and Rees, 2007) was based on 2001 Census data and took advantage of these data being made available with open access, as opposed to licensed distribution. This made the 2001 OAC the first UK classification to be entirely open, both in terms of data availability and the methodology used. Outside of academia, commercial organisations have been developing their own classifications over a number of decades. These are typically characterised by their use of ancillary sources in conjunction with census data (Sleight, 2004; Harris et al., 2005), thereby allowing topics not covered within the traditional census to be incorporated, such as income.

The release of data from the latest census provided an opportunity to build on the success of the 2001 OAC and produce a new classification for the UK. Work on this new classification was carried out at University College London (UCL) as a knowledge exchange activity co-sponsored by the Office for National Statistics (ONS). A requirement of ONS was that the new classification should be an evolution of the 2001 OAC, using the methodology outlined in Vickers and Rees (2007) as a guide. The direction from ONS was that, as the 2001 OAC had been well received by a broad user base, the retention of some of its methodological components would be prudent. Similar approaches are not uncommon with commercial geodemographic systems; thus, it would be correct to describe development of the new classification as evolutionary rather than revolutionary.

The design for the new classification aimed to take advantage of developments in geodemographics over the past decade, along with advances in computer software and processing power to address some methodological issues identified with the 2001 OAC (Vickers and Rees, 2011), whilst also providing methodological advances. Greater focus was placed upon user requirements, data selection and transformation issues, computational methods, and new methods of online visualisation and dissemination. Key to this was the testing and evaluation of multiple methodological approaches, with particular emphasis on exploring how interactions between different methods and techniques influenced the final cluster solutions; incorporating the needs of those who use open geodemographic systems, as voiced through a user engagement exercise; and utilising the expanded range of outputs provided by the 2011 Census. Predicating this was the decision to use only open source software and to release all outputs of the classification (such as code), once officially released by ONS.

Alongside the methodological advances, the overarching aims of this new classification, as with the 2001 OAC, remained: to describe the salient and multi-dimensional characteristics of small areas across the UK, as represented by the 2011 Census; to provide a usable classification, with a fully transparent and reproducible methodology; and to provide an alternative to commercial geodemographic classifications, with the final product being designated a key output of the 2011 Census by ONS.

17.2   Measuring socio-spatial structure

There are numerous methods of creating geodemographic classifications, differentiated by choice of input data, normalisation and standardisation procedures, clustering methods and development of associated descriptive 'pen portrait' materials. Building a geodemographic classification can be considered as both art and science (Harris et al., 2005), guided by user requirements, data content and coverage, and the choices and predilections of the classification builder (Webber, 1977; Harris et al., 2005). A wide range of methodological approaches are nevertheless successful in producing usable classifications, and we do not prescribe any particular general-purpose model: rather, we recommend user testing and evaluation of multiple classifications relative to application contexts (Brunsdon et al., 2011).

Although the 2011 Census recorded the status of the UK population as of 27 March 2011, it took two years before the small area statistics were released for England, Wales and Northern Ireland, and a further nine months before Scottish small area data were released, all following data assembly and checking. This period formed the planning phase for the 2011 Area Classification for Output Areas (2011 OAC), including a stakeholder consultation exercise and design evaluation, both carried out at UCL in collaboration with ONS. The results were used to guide the methodological approach, although specific details were only finalised after the data were released and exploratory empirical analysis had been conducted.

The first stage of the user consultation entailed a pilot study with ONS and the Demographics User Group, a consortium of retail organisations that, *inter alia*, lobbies government to open up access to data. An online self-completion questionnaire was also promoted through specific end-user websites and mailing lists. This comprised 20 questions on a range of topics about the format, methodology and outputs of a future classification for the 2011 Census. It was made publicly available for a six-week period from 17 February 2012 to 30 March 2012. The survey

Table 17.1 Summary of the OAC user survey responses (ONS, 2012a)

General

- Better discrimination *vis-à-vis* commercial geodemographics; especially within London and rural areas
- Some users are deterred from using commercial packages due to the cost
- The open source approach is viewed as a positive attribute within certain application areas

Methods

- Creating the new classification at the smallest areal level was preferable
- Integrating census data with wider open data sources was considered desirable
- No need for the new classification to be directly comparable with the 2001 OAC
- A general-purpose classification favoured
- No consensus on an appropriate geographic extent for the new classification

Outputs

- Better promotion of the new classification to stakeholder groups
- User-friendly outputs should be included, such as pictures and pen portraits
- Descriptive graphs and written material useful in providing greater understanding of the classification
- Interactive maps considered a useful enhancement to the classification
- Cluster naming very important
- A measure of uncertainty for cluster assignment desired as an additional product

generated 38 responses from a mixture of local and central government organisations, primary care trusts, other public sector organisations, consultancies, commercial organisations and academics. Summary findings are presented in Table 17.1, and the full survey was published by ONS in May 2012 (ONS, 2012a). The results of the survey helped to inform the classification methodology, and user requirements for an updated OAC.

The process of building the 2011 OAC involved selection of an initial set of input data, standardisation, refining of input data based upon correlation and sensitivity analysis and, finally, clustering to form the geodemographic classification. Once complete, the clusters were then described with a range of illustrative material.

This is a modification to the 2001 OAC methodology, with sensitivity analysis integrated into the variable selection process. It was felt that this was necessary to provide an improved assessment of the variables driving the geodemographic patterns in addition to reducing data redundancy. For the 2011 OAC, a total of 232,296 areal units required clustering, comprising: 181,408 output areas (OAs) in England and Wales; 46,351 OAs in Scotland; and 4,537 small areas (SAs) in Northern Ireland. The majority of the operations undertaken were performed in the statistical programming language *R* (*R* Development Core Team, 2011). The packages and code used are available through www.opengeodemographics.com.

17.3 Initial variable selection

A key requirement identified from the 2011 OAC user engagement exercise was that the new geodemographic classification should be based solely on 2011 Census data. Although there was recognition of the potential for additional open data sources to be utilised in conjunction with the census data, this was not pursued for a number of reasons. First, it was felt that the dearth of open data available at the OA or SA level would create too many compatibility issues with the census data, given that sources with complete UK coverage were only available at Lower Layer Super Output Area (LSOA) or equivalent, or higher levels of granularity. This decision should not, however, preclude use of open data in extended models, for example, in constrained geographic extents (e.g. a local authority) or bespoke domain specific applications (Singleton and Longley, 2009).

Outputs of the 2011 Census at the OA and SA level were released in stages by ONS for England and Wales, National Records of Scotland (NRS) for Scotland and the Northern Ireland Statistics and Research Agency (NISRA) for Northern Ireland. The outputs that were most appropriate for creating the 2011 OAC were those provided in univariate format. These data were less likely to be impacted by the perturbations linked with disclosure control methods applied to the 2011 Census (ONS, 2012b), where values were modified if the raw data could be used to identify specific households, families or businesses. For the 2011 Census data, these were identified by the statistical bodies as Key and Quick (univariate) Statistics, with Key Statistics providing summaries, and Quick Statistics being more detailed. ONS released 35 tables as Key Statistics and 73 tables as Quick Statistics for England and Wales, NISRA released 45 tables as Key Statistics and 58 tables as Quick Statistics for Northern Ireland, and, by December 2013, NRS had released 34 tables as Key Statistics and 59 tables as Quick Statistics for Scotland. The combined dataset for England and Wales contained 2,139 variables, in Scotland there were 1,326 and in Northern Ireland 1,378. Only variables that were consistent across the whole UK were considered for use in creating the 2011 OAC; however, within this reduced dataset, there were numerous cases of duplicated variables. For example, tables KS101EW and QS101EW contained identical data regarding the usual resident population of England and Wales.

A method by which high dimensionality has been managed elsewhere is through the application of principal components analysis (PCA), as in the post-1981 SuperProfiles classifications (Charlton et al., 1985). PCA is applied to a collection of individual variables, which are reduced to a series of linearly uncorrelated component scores. Historically, such processing has been necessary because the clustering of large multi-dimensional datasets was constrained by available computer technologies of the time. For the 2001 OAC, Vickers (2006) discounted using PCA because clustering from an entire dataset was seen as preferable, and advances in computational power meant that actual data points, rather than principal components, could be used. Following these same arguments, PCA was also not used for the 2011 OAC.

The objective of the 2011 OAC variable selection process was to obtain the smallest subset of variables that captured the main variations within the 2011 Census, with the same technique being used by other previous classifications (Bailey et al., 1999, 2000). As with the 2001 OAC, candidate variables were classified as belonging to one of five domains that aimed to best represent drivers of socio-spatial differentiation: demographic structure; household composition; housing; socio-economic; and employment (Vickers and Rees, 2007). Keeping the domains consistent with the 2001 OAC allowed for a similar variable structure to be maintained for the 2011 OAC, without restricting the final variable selection to being merely a replication of the previous classification. This also accommodates a key finding of the 2011 OAC user engagement exercise, which highlighted that, although broad comparability was desirable, a like-for-like replacement of the 2001 OAC was not. The output of this initial variable filter was 167 prospective variables that were used as the basis for the final attribute selection. This variable set aimed to assure coverage over the pre-identified domains. The 94 variables initially considered for the 2001 OAC (Vickers and Rees, 2007) guided the selection, whilst also accommodating the expanded number of outputs available from the 2011 Census, such as the increased number of ethnic group categories. Finally, the suitability of the 167 prospective variables to provide the basis of the new classification was assessed by ONS.

17.4   Attribute prevalence and standardisation

When building a geodemographic classification, it is important for all variables to be measured on the same scale, and to control for underlying population structure. This requires a process of rate and/or ratio calculation, followed by standardisation of these measures onto a single scale, as mixed scales can afford undue influence to attributes with larger ranges of values. Furthermore, as was the case with the 2001 OAC, a normalisation process may also be applied to all variables, if non-normality is viewed as counter to effective cluster formation.

The two methods of rate calculation that are most commonly used in the creation of a geodemographic classification are percentages or index scores. These present conceptually different approaches to classification: percentages compare areas based on rates for a particular attribute, whilst index scores compare areas based upon how different they are from the national average. A third option considered was the calculation of mean differences, to identify variables with the greatest deviation away from average characteristics. Our own view is that percentages are more appropriate when, as with the 2001 OAC, the objective is to compare similarities between places rather than to measure how places deviate from a global average, which is important where many attributes are known to be regionalised. Thus, percentages were calculated for all of the prospective attributes. Two additional candidate measures from the 2011 Census were added to the list of potential attributes: population density as a proxy to distinguish between urban and rural areas and a Standardised Illness Ratio (SIR) measure to compare observed illness counts within an area relative to expected values accounting for underlying age structure. The age structure of an area can have a significant impact on recorded illness rates. Areas that contain a high concentration of older individuals are more likely to be associated with high illness rates than areas containing a high proportion of younger people if values are not standardised.

Histograms were used to explore the degree to which the attribute data were normally distributed prior to clustering, with the majority of variables exhibiting varying degrees of skew. Kumar et al. (2014) note that the k-means clustering algorithm can be adversely affected by skewed data distributions, because Euclidean distances are calculated between points and cluster centroids, meaning the algorithm is optimised to find spherical groupings of cases with similar attributes (Jain, 2010). There are, however, differing views as to the optimal method. Harris et al. (2005) describe how commercial geodemographics seek to accommodate skewness through weighting, although the inherent subjectivity of weighting assignment led to the rejection of this approach with the 2001 OAC. Singleton and Spielman (2014) concur with this broad approach, but nevertheless caution that global normalisation may smooth away interesting local patterns, albeit that some such patterns may reflect the effects of outlying observations. The exploratory analysis conducted for the 2011 OAC identified a large number of outliers towards the high end of the value scale, notably concerning population density. Thus, a number of different transformations were investigated and applied to the prospective attributes, including log10, Box–Cox and inverse hyperbolic sine.

The Box–Cox and log10 transformations both require values to be positive and greater than 1 to effect transformation. There are different ways of managing this issue, the most common of which is to add a constant to all values prior to normalisation method (Osbourne, 2002). Log10 transformations artificially reduce the amount of variance to that of the normal distribution (Leydesdorff and Bensman, 2006). In essence, this method compresses the differences between the larger values whilst increasing those between the smaller values. A disadvantage of this approach is that the transformation is fixed across a dataset without sensitivity to different distributions that may appear between variables. An alternative method more sensitive to these issues is the Box-Cox transformation, which can be defined as:

$$x_i(\lambda) = \begin{cases} \left(x_i^\lambda - 1\right) / \lambda \left(\lambda \neq 0\right) \\ \log(y)(\lambda = 0) \end{cases} \tag{17.1}$$

An exponent, lambda (λ), transforms a variable (x) into a normal distribution (Box and Cox, 1964). Multiple values for lambda are tested, and the one that produces the most normal result is selected. There are numerous tests of normality that can be applied, and for the implementation here we used the Shapiro–Wilk test. A lambda value of 0 is mathematically equivalent to log10 transformation. As such, the implementation of the Box–Cox technique calculates a separate lambda value for each variable. The extent to which a variable is transformed will depend, however, on how skewed its distribution is, rather than a global skewness value (Dag et al., 2014).

A third approach does not require the addition of a constant and takes the form of the inverse hyperbolic sine (IHS), defined as $\log(x_i + (x_i^2 + 1)^{1/2})$. The IHS proposed by Johnson (1949) shares similarities with the standard log transformation, except that it can be defined at zero or for negative numbers (Burbidge et al., 1988). As a result, the technique is often favoured when transforming wealth datasets (Pence, 2006). One disadvantage relative to the Box–Cox transformation is that IHS, like log10, still maintains a single transformation applied uniformly to all attributes. In practice, our exploratory analysis revealed little difference in normality between the three procedures and, as such, IHS was adopted.

Finally, prior to final selection of the input attributes, the normalised dataset required standardisation, placing each variable onto the same scale. There are a number of different methods through which this can be achieved, and, as with the 2001 OAC, we evaluated z-scores, range standardisation and inter-decile range standardisation. Z-scores transform raw scores in terms of the number of standard deviations they are away from the mean, which can have the effect of assigning high leverage to outlying observations in the data. Our view was that it would not be helpful to the clustering procedure for a national general-purpose classification to comprise clusters formed from such outlying points and, as such, after extensive exploratory analysis, this approach was dropped.

Range standardisation compresses the values in a dataset into the range of 0 to 1. As a consequence, any outliers present within the dataset get compressed to fit within this 0 to 1 interval, potentially suppressing some interesting patterns within the data. Because the 2011 Census data contain numerous zero counts, this meant that most outliers in the dataset were found at the other end of the scale, but are confined to a much smaller range of values. The range standardisation method was used when creating the ONS 1991 classification of local authorities (Wallace and Denham, 1996) and for the 2001 OAC (Vickers and Rees, 2007). The inter-decile range standardisation method is a variant of range standardisation, in which the data are standardised over a smaller range, between the ninetieth percentile and the tenth percentile, in order to reduce the impact of outliers on the standardised data. Following extensive exploration of the 2011 Census data, we concurred with the finding of Vickers and Rees (2007) that this method gave too much weight to skewed variables.

In view of the issues with outliers associated with z-scores, and skewed variables within the inter-decile range, the 2011 OAC opted to follow the method of the 2001 OAC by implementing range standardisation on all variables. This and the decision to use percentages and IHS to create the 2011 OAC were informed by the experience of the classification building team. Such decisions were informed by exploratory analysis, and were used to understand how each proposed method of rate conversion, transformation and standardisation would influence the final classification and guided the judgements that were made.

17.5   Final attribute selection and testing

The previous section returned a dataset of standardised prospective variables generated through a process of percentage calculation, inverse hyperbolic sine and then range standardisation. Further analysis was required to select a final set of attributes prior to cluster analysis. The process of selecting the final variables took into consideration two key requirements. The first was for the 2011 OAC to be a general-purpose geodemographic classification. This required inclusion of a collection of variables that reflected the general characteristics of the population, whilst also emphasising characteristics that varied between OAs in order for distinct clusters to form. The second requirement was inclusion of the minimum number of variables in order to limit any potential weighting effect arising from co-linearity within the final list of inputs. The process of reducing the prospective variables to a final list of inputs was conducted with the overarching aim of retaining those variables that were likely to be the most important for the 2011 OAC, and removing those that would only add limited information to the final assignment of areas into clusters. Two empirical methods were utilised to guide the selection of final inputs for cluster analysis. The first explored variable correlation, and the second implemented a clustering-based sensitivity analysis technique based on total within cluster sum of squares (WCSS) values to give an indication of those variables that are important when forming output clusters.

It can be expected within any large multi-dimensional dataset that there will be an element of correlation between some variables. Correlated variables act in a similar way to a weight, giving added prominence to a shared dimension. Although most correlated variables were removed from the 2001 OAC, some which were considered highly correlated were retained for use. This was done in order to add predictive and descriptive power to the classification. There were three options available after identifying highly correlated variable pairs: remove the variable from the final selection; group it with other variable(s) to create a new composite variable; or ignore it. For the prospective variables, a Pearson correlation matrix was created, and those variables that had correlation of greater than 0.6 or less than -0.6 were examined. In total, 104 variables had correlations in excess of these values with at least one other variable within the dataset. Table 17.2

Table 17.2 Intercorrelation frequency of the 2011 OAC prospective variables

Variable name	Number of variables with intercorrelation values greater than 0.6 or less than -0.6
Households that only contain persons aged over 16 who are living in a couple and are married	17
Persons aged over 16 who are single	16
Persons aged over 16 who are married	16
Mean age	14
Persons whose country of birth is the United Kingdom	14
Households that only contain persons aged over 16 who are not living in a couple and are single (never married or never registered a same-sex civil partnership)	14
Households with two or more cars or vans	14
Median age	13
Persons who are White British and Irish	13
Households who have two or more rooms than required	13

is a summary of the ten variables that showed the greatest correlation with the other prospective variables. The three potential options to address these intercorrelations were carefully considered, with the most appropriate action dependent on the importance of each variable to the classification.

A second consideration before final attribute selection was identification of those variables that would have the greatest impact on cluster formation. Clustering of the 2011 OAC was completed using the previously discussed k-means algorithm (MacQueen, 1967) and maintains broad comparison with the 2001 OAC. This algorithm initialises with k 'seeds' randomly placed within the multi-dimensional attribute space of the input dataset, and the OAs and SAs are assigned to their closest seed, thereby creating an initial cluster assignment. Cluster centroids are then recalculated as the average of the attribute values for all data points assigned to each cluster. The data points are then reassigned if they become closer to the new cluster centroids. This process is repeated until no further data points move between clusters, thus meeting the convergence criterion.

This process aims to create clusters that are as homogenous as possible with variability within each cluster minimised, whilst the clusters themselves are more heterogeneous based on the initial seed assignment. Given the initialisation procedures, the k-means algorithm is stochastic and, as such, multiple runs are required with randomly allocated initial seeds. The within cluster (WCSS) and between cluster (BCSS) sums of squares are calculated as part of the clustering process. These values indicate how tightly clustered a particular dataset is. The WCSS and BCSS values signal how close objects within each cluster are to their centroids, thereby providing an indication of cluster homogeneity. The BCSS value measures the distances between the clusters and, as such, quantifies how similar they are to each other. In constructing the 2011 OAC, we were guided by the WCSS statistic, given our emphasis upon identifying homogenous groups within the UK's population rather than ensuring that clusters were as dissimilar to each other as possible.

At this stage, an optimal number of k clusters for the classification was not known, and k = 5 to k = 9 were selected for use in the sensitivity analysis as these corresponded to the most aggregate level of other UK geodemographic classifications, although these numbers were otherwise arbitrary. For each value of k, clustering was optimised (10,000 runs of k-means with the lowest WCSS selected) where a different variable was held back each time. This enabled identification of those individual attributes that impacted cluster performance by examining the WCSS and BCSS values. Although the tests were run for k = 5 to k = 9, the results identified similar variables having impact on the clustering performance. As an example, for k = 8, a summary of the variables that influence cluster homogeneity the most and least when removing each of the variables in turn from the 'percentage calculation, inverse hyperbolic sine and range standardisation' dataset is shown in Table 17.3. Where the removal of a variable resulted in a marked decrease in the WCSS value and a larger BCSS value, this suggested that the omission of that variable would result in more homogenous clusters and greater heterogeneity between clusters. Whilst this indicated that the variable should be discarded, the decision whether or not to do so was dependent on the importance of the variable in capturing key characteristics of the 2011 Census. Comparatively small decreases in the WCSS value and increases in the BCSS value for a variable suggested that the impact on the homogeneity of the clusters and heterogeneity between clusters was minimal, and therefore it was more likely that the variable would be retained.

The results from the correlation and sensitivity analysis provided a greater understanding of how individual variables would affect the final classification. However, these results were

Table 17.3 Variables that influence cluster homogeneity the most and least based on WCSS values in the percentage calculation, inverse hyperbolic sine and range standardisation dataset

Variables that influence cluster homogeneity the most

- Households who live in a terrace or end-terrace house
- Households who are private renting
- Persons whose main language is not English but can speak English well
- Households who live in a flat
- Households who live in a semi-detached house or bungalow
- Households who live in a detached house or bungalow
- Households with lone parent in part-time employment
- Employed persons aged between 16 and 74 who work in the information and communication industry
- Employed persons aged between 16 and 74 who work in the financial and insurance activities industries
- Households with different ethnic groups within partnerships (whether or not different ethnic groups between generations)

Variables that influence cluster homogeneity the least

- Day-to-day activities limited a lot or a little
- Persons aged over 16 who are married
- Persons aged 30 to 44
- Persons aged over 16 whose highest level of qualification is Level 1, Level 2 or Apprenticeship
- One person households: Aged 65 and over
- Median age
- Persons in fair health
- One person ethnic household
- Households who have one more room than required
- Persons aged between 16 and 74 who are economically active: Part-time employee

only a secondary consideration for the final variable selection, with the primary focus upon ensuring the relevant population and household structure was captured. This meant a variable that could be empirically shown to have a negative impact on the global homogeneity of clusters was still considered if it represented a key facet of the five census data domains identified by Vickers and Rees (2007). For example, Table 17.3 highlights the negative impact that certain housing variables have on the clustering process. It was, however, essential that a selection of these variables was included in the final classification to represent the built environment of the UK.

In total, from the 167 prospective variables: 86 were removed; 41 were retained without being combined with other variables; and 40 were merged to create 19 composite variables. This created a final set of 60 variables. Composite variables were created by combining variables which shared the same denominator, most commonly to reduce inter-correlation within the dataset. A full explanation of why variables were removed, merged or retained is given in Gale (2014).

The total of 60 variables represents a 46 per cent increase compared to the 41 variables used by the 2001 OAC, and offers additional dimensions to differentiate the UK's population. For example, an additional age variable is included to split those aged over 65 into two categories, to reflect the UK's ageing population (ONS, 2012c). Furthermore, the inclusion of a communal establishment indicator with the included age variables aims to make a distinction between areas where older members of society live independently *vis-à-vis* those who live within communal facilities such as residential care homes.

A greater range of demographic variables has also been included in the 2011 OAC. These provide a more in-depth perspective on individuals' ethnic background, country of birth and ability to speak the English language. Finally, more education variables have been incorporated, along with an expanded number of employment industry types. A full list of the 60 variables is given in ONS (2014), with additional accuracy of the finalised dataset assured through rigorous testing and validation by ONS prior to clustering.

17.6   Assessing similarity between areas

As with the 2001 OAC, the intention was to build a hierarchical classification to provide greater flexibility for potential applications. A top-down approach was adopted, whereby clusters were created for the most aggregate level of the classification, and then used to subdivide the input data, which were then successively clustered separately, forming the hierarchical levels of the classification. It was argued with the 2001 OAC that this method was favourable because the objects of study (OAs and SAs) are always clustered, rather than cluster centroids, as might be the case with 'bottom-up' methods implemented using alternate hierarchical algorithms such as Ward's clustering. This is important because cluster centroids will only rarely be representative of an entire cluster. Going up a hierarchical level and clustering using these centroids can create clusters containing objects with little in common, and thus result in clusters with low homogeneity.

The 2001 OAC adopted a three-tiered hierarchical approach, and the cluster frequency at each level was deemed satisfactory by the 2011 OAC user engagement. However, there was also an expectation that the 2011 classification would not necessarily have identical numbers of clusters at each level. The 2011 OAC therefore aimed to have similar, but not necessarily identical numbers in a three-tiered structure with the Supergroup, Group, and Subgroup terminology being retained.

There is no overall consensus as to how to structure a geodemographic classification. Singleton and Spielman (2014) point out that three-tier hierarchies tend to predominate in UK national classifications, albeit with different number of clusters per level. With no common methodological approach, the number of clusters required to best represent a population is unique to every classification; and deciding upon the optimum number of clusters is an inherently subjective process (Singleton and Longley, 2009). For the 2001 OAC, cluster numbers were decided upon following personal consultations with experts (cf. Vickers et al., 2005; Vickers, 2006). Decisions made with the 2001 OAC were therefore a balance between the ideal number of clusters identified by the experts, aligned with maximising cluster integrity and ensuring reasonably even cluster size.

Identifying the best solution for the 2011 OAC thus entailed exploratory analysis. Different k seeds were used to identify cluster solutions that differentiated between clusters, provided a scattered geographic distribution across the UK, and also gave an even assignment of OAs and SAs across all levels of the hierarchy. To ensure a similar hierarchical structure to the 2001 OAC, the top level of the hierarchy was examined for five to nine cluster options, whilst the middle and bottom levels had two to four. Each cluster analysis was repeated 10,000 times in line with other optimised geodemographics, and the iteration with the lowest WCSS, and therefore the most homogeneous clusters, was selected for each option (Singleton and Longley, 2009). The final selected classification had eight Supergroups, 26 Groups and 76 Subgroups.

17.7   Describing UK geodemographic patterns

Cluster names and descriptions are an important aspect of the user interface of geodemographic classifications, and are intended to represent the underlying complexity of the cluster compositions. Names and short 'pen portrait' descriptions were therefore developed for the final clusters making up each level of the 2011 OAC hierarchy. Vickers et al. (2005) noted that names and descriptions of clusters may be contentious, especially if they reinforce negative stereotypes. Additionally, the procedure of conflating individual variables with one another opens up the risk of ecological fallacy within the analysis (Robinson, 1950). However, past experience in both the private and public sectors confirms that this procedure does help end-users to identify with the names and descriptions given to local areas.

Given the predominant public sector usage that was anticipated in creating the 2011 OAC, words and phrases that might be construed as pejorative were not used, and all descriptors had strong and literal links to the underlying distributions revealed by the data. Words that were overtly negative or positive were avoided where possible, and the abiding sense of the descriptors was to present the characteristics of each cluster as consequences of factors that have happened to areas rather than the consequences of human agency. This was consistent with avoiding value judgements when assigning cluster names and descriptions.

Names and descriptions of clusters are, of course, based on the characteristics of their centroids, and the workings of the cluster assignment procedure. This means that specific areas differ in the degrees to which they conform to these average characteristics. Attribute range statistics were thus used to avoid descriptors that pertained only to the OAs and SAs that were closest to their assigned cluster's centroid. This consideration was balanced by avoiding names that are too broad and too vague to offer any insight into an area's attributes. Additional checks were undertaken to ensure that names, where possible, did not duplicate those used by previous commercial and non-commercial geodemographic classifications, and final approval of the cluster names was sought from ONS, who conducted their own internal review and consultation.

The final names for the 2011 OAC are shown in Table 17.4, and Figures 17.1 and 17.2 present an illustration of the 2011 OAC for the city of Brighton and Hove. The mapping of clusters allowed for internal validation, where names were checked against local knowledge of areas. For example, Brighton and Hove has one of the largest concentrations of 'Cosmopolitans' in any urban area in the UK. In total, 44 per cent of its OAs belong to that Supergroup. The spatial distribution of 'Cosmopolitans' shown in Figure 17.1 is, however, limited to the central areas of the city, with an exception being the University of Sussex campus to the north east. The population characteristics of those assigned to the 'Cosmopolitans' Supergroup suggest these areas are more likely to contain a mixture of affluent workers and students, which corresponds to areas of Brighton and Hove known for containing high concentrations of such individuals.

At the Group level of the 2011 OAC, however, it is evident there is a clear spatial divide between the student and non-student populations. Figure 17.2 illustrates the Groups derived from the 'Cosmopolitans' Supergroup and shows student populations are, unsurprisingly, more likely to be found in close proximity to the campuses and buildings of the Universities of Sussex and Brighton. The non-student populations are more likely to be found in affluent areas of the city, notably Brighton Marina and areas of Hove. Again, the concentrations of these populations in these areas is a reflection of reality. The example of Brighton and Hove, with similar exercises undertaken in London, Southampton and Bristol, indicates known neighbourhood characteristics can be identified with the 2011 OAC. However, a fuller validation exercise, similar to that undertaken by Vickers and Rees (2011), would be required to evaluate the true representation that the 2011 OAC provides of the UK population.

223

Table 17.4 2011 OAC cluster names and hierarchy

Supergroups	Groups	Subgroups
1 – Rural Residents	1a – Farming Communities	1a1 – Rural Workers and Families
		1a2 – Established Farming Communities
		1a3 – Agricultural Communities
		1a4 – Older Farming Communities
	1b – Rural Tenants	1b1 – Rural Life
		1b2 – Rural White-Collar Workers
		1b3 – Ageing Rural Flat Tenants
	1c – Ageing Rural Dwellers	1c1 – Rural Employment and Retirees
		1c2 – Renting Rural Retirement
		1c3 – Detached Rural Retirement
2 – Cosmopolitans	2a – Students Around Campus	2a1 – Student Communal Living
		2a2 – Student Digs
		2a3 – Students and Professionals
	2b – Inner-City Students	2b1 – Students and Commuters
		2b2 – Multicultural Student Neighbourhoods
	2c – Comfortable Cosmopolitans	2c1 – Migrant Families
		2c2 – Migrant Commuters
		2c3 – Professional Service Cosmopolitans
	2d – Aspiring and Affluent	2d1 – Urban Cultural Mix
		2d2 – Highly Qualified Quaternary Workers
		2d3 – EU White-Collar Workers
3 – Ethnicity Central	3a – Ethnic Family Life	3a1 – Established Renting Families
		3a2 – Young Families and Students
	3b – Endeavouring Ethnic Mix	3b1 – Striving Service Workers
		3b2 – Bangladeshi Mixed Employment
		3b3 – Multi-Ethnic Professional Service Workers
	3c – Ethnic Dynamics	3c1 – Constrained Neighbourhoods
		3c2 – Constrained Commuters
	3d – Aspirational Techies	3d1 – New EU Tech Workers
		3d2 – Established Tech Workers
		3d3 – Old EU Tech Workers
4 – Multicultural Metropolitans	4a – Rented Family Living	4a1 – Social Renting Young Families
		4a2 – Private Renting New Arrivals
		4a3 – Commuters with Young Families
	4b – Challenged Asian Terraces	4b1 – Asian Terraces and Flats
		4b2 – Pakistani Communities
	4c – Asian Traits	4c1 – Achieving Minorities
		4c2 – Multicultural New Arrivals
		4c3 – Inner-City Ethnic Mix

Supergroups	Groups	Subgroups
5 – Urbanites	5a – Urban Professionals and Families	5a1 – White Professionals 5a2 – Multi-Ethnic Professionals with Families 5a3 – Families in Terraces and Flats
	5b – Ageing Urban Living	5b1 – Delayed Retirement 5b2 – Communal Retirement 5b3 – Self-Sufficient Retirement
6 – Suburbanites	6a – Suburban Achievers	6a1 – Indian Tech Achievers 6a2 – Comfortable Suburbia 6a3 – Detached Retirement Living 6a4 – Ageing in Suburbia
	6b – Semi-Detached Suburbia	6b1 – Multi-Ethnic Suburbia 6b2 – White Suburban Communities 6b3 – Semi-Detached Ageing 6b4 – Older Workers and Retirement
7 – Constrained City Dwellers	7a – Challenged Diversity	7a1 – Transitional Eastern European Neighbourhoods 7a2 – Hampered Aspiration 7a3 – Multi-Ethnic Hardship
	7b – Constrained Flat Dwellers	7b1 – Eastern European Communities 7b2 – Deprived Neighbourhoods 7b3 – Endeavouring Flat Dwellers
	7c – White Communities	7c1 – Challenged Transitionaries 7c2 – Constrained Young Families 7c3 – Outer City Hardship
	7d – Ageing City Dwellers	7d1 – Ageing Communities and Families 7d2 – Retired Independent City Dwellers 7d3 – Retired Communal City Dwellers 7d4 – Retired City Hardship
8 – Hard-Pressed Living	8a – Industrious Communities	8a1 – Industrious Transitions 8a2 – Industrious Hardship
	8b – Challenged Terraced Workers	8b1 – Deprived Blue-Collar Terraces 8b2 – Hard-Pressed Rented Terraces
	8c – Hard-Pressed Ageing Workers	8c1 – Ageing Industrious Workers 8c2 – Ageing Rural Industry Workers 8c3 – Renting Hard-Pressed Workers
	8d – Migration and Churn	8d1 – Young Hard-Pressed Families 8d2 – Hard-Pressed Ethnic Mix 8d3 – Hard-Pressed European Settlers

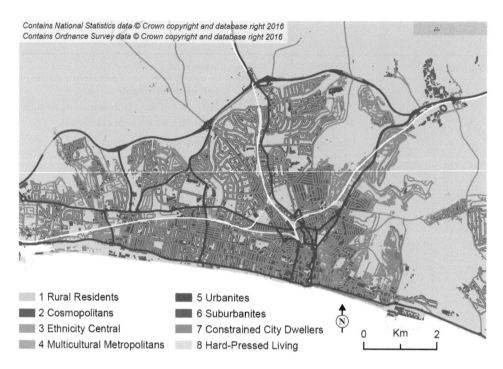

1 Rural Residents   5 Urbanites

2 Cosmopolitans   6 Suburbanites

3 Ethnicity Central   7 Constrained City Dwellers

4 Multicultural Metropolitans   8 Hard-Pressed Living

N   0   Km   2

Figure 17.1 The 2011 OAC Supergroups in Brighton and Hove and the surrounding areas

Source: National Statistics and OS data © Crown copyright and database right 2016.

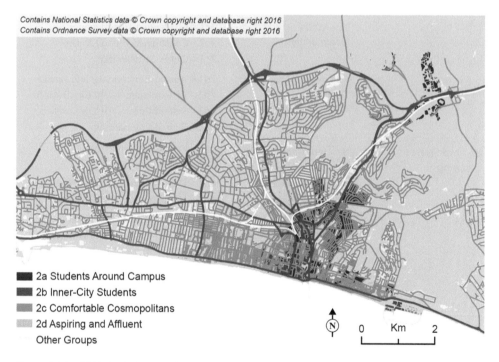

2a Students Around Campus

2b Inner-City Students

2c Comfortable Cosmopolitans

2d Aspiring and Affluent

Other Groups

N   0   Km   2

Figure 17.2 2011 OAC Groups derived from the 'Cosmopolitans' Supergroup in Brighton and Hove and the surrounding areas

Source: National Statistics and OS data © Crown copyright and database right 2016.

17.8   Conclusion

This chapter has outlined the process that underpinned the creation of the 2011 OAC. As with all geodemographic classifications, it is inevitable that subjective decisions are made based upon our own accumulated experience, and in this context 'best' should be thought of in terms of creating a general-purpose classification that fulfils utilitarian objectives – of use to the greatest number of people and the largest number of applications.

Creating geodemographic classifications can lean heavily on past methodologies or be created with new techniques utilising methods and procedures never previously applied in the field. Ultimately, however, whatever the techniques used, the aim is the same: to create a geodemographic classification that summarises the varying characteristics of the study population and built environment. The methodology for the 2011 OAC is an evolution of that used for the 2001 OAC, with focus on the testing and evaluation of multiple methodological approaches forming a core component of this. This takes on board the views expressed by current and past users of the 2001 OAC in terms of the structure, outputs and general characteristics of the classification. However, the aspirations of openness with the 2011 OAC far exceed that of its predecessor. This has included the use of open source software such as R and the dissemination of data and materials, including the code created being uploaded to GitHub.[1] It is hoped that this resource will aid others to build their own bespoke classifications, such as that already done for London (Longley and Singleton, 2014), and with workplace data (Cockings et al., 2015).

The practices of geodemographic classification continue to evolve. The primary motivation for the 2001 OAC was to demonstrate the feasibility of creating a free open geodemographic classification at a fine level of spatial granularity. The 2011 OAC extends this methodology whilst incorporating a greater level of openness and transparency than its predecessor. As such, the 2011 OAC can be considered an advance in open geodemographics in the UK.

Acknowledgements

This research was sponsored by the Office for National Statistics under a UCL 'Impact' PhD studentship, with further analysis undertaken under EPSRC grants EP/J004197/1 (Crime, Policing and Citizenship (CPC) – Space–Time Interactions of Dynamic Networks) and EP/J005266/1 (The Uncertainty of Identity: Linking Spatiotemporal Information between Virtual and Real Worlds) and ESRC grants ES/K004719/1 (Using Secondary Data to Measure, Monitor and Visualise Spatiotemporal Uncertainties in Geodemographics) and ES/L011840/1 (Retail Business Datasafe).

Note

1   http://geogale.github.io/2011OAC/.

References

Adnan, M., Longley, P.A., Singleton, A.D. and Brunson, C. (2010) Towards real-time geodemographics: Clustering algorithm performance for large multidimensional spatial databases. *Transactions in GIS*, 14(3): 283–297.

Bailey, S., Charlton, J., Dollamore, G. and Fitzpatrick, J. (1999) Which authorities are alike? *Population Trends*, 98: 29–41.

Bailey, S., Charlton, J., Dollamore, G. and Fitzpatrick, J. (2000) Families, groups and clusters of local and health authorities of Great Britain: Revised for authorities in 1999. *Population Trends*, 99: 37–52.

Blake, M. and Openshaw, S. (1994) *GB Profiles: A User Guide*. School of Geography, University of Leeds, Leeds.

Blake, M. and Openshaw, S. (1995) Selecting variables for small area classifications of 1991 UK census data. *Working Paper 95-2*, School of Geography, University of Leeds, Leeds. Available at: http://www.geog.leeds.ac.uk/papers/95-2/.

Box, G.E.P. and Cox, D.R. (1964) An analysis of transformations. *Journal of the Royal Statistical Society: Series B (Methodological)*, 26(2): 211–252.

Brunsdon, C., Longley, P., Singleton, A. and Ashby, D. (2011) Predicting participation in higher education: A comparative evaluation of the performance of geodemographic classifications. *Journal of the Royal Statistical Society: Series A (Statistics in Society)*, 174(1): 17–30.

Burbidge, J. B., Magee, L. and Robb, A.L. (1988) Alternative transformations to handle extreme values of the dependent variable. *Journal of the American Statistical Association*, 83(401): 123–127.

Charlton, M., Openshaw, S. and Wymer, C. (1985) Some new classifications of census enumeration districts in Britain: A poor man's ACORN. *Journal of Economic and Social Measurement*, 13(1): 69–96.

Cockings, S., Martin, D. and Harfoot, A. (2015) A Classification of Workplace Zones for England and Wales (COWZ-EW). University of Southampton. Available at: http://cowz.geodata.soton.ac.uk/download/files/COWZ-EW_UserGuide.pdf.

Dag, O., Asar, O. and Ilk, O. (2014) A methodology to implement Box–Cox transformation when no covariate is available. *Communications in Statistics – Simulation and Computation*, 43(7): 1740–1759.

Gale, C.G. (2014) Creating an open geodemographic classification using the UK Census of the Population. Doctoral Thesis, University College London, Department of Geography. Available at: http://discovery.ucl.ac.uk/1446924/.

Harris, R., Sleight, P. and Webber, R. (2005) *Geodemographics, GIS and Neighbourhood Targeting*. Wiley, London.

Jain, A.K. (2010) Data clustering: 50 years beyond K-means. In *Pattern Recognition Letters, Award Winning Papers from the 19th International Conference on Pattern Recognition (ICPR)*, 31(8): 651–666.

Johnson, N.L. (1949) Systems of frequency curves generated by methods of translation. *Biometrika*, 36(1–2): 149–176.

Kumar, N. S., Rao, K.N., Govardhan, A., Reddy, K.S. and Mahmood, A.M. (2014) Undersampled K-means approach for handling imbalanced distributed data. *Progress in Artificial Intelligence*, 3(1): 29–38.

Leydesdorff, L. and Bensman, S. (2006) Classification and powerlaws: The logarithmic transformation. *Journal of the American Society for Information Science and Technology*, 57(11): 1470–1486.

Longley, P.A. and Singleton, A.D. (2014) London Output Area Classification: Final Report. Greater London Authority. Available at: https://londondatastore-upload.s3.amazonaws.com/Vik%3D2011+LOAC+Report.pdf.

MacQueen, J.B. (1967) Some methods for classification and analysis of multivariate observations. In *Proceedings of 5th Berkeley Symposium on Mathematical Statistics and Probability*. University of California Press, Oakland, CA, pp. 281–297.

ONS (2012a) User engagement on a new United Kingdom Output Area Classification – summary of responses. Available at: http://www.ons.gov.uk/ons/guide-method/geography/products/area-classifications/ns-area-classifications/new-uk-output-area-classification/user-engagement-on-uk-output-area-classification.pdf.

ONS (2012b) Statistical disclosure control for 2011 Census. Available at: http://www.ons.gov.uk/ons/guide-method/Census/2011/the-2011-Census/processing-the-information/statistical-methodology/statistical-disclosure-control-for-2011-Census.pdf.

ONS (2012c) Population ageing in the United Kingdom, its constituent countries and the European Union. Available at: http://www.ons.gov.uk/ons/dcp171776_258607.pdf.

ONS (2014) Methodology note for the 2011 area classification for output areas. Available at: http://www.ons.gov.uk/ons/guide-method/geography/products/area-classifications/ns-area-classifications/ns-2011-area-classifications/methodology-and-variables/methodology-oa.pdf.

Osbourne, J.W. (2002) Notes on the use of data transformation. *Practical Assessment, Research & Evaluation*, 8(6). Available at: http://pareonline.net/getvn.asp?v=8&n=6.

Pence, K. (2006) The role of wealth transformations: An application to estimating the effect of tax incentives on saving. *Contributions to Economic Analysis & Policy*, 5(1): 1430–1430.

R Development Core Team (2011) *R: A Language and Environment for Statistical Computing*. The R Foundation for Statistical Computing, Vienna, Austria. Available at: http://www.R-project.org.

Robinson, W.S. (1950) Ecological correlations and the behavior of individuals. *American Sociological Review*, 15(3): 351–357.

Singleton, A.D. and Longley, P.A. (2009) Creating open source geodemographics – Refining a national classification of census output areas for applications in higher education. *Papers in Regional Science*, 88(3): 643–666.

Singleton, A.D. and Spielman, S. (2014) The past, present and future of geodemographic research in the United States and United Kingdom. *Professional Geographer*, 66(4): 558–567.

Sleight, P. (2004) *Targeting Customers: How to Use Geodemographic and Lifestyle Data in Your Business.* Third edition. World Advertising Research Center Limited, Henley-on-Thames.

Vickers, D.W. (2006) Multi-level integrated classifications based on the 2001 Census. Doctoral Thesis, School of Geography, University of Leeds, Leeds.

Vickers, D.W. and Rees, P.H. (2007) Creating the UK National Statistics 2001 Output Area Classification. *Journal of the Royal Statistical Society: Series A (Statistics in Society)*, 170(2): 379–403.

Vickers, D.W. and Rees, P.H. (2011) Ground-truthing geodemographics. *Applied Spatial Analysis and Policy*, 4(1): 3–21.

Vickers, D.W., Rees, P.H. and Birkin, M. (2005) Creating the National Classification of Census Output Areas: Data, methods and results. *Working Paper 05/2*, School of Geography, University of Leeds, Leeds. Available at: http://www.geog.leeds.ac.uk/fileadmin/documents/research/csap/wpapers/05-2.pdf.

Wallace, M. and Denham, C. (1996) *The ONS Classification of Local and Health Authorities of Great Britain.* Studies on Medical and Population Subjects, HMSO, London.

Webber, R.J. (1977) An introduction to the national classification of wards and parishes. *Planning Research Applications Group Technical Paper*, Centre for Environmental Studies, London.

Webber, R.J. and Craig, J.A. (1976) Which local authorities are alike? *Population Trends*, 5: 13–19.

Webber, R.J. and Craig, J.A. (1978) *Socio-Economic Classification of Local Authority Areas.* Studies on Medical and Population Subjects, Office of Population Censuses and Surveys, Titchfield.

18

Uneven family geographies in England and Wales

(Non)traditionality and change between 2001 and 2011

Darren Smith and Andreas Culora

18.1 Introduction

The first release of 2011 Census data for England and Wales sparked the national mainstream media to widely report dramatic population changes and pose challenging questions about the salience of current social policy and ongoing welfare reforms. For example, a *Telegraph* (2012) headline claimed "Census 2011 'shows the changing face of Britain'". Likewise, *The Guardian* (2012) stressed that "the main story is surely that this country has undergone a radical transformation in this last decade", citing, for instance, the effects of unprecedented immigration, changing household and living arrangements, and the proliferation of mixed-ethnicity households in a more multi-cultural Britain. Comparatively, the release of the 2001 Census data, one decade earlier, did not herald such extreme articulations from the national media (Boyle and Dorling, 2004) or create entrenched representations of a profoundly altered British population, despite widening polarisation and marginalisation within British society between 1991 and 2001 (Dorling and Rees, 2003, 2004; Dorling and Thomas, 2004).

Surprisingly, given the relatively high profile within the media of emerging demographic trends (e.g. rising birth rates and ageing society) from the 2011 Census, as well as the flagging-up of ethnic and racial, housing and labour market-related restructuring (which have been substantiated by recent academic studies, e.g. Stillwell and Dennett, 2012), there has been a general lack of attention to how family formations have changed between 2001 and 2011. This is despite assertions, just before the launch of the 2011 Census, that "the stereotypical family image – mother, father and two children in a detached or semi-detached house – is fast becoming a myth" (*The Guardian*, 2011). Such views are in close alignment with prominent academic debates, such as Edwards and Gillies' (2012) treatise of 'farewell to the family', and in line with common understandings of the growing diverse makeup of family life in twenty-first-century Britain (Williams, 2004).

Contrarily, narrow representations of the 'family' and 'family life' have recently become even more highly politicised in Britain, with the virtues of the 'traditional family' widely espoused by the previous Prime Minister, David Cameron, such as:

> For me, nothing matters more than family. It's at the centre of my life and the heart of my politics. As a husband and a father I know how incredibly lucky I am to have a wonderful wife and to have had 4 amazing children . . . It's family that brings up children, teaches values, passes on knowledge, instils in us all the responsibility to be good citizens and to live in harmony with others. And so for someone from my political viewpoint who believes in building a stronger society from the bottom up, there is no better place to start than with family.
>
> *(Cameron, 2014)*

These statements are paradoxically delivered against the backdrop of academic scholarship which identifies that modern families are increasingly deviating away from this ideal of the traditional family (e.g. Wilkinson, 2013), and that there is a spatial unevenness to family geographies in Britain (McDowell et al., 2014). Although knowledge of the social, economic, cultural and political processes (e.g. civil partnerships, dual-residence couples, changing benefits) that are reshaping family formations in the UK (e.g. Chambers, 2012) is advancing, complete understanding of the different sub-national geographies of family formations is seriously lacking. Indeed, it can be argued that sub-national family geographies are under-researched, and there is a current paucity of empirical studies of the geographic distribution of different types of family.

The main aim of this chapter is thus to map some different dimensions of family geographies in England and Wales, and to examine the changing and enduring patterns of family geographies using 2001 and 2011 Census data. We analyse census data at local authority district (LAD) level in England and Wales, and our methodology adopts six measures from Duncan and Smith's (2002) indices of family formations, to explore the divergence to and from the normative male breadwinner/female homemaker model. The chapter is divided into four main sections. The next two sections briefly outline some key findings from relevant recent academic scholarship on family and population change, and then describe the methods to directly explore the uneven spatiality of six themes of family change. Sections 18.4 and 18.5 provide descriptive analyses of our mapping of the measures of family formations in 2011, and then examine changes in family formations between 2001 and 2011. Section 18.6 provides some brief concluding remarks.

18.2   Changing family formations

There is a substantial and well-established social science scholarship documenting the ways that family formations have changed during the last few decades (e.g. Weston, 2013), which provides theoretical, conceptual and empirical groundings to our understanding in this field of study (Cannan, 2014). One exemplar here is the current flagship Economic and Social Research Council (ESRC) Research Centre for Population Change which, during the last decade, has delivered an impressive stream of outputs on contemporary family life in the UK. This work serves to demonstrate some of the key ways in which family formations are being reconfigured, revealing both how and why notions of the traditional family are increasingly disrupted and complicated by more diverse and dynamic forms of family formation. This work consolidates earlier original findings from the ESRC Care, Values and the Future of Welfare (CAVA) Research Project on Care, Values and the Future of Welfare (e.g. Williams, 2004). Six key themes are particularly important for this chapter (see Table 18.1), and are emblematic of the changing context of family formations; they form the focus of our investigations in the following sections.

Table 18.1 Changing family formations and findings from ESRC Research Centre for Population Change

Themes of changing family formations	*Evidence from ESRC Research Centre for Population Change*
1. Postponement/ rejection of formal (marriage) and informal (cohabiting) heterosexual partnership unions, and rise of solo/multi-person household living	Stone et al. (2014) draw attention to the increasing 'boomeranging' returns of young people to their parental homes following university study, dissolution of partnerships, and/or more precarious employment conditions (Berrington et al., 2014), and influenced by the lack of affordable housing for young adults (Berrington and Stone, 2014). The implications of this trend on the rate and speed of the formation of new families and reshaping established families (i.e. reduction of empty-nest households) is noteworthy.
2. De-formalisation of childrearing by co-residence partners	Berrington and McGowan (2014, p.32) contend that "the likelihood of becoming a lone mother, either through experiencing a birth prior to any coresidential partnership, or through the experience of partnership dissolution, may have slowed", although it is noted that this may not reduce the total numbers of lone parents in the UK.
3. Increase of partnership dissolution and re-partnering practices	Demey et al. (2013) stress that the overall increase of solo living is associated with both young adults and mid-life adults (see also Falkingham et al., 2012; Dieter et al., 2013; Demey et al., 2014a). Demey et al. (2014b, p. 1) also note that, for relatively large numbers of adults both in childbearing and mid-life phases of their lifecourse, "repartnering is steadily turning into a common life experience for many as more and more enter a second or higher-order co-residential union". This clearly disrupts the boundaries of the conventional uni-residential family unit, and demonstrates one of the key ways that contemporary family units straddle multiple household and home spaces. The CAVA work of Duncan and colleagues on the growth of couples living apart together (LATs), estimated to represent 10% of adults in the UK (Duncan et al., 2012), may be pertinent to this last point (Duncan et al., 2013, 2014; Duncan, 2015).
4. Changing ideas of gendered role allocation about breadwinner and domesticity/homemaker responsibilities	Stone et al. (2015) construct a novel taxonomy of women's lifecourse economic activity trajectories based on their experiences between ages 16 and 64 years, to identify the diverse combinations of ways that women balance different gendered paid work/employment and domestic roles (see also Roberts et al., 2014).
5. Changing normative ideas of motherhood and fatherhood and normative career/employment aspirations	Berrington and Pattaro (2014) assert that the traditional relationships between fertility intentions/outcomes are changing, and these cross-cut with changing partnership, educational attainment and employment practices. Key factors here are linked to changing flows and rates of immigration into the UK (see Waller et al., 2014; Robards and Berrington, 2015), as well as the postponement of childrearing by well-educated women (Berrington et al., 2015b, 2015c).
6. Decoupling of normative connections between marriage and childbirth	Berrington et al. (2015a) describe the rise of a "de-standardized life course", with the rising postponement of marriage and growth of cohabitation (see also Perelli-Harris et al., 2014). Tied to these trends is the weakening of ties between childbearing and marriage, particularly in light of new meanings of cohabitation and public displays of personal commitment via cohabitation (e.g. shared mortgages and childrearing). Also influential here are re-envisaged meanings and symbolisms of weddings (for example, see Carter and Duncan, 2017) within society.

18.3 Methods

To explore the effects of the above changing processes of family formation, and to consider how these facets of change are expressed spatially within family geographies in England and Wales at a sub-national level in 2001 and 2011, the six emblematic themes from Table 18.1 are directly matched to six comparative measures of family formations drawn from Duncan and Smith's (2002) earlier study of family formations.

First, aggregated 2011 Census datasets were accessed from the Office for National Statistics (ONS) to extract census data to reconstruct four measures of family formation that Duncan and Smith used to examine the first four themes outlined in Table 18.1. These include:

- One person, multi-person (more than two unrelated people living together) or same-sex civil partnership households with dependent or no dependent children as a percentage of all households. The data were extracted from the Quick Statistics dataset (Table QS116EW: Household Type). The measure represents an indication of the relative (re)alignment to the normative model of heterosexual partnership forming and living, and the adoption of alternative forms of partnership forming and living.
- Lone parent families (aged 16–74) with dependent children as a percentage of all families with one or more dependent children. The data were extracted from the Key Statistics dataset (Table KS107EW: Lone Parent Households with Dependent Children). The measure indicates adherence to childrearing and non-co-residence of partners.
- All usual residents (aged 16 and over) who are divorced and widowed as a percentage of total usual residents (aged 16 and over). These data were extracted from the Key Statistics dataset (Table KS103EW: Marital and Civil Partnership Status). The measure is an indication of the de-alignment of marriage and lifelong partnership connections.
- Married women that are economically inactive in the formal labour market as a percentage of total married women. Duncan and Smith (2002) referred to these females as 'married domestic workers', which represents an indication of traditionality in households and the marriage contract, and gendered role allocations of caring, domestic work and household reproduction.

Second, we reconstructed the Motherhood Employment Effect (MEE), which provides a standardised measure of the relative adherence to the so-called traditional male breadwinner and female homemaker family model, using individual person records from the 2011 Census microdata. Here we are exploring the relativity of the withdrawal of mothers from full-time and part-time paid employment in the formal labour market (termed economic inactivity in the census). This is an index of the difference between the full-time employment rates of partnered mothers with one or more dependent children and partnered non-mothers. Unfortunately, the age range bands between the 2001 Individual Sample of Anonymised Records (SARs) and the 2011 non-regional safeguarded Individual file (5 per cent sample) are broken down in different ways, and we have to compare different so-called 'prime motherhood' ages of 20–45 years in 2001 and 24–49 years in 2011 respectively (see Duncan and Smith (2002) for discussion of some weaknesses of this index). Although this is not ideal, it does allow some crude indications to be drawn.

Third, we replicated the construction of the Family Conventionality Index (FCI), drawing upon birth registration datasets from population and vital statistics (and accessed from the ONS). Duncan and Smith used data for 1997; we use comparative data for 2014. Here we capture the ratio of births to married couples (including within marriage and civil partnerships in 2014) and births to non-married (cohabiting) couples (joint registrations at same address in 2014).

This is an indication of (less) conventionality of parenting practices. We exclude births to lone parents given the geographic clustering of this phenomenon.

18.4   Uneven family geographies in 2011

In analyses of 2001 Census data, Duncan and Smith (2002) argued that the well-known North–South and urban–rural divides, deeply embedded in the national consciousness, do not wholly explain the uneven geographies of family formations in Britain. Instead, it was argued that "different areas show different norms in terms of their relative adherence to the male breadwinner family" (p. 490), which are influenced by: a cross-cutting gamut of localised and regional histories of gendered household and work-place divisions of labour; diverse geographies of social class, religion and ethnicity/race; and local and regional normative ideas of good partnering and parenting. Duncan and Smith thus conclude that "there has never been a standard geographical family at any one time" (2002, p. 490). In other words, this can be interpreted as the geography of families has been and will always be plural – more effectively captured by the term 'uneven family geographies'. To some degree, Figures 18.1–18.4 concur with this need to more fully recognise the unevenness of family formations in the UK, which, as we illustrate, is particularly pertinent to family geographies in 2011.

First, both the maps in Figure 18.1 show the distribution, in quartiles, of less-conventional households (single, multi-person and civil partnership same-sex couples) and lone parents with dependent children, respectively. Strikingly, there are some similarities between the patterns of

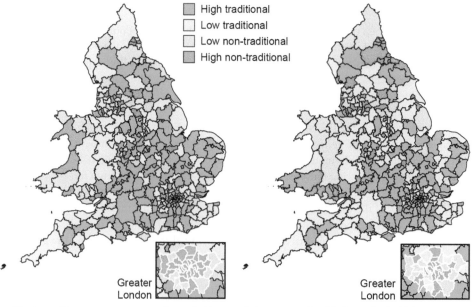

a. Single, multi-person or same sex households   b. Lone parents with dependent children

Figure 18.1   Distribution of less-conventional households and lone parents with dependent children by LAD, England and Wales, 2011

Source: ONS (2016) 2011 Census aggregate data. UK Data Service (Edition: June 2016). DOI: http://dx.doi.org/10.5257/census/aggregate-2011-1.

these phenomena in England and Wales at LAD level, with particularly high concentrations in the inner boroughs of London (see map inset), in part, likely to be influenced by the in-migration of young adults stepping onto the metaphorical escalator for employment/career opportunities and upward social mobility (see Champion, 2012; Gordon et al., 2015). This is in contrast to the outer suburban boroughs of London, where there are much lower levels of less-conventional households.

Concentrations of less-conventional households are also relatively high in university towns and cities, expressing the high number of young single adults attending higher education institutions and living within intensifying studentified neighbourhoods (Smith and Hubbard, 2014) and graduates staying after graduation, as well as less-conventional households living in so-called 'alternative' neighbourhoods of university towns and cities (such as Jericho in Oxford). It is also notable that there are high concentrations of less-conventional households in many coastal resorts (e.g. Margate, Kent), probably tied to the high supply of private sector housing for benefit recipients (Smith, 2012; Ward, 2015) and single adults seeking 'escape' areas.

A noteworthy difference between the two maps in Figure 18.1 is the high number of lone parents with dependent children in South Wales, North East, and South Manchester/Merseyside, pointing to an alignment between high levels of socio-economic deprivation and lone parenthood in these locations. The maps in Figure 18.1 also serve to demonstrate swathes of high traditionality (i.e. low levels of less-conventional households) in the South East (Hampshire, Sussex), South West (Mid Devon, Mid Dorset), Surrey/Buckinghamshire, M11 corridor (Cambridgeshire up to North Norfolk), Suffolk, Cotswolds and North Yorkshire. This is in line with the findings of Duncan and Smith (2002) and may be influenced by the out-migration of family-forming couples from London, seeking more rural and semi-rural locations for childrearing and high-quality education for their children (Smith and Higley, 2012).

Figure 18.2 maps the distribution of the percentage of widowed and divorced adults in England and Wales. On the whole, it can be seen that there is a general 'donut effect' to the mapping

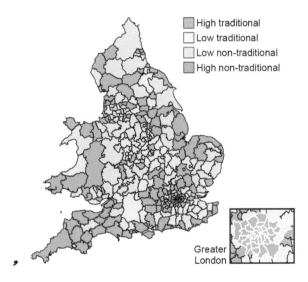

Figure 18.2 Total adults divorced and widowed by LAD, England and Wales, 2011

Source: ONS (2016) 2011 Census aggregate data. UK Data Service (Edition: June 2016). DOI: http://dx.doi.org/10.5257/census/aggregate-2011-1.

of this measure in England and Wales, with a concentrated core of high traditionality in the South East and Midlands (i.e. low levels of divorce and widowhood). The areas of less-traditionality may conflate different social processes using this measure. For instance, previous flows of (pre-) retirement migration to the South East coast (e.g. Eastbourne and Bournemouth) and Devon/Cornwall may have influenced the relatively high number of widowed individuals within established retirement hotspots. In a different way, the appeal of some coastal towns (such as Blackpool) as 'escape areas' may have influenced the relatively high number of divorced individuals in some coastal areas.

Figure 18.3a presents the mapping of the percentage of married domestic workers (i.e. married women who are economically inactive). Strikingly, this map generally divides England and Wales along an imaginary line from the Wash to the Severn Estuary, with some additional contrast between urban–rural in the North. In the vast majority of South East and East England, non-traditionality predominates with married women having a higher propensity to be economically active when compared to their northern counterparts, probably influenced by higher numbers of dual-earning couples in the South East *per se*, and possibly higher labour market opportunities for female workers in the South East. Clearly, the exception to this rule is Devon and Cornwall, where high traditionality would appear to be prevalent (with higher numbers of married domestic workers), perhaps influenced by the more rural labour markets of Devon and Cornwall. In contrast to the South of England, the more northerly regions of England and Wales are characterised by traditionality (i.e. high numbers of economically inactive married women), with the notable exception of Lancashire and Birmingham. This latter finding concurs

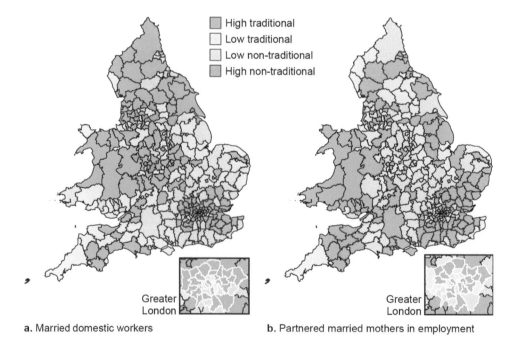

a. Married domestic workers   b. Partnered married mothers in employment

Figure 18.3 Distribution of married domestic workers and partnered married mothers in employment by LAD, England and Wales, 2011

Source: ONS (2016) 2011 Census aggregate data. UK Data Service (Edition: June 2016). DOI: http://dx.doi.org/10.5257/census/aggregate-2011-1.

with Duncan and Smith's (2002) view of enduring and historically 'independent women' in the former cotton towns of Lancashire who have a high propensity to work in the formal labour market.

To some degree, the map in Figure 18.3b, which presents findings from the motherhood employment effect, is in general alignment with Figure 18.3a. However, representations of traditionality versus non-traditionality are not as marked, although there is a noteworthy North–South dividing line, again. The main differences between the two maps are the areas of less-traditionality (i.e. high numbers of partnered mothers with dependent children in paid work) in the M5 corridor (Devon, Somerset), parts of Shropshire, the East Midlands, and the Birmingham City Region. This may point to the effects of commuting to larger metropolitan centres (i.e. Bristol, Leicester, Nottingham and Birmingham) by partnered mothers that reside in more rural and semi-rural locations (see Brown et al., 2015), and may point to the relatively high uptake of childcare. On the other hand, it is also noteworthy that there are pockets of traditionality (i.e. high numbers of partnered mothers with dependent children that are not in paid work) along the North Norfolk coast, East Kent, and parts of the South East coast. This may be influenced by the rural and coastal labour markets in these locations.

There is also some general alignment between the map in Figure 18.4 and those in Figure 18.3. Again, the line of division from the Wash to the Severn estuary is notable. South of the line of division is marked by areas of less-traditionality, characterised by relatively high levels of births outside marriage, including the M5 corridor in Somerset and Devon. The main divergences here include Norfolk (with the exception of Norwich), parts of Suffolk, North and East Kent, and most of the South East coast. In these more rural and coastal parts of the margins of the South East, there is a relatively high proportion of births to married couples. This is in line with the vast majority of LADs to the north of the line of division, which are characterised by traditionality. The exceptions

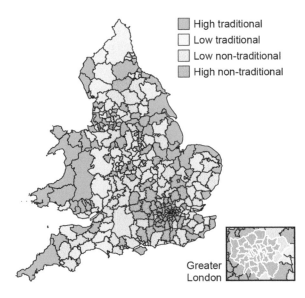

Figure 18.4 Ratio of births inside marriage to births outside marriage by LAD, England and Wales, 2011

Source: ONS (2016) 2011 Census aggregate data. UK Data Service (Edition: June 2016). DOI: http://dx.doi.org/10.5257/census/aggregate-2011-1.

to this rule include areas of less-traditionality in the metropolitan labour market areas of the Leeds City Region extending into North Yorkshire, Manchester/South Manchester, and South Birmingham City Region.

18.5   Change between 2001 and 2011

The maps in Figure 18.5 show how the six measures of family change between 2001 and 2011. We focus here on LADs that were in either the top or bottom quartile in both 2001 and 2011, and that have moved into the top and bottom quartile between 2001 and 2011 respectively. We do not focus on the 'middling' locations that were not in the top or bottom quartiles in 2011.

Figure 18.5a, expressing LADs with relatively high and low proportions of less-conventional family structures, identifies that concentrations are highest in university towns and cities (single students), deprived coastal towns (housing benefit recipients), and coastal Wales (perhaps influenced by university towns of Aberystwyth and Bangor). By contrast, less-conventional family structures are low in more semi-rural, small town and suburban locations in Surrey, Sussex, Berkshire, Wiltshire, Norfolk, Suffolk, Cambridgeshire and the West/East Midlands. This may point to important lifecourse differentials with family forming and rearing households more likely to reside in these locations, when compared to the dominance of solo, multi-person households living within university and coastal towns. These different population dynamics will clearly have a bearing on the patterns of (non)traditionality outlined in Figure 18.1.

The distribution shown in Figure 18.5b will have some connection to the interpretation of the map in Figure 18.5a. It can be seen that locations with the highest proportions of lone parents with dependent children are found in deprived coastal areas such as Margate in Kent, and Hastings in Sussex; and in relatively deprived locations in Lancashire and South Wales, and some inner boroughs of London. It is noteworthy that high proportions of lone parents with dependent children have become more entrenched in the North East of England between 2001 and 2011. Concentrations of high proportions of lone parents with dependent children are also high in the majority of urban provincial towns and cities (e.g. Bristol), perpetuated between 2001 and 2011.

Figure 18.5c reveals that the geography of divorced and widowed adults tends to be a coastal phenomenon in England and Wales, associated with 'escape areas' such as Blackpool. Between 2001 and 2011 there has been a marked amplification of this geography, with growing patterns along the South East coast, South Wales, Devon/Somerset/Cornwall, Norfolk, Lake District and, to a lesser extent, Yorkshire. This trend may be tied to an ageing of previous flows of retirement in-migrants, with one partner perhaps subsequently passing away in the place of retirement. Thus, it is possible that map c may conflate different geographies of widowhood and divorce. Nevertheless, geographies of widowhood and/or divorce have become more widespread between 2001 and 2011. By contrast, widowhood and/or divorce are relatively absent from the more conventional family-oriented semi-rural, small town and suburban locations in Surrey, Sussex, Berkshire, Wiltshire, Norfolk, Suffolk, Cambridgeshire and the West/East Midlands. These patterns of the absence of divorced and widowed individuals in semi-rural, small town and suburban have endured between 2001 and 2011.

Figure 18.5d shows that areas becoming more traditional, as identified by the measure of married domestic workers, are evident in large swathes of North Yorkshire and Northumbria,

a. Single, multi-person or same sex households

London

b. Lone parents with dependent children

London

c. Adults divorced and widowed

London

d. Married domestic workers

London

e. Motherhood employment effect

London

f. Ratio of births inside marriage to outside marriage

London

☐ Middling　■ Traditional　☐ Becoming more traditional　☐ Becoming less traditional　■ Less traditional

Figure 18.5　Changes in measures of family formation by LADs in England and Wales, 2001–11

Source: ONS (2016) 2011 Census aggregate data. UK Data Service (Edition: June 2016). DOI: http://dx.doi.org/10.5257/census/aggregate-2011-1.

the Lake District, North/Mid Wales, West Midlands, South Devon and the Dorset coast. These areas may be experiencing processes of rural gentrification, and witnessing the in-migration of childrearing couples, where the female partner is perhaps stepping out of the labour market in line with dominant representations of family life and the rural idyll. By contrast, Figure 18.5d reveals that clustered parts of South Manchester and large areas of the South East are becoming less traditional, perhaps pointing to the increase of more dual-career couples commuting into the metropolitan centres for work from the margins of labour market city regions.

Figure 18.5e identifies the perpetuation of high and low scoring MEEs between 2001 and 2011, although caution needs to be noted here given the measures are not directly comparable due to changing age band ranges between the censuses. The region with the highest MEE scores is the Greater South East, particularly in locations encircling the M25. This pattern has become more intensified between 2001 and 2011, with MEE scores becoming more entrenched in Hampshire, Sussex, Kent, Essex and Suffolk. In essence, the wider Greater South East has become a hotspot of high MEE scores between 2001 and 2011, with higher proportions of mothers not withdrawing from the formal labour market. There would also appear to be an increasing pattern of high MEE scores in the provinces, with increases in Bristol and Bath, the Cotswolds and South West M5 corridor, and the East Midlands. This suggests that the phenomenon of mothers not withdrawing from the formal labour market is tied to metropolitan areas and university towns (Cambridge and Oxford), perhaps due to a combination of choice (e.g. changing expectations of motherhood, employment and parenting; affordable childcare relative to income) and constraint (i.e. mortgage repayments, financial commitments), as well as the possible rolling-out of more family-friendly working practices of employers (i.e. flexitime, shared posts) and the wider uptake of technological developments (e.g. potential for homeworking and more mobile employment practices, e.g. Skype/FaceTime).

By contrast, Figure 18.5e reveals the reproduction of regions with low MEE scores between 2001 and 2011. This pattern is dominated by rural Wales, and rural parts of Lincolnshire, Norfolk, Northumbria and North Yorkshire. Traditional familial and gender relations within agricultural households and communities may be important here. It is also notable that there is a growing prominence of locations across the North West with low MEE scores, likely to be tied to the traditional familial cultures of Muslim populations in locations such as Oldham, Rochdale and Bury. This may also explain the growing pattern of low MEE scores in some outer suburban London boroughs, which have witnessed the in-migration of second or third generations of Muslim families between 2001 and 2011 (Stillwell and Dennett, 2012).

Figure 18.5f illuminates some general overlaps with Figure 18.5e, in terms of (non)traditionality, yet there are some important, subtle differences which are noteworthy. Alignment to traditionality (i.e. high proportions of births within marriage) is prominent and becoming more entrenched in rural Wales (contrary to urban locations of Swansea and Cardiff), North West and Lake District, Norfolk and Lincolnshire. The high concentration of locations with relatively low levels of births outside marriage in South Yorkshire is also notable, perhaps reflecting entrenched notions of traditional partnering and parenting practices within former industrial and mining communities. It is also interesting here to compare this interpretation to rural Kent in the South East, and past associations with the coal mining industry (e.g. Aylesham). Kent clearly contrasts with the majority of the rest of the South East, where births outside marriage are relatively high in London, and the commuting corridors of Berkshire, Buckinghamshire, Cambridgeshire, and the ring of the M25 (Hertfordshire, Surrey, Sussex, Kent). These patterns have been amplified between 2001 and 2011.

At the same time, the growing trend of non-traditionality towards non-withdrawal of mothers from the formal labour market (outlined in Figure 18.5e) in the provinces of Bristol and Bath, the Cotswolds and South West M5 corridor, and the East Midlands, is not matched by non-traditionality of births outside marriage. This may suggest that the non-withdrawal of mothers from the formal labour market in these provincial locations may not be emblematic of a growing propensity towards non-traditional familial and gender relations. Rather, it may represent rational economic household decision making, and negotiations including relationships between female income and childcare costs.

18.6 Conclusion

The starting point for this chapter was recognition of the national media stressing the profound population changes identified by 2011 Census data when first released in late 2012. Overall, the findings presented in this chapter suggest that there may be some resonance to this viewpoint in the context of changing patterns of family formation. However, we would argue that equally, if not more, important are the enduring regional patterns of family formations that we have identified between 2001 and 2011. In Duncan and Smith's (2002) analyses of 1991 and 2001 Census data, distinct sub-national geographies of family formations were mapped that revealed an uneven and, arguably, divided UK. In part, it was asserted that this 'geographical difference' was tied to a combination of: regional (gender) cultures and different socio-economic and gendered work and domestic histories; contemporary geographic contingencies and spatial divisions of labour; and different normative expectations of partnering, parenting, motherhood and fatherhood.

Our main findings generally concur with this interpretation, yet we would argue that these socio-spatial divisions would appear to have become more entrenched and amplified between 2001 and 2011. The unevenness of family geographies, often based on important regional differences, have become more intense, with adjoining LADs seemingly becoming members of regional clubs of (non)traditionality between 2001 and 2011. What would appear to be happening is a sub-national divergence of family geographies in England and Wales, including some notable internal anomalies within the two general parts of the widening duality. For instance, Devon and Cornwall seem to have more in common with Wales and Northern England, than their Southern England counterparts. Equally, the metropolitan centres of Manchester, Leeds and Birmingham seem to have some commonalities with the non-traditional swathes of Southern England.

Although there is clearly an underlying North–South and urban–rural influence to the lines of division, this factor only provides a partial understanding of the widening divergence between more high traditional and non-traditional parts of England and Wales. As previously noted, the uneven patterns of family formations are shaped by the effects of different gendered cultures of motherhood, female partnering and female employment practices. Lifecourse effects would also appear to be a major influential factor in the distribution of different family formations, with particular types of location such as university towns and cities perpetuating such geographies through the expansion of higher education and processes of studentification. Likewise, coastal towns would appear to be a magnet for less-conventional family formations, and individuals at stages of their lifecourse which are often characterised by single or solo living (i.e. higher education or post-student lifestyles), and relatively high levels of divorce and widowhood are prevalent in coastal locations for different reasons.

At the same time, it would appear that there are well-established hotspots for heterosexual couples to raise children, and these bastions of traditionality would appear to have been extended into adjacent neighbouring LADs between 2001 and 2011, perhaps pointing to the spread of semi-rural and rural gentrification in middle England (Smith and Higley, 2012).

All of these diverse geographies of family formation will have important implications for social policies, welfare budgets, and demands on public and private services, for instance different needs for childcare, marriage counselling, nurseries and schools and health services. Different normative ideas about what constitutes the 'right family' and the 'right familial relations' will also impact on personal senses of belonging and attachment, quality of life, stresses and strains, and the accepted routines of everyday life. More fully understanding the uneven geographies of family formation is therefore important for the wider well-being of society, and will have resource implications for public and voluntary sector organisations.

It is also important to stress, in conclusion, that our descriptive analyses are based on numerical statistical aggregates at relatively broad geographical units using cross-sectional data. Although these broad representations of families and family life in particular localities and regions may undoubtedly act as powerful structural conditions that shape perceptions of what constitutes good partnering, parenting, motherhood, fatherhood, and so on, we have not explored how more micro-level geographies of family formations are hidden within the broader geographical resolution of LADs. Analyses at the levels of output area, lower super output area, middle super output area or census wards, for example, may have borne very different results, and perhaps captured the tangible effects of neighbourhoods and/or streets on local geographies of family formation. At the same time, our analyses have sought to shed light on the 'where' questions of family formation, and we have not been able to grapple with the 'why' and 'how' questions that underpin the formation, perpetuation and/or transformation of family formation.

Finally, it is valuable to emphasise that the enduring and changing family geographies identified in this chapter have unfolded during a decade that was marked by a severe global economic recession, and the slowing-down of internal migration flows in Britain (Smith and Sage, 2014; Champion and Shuttleworth, 2015). Given sub-national migration flows, both short- and long-distance, are arguably fundamental to the replenishment and/or reconfiguration of spatial aggregations of distinct family formations within specific places and regions (Smith, 2011), it is possible that the geographic patterns of families that we have mapped and analysed may have been more pronounced if the global economic recession of the 2000s had not acted as a brake on sub-national population redistribution. At the same time, some of the changing patterns of family geographies may be influenced by recent immigration flows and losses of population due to emigration. There is an urgent need to more effectively connect together the demographic and migration components of population change to geographies of family formations in Britain, particularly at a time of flux and uncertainty.

References

Berrington, A. and McGowan, T. (eds.) (2014) The changing demography of lone parenthood in the UK. *Working Paper Series 48*, ESRC Centre for Population Change, Southampton.

Berrington, A. and Pattaro, S. (2014) Educational differences in fertility desires, intentions and behaviour: A life course perspective. *Advances in Life Course Research*, 21(1): 10–27.

Berrington, A.M. and Stone, J. (2014) Young adults' transitions to residential independence in Britain: The role of social and housing policy. In Antonucci, L., Hamilton, M. and Roberts, S. (eds.) *Young People and Social Policy in Europe: Dealing with Risk, Inequality and Precarity in Times of Crisis*. Palgrave MacMillan, London, pp. 210–235.

Berrington, A., Perelli-Harris, B. and Trevena, P. (2015a) Commitment and the changing sequence of cohabitation, childbearing and marriage: Insights from qualitative research in the UK. *Demographic Research*, 1(1): 1–46.

Berrington, A., Stone, J. and Beaujouan, E. (2015b) Educational differences in timing and quantum of childbearing in Britain: A study of cohorts born 1940–1969. *Demographic Research*, 33(1): 733–764.

Berrington, A., Stone, J., Beaujouan, E., McGowan, T. and Henderson, S. (2015c) Educational differences in childbearing widen in Britain. *Working Paper Series 29*, ESRC Centre for Population Change, Southampton.

Berrington, A., Tammes, P., Roberts, S., McGowan, T. and West, G. (2014) Measuring economic precarity among UK youth during the recession. *Working Paper Series 22*, ESRC Centre for Population Change, Southampton.

Boyle, P. and Dorling, D. (2004) Guest editorial: The 2001 UK Census: Remarkable resource or bygone legacy of the 'pencil and paper era'? *Area*, 36(2): 101–110.

Brown, D.L., Champion, T., Coombes, M. and Wymer, C. (2015) The migration-commuting nexus in rural England: A longitudinal analysis. *Journal of Rural Studies*, 41(1): 118–128.

Cameron, D. (2014) David Cameron on families. Speech at the Relationships Alliance Summit, held at the Royal College of GPs, on putting families at the centre of domestic policy-making, 18 August. Available at: www.gov.uk/government/speeches/david-cameron-on-families.

Cannan, C. (2014) *Changing Families*. Routledge, London.

Carter, J. and Duncan, S. (2017) Wedding paradoxes: Individualized conformity and the 'perfect day'. *The Sociological Review*, 65(1): 3–20.

Chambers, D. (2012) *A Sociology of Family Life*. Polity, Cambridge.

Champion, T. (2012) Testing the return migration element of the 'escalator region' model: An analysis of migration into and out of South-East England, 1966–2001. *Cambridge Journal of Regions, Economy and Society*, 5(2): 255–270.

Champion, T. and Shuttleworth, I. (2015) Are people moving home less? An analysis of address changing in England and Wales, 1971–2011, using the ONS Longitudinal Study. *SERC Discussion Papers, SERCDP0177*, Spatial Economics Research Centre, London.

Demey, D., Berrington, A., Evandrou, M. and Falkingham, J. (2013) Pathways into living alone in mid-life: Diversity and policy implications. *Advances in Life Course Research*, 18(3): 161–174.

Demey, D., Berrington, A., Evandrou, M. and Falkingham, J. (2014a) Living alone and psychological well-being in mid-life: Does partnership history matter? *Journal of Epidemiology and Community Health*, 68(5): 403–410.

Demey, D., Berrington, A., Evandrou, M. and Falkingham, J. (2014b) The determinants of repartnering in mid-life in the United Kingdom. Paper presented at the European Population Conference, Budapest, Hungary, 25–28 June.

Dieter, D., Berrington, A., Evandrou, M. and Falkingham, J. (2013) Who is living alone in mid-life and why does it matter? *Understanding Society News*. Available at: www.understanding society.ac.uk/2013/12/09/who-is-living-alone-in-mid-life-and-why-does-it-matter?

Dorling, D. and Rees, P. (2003) A nation still dividing: The British census and social polarisation 1971–2001. *Environment and Planning A*, 35(7): 1287–1313.

Dorling, D. and Rees, P. (2004) A nation dividing? Some interpretations of the question. *Environment and Planning A*, 36(2): 369–374.

Dorling, D. and Thomas, B. (2004) *People and Places: A 2001 Census Atlas of the UK*. Policy Press, Bristol.

Duncan, S. (2015) Women's agency in living apart together: Constraint, strategy and vulnerability. *The Sociological Review*, 63(3): 589–607.

Duncan, S. and Smith, D. (2002) Geographies of family formations: Spatial differences and gender cultures in Britain. *Transactions of the Institute of British Geographers*, 27(4): 471–493.

Duncan, S., Carter, J., Phillips, M., Roseneil, S. and Stoilova, M. (2012) Legal rights for people who 'live apart together'. *Journal of Social Welfare and Family Law*, 34(4): 443–458.

Duncan, S., Carter, J., Phillips, M., Roseneil, S. and Stoilova, M. (2013) Why do people live apart together? *Families, Relationships and Societies*, 2(3): 323–338.

Duncan, S., Phillips, M., Carter, J., Roseneil, S. and Stoilova, M. (2014) Practices and perceptions of living apart together. *Family Science*, 5(1): 1–10.

Edwards, R. and Gillies, V. (2012) Farewell to family? Notes on an argument for retaining the concept. *Families, Relationships and Societies*, 1(1): 63–69.

Falkingham, J., Demey, D., Berrington, A. and Evandrou, M. (2012) The demography of living alone in mid-life: A typology of solo-living in the United Kingdom. Paper presented at the European Population Conference, Stockholm, Sweden, 13–16 June.

Gordon, I., Champion, T. and Coombes, M. (2015) Urban escalators and interregional elevators: The difference that location, mobility, and sectoral specialisation make to occupational progression. *Environment and Planning A*, 47(3): 588–606.

Perelli-Harris, B., Mynarska, M., Berghammer, C., Berrington, A., Evans, A., Isupova, O. and Vignoli, D. (2014) Towards a deeper understanding of cohabitation: Insights from focus group research across Europe and Australia. *Demographic Research*, 31(4): 1043–1078.

McDowell, L., Rootham, E. and Hardgrove, A. (2014) Precarious work, protest masculinity and communal regulation: South Asian young men in Luton, UK. *Work, Employment & Society*, 28(6): 847–864.

Robards, J. and Berrington, A. (2015) The fertility of recent migrants to England and Wales: Interrelationships between migration and birth timing. *Demographic Research*, 34(4): 1037–1052.

Roberts, S., Berrington, A., Tammes, P., McGowan, T. and West, G. (2014) Educational aspirations among UK young teenagers: Exploring the role of gender, class and ethnicity. *Working Paper Series 21*, ESRC Centre for Population Change, Southampton.

Smith, D.P. (2011) Geographies of long-distance family migration: Moving to a 'spatial turn'. *Progress in Human Geography*, 35(5): 652–668.

Smith, D.P. (2012) The social and economic consequences of housing in multiple occupation (HMO) in UK coastal towns: Geographies of segregation. *Transactions of the Institute of British Geographers*, 37(3): 461–476.

Smith, D.P. and Higley, R. (2012) Circuits of education, rural gentrification, and family migration from the global city. *Journal of Rural Studies*, 28(1): 49–55.

Smith, D.P. and Hubbard, P. (2014) The segregation of educated youth and dynamic geographies of studentification. *Area*, 46(1): 92–100.

Smith, D.P. and Sage, J. (2014) The regional migration of young adults in England and Wales (2002–2008): A 'conveyor-belt' of population redistribution? *Children's Geographies*, 12(1): 102–117.

Stillwell, J. and Dennett, A. (2012) A comparison of internal migration by ethnic group in Great Britain using a district classification. *Journal of Population Research*, 29(1): 23–44.

Stone, J., Berrington, A. and Falkingham, J. (2014) Gender, turning points, and boomerangs: Returning home in young adulthood in Great Britain. *Demography*, 51(1): 257–276.

Stone, J., Evandrou, M., Falkingham, J. and Vlachantoni, A. (2015) Women's economic activity trajectories over the life course: Implications for the self-rated health of women aged 64+ in England. *Journal of Epidemiology and Community Health*, 69(1): 873–879.

The Guardian (2011) Census 2011: The typical family is not what it used to be. 27 March. Available at: www.theguardian.com/uk/2011/mar/27/census-family-housing-ageing-population.

The Guardian (2012) Census shows a changing of the guard in Britain. 11 December. Available at: www.theguardian.com/commentisfree/2012/dec/11/census-2011-england-wales.

The Telegraph (2012) Census 2011 'shows the changing face of Britain'. 11 December. Available at: www.telegraph.co.uk/news/uknews/immigration/9737874/Census-2011-shows-the-changing-face-of-Britain.html.

Waller, L., Berrington, A. and Raymer, J. (2014) New insights into the fertility patterns of recent Polish migrants in the United Kingdom. *Journal of Population Research*, 31(2): 131–150.

Ward, K.J. (2015) Geographies of exclusion: Seaside towns and houses in multiple occupancy. *Journal of Rural Studies*, 37(1): 96–107.

Weston, K. (2013) *Families We Choose: Lesbians, Gays, Kinship.* Columbia University Press, New York.

Wilkinson, E. (2013) Learning to love again: 'Broken families', citizenship and the state promotion of coupledom. *Geoforum*, 49(1): 206–213.

Williams, F. (2004) *Rethinking Families.* Calouste Gulbenkian Foundation, London.

19

Using census data in microsimulation modelling

Mark Birkin, Michelle Morris,
Tom Birkin and Robin Lovelace

19.1 Introduction

The meaning of the term 'microsimulation' can be identified by looking at its component parts. 'Micro' refers to small (usually individual-level) units of analysis and 'simulation' refers to the process of modelling these individual units to represent some component of reality that one is trying to understand. However, it is important to understand that microsimulation is not a single discipline or method, but an approach which contains a variety of practices and conventions. Let us take three examples where the meaning of the term is more distinct: (i) traffic microsimulation usually refers to modelling the movement of individual vehicles on the transport network;[1] (ii) economic microsimulation is often used in the context of estimating the distributional impacts of new policies, as showcased in EUROMOD, probably the largest microsimulation modelling project on earth (and heavy user of census data) (Sutherland and Figari, 2013); and (iii) spatial microsimulation is "the creation, analysis and modelling of individual level data allocated to geographic zones" (Lovelace and Dumont, 2016, p. 31).

In the context of this chapter, we use microsimulation to mean 'the modelling of individual-level data' – and the associated behaviours – with an emphasis on spatial microsimulation. The methods are applicable to many fields. The purpose of the chapter is to show how the methods work: the user can apply them to their own fields as they see fit. We will illustrate the message with case studies from holiday-making, diet and urban travel. It is important to note that microsimulation is still an emerging field, and there is much scope for future innovation. The aim of this chapter is to provide pointers to the key methods that fall under the microsimulation banner rather than to provide a definitive guide on the technicalities. It also aims to demonstrate the crucial role of census data when using spatial microsimulation methods in these contexts.

19.2 Spatial microsimulation methods

The concept of microsimulation was introduced in the 1950s as a means to evaluate the distributional consequences of economic policy on a heterogeneous population (Orcutt, 1957). The essence of the method is the application of well-defined rules according to the specific attributes of a population of individuals or households. For example, in the case of child benefit, then the rule might be that a payment of £20 per month is attributed for each household member aged 16 or less. If a list of households can be provided with the number and age of household members, the application of this rule entails a count of children under 16 and a multiplication of £20 per month. This technique of 'list processing' is a staple element of the microsimulation approach, and is a straightforward but powerful means for representing diversity, as well as for understanding the effect of policy changes. Thus, if child benefit is reduced from £20 to £15, the impact of this change can easily be seen in relation to any other attributes which might be held in the original list; for example, to compare single parents against married couples.

In the event that the list can be augmented by the addition of a location attribute, this leads towards a spatial microsimulation approach which has been popular with geographers and regional scientists, not least in view of the efficient representation which is important when multiple places are connected by spatial interactions, for example between residential properties and retail or workplace destinations. Conventional applications of spatial microsimulation modelling (MSM) have used synthetic populations for the evaluation of policy problems in domains such as health care, education, transport, crime, housing and regional economic development (e.g. Morrissey et al., 2012; Kavroudakis et al., 2013; Lovelace et al., 2014; Malleson and Birkin, 2012; Tanton, 2014).

The emergence of new forms of (big) data has important consequences for spatial MSM; for example, by providing opportunities for simulation using real individuals in preference to synthetic ones, or by extracting detailed evidence on the behaviour or attitudes of members of the population according to their characteristics and/or locations. In this chapter we will also explore the idea that spatial MSM may have important consequences for big data analytics. In particular, spatial MSM is typically based around a complete enumeration of the population, so that the process of linkage between the members of the model and the representatives of a big dataset could supply a powerful framework for addressing the biases in big data that many authors have noted. Furthermore, the adoption of synthetic representations could sidestep many of the concerns with privacy, anonymisation and ethical use of big data.

This chapter will articulate the approach with three case study examples. Section 19.3 will lay the foundations by outlining the ways in which synthetic microdata can be generated using data from the census and other sources. Some essential techniques for the manipulation and exploitation of microdata will be discussed. The case studies are then introduced in sections 19.4, 19.5 and 19.6. For each example, we seek to clarify the purpose and approach to the problem, to outline some indicative results, and review any obvious difficulties or limitations of the work. Finally, in section 19.7, the broader ramifications of the entire portfolio will be reviewed, we will offer some comments on the significance of census data within this context, and put forward some suggestions for important and relevant next steps.

19.3   The creation of synthetic microdata

Rationale

Individual-level data are rarely released at all, let alone with geo-references, for data security reasons. Although Samples of Anonymous Records (SARs) are produced by the Census Offices, they contain only a small subset of the population (albeit representative) and do not include detailed home location (see Chapter 8). In this context, MSM can be used to "combine the advantages of individual level data with the geographical specificity of geographical data . . . to provide 'the best of both worlds' of available data by combining them into a single analysis" (Lovelace and Dumont, 2016, p. 32). This provides a powerful rationale for creating synthetic microdata, allowing for more advanced modelling techniques than can be conducted with aggregate data alone such as intra-zonal inequality analysis and agent-based modelling. Thus, overcoming (often justified) data security barriers to data access and a foundation for sophisticated models provide a powerful rationale for the creation of synthetic microdata.

Method

The synthetic population used in this research was generated using open data from the 2011 Census for England and Wales. The census surveyed all 56,075,912 individuals in England and Wales on 27 March 2011. The census provides a snapshot of the population and captures many characteristics ranging from age and sex to general health and car ownership. The census data on each of these individuals and where they reside are not available in their entirety for research purposes due to data protection and for ethical reasons.

The Office for National Statistics (ONS) released a 2011 Sample of Anonymised Records (SAR) teaching file, containing a 1 per cent sample of the population for England and Wales, which can be used for a number of purposes including the most relevant concern here which is to "encourage wider use of census data by providing a way of examining census data beyond the standard tables" (ONS, 2014, p. 3). Data for the other countries of the UK, namely Northern Ireland and Scotland, are collected and managed by separate organisations and are not included in this model. This sample contains 17 variables relating to 569,741 individuals with a regional (large geographical unit) identifier included, of which there are nine in England and one for Wales. These data serve as the sampling population from which to generate the complete synthetic population using spatial MSM. The term 'sampling population' is used to refer to a sample set of individuals, members of which are selected randomly, multiple times and copied to form a full population. Within the sampling population there were 8,701 individuals whose usual residence was outside of England and Wales so were removed from the model. A further 74 individuals were removed as they had not completed the question relating to general health (a variable which was used for external validation of this model), which left a sampling population of 560,996.

Six variables were selected to be included in the sampling population for the modelling process. The general health variable was added as an additional constraint table and therefore there were seven variables in the final synthetic population. General health was used as a marker for validation of the synthetic population model. The six sampling population variables were chosen according to the strength of association with general health using Cramer's V statistic (Cramer, 1946). These variables are: age, marital status, hours worked, occupation, economic activity and sex.

Constraint tables include counts for the number of individuals within each category of the six chosen variables at a small area geographical unit of middle layer super output area (MSOA). There are 7,201 MSOAs in England and Wales. The average usual resident population of an MSOA is 7,787 individuals, with a range of 2,203 to 16,342. Considerable cleaning of the constraint tables was required in order that each file contained the same number of rows as the final population. For example, for tables relating to economic activity where ages refer to those aged 16–74, count data for those below and above working age were added to the table in order to complete the population. Only univariate constraint tables were included in the generation of this model.

This spatial MSM used the mathematical method of combinatorial optimisation to effectively 'clone' the individuals in the sampling population until they matched the aggregate counts of the constraint tables. This was completed using the *Flexible Modelling Framework* (*FMF*) application, an open source software package developed at the University of Leeds (Harland, 2013). The metric used to assess how well each individual was 'cloned', taking into account each combination of all the variables of interest, was total absolute error (TAE). TAE sums the absolute differences between the input and output counts across all the categories within the tables. The *FMF* re-iterated the optimisation process until TAE was equal to zero. For this spatial MSM, 91 per cent of MSOA zones completed with one iteration of fitting, 8 per cent required two iterations and the remaining 1 per cent needed between three and 11 *FMF* iterations. The final synthetic population contained 56,075,912 individuals all with TAE equal to zero.

The synthetic population was validated using both internal and external validation methods. Internal validation involved testing for difference between characteristics in both the sampling population and the resultant synthetic population. No statistically significant differences were observed (*t* test $p > 0.1$). External validation used general health, which was not included in the sampling population. Again, no statistically significant difference was observed (*t* test $p > 0.1$). As univariate constraint tables were used in the model, further validation was carried out through comparison of cross-tabulations of the synthetic population with the constraint tables. Results were identical to one decimal place. For example, the percentage of 25–34 year old males in the constraint files was 50.29 per cent and 50.32 per cent in the synthetic population. Generating this synthetic population of 56 million people was computationally intensive, but was possible using *FMF* on a desktop PC with 32 GB of RAM and took 13 minutes and 27 seconds of processing time.[2]

Potential applications of synthetic microdata

In a world with increasing volumes of 'big data', it is fair to ask what value synthetic microdata add to this wealth of data. One function is that synthetic microdata offer a method for understanding the biases observed in other types of data. Through utilisation of a complete and representative population of England and Wales – the census – a population can be generated which includes all individuals.

In other sources of data, we often see systematic missingness in the data. For example, those for whom English is not their first language will find difficulty engaging with services for which English is required; data collected via smart phone apps will not select from those who do not have a smart phone or, more specifically, do not use that app; and supermarket loyalty card data will only include the customers of a given supermarket who have chosen to enrol for a loyalty card. These differences are not at random; they often occur across a socio-economic gradient. Through use of synthetic microdata, we can understand both the population for whom we can find data and for those who we cannot, thereby adding valuable insight to research in many disciplines.

Synthetic microdata will also have valuable application in research where ethics and governance challenges around the possible identification of individuals in other data sources prevent those data being used for research. With confidence in the qualities of synthetic microdata, research can continue using the 'synthetic' individuals as the subjects, which in turn will offer insight into a range of real-world problems.

19.4 Case study A: application to the UK holiday market

Data source

Evidence generated by commercial organisations relating to the choices of individual customers is starting to provide new opportunities for social research in the 'big data' age. Case study A involves data obtained through access to customer records containing the destinations and personal characteristics of holiday-makers from the Leeds area in the calendar year 2014. These data were collected from customer travel questionnaires which are usually filled out by individuals on buses or aircraft and in airports on the return leg of their holidays, and typically address satisfaction levels and attitudes to the holiday experience. In addition to basic details such as the date and location of travel, the questionnaires contain measures of satisfaction on a Likert scale. For each respondent, sex, age, occupation, income, home postcode, and the identity of travel companions are all recorded. The total number of respondents from Leeds is 16,337.

The data are an extract from a national dataset and are restricted to residents whose home postcode is in the Leeds (LS) postal area. They represent a sample of about 1 per cent of the total customer file for Leeds, which has a generally similar demographic profile to the UK population as a whole (Unsworth and Stillwell, 2004).

Data linkage

The specific challenge which we address here is how to allocate a holiday destination for each member of the synthetic population. This step requires linkage/merger of records between the MSM and the holiday records. Since the number of holiday records is small relative to the size of the synthetic population this is a 'one to many' mapping from the former to the latter, i.e. each holiday-maker typically has a large number of lookalikes in the simulation. The linkage was executed according to a common set of attributes in the component datasets. Sex and age are relatively straightforward; employment variables were a little more complicated and involved the preliminary step of estimating employment status in the travel file based on income as a proxy variable. The final attributes for linkage are sex, age, economic activity, employment status and general health.

Although there are a number of examples of linking records in the MSM literature, there are no generally accepted methods and algorithms for this procedure. A difficulty in many situations is that the number of possible states in each population may be large, limiting the variety or likelihood of a direct match. In this application, there are only two possible states for sex, but six for age, three for employment status, three for economic activity, five for health, and so on. When we multiply these up we get a total of 540 possible states with the potential for low membership, which could lead in turn to significant under- or over-representation of customer groups in the matching process. The challenge was to aggregate states into a smaller number of combinations – or 'clusters' – with a target population of 100 or more observations in each cluster to provide some degree of statistical robustness to this process. We tackled this problem using a decision tree. Starting with sex, the travel data were split into two groups. Then we used age to split into 12 groups. We continued, using further characteristics, but each branch of the tree was terminated as the number of observations approached 100.

In this way, 49 customer 'segments' were identified. The trees associated with males and females are shown in Figure 19.1. The linkage procedure is now applied as follows. For each individual in both files, we allocate a segment based on the decision tree. Individual members of the MSM file are considered sequentially, then matched at random to the travel file (i.e. we take repeated random draws from the travel file until we find an individual with the same segment). When a match is found the destination of the individual in the travel file is mapped to the individual in the MSM.

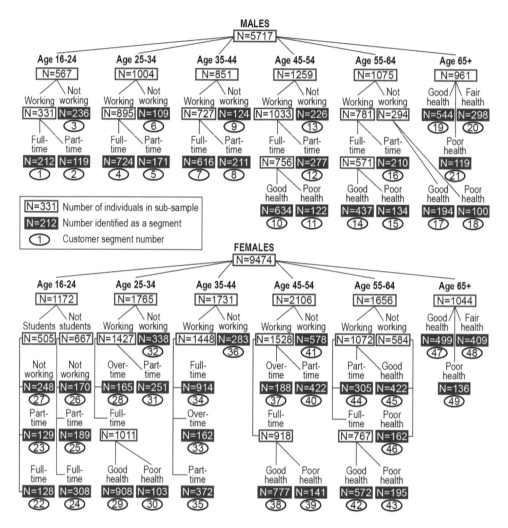

Figure 19.1 Decision trees for males and females for travel destinations

Socio-demographics of UK holiday-makers

The segmentation approach outlined above can be used to evaluate the holiday choices of different population groups. Indexing for shares, we compare the propensity for each of several hundred destinations relative to a 'par score' representing the average for all segments:

$$T^{km} = 100 \{O^{km} / [(O^{k*} / O^{**})(O^{*m} / O^{**})O^{**}]\} \tag{19.1}$$

where T^{km} is the index for destination m and segment k, O^{km} are the observed trips from k to m and the asterisk denotes summation over the missing superscript.

This method is widely used in target marketing and provides the basis for profiling segments, for example in geodemographic classification (Brown, 1991). In general, the older age segments have a preference for short-haul destinations in the Mediterranean (e.g. Malaga, La Palma and Madeira all have a clientele with an average age above 55, compared to 47 for the whole file); the more affluent segments have a preference for exotic destinations, with the islands of Aruba and the Maldives, and the African resorts of Mombasa and Taba showing the strongest concentration of high-income customers.

Modelling travel behaviour

Once the data have been merged, the synthetic population can be deployed to investigate an applied problem. The contribution of international travel to the use of fossil fuels is well known, and the behaviour of holiday-makers is therefore an important consideration in assessment of the long-term impact of human behaviour on the environment and sustainability of current social and economic practices. Druckman and Jackson (2009) found that 'recreation and leisure' contributes 26 per cent of the CO_2 emissions of an average UK household, following 195 per cent growth in these emissions between 1990 and 2004. The combination of MSM with consumer data will be used to estimate the 'carbon footprint' of population groups on the environment, and variations will be considered for a simple scenario of demographic change.

Model baseline: carbon footprint of holiday choices

We used the straight-line distance between origin and destination airports as a simple but robust estimate of the impact of holiday travel on the environment. For each segment in the enhanced MSM, the average travel distance was determined as a trivial list-processing computation. The resulting values are shown in Figure 19.2. For convenience of presentation, we index the values around a standardised mean of 100, and observe considerable variations from 17 per cent below the mean (e.g. Group 47) to 18 per cent above (e.g. Group 30). Some indicative examples are shown in Table 19.1, which tend to endorse the earlier finding that elderly holiday-makers have a tendency to stay closer to home.

The MSM provides a powerful means for reweighting the original sample. For example, segment 3 represents 5.1 per cent of the population, but only 1.4 per cent of the sample. The importance of this segment will be magnified considerably from the travel survey to the linked data. Segment 12 is 1.8 per cent of the survey but only 0.6 per cent of the population. The influence of this segment will be much less in the linked data. The overall impact of this reweighting process

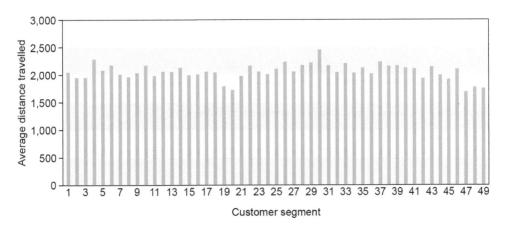

Figure 19.2 Average distance travelled by segment (see segment definitions in Figure 19.1)

Table 19.1 Illustrative profiles and travel patterns

Segment	Description	Distance (km)	Index
4	Males aged 25–34 in full-time employment	2,292	111
10	Males aged 45–54, in full-time employment with good health	2,170	105
20	Males aged 65+ in reasonable health	1,733	84
30	Females aged 25–34, in full-time employment with poor health	2,458	118
47	Females aged 65 and over in good health	1,702	83

is limited as the average distance travelled in the survey is 2,054 miles compared to 2,035 miles in the linked MSM. So there is a systematic over-representation of customers in the survey who travel longer distances, but the difference is relatively slight. The effect on individual destinations is a little larger – mass market resorts including Alicante and Marrakech are over-represented by more than 10 per cent in the sample relative to the synthetic population; travel operators should be wary of allowing too much capacity at these destinations based on the survey results. Niche destinations like Malé and Oranjestad are under-represented by more than 20 per cent in the survey, so it could be that low provision of flights or hotel spaces in these locations contributes to their low popularity.

Model scenario: holiday destinations with an ageing population

It is well known that the population of the UK is becoming more elderly at quite a rapid rate. For example, the population aged 75 and over is expected to almost double from 5.2 to 9.8 million by 2039 (ONS, 2015). We introduced a 'what if?' simulation in which the average age of the population increases by ten years as a 'thought experiment' to understand the potential impact of an ageing population on holiday patterns.

The main beneficiaries from a demographic shift towards a more elderly population are mainland Spain, the Balearics and Canary Islands, for example Tenerife (4,884 new holiday-makers from Leeds), Malaga (3,623), Palma de Mallorca (3,201) and Lanzarote (2,073). The big losers are long-haul and exotic destinations such as Enfidha (a loss of 5,452 visitors), Sharm-el-Sheikh (4,584) and Cancun (3,450). Scaled across the national population (which would be a relatively straightforward application for the linked MSM), these changes would have far-reaching implications for the travel sector. While the specifics of the simulation described here are somewhat stylised, the technique of spatial MSM is also extremely well suited to the generation of more robust demographic forecasts (e.g. Wu et al., 2008, 2011). The analysis of long-term demographic impacts on UK infrastructure has been explored by Zuo et al. (2014). A wide variety of applications for social policy have been presented by Ballas et al. (2006).

19.5 Case study B: application using UK diet and lifestyle data

Data source

In the previous section, our application explored the linkage of records from a large but somewhat distorted customer file. Established approaches to survey design have addressed the problem of bias by stratified (random) sampling, so that population sub-groups are evenly represented regardless of the sample size. This allows detailed investigations to be undertaken at reasonable cost to support whole country analysis, for example within the National Travel Survey, Family Spending and so on. A limitation of these samples is that they are unable to support local analysis due to a combination of lack of data and privacy restrictions on small respondent numbers. In these situations, the merger of survey data to small area census populations can facilitate insights down to a neighbourhood scale.

In the UK, comprehensive dietary data are collected in the National Diet and Nutrition Survey (NatCen Social Research, MRC Human Nutrition Research, University College London. Medical School, 2015). This survey is a rich and valuable asset. The survey has been designed to be nationally representative, sampling from 799 primary sample units. The sample size is relatively small with approximately 1,000 individuals being surveyed each year of a rolling programme with data available from four years. This annual sample represents both adults and children and, as such, 3,450 adults and 3,378 children are now included in the sample.

Dietary data from the NDNS are presented at a regional geographic scale for the UK (NDNS, 2014). Much evidence suggests that small area variations in diet and subsequent health exist (Morris et al., 2016; Hughes et al., 2012; Moon et al., 2007). Therefore, value exists in understanding small area variations in diet. Such information can inform the implementation of specific interventions for areas most in need. Using the spatial MSM of the England and Wales population described above, it is possible to estimate small area diet and lifestyle patterns in the UK.

Data linkage

Dietary data from the NDNS have been linked to the spatial MSM population using key demographic characteristics which appear in both the NDNS survey and the spatial MSM population. These include: age category, general health, marital status, hours worked and sex. In addition to these variables which appear in both datasets, we were able to link on occupation following some rearrangement of the economic activity variable in the synthetic population to align with the Standard Occupational Classification used by the NDNS. The MSOA geographies were aggregated to regions for the purpose of linking via the region in the NDNS data.

The data matching algorithm for this case study uses a modified Gower distance to exploit the similarity between the survey individuals and the synthetic population. The Gower distance is a dissimilarity measure which can accommodate both nominal and categorical variables. In essence, for a given attribute with n possible values, the Gower distance is zero between a pair with a shared state, and $1/n$ for any pair with distinct states. For example, suppose the attribute is marital status in which three possible states are 'married', 'single' and 'widowed/divorced'. If the objective is to match a single individual from the survey to a synthetic individual in the MSM who is also single, the Gower distance is zero. However, if the synthetic individual is not single, the Gower distance would be 1/3 whether that individual is married, widowed or divorced. The overall similarity between pairs i and j is therefore:

$$X_{ij} = \sum_k \frac{1}{n^k} \delta_{ij}^k \tag{19.2}$$

where k is the number of attributes, n^k is the number of states for attribute n, and δ_{ij}^k is zero for any attributes k in which i and j share the same state, and 1 otherwise.

In this example, we found better results with a modified Gower distance in which the age category is 'distance weighted' to introduce a penalty for age categories which are not tangential, and by a reduced weighting for region mismatches. The effect of this adjustment is to promote an exact match on age in all cases while encouraging the selection of similar individuals from different regions in preference to dissimilar individuals from the same region.

Visualisation of results

Once the population is generated, it is possible to analyse dietary components, reported in the NDNS for the full population of England and Wales. These can be represented geographically in the maps in Figure 19.3, which show spatial variation in physical activity level (PAL), energy intake (Kcal) and prevalence of obesity respectively.

These maps show how variation in energy expenditure (PAL), energy intake (Kcal) and energy imbalance (obesity) vary at a small area geography in England and Wales. Such data could provide valuable insights for health professionals and policy makers in local area geographies. Differences in PAL and Kcal intake are less pronounced than patterns in obesity. This may be due to misreporting bias in the survey self-reported data on PAL and dietary intake. Obesity values were generated using body mass index (BMI) calculated from nurse measured height and weight in the NDNS population. Such differences highlight that the resulting synthetic population is only as good as the data which were used to generate it. For models including self-reported data like diet and physical activity, bias may be expected. This synthetic population could be improved through addition of more objective measures such as transaction records from supermarket loyalty cards.

The NDNS case study presents variation in diet, activity and obesity. Using additional data sources, such as the Health Survey for England or aggregated Hospital Episode Statistics to enrich the population further, it would be possible to investigate changes in a wealth of health outcomes

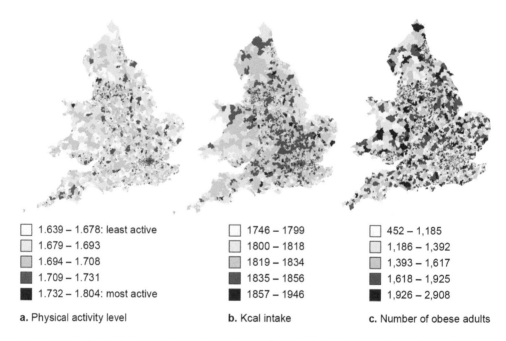

☐ 1.639 – 1.678: least active	☐ 1746 – 1799	☐ 452 – 1,185
☐ 1.679 – 1.693	☐ 1800 – 1818	☐ 1,186 – 1,392
▨ 1.694 – 1.708	▨ 1819 – 1834	▨ 1,393 – 1,617
▓ 1.709 – 1.731	▓ 1835 – 1856	▓ 1,618 – 1,925
■ 1.732 – 1.804: most active	■ 1857 – 1946	■ 1,926 – 2,908

a. Physical activity level   **b.** Kcal intake   **c.** Number of obese adults

Figure 19.3 Physical activity levels in adults, daily Kcal intake in adults and prevalence of obese adults by MSOA in England and Wales

through 'what-if' scenarios which change demographic characteristics or lifestyle behaviours. Such 'what-if' scenarios could be used to mimic planned public health interventions or policy changes. A topical example would be to estimate the impact of a 'sugar tax' on health outcomes (*The Guardian*, 2016).

19.6 Case study C: urban mobility

As we argued in Section 19.2, the use of rule-based procedures is a classic application for MSM. In an early illustration, Birkin and Clarke (1989) showed how to isolate 'dinkies' (dual income, no kids; a household category much beloved by estate agents and tabloid newspapers in the 1980s) in an urban area by synthesising household composition for small areas from the census with national income data from the New Earnings Survey. In a similar way, the third case study explores how 'oil vulnerability' (the likelihood of being negatively affected by oil price shocks) varied by individual-level characteristics (e.g. class) and spatial location (Lovelace and Philips, 2014). In contrast to previous approaches which typically assess vulnerability in relation to a single measure, here spatial MSM is used to drill down into the complex factors leading to vulnerability, including car accessibility, distance travelled to work and family size.

In this example, the component data sources were linked using the method of iterative proportional fitting (IPF), which facilitates the estimation of conjoint distributions from linked multi-dimensional input tables. Suppose, for example, that the objective is to estimate (for any neighbourhood) distance travelled to work on the basis of family size and commuting mode (a three-way cross-tabulation) and that what is known *a priori* from census data is distance travelled by family size, commuting mode by family size and distance by mode (i.e. a series of two-way cross-tabulations). IPF would be used to generate the three-way distribution of family size by mode by distance from the input tables. The synthetic population would then be estimated by assigning probabilities to each cell in the modelled output table and Monte Carlo sampling (random assignment) from these probabilities. This procedure generates results which are qualitatively similar to the *FMF* (see Section 19.3) with multi-dimensional input tables (Harland et al., 2012).

In order to calculate measures of oil vulnerability, data were again linked to a third party dataset, in this case Understanding Society.[3] This allows energy use and income to be added to the MSM using rules based on the household attributes. Figure 19.4 shows variations in energy use, distance travelled and the number of trips generated by this method for zones in West Yorkshire. In this diagram, the 'boxes' show the average proportions across the whole region and the 'whiskers' show the range of observations for small areas (MSOAs).

Figure 19.4 illustrates the wide range of outputs that can be generated when using individual-level input data; aggregate models would be able to provide estimates of one or two of the variables illustrated, but the use of individual-level data with MSM facilitates a more detailed understanding of the energy costs of transport systems. The spatial dimension of this problem is crucial in view of the importance of cars in Figure 19.4 and the variability of commuting patterns across urban areas and their hinterlands.

In further analysis, a suite of four indices was created with the MSM model to profile areas and their constituent households including measures of household spending on commuting and energy, the relative importance of energy use for commuting, and a compound measure of energy poverty (Lovelace and Philips, 2014). An example of the spatial distribution of one of these variables (Vcfp, vulnerability as commuter fuel poverty) is shown in Figure 19.5.

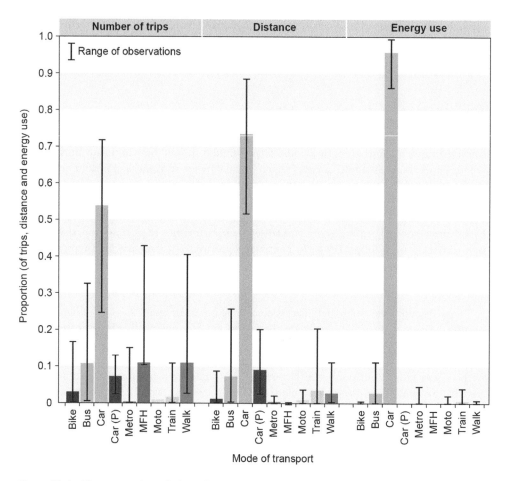

Figure 19.4 The proportion of trips taken, distance travelled and energy consumed by different modes in West Yorkshire, an output from a microsimulation model of urban travel

Source: Adapted from Lovelace and Philips (2014).

Note: Moto and MFH refer to commuting by motorbike and working 'mainly from home'.

Thus, MSM has already been used to greatly enhance our understanding of urban mobility and there is much potential for future research in this area (Lovelace et al., 2014). Historically, there has been a disconnect between MSM methods used by the transport modelling community (who have developed their own open source software such as *MATSIM*,[4] and even terminology for MSM, such as the term 'population synthesis' and the use of the term MSM itself to refer to vehicle movements) and other social scientists. While transport modellers tend to use a few mature methods, other social scientists tend to be more promiscuous in their use of methods, data and software, with the R packages *simPop*[5] and *ipfp*[6] providing examples. There are signs of convergence, however. More powerful computers running more efficient and clever code (such as *UrbanSim*, an open source city simulator) provide access to MSM methods for urban mobility researchers of all stripes (Waddell, 2002). The literature is large and rapidly growing.

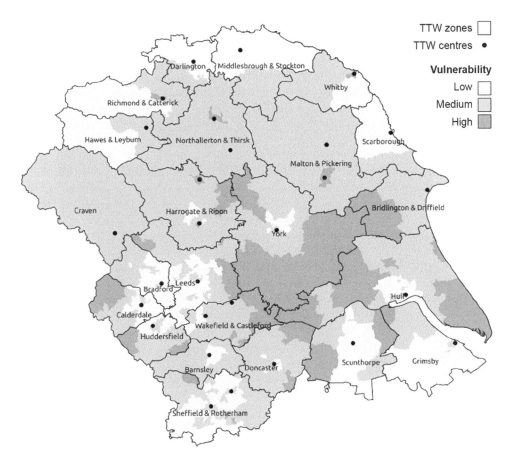

Figure 19.5 Vulnerability in terms of commuter fuel poverty in Yorkshire and the Humber

Source: Adapted from Lovelace and Philips (2014).

Note: TTW zones are Travel To Work zones.

19.7 Conclusions

The examples described in this chapter have shown how the adoption of spatial MSM can provide a powerful framework for analysis of applied problems. In the instance of consumer data, we have shown that a set of holiday destination choices can be modelled across an entire population when synthetic microdata are linked to consumer records from an emerging 'big data' source. In the second case study, the methodology is similar, but population surveys on the diet and nutrition of the nation are used to generate small area lifestyle and behaviour profiles. When urban transportation is considered, the challenge is to reconnect a level of cross-sectional detail between household demographics, location and travel preferences which are presented by the census area and interaction statistics, but in which the crucial detail for behavioural insight and policy insight has been lost through prior aggregation and the separation of domains.

Various elements hinting at the scope of the MSM method are embedded within one or more of these applications. The importance of MSM as a means for efficient representation of combinations of attributes is demonstrated to some degree by all of the applications, and this logic becomes even more compelling when a wide variety of choices are also incorporated. Thus, in the case of holidays, we considered outcomes for six age groups by sex, with three levels of health status, three layers of employment status, and four occupations for each of 108 small areas (MSOAs) in Leeds, travelling to 78 holiday destinations. The number of possible individual states is more than 3.6 million, which exceeds the number of households in Leeds by a factor of around 12; hence MSM is the only sensible way to handle such diversity, as has long been recognised by, for example, van Imhoff and Post (1998). In the case studies, we have also seen how complex choices on energy consumption and lifestyle can be represented efficiently in the MSM.

The spatial component of the models is essentially twofold. On the one hand, location plays an important role in determining variations in demand through the underlying demographic structure. Thus, in affluent suburban areas, incomes are high and individuals will tend to select exotic holiday destinations, have high rates of car ownership, and consume plenty of fruit and vegetables. These factors could ultimately determine the range of flights and package holidays offered through different airports, the provision of hospital beds and health care services, or the introduction of new transport infrastructure such as 'High Speed Rail' or dedicated cycleways. In a sense the deeper question is how location affects behaviour through environment, accessibility and service provision, which could manifest itself in various ways. A simple example could be that car ownership is higher in rural areas and people may travel longer distances because the provision of infrastructure is sparser. A more subtle case might be the possibility of spatial clustering in the preference for holiday destinations, for example due to shared experience in social networks. These kinds of micro-scale trends will be difficult or impossible to detect in survey data, but the increasing availability of big data offers the prospect of many more intensive and appealing investigations.

The impact of MSM increases in accordance with the range and diversity of the model scope. An important means of enhancing the models is therefore through data linkage, and we have illustrated two different approaches through the holiday and health examples. In each of these cases the linkage process involves synthetic data, and adds value in at least two distinct ways. First, linkage allows us an improved understanding of the interaction between risk factors and system drivers – for example, combining demographics with health, behaviour and lifestyle begins to provide important insights into problems of obesity, fitness and morbidity (e.g. in relation to retail provision by Aggarwal et al., 2014). Second, linkage of a survey to a complete synthetic database provides a means to reweight and adjust for sample bias. In view of debates about the robustness and reliability of new big data sources this could be even more of an issue for large datasets than smaller and well-stratified surveys.

As with all forms of modelling and analysis, the value of the research increases with the quality of the data. In relation to linkage, this is important as, in many cases, establishing a link between real individual data records is an exciting holy grail for the future. Combining health records with consumer data is a perfect example of this proposition – thus, being able to link through from individual consumption records (e.g. from till rolls or loyalty cards) to specific health outcomes (medication, illness or surgery) would provide a fantastic basis for a more definitive understanding of risks associated with lifestyle, demographics and behaviour. We expect MSM to move increasingly away from synthetic data in the direction of linking and modelling real data, but there are also many challenges, not least around consent, confidentiality and ownership of data and research ethics more generally. MSM of synthetic, anonymised or aggregate data could be ideal middle ground in exploiting big data within a robust ethical framework.

All of the examples in this chapter are predicated on the use of small area census data as a basis for the MSM models. As we have seen from the detailed discussion, this has involved core distributions of age, occupation, health and economic activity. Exploiting these census data within the MSM framework has suggested ways in which exciting data sources on lifestyles, recreation and consumption can be enhanced, and therefore provide a robust basis for diagnosis, forecasting and predictive analytics. It is hard to see how one might generate background data of similar reliability in the immediate future, although some solution looks necessary given the possible demise of the census in the medium-term. The use of MSM models to synthesise activity data, for example on lifestyle and consumption, with other sources of background population data ranging from the electoral roll to housing market data, satellite imagery or employment statistics could provide one of the more promising approaches to this problem. More studies of this type can be expected in the future, but well-designed tests of the validity of these methods and the suitability of the underlying data are still urgently needed.

Acknowledgements

This work was supported by the Consumer Data Research Centre, Economic and Social Research Council (grant number ES/L011891/1). The authors would like to thank the Consumer Data Research Centre data partners who provided data for use in this chapter and also all the National Diet and Nutrition Survey participants who agreed to take part in the survey.

Notes

1 See, for example, www.sias.com/paramics.
2 The synthetic population is available for research purposes via the Consumer Data Research Centre at https://data.cdrc.ac.uk/dataset/synthetic-population.
3 www.understandingsociety.ac.uk.
4 See the user manual at www.matsim.org/the-book.
5 https://cran.r-project.org/web/packages/simPop/index.html.
6 https://cran.r-project.org/web/packages/ipfp/index.html.

References

Aggarwal, A., Cook, A., Jiao, J., Seguin, R., Moudon, A., Hurvitz, P. and Drewnowski, A. (2014) Access to supermarkets and fruit and vegetable consumption. *American Journal of Public Health*, 104(5): 917–923.

Ballas, D., Clarke, G., Dorling, D. and Rossiter, D. (2006) Using SimBritain to model the geographical impact of national government policies. *Geographical Analysis*, 39(1): 44–77.

Birkin, M. and Clarke, M. (1989) The generation of individual and household incomes at the small area level using synthesis. *Regional Studies*, 23: 535–548.

Brown, P. (1991) Exploring geodemographics. In Masser, I. and Blakemore, M. (eds.) *Handling Geographical Information: Methodology and Potential Applications.* Longman Group UK Limited, Avon, pp. 221–258.

Cramer, H. (1946) *Mathematical Methods of Statistics.* Asia Publishing House, London.

Druckman, A. and Jackson, T. (2009) The carbon footprint of UK households 1990–2004: A socio-economically disaggregated, quasi-multiregional input–output model. *Ecological Economics*, 68(7): 2066–2077.

Harland, K. (2013) Microsimulation Model User Guide (Flexible Modelling Framework). Available at: http://eprints.ncrm.ac.uk/3177/.

Harland, K., Heppenstall A., Smith, D. and Birkin, M. (2012) Creating realistic synthetic populations at varying spatial scales: A comparative critique of population synthesis techniques. *Journal of Artificial Societies and Social Simulation*, 15(1): 1–15.

Hughes, R.J., Edwards, K.L., Clarke, G.P., Evans, C.E., Cade, J.E. and Ransley, J.K. (2012) Childhood consumption of fruit and vegetables across England: A study of 2306 6–7-year-olds in 2007. *British Journal of Nutrition*, 2012: 1–10.

Kavroudakis, D., Ballas, D. and Birkin, M. (2013) Using spatial microsimulation to model social and spatial inequalities in educational attainment. *Applied Spatial Analysis and Policy*, 6(1): 1–23.

Lovelace, R. and Dumont, M. (2016) *Spatial Microsimulation with R.* CRC Press, London.

Lovelace, R. and Philips, I. (2014) The 'oil vulnerability' of commuter patterns: A case study from Yorkshire and the Humber, UK. *Geoforum*, 51: 169–182.

Lovelace, R., Ballas, D. and Watson, M. (2014) A spatial microsimulation approach for the analysis of commuter patterns: From individual to regional levels. *Journal of Transport Geography*, 34: 282–296.

Malleson, N. and Birkin, M. (2012) Analysis of crime patterns through the integration of an agent-based model and a population microsimulation. *Computers, Environment and Urban Systems*, 36(6): 551–561.

Moon, G., Quarendon, G., Barnard, S., Twigg, L. and Blyth, B. (2007) Fat nation: Deciphering the distinctive geographies of obesity in England. *Social Science and Medicine*, 65: 20–31.

Morris, M.A., Clarke, G.P., Edwards, K.L., Hulme, C. and Cade, J.E. (2016) Geography of diet in the UK Women's Cohort Study: A cross-sectional analysis. *Epidemiology*, 1(1): 20–32.

Morrissey, K., Ballas, D., Clarke, G., Hynes, S. and O'Donoghue, C. (2012) Spatial access to health services. In O'Donoghue, C., Ballas, D., Clarke, G.P., Hynes, S. and Morrissey, K. (eds.) *Spatial Microsimulation for Rural Policy Analysis*. Springer, Dordrecht, pp. 213–230.

NatCen Social Research, MRC Human Nutrition Research, University College London. Medical School (2015). *National Diet and Nutrition Survey Years 1–4, 2008/09–2011/12*. [data collection]. Seventh edition. UK Data Service. SN: 6533. Available at: http://doi.org/10.5255/UKDA-SN-6533-6.

NDNS (2014) National Diet and Nutrition Survey. Available at: www.gov.uk/government/statistics/national-diet-and-nutrition-survey-results-from-years-1-to-4-combined-of-the-rolling-programme-for-2008-and-2009-to-2011-and-2012.

ONS (2014) 2011 Census Microdata Teaching File User Guide. ONS, Titchfield. Available at: www.ons.gov.uk/census/2011census/2011censusdata/censusmicrodata/microdatateachingfile/microdatauserguide.

ONS (2015) 2014-based National Population Projections. 29 October. Available at: www.ons.gov.uk/peoplepopulationandcommunity/populationandmigration/populationprojections/datasets/2014basednationalpopulationprojectionstableofcontents.

Orcutt, G. (1957) A new type of economic system. *Review of Economics and Statistics*, 58: 773–797.

Sutherland, H. and Figari, F. (2013) EUROMOD: The European Union tax-benefit microsimulation model. *EUROMOD Working Paper No. EM8/13*. Available at: www.iser.essex.ac.uk/research/publications/working-papers/euromod/em8-13.pdf.

Tanton, R. (2014) A review of spatial microsimulation methods. *International Journal of Microsimulation*, 7(1): 4–25.

The Guardian (2016) George Osborne backs sugar tax and £3.5bn of Whitehall cuts. 16 March. Available at: http://www.theguardian.com/uk-news/2016/mar/16/george-osbornes-sugar-tax-economic-fears-budget.

Unsworth, R., and Stillwell, J. (eds.) (2004) *Twenty-first Century Leeds: Geographies of a Regional City*. Leeds University Press, Leeds.

van Imhoff, E. and Post, W. (1998) Microsimulation methods for population projection. *Population*, 10(1): 97–136.

Waddell, P. (2002) UrbanSim: Modeling urban development for land use, transportation, and environmental planning. *Journal of the American Planning Association*, 68: 297–314.

Wu, B., Birkin, M. and Rees, P. (2008) A spatial microsimulation model with student agents. *Computers, Environment and Urban Systems*, 32: 440–453.

Wu, B., Birkin, M. and Rees, P. (2011) A dynamic microsimulation model with agent elements for spatial demographic forecasting. *Social Science Computing Review*, 29(1): 145–160.

Zuo, C., Birkin, M. and Malleson, N. (2014) Spatial microsimulation modelling for residential energy demand of England in an uncertain future. *GeoSpatial Information Science*, 17: 153–169.

20

Local ethnic inequalities and ethnic minority concentration in England and Wales, 2001–11

Kitty Lymperopoulou, Nissa Finney and Gemma Catney

20.1 Introduction

Ethnic residential segregation has been studied for almost a century and has been seen as a social and political concern for modern democracies. The connection between a group's spatial position in society and the socio-economic well-being of its members has received attention in academic and political debates, particularly in relation to its impact on social cohesion. Following the aftermath of the 'riots' in northern cities in England in the early 2000s, ethnic residential segregation was increasingly viewed as problematic, promoting 'parallel lives' and perpetuating the socio-economic deprivation of ethnic minority groups and their unequal position in British society. While there was little evidence on the social and economic consequences of ethnic segregation, in political discourses 'ethnic' areas were seen as synonymous with high levels of socio-economic disadvantage (Phillips, 2006).

This chapter examines the association between the degree to which an ethnic group is unevenly spread across neighbourhoods within a particular district and the socio-economic disadvantage it experiences in relation to the White British group within the same district. Specifically, we seek to answer three research questions:

1   What relationships exist between ethnic residential segregation and ethnic inequality in terms of education, employment, health and housing in districts in England and Wales?
2   How do the relationships between ethnic segregation and ethnic inequality differ between 2001 and 2011?
3   Are districts that have experienced increasing ethnic residential segregation over the 2000s associated with the highest levels of ethnic inequalities?

Residential segregation and ethnic inequalities have been examined separately in many studies, with little attention paid to how they relate to each other. In the next sections we review the literature on the linkages between ethnic concentration and ethnic inequalities and discuss the data and methods we use to examine the relationships between them. We then present evidence on patterns of residential segregation and ethnic inequality in districts of England and Wales in

2011, the association between within-district residential segregation and ethnic inequalities, and how these have changed over time. The chapter concludes with a discussion of the implications of the results.

20.2 Ethnic concentration and ethnic inequalities: what are the links?

In recent decades, in addition to the development of sophisticated datasets and innovative methods to monitor segregation, interest has turned towards understanding the impacts of segregation. While high levels of residential segregation have been associated with socio-economic inequalities between ethnic minority and majority groups (Massey, 2001), concurrently, arguments have been made for the positive effects of the maintenance of residential clusters of ethnic groups. Peach (1996) and Phillips (2015) both argue, in a UK context, for the benefits provided to minority groups in terms of resources, support and sense of belonging. Merry (2013) argues that 'voluntary separation' can improve equality and citizenship for minority groups and the population as a whole, in liberal democratic contexts.

These developments in the study of ethnic residential segregation trace interests in the broader social sciences in the meaning, significance and distribution of social capital. Influential arguments were made in the 2000s that ethnically mixed neighbourhoods improved social capital and social cohesion by increasing relations and trust within and between ethnic groups (Putnam, 2007). A large body of work has assessed these assertions, including quantitative spatial studies that can be grouped as scholarship on 'neighbourhood effects' (van Ham et al., 2012).

Explanations of the mechanisms through which ethnic concentration impacts on ethnic inequalities emphasise the role of intergroup attitudes and intragroup solidarity (Quillian, 2014). The 'racial threat' hypothesis (Blumer, 1958; Blalock, 1967) postulates that as a minority group in an area increases, this fuels prejudice and hostility by some members of the majority group, who feel threatened by this demographic change. From this, it follows that if higher ethnic segregation results in lower inter-ethnic group contact and increased prejudices, hostile attitudes and discrimination in an area (for example, by potential employers, housing providers and school funding bodies), ethnic minority groups will have poorer socio-economic outcomes and ethnic inequalities will increase. When there are barriers to spatial mobility by prejudice and discrimination, residential segregation will be expected to persist over time and this will act as a barrier to social mobility (Massey, 2001), further perpetuating ethnic inequalities and their spatial manifestations.

A contrasting theory, the 'intragroup solidarity' hypothesis, posits that ethnic concentration aids the formation of group social capital via dense co-ethnic social networks and neighbourhood institutions which facilitate information exchange and resource sharing (Quillian, 2014). In this case, residential segregation can improve (minority) group socio-economic outcomes and reduce ethnic inequalities.

Neighbourhood effects studies have produced mixed findings about the effect of neighbourhood ethnic concentration and diversity. Bécares and Nazroo (2013) find a protective effect of 'ethnic density' on the physical and mental health outcomes of minority groups in the UK and USA. Nelson (2013), however, finds segregation to have a negative effect on the self-rated health of Hispanics and Mexicans in the USA. On socio-economic outcomes, Steil et al. (2015, p. 1552) find segregation to be "consistently associated with lower levels of educational attainment and labor market success for both African American and Latino young adults compared with whites". They posit that this is due to minority groups in segregated neighbourhoods having lower exposure to residents with higher education degrees than their White counterparts. So, the negative effects of segregated neighbourhoods are felt more by minority groups living within them than by Whites.

In terms of the effects of neighbourhood ethnic minority concentration on housing inequalities, higher levels of segregation would be expected to be associated with higher levels of ethnic inequality in housing. This is for three reasons. First, the correlation of neighbourhoods of high concentration of ethnic minority groups and high levels of deprivation (Jivraj and Simpson, 2015) indicates that neighbourhoods with high co-ethnic concentration are characterised more than other areas by housing deprivation, and we can assume this will affect minority groups within those areas more than the White population (Finney and Harries, 2015). Second, according to the intergroup contact hypothesis, racism and discrimination in housing processes are more evident in 'segregated' areas, meaning minority groups are more constrained than Whites in their housing choice within areas with poorer housing quality (Ratcliffe, 2009; Markkanen and Harrison, 2013). Third, areas with higher segregation levels may provide minority groups with fewer sources of information about housing opportunities and rights.

Assessing the consequences of living in a neighbourhood with certain characteristics is far from straightforward. One major challenge is how to tease out the effects of place from the (self-)selection of individuals with particular characteristics into particular neighbourhoods via internal migration (Darlington, 2015; Sampson and Sharkey, 2008; Sharkey, 2014). In addition, the 'effects' of neighbourhood segregation are neither static nor unidirectional. Indeed, in contrast to a focus on the impact of ethnically diverse neighbourhoods on socio-economic inequalities, it is plausible to hypothesise that within-district inequality drives within-district residential segregation. For example, if minority groups are disadvantaged relative to Whites in education and employment, we might expect their residential options to be restricted to the cheapest (and, possibly, least desirable) neighbourhoods within districts. The impact of inequality has tended to be overlooked in neighbourhood research.

At a national scale, however, arguments have been made for the negative effects of inequality on a number of social indicators including health (Wilkinson and Pickett, 2009) and counterarguments have also been put forward (Snowdon, 2010). At the sub-national level, (socio-economic) inequalities have also been found to have negative impacts on health, possibly operating through mechanisms of social comparison and positioning: an individual's self-rated health is better when their residential contexts do not allow them to make extreme contrasts upwards (Marshall et al., 2014). This idea could be translated to the connection between within-neighbourhood ethnic inequalities and residential segregation: minority groups living in areas of high inequality may perceive themselves (more than their contemporaries in less unequal neighbourhoods) to be faring less well in the social hierarchy. This could, in turn, impact residential decision making by engendering a more restricted perspective on the (types of) neighbourhoods perceived as viable residential options, thereby encouraging clustering and segregation.

This chapter explores these theoretical propositions by examining the relationship between within-district ethnic group residential segregation and spatial patterns of ethnic inequality in education, employment, housing and health in districts of England and Wales. The analysis in this chapter is possible only due to the unique and rich qualities of UK census data that allow examination of socio-economic outcomes and residential patterns for ethnic groups for small areas and across time.

20.3 Data and methods

This chapter uses ethnic inequality and residential segregation measures constructed with data from the 2001 and 2011 Censuses for the 348 districts in England and Wales in 2011.[1] The decennial UK census of the population provides a robust data source to explore ethnic inequalities across different dimensions and across local areas, and how they are related to residential segregation.

Ethnic inequalities are examined in four dimensions: education, employment, housing and health. In these dimensions, ethnic minority groups have been shown to experience persistent disadvantage and they represent key areas of policy interest in relation to social and spatial inequality. Education and employment are central aspects of life chances and equality (Modood et al., 1997; Heath and Cheung, 2006); poor education and employment outcomes have negative consequences for well-being and participation in society (Machin and McNally, 2006); similarly, housing characterised by poor living conditions can create stress in the home and affect well-being, particularly that of children (Solari and Mare, 2012); while the well-established relationship between social deprivation and ill-health (Bécares et al., 2009) indicates that these are also important aspects of social inequality.

Ethnic inequality measures for districts (Table 20.1) are calculated for each ethnic minority group in 2001 and 2011 as the percentage point difference in the proportion of the White British group and each ethnic minority group, for those aged 16–24 with no qualifications (education); aged 25 and over who are unemployed (employment); with a (indirectly age standardised) limiting long-term illness (LLTI) (health); and living with an occupancy rating of -1 or below, indicating overcrowding (housing). Positive ethnic inequality scores indicate ethnic minority disadvantage on a particular dimension (e.g. a higher unemployment rate for an ethnic minority group compared with the White British group), while negative values indicate ethnic minority advantage (e.g. a lower unemployment rate for an ethnic minority group compared with the White British group). Segregation is measured using a version of the commonly applied Index of Dissimilarity (D), which tells us about the spatial spread of an ethnic group across an area (for a recent application see Catney, 2016).

In this chapter, each ethnic group is compared to the rest of the population. The proportion of an ethnic group in England and Wales who live in a given output area (OA) (the smallest of census areas for which ethnic group data are reported) is subtracted from the proportion of the rest of the population who live in that OA. The absolute differences are summed for all OAs in their respective district. The product is then multiplied by 50, giving a D value for each district in England and Wales which can range between 0 and 100. If the percentages of the two groups are the same in all OAs (e.g. all have 60 per cent group 1 and 40 per cent group 2), the population is evenly spread ($D = 0$), while if each zone comprises members of only one ethnic group or only

Table 20.1 Dimensions, indicators and data sources for ethnic inequalities and segregation

Dimension	Indicator	Data source
Education	% with no qualifications out of those aged 16–24	Census 2001 (ST117); Census 2011 (CT0260)
Employment	% unemployed out of economically active for those aged 25 and over	Census 2001 (ST108); Census 2011 (DC6201)
Health	% with LLTI (indirectly age standardised)	Census 2001 (ST16, ST65, ST101 and ST107); Census 2011 (DC2101, CT0261 DC3402, LC3205)
Housing	Overcrowding: % with occupancy rating -1 or lower	Census 2001 (ST124); Census 2011 (DC4205)
Segregation	Index of Dissimilarity	Census 2001 (KS006); Census 2011 (KS201EW)

members of other groups, the population is unevenly spread ($D = 100$). The D index is calculated for 2001 and 2011, aligned to 2011 district boundaries.[2] While the ethnic inequality measures are calculated as a difference from the White British population, for residential segregation we compare each ethnic group to *all other* groups, since the concern is with the impact of segregation in neighbourhoods *per se*, rather than their segregation from the majority group only. This research offers a fresh direction in segregation research, by *combining* hitherto separate analyses of segregation for OAs within districts (see Catney, 2016) and analyses of geographical inequalities across districts (Finney and Lymperopoulou, 2014). Ethnic concentration and inequality measures have been calculated for nine ethnic group categories for which data were available in both 2001 and 2011, to enable comparisons across time: White Irish, White Other, Mixed, Indian, Bangladeshi, Pakistani, Chinese, Black African and Black Caribbean groups.

We examine ethnic segregation and inequality patterns at the district level for a number of reasons. Local authority districts (LADs) are commonly used as approximations of the wider areas of socio-economic opportunity within which neighbourhoods sit (Kearns and Parkinson, 2001) and since they correspond to administrative local government boundaries, they are relevant in terms of structural processes — for example, planning and housing, education and labour market policies — that shape both segregation and inequality patterns. LADs provide the broader context outside an ethnic group's neighbourhood of residence such as schools, workplaces and public spaces, where everyday interactions take place and where attitudes towards ethnic minority groups underlying socio-economic inequalities are formed (Quillian, 2014). LADs also have sufficient ethnic minority population size to derive statistically meaningful ethnic inequality measures.

20.4 Patterns of ethnic inequality in education, employment, housing and health and residential segregation, 2011

Analysis of 2011 Census data shows that even though inequalities in education, employment, health and housing vary by ethnic group, they remain widespread in England and Wales. Figure 20.1 shows, respectively, the variation in ethnic inequality scores for LADs for the education, employment, health and housing inequality and segregation indicators. Overall, while inequalities in housing and employment are severe for ethnic minority groups in most districts, most ethnic minority groups fared better in 2011 than the White British group in terms of education and health. As discussed in the next sections, these LADs vary in geographical location, and include urban districts with large ethnic minority populations as well as semi-rural districts outside traditional (urban) ethnic minority settlement areas.

The degree of inequality in education varies greatly across LADs for each ethnic group. The Indian, White Irish, Chinese and Pakistani ethnic groups were advantaged educationally compared with White Britons in three-quarters (or more) of LADs in 2011, while the Black African, Black Caribbean and Bangladeshi groups fared better than the White British group in around two-thirds of LADs. The Other White and Mixed groups, in contrast, were disadvantaged educationally compared with White Britons in three-quarters and two-thirds of LADs in England and Wales respectively.

In contrast, most ethnic minority groups were disadvantaged in terms of employment in 2011, with the most disadvantaged groups being the Black groups (African and Caribbean). In the overwhelming majority of LADs in England and Wales, the Mixed, Bangladeshi, Pakistani, Black African and Black Caribbean groups fared worse in terms of employment compared with the White British group. The range in levels of employment inequality

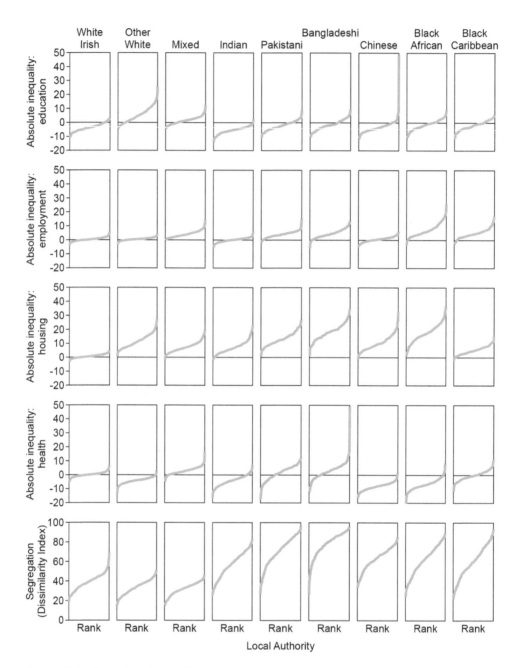

Figure 20.1 Plots of ranked LAD ethnic inequality in education, employment, housing and health and residential segregation, 2011

Source: 2011 Census, as indicated in Table 20.1.

Note: Ethnic inequality is the absolute difference between the proportion of people from the White British group and each ethnic group who are disadvantaged on each indicator (Table 20.1) in 2011. Positive ethnic inequality scores indicate ethnic minority disadvantage on a particular dimension.

between LADs for the Black African ethnic group were marked, with some LADs experiencing extreme inequality on this dimension. In contrast, in around half of all districts the Indian and White Irish group fared worse than the White British. The Other White group experienced employment inequality compared with the White British group in two-thirds of LADs in England and Wales.

In the majority of districts, the Indian, Other White, Chinese and Black African groups fared better than the White British group in terms of health (measured by the age-standardised LLTI rate). The Pakistani and Bangladeshi groups had poorer health than the White British group in around two-thirds of districts but the Mixed group was the most disadvantaged ethnic minority group in terms of health, with LLTI rates being higher in over three-quarters of districts in England and Wales in 2011.

Ethnic inequality in housing in 2011 was widespread across LADs in England and Wales. In the majority of districts, people from ethnic minority groups had higher levels of overcrowding than the White British. The largest inequalities in terms of housing across LADs in 2011 were seen for the Other White, Black African and Bangladeshi groups, while housing inequality was less severe for the White Irish group.

At the same time, the measure of segregation employed (the Index of Dissimilarity, D) shows marked variation between ethnic minority groups in the extent of residential clustering in neighbourhoods within LADs. The Bangladeshi and Pakistani groups exhibited on average higher levels of segregation than other groups in 2011, while the Mixed and White groups had lower segregation levels. The LADs with higher levels of ethnic segregation for most ethnic minority groups were more likely to be rural than urban districts, where the populations of the respective minority groups were small. For example, less than a quarter of LADs in the upper quartile of the D index for the Chinese, Mixed and Indian groups were urban districts. In contrast, there were more urban districts among those with higher levels of segregation for the White groups, particularly for the Other White group, where just over half of the LADs with high levels of segregation were urban districts.

20.5 Relationships between ethnic inequality in education, employment, housing and health, and residential segregation, 2001–11

The central question for consideration in this chapter is whether there is a relationship between ethnic socio-economic inequality and segregation. To answer this question we examine correlations between ethnic segregation within neighbourhoods across districts and district-level ethnic inequality scores in education, employment, health and housing in 2001 and 2011 (Table 20.2). The evidence on the relationship between residential segregation and ethnic inequality appears to be rather mixed: LADs with higher segregation are associated with higher inequality for some ethnic groups and lower inequality for other groups, and the relationship between inequality and segregation varies by the dimension of inequality chosen. However, it is possible to extract some general patterns from our results.

The correlations between educational inequality and segregation were positive (though small) for most ethnic groups, indicating an association between higher levels of educational inequality and higher levels of residential segregation within districts. Conversely, the correlations are on the whole negative for the relationship between segregation and inequality in employment, health and housing: higher residential segregation is associated with lower levels of ethnic inequality (and *vice versa*) for these domains.

Table 20.2 Coefficients of correlation between ethnic inequality in education, employment, housing and health and ethnic segregation, 2011 (2001 coefficients shown in parenthesis)

	Education inequality	*Employment inequality*	*Housing inequality*	*Health inequality*
White Irish	−.456** (−.511**)	.005 (.107*)	−.025 (−.232**)	−.328** (−.187**)
Other White	.388** (−.045)	−.068 (0.071)	.118* (−.298**)	−.245** (.100)
Mixed	.128* (−.104)	−.059 (−0.091)	−.598** (−.597**)	.150** (.009)
Indian	−.207** (−.101)	−.398** (−.213**)	−.100 (−.262**)	−.433** (−.363**)
Pakistani	.215* (.405**)	−.157* (−.127)	−.412** (−.346**)	−.436** (−.269**)
Bangladeshi	.181 (.316**)	−.322** (−.215)	−.494** (−.396**)	−.222** (−.172*)
Chinese	.204* (.147)	−.116 (−.300**)	−.116 (−.194*)	−.155** (−.048)
Black African	−.074 (−.425**)	−.222** (−.400**)	−.622** (−.709**)	−.302** (−.005)
Black Caribbean	.024 (−.006)	−.198** (−.342**)	−.262** (−.516**)	−.135* (−.016)

** Significant at 0.01 level; * Significant at 0.05 level.

In considering these observations in more depth, we can see that the relationship between residential segregation and educational inequality varies somewhat between ethnic groups.

There is a moderate positive relationship between ethnic inequality in education and segregation for the Other White group, indicating that LADs with higher segregation levels for this group also have higher ethnic inequality in education. There is also a weak positive relationship between education inequality and segregation for the Mixed, Pakistani and Chinese groups. For the Indian and White Irish groups the relationship is negative: higher residential segregation is associated with less difference between the minority group and White British in the proportion of young people with no educational qualifications.

For most ethnic minority groups there are weak to moderate negative associations between segregation and inequality in employment, housing and health. There is a strong negative relationship between segregation and housing inequality for the Black African group, indicating, somewhat surprisingly, that LADs with higher levels of overcrowding for the Black African group (in comparison with the White British group) are districts with lower Black African residential segregation. For the Pakistani and Bangladeshi groups (who exhibit higher levels of segregation in neighbourhoods within districts), higher levels of housing and employment inequalities are moderately associated with lower levels of segregation. The relationship between employment inequality and residential segregation in 2011 is negative for all ethnic groups apart from the White Irish; the association is statistically significant, though small, for the Asian and Black groups. That is, higher levels of residential segregation are associated with less difference in unemployment rates between White and ethnic minority populations of a district. Associations between segregation and health inequalities are also negative (and significant) for each group except the Mixed group and the associations are strongest for the Indian and Pakistani groups. In LADs with higher levels of residential segregation of Indian and Pakistani populations there is less difference between the health of Indians and Pakistanis and the White population than in districts where Indians and Pakistanis are less concentrated.

The second question under consideration is whether the relationship between ethnic inequality and segregation has changed over time. A comparison of the correlation coefficients for segregation and inequalities in 2001 and in 2011 (Table 20.2) shows that the associations between education inequality and segregation for the Other White and Mixed groups became positive and significant in 2011 (although they were not significant in 2001). The positive relationship between educational inequality and residential segregation became stronger between 2001 and 2011 for the Chinese group, but became less strong over the decade for the Pakistani and Bangladeshi groups. The negative relationships between segregation and employment, health and housing inequalities observed in the main in 2011 were also observed in 2001, suggesting persistence in these relationships. For most of the ethnic groups where the relationship is statistically significant at both time points, its strength increased over the decade. This is particularly the case for the relationship between health inequality and residential segregation.

The third question addressed by this chapter is whether increasing ethnic residential segregation in districts is associated with higher levels of within-district ethnic inequalities. To investigate this we examined the correlation coefficients for inequality scores in 2011 and change in the D index between 2001 and 2011. As shown in Table 20.3, the correlation coefficients of the inequality and segregation indicators present consistent results: positive correlations indicate an association between increasing residential segregation and higher levels of ethnic inequalities within districts in England and Wales. In highlighting this result, two points should be noted. First, the positive correlations are mostly small or moderate. Second, since segregation decreased in the majority of LADs in England and Wales for most ethnic minority groups, (see next section) the positive association between inequalities and segregation change also suggests that districts with higher ethnic inequality experienced lower reductions in residential segregation for ethnic minority groups.

This can be seen in Figure 20.2, which plots district ethnic inequality scores against change in the D index for selected ethnic minority groups. For example, LADs with increasing levels of segregation for the Other White group tend to be districts where this group is most disadvantaged in terms of housing and (to a lesser extent) education. On the other hand, LADs with the highest levels of inequality in housing, employment and health for the Mixed group are districts where generally this group saw lower reductions in segregation.

Table 20.3 Coefficients of correlation between ethnic inequality in education, employment, housing and health, 2011, and change in ethnic segregation, 2001–11

	Education inequality	Employment inequality	Housing inequality	Health inequality
White Irish	.202	−.049	−.014	.170**
Other White	.218**	−.054	.334**	.119*
Mixed	−.091	.217**	.564**	−.125*
Indian	.111	.351**	−.033	.481**
Pakistani	.293**	.227**	.222**	.281**
Bangladeshi	.277*	.296**	.030	.249**
Chinese	−.055	.117*	.231**	−.008
Black African	.094	.195**	.450**	.439**
Black Caribbean	.274*	.363**	.153*	.280**

** Significant at 0.01 level; * Significant at 0.05 level.

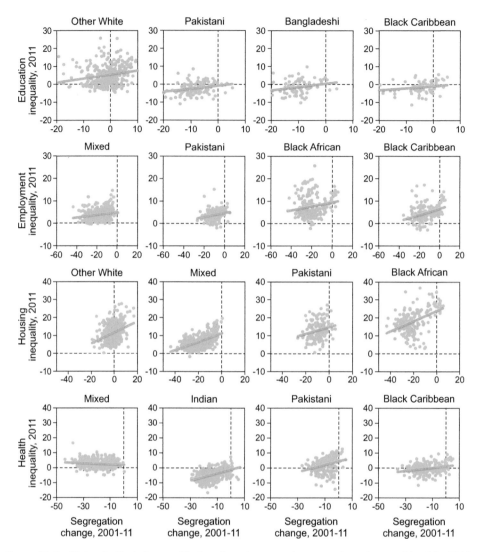

Figure 20.2 Plots of ethnic inequality in education, employment, housing and health, 2011, against segregation change, 2001–11, for selected ethnic minority groups

Sources: 2001 and 2011 Censuses, as indicated in Table 20.1.

Note: Positive ethnic inequality scores indicate ethnic minority disadvantage on a particular domain in 2011 while positive segregation values indicate increasing segregation in 2001–11.

20.6 Patterns of inequality in education, employment, housing and health, 2011 and changes in segregation, 2001–11

This section is concerned with the geography of the relationships described previously. To explore which districts with increasing ethnic residential segregation are associated with higher levels of ethnic inequalities, we examine 2011 LAD population cartograms (Dorling and Thomas, 2011) showing ethnic inequality scores in 2011 and changes in the *D* index in 2001–11.

Inequalities for ethnic minority groups in 2011 were more severe in urban (and often deprived) districts with larger ethnic minority populations, as well as (largely rural) districts with low ethnic minority concentrations, although there is variation in the geographies of inequalities by socio-economic dimension and ethnic group. In terms of education, the most marked inequalities exist outside major conurbation centres, and in rural and semi-rural districts, although most ethnic minority groups had better educational attainment than their White British counterparts in 2011, particularly in London boroughs. In contrast, for most ethnic minority groups, employment and housing inequality was most severe in urban conurbations, particularly in London, parts of the Midlands and the north of England, despite these districts experiencing decreasing segregation levels. Health inequalities were particularly marked for the Pakistani, Bangladeshi and Mixed groups in inner London boroughs.

Analysis of change in neighbourhood segregation within districts between 2001 and 2011 shows that in the majority of LADs in England and Wales, residential segregation decreased for all ethnic minority groups. In nearly all LADs, segregation levels were lower in 2011 than in 2001 for the Mixed group, and there were decreases in segregation levels in the majority of districts for the Black African, Black Caribbean, Chinese, Pakistani and Indian groups. In contrast, segregation levels increased in more than two-fifths of districts for the White Irish and Other White groups. The largest decreases in segregation for most ethnic minority groups were observed in urban districts, although for the White Irish, Mixed and Indian groups there were more rural districts among those with the largest decreases in segregation. The large decreases in ethnic minority segregation can be partly attributed to migration away from ethnic minority settlement areas linked to suburban housing aspirations and lifecourse changes. Few LADs saw increases in segregation, and these were mainly rural districts with small ethnic minority populations, reflecting the movement of small numbers of ethnic minority people into a small number of areas (Catney, 2016).

The ethnic inequality cartograms shown in Figure 20.3 show LAD inequality patterns in 2011 and change in segregation in 2001–11 for the Other White, Black African and Pakistani groups (other groups are not included given space constraints). The maps show the variation in inequality and neighbourhood segregation change across districts and the ways the association between segregation and inequality manifests for different ethnic minority groups.

Segregation increased in nearly half of all LADs where the Other White group experienced inequality in terms of education inequality in 2011. Housing disadvantage for the Other White group was most severe in both urban and rural districts, although segregation increased the most in rural districts which ranked among the most unequal in terms of housing and education inequality, such as South Holland and Boston in the East Midlands and Fenland in the east of England. These districts experienced large population increases for the Other White ethnic minority group in 2001–11, and have among the largest EU Accession migrant populations in the country.

The Black African group (which experienced more severe inequalities in terms of employment and housing in 2011 compared to other groups) was particularly disadvantaged in London and the large urban conurbations. While segregation levels for the Black African group increased in some districts in London, this group experienced decreasing levels of segregation in the majority of LADs in England and Wales.

The Pakistani group experienced housing and employment inequality across LADs in England and Wales, and housing inequality was particularly marked in London and the south of England. With just one exception, segregation for the Pakistani group increased in districts with among the highest inequalities in terms of employment and housing in 2011 for the Pakistani group. Segregation decreased in all other districts in England and Wales.

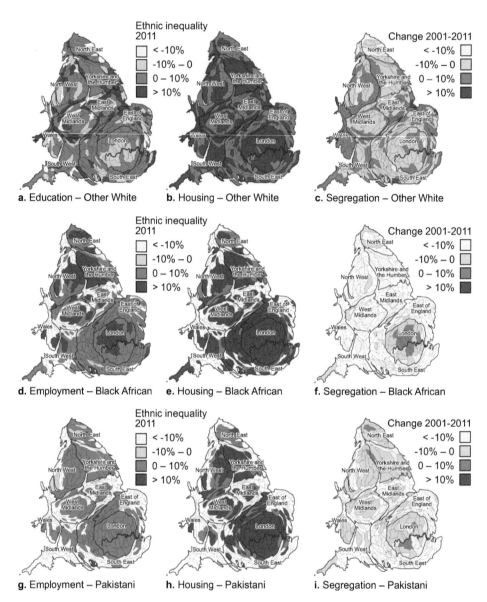

Figure 20.3 Ethnic inequalities, 2011, and segregation change, 2001–11, selected ethnic minority groups

Sources: 2001 and 2011 Censuses, as indicated in Table 20.1.

Note: Inequality and segregation scores are only shown if the district ethnic population at risk is at least 100.

There are several possible reasons for the variation in the geography of the relationship between socio-economic inequality and segregation across ethnic minority groups. Earlier we suggested the different mechanisms through which residential segregation can affect the socio-economic outcomes of ethnic minority groups. In the next and final section we review these in relation to our main findings.

20.7   Discussion and conclusions

There has been a longstanding debate about the consequences of ethnic residential segregation for the life chances of ethnic minority groups. This chapter made use of the unique characteristics of census data to examine the relationship between ethnic residential segregation and ethnic inequality in education, employment, health and housing and how it has changed over time. The findings point to both beneficial and harmful effects of ethnic residential segregation on ethnic minority socio-economic outcomes. Although we have not explored the direction of causality between residential segregation and ethnic inequality, what evidence exists suggests causal processes of segregation on a range of ethnic minority socio-economic outcomes (Quillian, 2014). The discussion in this section concentrates on three main findings of significance.

First, in general, there is a positive relationship between residential segregation within districts and ethnic educational inequality in 2011. The connection here may be via deprivation; ethnic minority groups have higher poverty rates than the White British group (Platt, 2011) and are more likely to live in deprived neighbourhoods (Jivraj and Khan, 2015) and it is likely that this is compounded by those (minority groups) without educational qualifications being restricted in their residential choice to the most deprived (segregated) neighbourhoods within LADs. This result also raises questions about differential performance of educational institutions according to the diversity of their catchments.

To understand the results more thoroughly it is useful to consider the experience of specific ethnic groups. The positive association between residential segregation and educational inequality found is strongest for the Other White group which grew mostly from immigration over the inter-censal decade (Simpson and Jivraj, 2015). The educational disadvantage of the Other White group, particularly in rural areas, is likely to reflect the lower educational attainment of younger EU Accession migrants (Lymperopoulou and Parameshwaran, 2015). Recently arrived migrants are more likely to co-locate with co-ethnics upon arrival in the receiving country, particularly those with poor educational attainment (McDonald, 2004), and may face barriers to obtaining qualifications, for example as a result of lower English language ability. Recent migrants are also more likely to under-report lower-level qualifications in the census (ONS, 2012), which may partially explain why young people from the Other White group have lower qualifications than their counterparts from other ethnic groups.

Our second notable finding is that, in general, there is a negative relationship between residential segregation and ethnic inequalities in employment, health and housing in 2011; that is, high residential segregation is associated with low levels of inequality in these domains. This relationship has generally become stronger between 2001 and 2011. This finding supports the proposition that there are protective effects with intra-group solidarity: dense co-ethnic social networks and neighbourhood institutions facilitate information exchange and resource sharing and protect ethnic minority groups from alienation and discrimination (Bécares and Nazroo, 2013; Quillian, 2014). In established ethnic minority settlement areas minority groups may also have greater political power and greater influence on the provision of public services and community infrastructures, which can diminish the socio-economic disadvantage of ethnic minority groups (Stafford et al., 2010).

Higher levels of segregation are significantly associated with lower levels of inequality in health for all ethnic minority groups, which supports the literature which evidences the protective effects of ethnic density on health, for example as a result of lower exposure to racism and discrimination (Bécares et al., 2009). There is also some evidence of the association between ethnic community infrastructures and health outcomes for ethnic minority groups, although the direction of causality is not clear (Stafford et al., 2010). Similarly, in ethnically dense neighbourhoods minority groups can have access to ethnic networks which may improve access to information about employment opportunities, which may result in higher ethnic minority employment rates, although these are likely to be in low pay and precarious occupations (Waldinger, 2005). This may explain the association between higher levels of segregation and lower levels of inequality in employment for the Indian, Pakistani, Bangladeshi, Black African and Black Caribbean groups found in this study.

Higher levels of segregation are also significantly associated with lower levels of inequality in housing. For some ethnic minority groups, particularly those with longer immigration histories in the UK, this may reflect an advantage of information networks about housing opportunities and rights in established ethnic minority settlement areas compared to other neighbourhoods, together with the better provision of housing (including social housing) to suit the diverse preferences of ethnic groups. For the Other White group, the positive association between segregation and housing inequality may reflect their more recent migration history and the preference of recent immigrants to co-locate in areas with co-ethnics, together with their initial disadvantage as a result of lack of knowledge of local housing markets and the limited financial resources they have upon arrival to the UK.

Our third important result is that the relationship between ethnic inequality in 2011 and change in ethnic segregation between 2001 and 2011 indicates that districts with higher ethnic inequality in education, employment, housing and health in 2011 are districts where levels of segregation increased or where there were lower reductions in segregation between 2001 and 2011. Segregation levels increased the most in rural districts and ethnic inequalities in housing and education in these districts were more severe than elsewhere in England and Wales. Rural districts have been associated with higher experiences of racism among ethnic minority groups (Chakraborti and Garland, 2004) and the absence of dense ethnic social networks, community resources and associated support which characterise these areas may help explain the existence of higher ethnic inequalities in rural districts with increasing levels of ethnic clustering.

The evidence presented in this chapter, using census data which provide a unique opportunity to explore these issues, raises many questions about the relationship between segregation and ethnic inequalities, the way it varies by dimension of socio-economic inequality, and the ways in which it manifests across districts for different ethnic minority groups. This provides considerable impetus for further research which could examine the effect of residential segregation on ethnic inequality in different socio-economic domains while controlling for ethnic group and district characteristics, and focus on providing a better understanding about the direction of the relationship(s) between segregation and ethnic inequality.

Notes

1 The data from the 2001 Census were matched to the 2011 local authority district boundaries to make them consistent. Office for National Statistics (ONS) disclosure controls have resulted in the merging of two pairs of districts: Cornwall has been merged with the Isles of Scilly and Westminster with the City of London.

2 The segregation results presented in this chapter differ from those published in Catney (2016) since in the latter 2001 segregation values are computed using revised ethnic group population estimates (see Sabater and Simpson, 2009) rather than the 2001 Census data as published, and they use a different population threshold.

References

Bécares, L. and Nazroo, J. (2013) Social capital, ethnic density and mental health among ethnic minority people in England: A mixed-methods study. *Ethnicity & Health*, 18(6): 544–562.

Bécares, L., Nazroo, J. and Stafford, M. (2009) The buffering effects of ethnic density on experienced racism and health. *Health and Place*, 15: 670–678.

Blalock, H.M. (1967) *Toward a Theory of Minority-group Relations*. John Wiley and Sons, New York.

Blumer, H. (1958) Race prejudice as a sense of group position. *Pacific Sociological Review*, 1: 3–7.

Catney, G. (2016) Exploring a decade of small area ethnic (de-)segregation in England and Wales. *Urban Studies*, 53(8): 1691–1709.

Chakraborti, N. and Garland, J. (2004) *Rural Racism*. Willan, Cullompton.

Darlington, F. (2015) Ethnic inequalities in health: understanding the nexus between migration, deprivation change and social mobility. PhD thesis, University of Leeds. Available at: http://etheses.whiterose.ac.uk/12108/.

Dorling, D. and Thomas, B. (2011) *Bankrupt Britain: An atlas of social change companion website*. Available at: http://sasi.group.shef.ac.uk/bankruptbritain/index.html.

Finney, N. and Harries, B. (2015) Which ethnic groups are hardest hit by the 'housing crisis'? In Jivraj, S. and Simpson, L. (eds.) *Ethnic Identity and Inequalities in Britain: The Dynamics of Diversity*. Policy Press, Bristol, pp. 141–160.

Finney, N. and Lymperopoulou, K. (2014) *Local Ethnic Inequalities: Ethnic Differences in Education, Employment, Health and Housing in Districts of England and Wales, 2001–2011*. Runnymede, London.

Heath, A. and Cheung, S.Y. (2006) *Ethnic Penalties in the Labour Market: Employers and Discrimination*. Department of Work and Pensions, Leeds.

Jivraj, S. and Khan, O. (2015) How likely are ethnic minorities to live in deprived neighbourhoods? In Jivraj, S. and Simpson, L. (eds.) *Ethnic Identities and Inequalities in Britain: The Dynamics of Diversity*. Policy Press, Bristol, pp. 199–213.

Jivraj, S. and Simpson, L. (2015) *Ethnic Identity and Inequalities in Britain: The Dynamics of Diversity*. Policy Press, Bristol.

Kearns, A. and Parkinson, M. (2001) The significance of neighbourhood. *Urban Studies*, 38: 2103–2110.

Lymperopoulou, K. and Parameshwaran, M. (2015) Is there an ethnic group educational gap? In Jivraj, S. and Simpson, L. (eds.) *Ethnic Identity and Inequalities in Britain: The Dynamics of Diversity*. Policy Press, Bristol, pp. 181–198.

Machin, S. and McNally, S. (2006) *Education and Child Poverty*. Joseph Rowntree, York.

Markkanen, S. and Harrison, M. (2013) Race, deprivation and the research agenda: Revisiting housing, ethnicity and neighbourhoods. *Housing Studies*, 28(3): 409–428.

Marshall, A., Jivraj, S., Nazroo, J., Tampubolon, G. and Vanhoutte, B. (2014) Does the level of wealth inequality within an area influence the prevalence of depression amongst older people? *Health and Place*, 27: 194–204.

Massey, D.S. (2001) Residential segregation and neighborhood conditions in U.S. metropolitan areas. In Wilson, W. (ed.) *America Becoming: Racial Trends and Their Consequences*. National Academy Press, Washington, DC, pp. 391–434.

McDonald, J.T. (2004) Toronto and Vancouver bound, the location choice of new Canadian immigrants. *The Canadian Journal of Urban Research*, 13: 85–101.

Merry, M. (2013) *Equality, Citizenship and Segregation: A Defence of Separation*. Palgrave Macmillan, Basingstoke.

Modood, T., Berthoud, R., Lakey, J., Nazroo, J., Smith, P., Virdee, S. and Beishon, S. (1997) *Ethnic Minorities in Britain: Diversity and Disadvantage*. Policy Studies Institute, London.

Nelson, K.A. (2013) Does residential segregation help or hurt? Exploring differences in the relationship between segregation and health among US Hispanics by nativity and ethnic subgroup. *Social Science Journal*, 50(4): 646–657.

ONS (2012) 2011 Census: User Guide, 2011–2001 Census in England and Wales Questionnaire Comparability. Available at: www.ons.gov.uk/ons/guide-method/census/2011/census-data/2011-census-user-guide/comparability-over-time/2011-2001-census-questionnaire-comparability.pdf.

Peach, C. (1996) Good segregation, bad segregation. *Planning Perspectives*, 11(4): 379–398.

Phillips, D. (2006) Parallel lives? Challenging discourses of British Muslim self-segregation. *Environment and Planning D: Society and Space*, 24(1): 25–40.

Phillips, D. (2015) Race, community and ongoing conflict. *Ethnic and Racial Studies*, 38(3): 391–397.

Platt, L. (2011) *Inequality Within Ethnic Groups*. Joseph Rowntree Foundation, York.

Putnam, R. (2007) E Pluribus Unum: Diversity and community in the twenty-first century. The 2006 Johan Skytte Prize Lecture, *Scandinavian Political Studies*, 30(2): 137–174.

Quillian, L. (2014) Does segregation create winners and losers? Residential segregation and inequality in educational attainment. *Social Problems*, 61: 402–426.

Ratcliffe, P. (2009) Re-evaluating the links between 'race' and residence, *Housing Studies*, 24(4): 433–450.

Sabater, A. and Simpson, L. (2009) Enhancing the population census: A time series for sub-national areas with age, sex, and ethnic group dimensions in England and Wales, 1991–2001. *Journal of Ethnic and Migration Studies*, 35(9): 1461–1477.

Sampson, R.J. and Sharkey, P. (2008) Neighborhood selection and the social reproduction of concentrated racial inequality. *Demography*, 45(1): 1–29.

Sharkey, P. (2014) Spatial segmentation and the Black middle class. *American Journal of Sociology*, 119(4): 903–954.

Simpson, L. and Jivraj, S. (2015) Why has ethnic diversity grown? In Jivraj, S. and Simpson, L. (eds.) *Ethnic Identity and Inequalities in Britain: The Dynamics of Diversity*. Policy Press, Bristol, pp. 33–48.

Snowdon, C. (2010) *The Spirit Level Delusion: Fact Checking the Left's New Theory of Everything*. Little Dice, London.

Solari, C.D. and Mare, R.D. (2012) Housing crowding effects on children's wellbeing. *Social Science Review*, 41: 464–476.

Stafford, M., Bécares, L. and Nazroo, J. (2010) Racial discrimination and health: Exploring the possible protective effects of ethnic density. In Stillwell, J. and van Ham, M. (eds.) *Ethnicity and Integration: Understanding Population Trends and Processes – Volume 3*. Springer, Dordrecht, pp. 225–250.

Steil, J., De la Roca, J. and Ellen, I.G. (2015) Desvinculado y desigual: Is segregation harmful to Latinos? *Annals of the American Academy of Political and Social Science*, 660(1): 57–76.

van Ham, M., Manley, D., Bailey, N., Simpson, L. and Maclellan, D. (2012) *Neighbourhood Effects Research: New Perspectives*. Springer, Dordrecht.

Waldinger, R. (2005) Networks and niches, the continuing significance of ethnic connections. In Loury, G.C., Modood, T. and Teles, S.M. (eds.) *Ethnicity, Social Mobility and Public Policy, Comparing the U.S. and U.K.* Cambridge University Press, Cambridge, pp. 342–362.

Wilkinson, R. and Pickett, K. (2009) *The Spirit Level: Why Equality is Better for Everyone*. Penguin, London.

Using the 2001 and 2011 Censuses to reconcile ethnic group estimates and components for the intervening decade for English local authority districts

Philip Rees, Stephen Clark, Pia Wohland,
Nik Lomax and Paul Norman

21.1 Introduction

Periodic censuses, annual population estimates and projections all provide important data for monitoring the state of the UK's population, where the aim is to count, estimate or project everybody. These datasets can be supplemented by survey data and administrative records which cover sub-sets of the population. Integration of these elements is currently being explored (Office for National Statistics (ONS), 2015a). UK census tables provide detailed population data down to small area scale (e.g. output areas); the mid-year official estimates provide populations by age and sex for local authority districts (LADs) and sub-LAD areas (e.g. super output areas, wards); official projections provide population tables by age and sex and households for LADs for the next 25 years. Compared with full censuses, the estimates and projections are very 'information-lite'. Should we be interested in reliable information beyond age and sex for the whole population, reliance must be placed on the census.

Population estimates and projections have been developed for various attributes beyond just age and sex. Ethnicity has been estimated at LAD scale by ONS (2011) and the ethnic composition of the population has been projected at LAD scale in England by Wohland et al. (2010) and Rees et al. (2011, 2012) and for London boroughs by the Greater London Authority (GLA, 2015). Forecasts of health status combining census, survey and population projection data for older people in English LADs have been generated by Clark (2015).

UK population estimates and projections are based on rolling forward populations from census counts using estimates of the components of change based on register and administrative data in the short term and assumptions about future behaviour of the rates in the long term. Evaluation of LAD population estimates by ethnicity rolled forward from the 2001 Census to mid-2009 (ONS, 2011) and the ETHPOP (ETHnic POPulation) projections by ethnicity from mid-2001 to mid-2011 revealed serious departures from the results of the 2011 Census (Rees et al., 2016).

The aim of this chapter is to explain how we estimated the populations and components of change by ethnicity, age and sex for the LADs of England together with Wales, which is treated as a single zone. Because population-level information (rather than sample based or proxy administrative information) on ethnicity is virtually absent in UK demographic statistics, the methods used here rely on the 'book ends' of the 2001 and 2011 Censuses. The ethnic group estimates are constrained so that they sum to the official mid-year population estimates and the mid-year to mid-year components of change. The time series of component rates by ethnicity provides the basis for setting assumptions for projections with greater intelligence.

The chapter focuses on methods used; empirical results are available from the UK Data Service (Wohland, 2016). The methods, although detailed and perhaps hard to follow for those not used to algebra using subscripted variables, have general applicability beyond the estimation of the inter-census populations and their components. We provide flow diagrams to show how the different pieces of information are combined which readers may find easier to follow. The methods are also applicable in longitudinal studies which follow birth cohorts of people over time. They also show how important census information is for estimating demographic time series for geographical areas when the population group is only recorded in national surveys between censuses.

21.2 Estimating ethnic populations and components between censuses

One technique, employed since censuses have been taken and where good registers of births and deaths exist, is to infer net migration for different populations by subtracting natural change from population change between censuses. When the populations are disaggregated by age, the calculation simplifies into a subtraction of deaths from population change. This technique is best applied to age bands with intervals equal to the inter-census time interval. It has not been used to estimate detailed annual components of change for populations classified by ethnicity, although Finney and Simpson (2008) provide estimates for 1991–2001 of natural change and net migration by ethnicity for districts in Great Britain.

Where estimates of inter-census populations by ethnicity are made, a cohort-component method is used. For example, the ONS produces roll forward ethnic population estimates for LADs in England (Large and Ghosh, 2006a, 2006b; ONS, 2011). The method depends on estimation of the components of change (births, deaths, internal and international migrations) for each year intervening between two censuses. These estimates assumed that mortality rates were the same for all ethnicities, concealing important variation (Wohland et al., 2015) and comparison with estimates based on Annual Population Survey (APS) data suggested considerable differences.

Bryant and Graham (2013) recognise that both population estimates and projections can drift away from the later 'true' figures. They construct a method for sub-national population estimation using a formal Bayesian framework. This enables them to develop credible intervals around estimates and demonstrate how uncertainty can be controlled for estimates between censuses. This uncertainty, however, can escalate without the constraint of a second census. At the heart of the estimation is a demographic account, providing a complete description of the population system. The relationship between the account and data is described by an 'observation' model which involves simulation carried out using Markov Chain–Monte Carlo methods (Bryant and Graham, 2013). We considered but did not use the Bryant–Graham methods, because of the computational challenge of processing 12 ethnic groups and 324 LADs in England compared with four ethnic groups and 67 territorial authorities in New Zealand.

21.3 Getting the census populations into shape

In order to use population data from the UK 2001 and 2011 Censuses to estimate populations and component rates by ethnicity for mid-year intervals between mid-2001 and mid-2011, it is necessary to harmonise key variable definitions for ethnic groups and LADs. Harmonisation is needed because the ethnic classifications differ between censuses and between the different 'census' countries: England and Wales, Scotland and Northern Ireland. The ONS harmonised classification uses ten groups but several features are problematic. The Gypsy/Traveller/Irish Traveller category is very small and any component estimates are likely to be poor. The absorption of the White Other group into a general White category means that significant new population flows in the decade cannot be tracked: the immigration from eastern European states that joined the EU in May 2004 and the immigration from southern European states with high unemployment. Finally, in the ONS harmonised classification, the three Black groups, distinguished in both censuses with different ethnic origins and demographic profiles, are merged into one. So, in the classification used in our analysis we move the Gypsy/Traveller/Irish Traveller population into the White British and Irish group and retain a separate White Other group. We accept the ONS merger of the different mixed or multiple ethnicities into one Mixed group. However, we retain three distinct Black groups. None of these decisions causes problems for LADs in England.

Harmonisation of geographical areas is essential in any analysis of population change. The principle we adopted was to employ the most recent definitions of LADs (lowest tier) in each home country of the UK. The current analysis is focused on English LADs, which were re-organised in 2009 through the creation of new unitary authorities (UAs) in place of previous shire counties and districts. Where this occurred, 2001 Census shire districts were aggregated to their successor UAs. We used two mergers of LADs with small populations into larger neighbours from previous analysis: the Isles of Scilly were merged with the Cornwall UA and the City of London with the City of Westminster. In this analysis, we used 324 English LADs and Wales is treated as a single zone.

21.4 The time and age–time frameworks for inter-census estimation

We need a clear framework for reconciling populations by age and ethnicity at the two censuses with the intervening components of change. By reconciliation we mean that the end population defined by the second census is obtained by inputting the components of change for the decade. To avoid complexity, we adjust the LAD populations at the census by ethnicity, age and sex, counted on 29 April 2001 or 27 March 2011, to agree with the following mid-year LAD population by age and sex. UK demographic statistics use mid-years as the time points at which the population is estimated or projected. This means we have an exact ten-year interval between the mid-year estimates directly informed by the first and second censuses. We label these adjusted counts 'census-based book end' (CBBE) populations and CBBE components.

There are two parts to this framework: an identification of the time points and intervals for which data are available; and a specification of the relationship between age and time. The

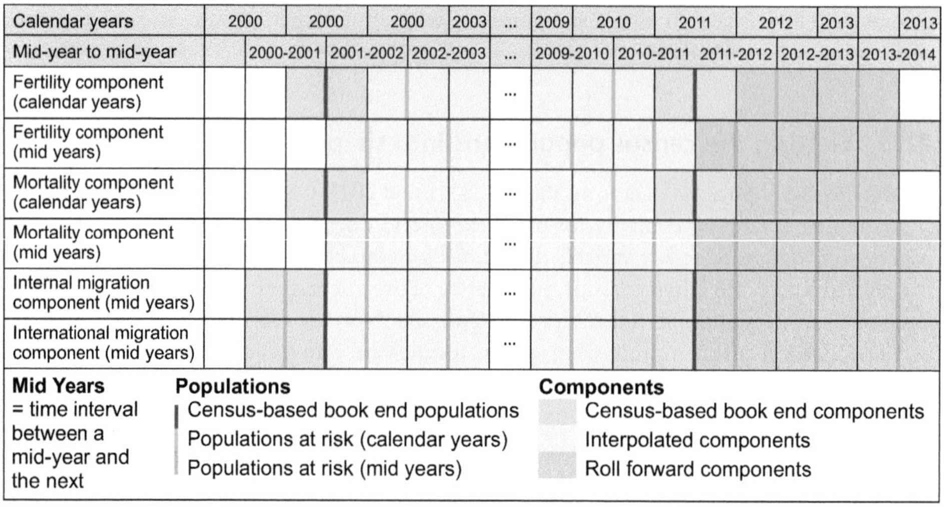

Calendar years	2000	2000	2000	2003	...	2009	2010	2011	2012	2013		2013
Mid-year to mid-year		2000-2001	2001-2002	2002-2003	...	2009-2010	2010-2011	2011-2012	2012-2013	2013-2014		
Fertility component (calendar years)					...							
Fertility component (mid years)					...							
Mortality component (calendar years)					...							
Mortality component (mid years)					...							
Internal migration component (mid years)					...							
International migration component (mid years)					...							

Mid Years
= time interval between a mid-year and the next

Populations
| Census-based book end populations
| Populations at risk (calendar years)
| Populations at risk (mid years)

Components
Census-based book end components
Interpolated components
Roll forward components

Figure 21.1   A time framework for estimation and projection of populations and components by ethnicity

columns in Figure 21.1 indicate the time intervals and points for which demographic data are estimated, either for calendar years or for mid-year to mid-year intervals. The rows represent the four components of change. Fertility and mortality are assigned two rows because the available input data refer to calendar years while the data needed for the reconciliation exercise refer to mid-year to mid-year intervals. The mid-year to mid-year demographic flows are estimated from the calendar-year data for births and deaths. The internal migration data are published for mid-year to mid-year intervals, while international migration data are provided for a variety of annual periods and are converted into mid-year to mid-year estimates.

Figure 21.1 represents populations counted or estimated for a point in time as vertical lines in different colours. It shows components (events or transitions that change the population) as horizontal bars, shaded in different colours. The CBBE adjusted counts are represented as red double lines, positioned at mid-2001 and mid-2011. The time intervals for component flows and rates extend from one mid-year to the next. However, often component flows are reported only for calendar years. Data from two successive years need to be combined, usually through averaging. CBBE components are identified in a pink shade. We sum the initial estimates of the LAD components by ethnicity, age and sex and then adjust these totals to agree with the ONS LAD components.

Mid-year LAD populations by ethnicity, age and sex are shown as vertical lines in the figure, from which populations at risk (PAR) are computed. The PAR for fertility and mortality rates are the mid-year populations of the calendar year, shown as vertical black double lines. The PAR for internal and international migration are an average of mid-year populations in the two years making up the mid-year to mid-year intervals and are shown as vertical green double lines. The PAR are multiplied by the component rates to yield first estimates of ethnic-specific flows, which are adjusted to the all-person estimates produced by ONS. The migration statistics, which refer to the year prior to the census, are used in combination with migration statistics for the mid-year to mid-year interval preceding the CBBE population. For international migration the interpolation is carried out on flows rather than rates because we use assumptions about future flows rather than rates in subsequent projections.

The second part of the framework is provided by the age–time diagram, a graph in which age is plotted on the vertical axis and time on the horizontal (Figure 21.2). The age bandwidths and time intervals in an age–time diagram must be equal in a roll-forward population estimation or projection model. Figure 21.2a sets out the standard case and Figure 21.2b the case for new-borns. The vertical axis is marked into equal sized bands of single years of age and labelled with age at last birthday. There are ten annual time (mid-year to mid-year) intervals on the horizontal axis and 20 ages (at last birthday) shown on the vertical axis. Horizontal bands refer to the same age over time intervals that change, while vertical bands refer to the same time intervals over ages that change. Diagonal bands in the graph refer to persons belonging to the same birth cohort. Each parallelogram in the graph refers to a period-cohort which tracks the changes in a cohort population in an annual time interval.

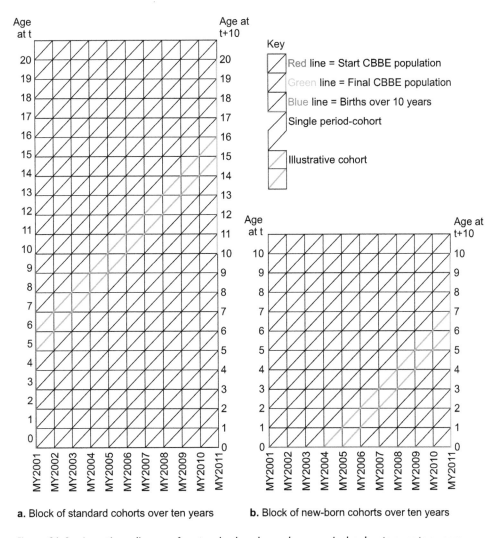

a. Block of standard cohorts over ten years   b. Block of new-born cohorts over ten years

Figure 21.2   Age–time diagram for standard and new-born period-cohorts over ten years

In Figure 21.2a, selected lines have been highlighted by colour, marking off a block of period-cohort spaces. Consider the space (a parallelogram) which shows the progression of persons aged 0–9 at mid-year 2001, the starting CBBE population which is highlighted in red, to being aged 10–19 at mid-2011, the finishing CBBE population, highlighted in green. The brown lines in the diagram delimit the age-time trajectories of a single cohort, born between mid-year 1995 and mid-year 1996. Members of this birth cohort are aged 5 at mid-year 2001 and aged 15 at mid-year 2011. What we need to do is to find the set of period-cohort component flows which, over the ten years (mid-year to mid-year intervals), change the starting population into the finishing population. The standard age–time framework applies to ages 0–100+ at mid-year 2001. Babies are born each year and require special treatment (Figure 21.2b): births replace the 2001 CBBE population as the starting book-end population and the estimates are made over time intervals which shrink from 9.5 years to 0.5 over the 2001–11 decade.

21.5 The estimation of mid-year populations and populations at risk

Before proceeding with the estimates of mid-year populations, the information available on ethnicity for mid-2002 to mid-2010 was reviewed for use in the estimation. The candidates for proxy populations were: (i) the Trend and/or the Emigration Rate projections for 2001 to 2011 from the 2001 Census-based ETHPOP projections (Wohland et al., 2010; Rees et al., 2011, 2012); (ii) the Population Estimates by Ethnic Group (PEEG; ONS, 2011); (iii) APS estimates (ONS, 2016); and (iv) linear interpolation between CBBE populations. Following evaluation of each alternative (Rees et al., 2016), linear interpolation between CBBE populations for LADs by gender, age and ethnicity was implemented.

Figure 21.3 sets out the scheme used in computing the ethnic-specific mid-year populations, in this case for 2000–01 and 2010–11. The top row of boxes indicate the inputs to the population estimation. The boxes shaded in blue refer to the census populations aggregated into our 12 harmonised ethnic groups. The boxes shaded in grey contain the mid-year population estimates by gender and age, reconciled by ONS between the 2001 and 2011 Censuses (ONS, 2013). The left-most box refers to the mid-2000 population reconciled to the 1991 and 2001 Census populations. The second row contains the CBBE populations in pink shaded boxes for 2001. The mid-year populations by gender, age and ethnicity are linearly interpolated between the CBBE populations in 2001 and 2011 as follows:

$$P^{i,2001+y}_{g,x+y,e} = P^{i,2001}_{g,x,e} + \left(\frac{y}{10}\right) \times \left[P^{i,2011}_{g,x+10,e} - P^{i,2001}_{g,x,e}\right] \tag{21.1}$$

where P stands for population, i for LAD, g for gender, x for age (0, 100+) and e for ethnicity; y (= 1, 9) is the mid-year number counting forward from year zero (mid-year 2001). A tenth of the difference between the 2001 and 2011 CBBE populations in the same cohort, 10 years of age apart, is multiplied by the year index and added to the mid-2001 population. This equation applies to the cohorts aged 0–100+ at mid-year 2001. For those ages where the population is born during the period, we implement a scheme for people starting life in the interval between the CBBEs for birth years MY2001–MY2002 to MY2010–MY2011:

$$P^{i,2001+(10-y)+x}_{g,x,e} = B^{i,2001+(10-y)}_{g,e} + \left(\frac{x+0.5}{y+0.5}\right) \times \left[P^{i,2011}_{g,y,e} - B^{i,2001+(10-y)}_{g,e}\right]$$

$$\forall y \in [9,0], x \in [0, y-1] \tag{21.2}$$

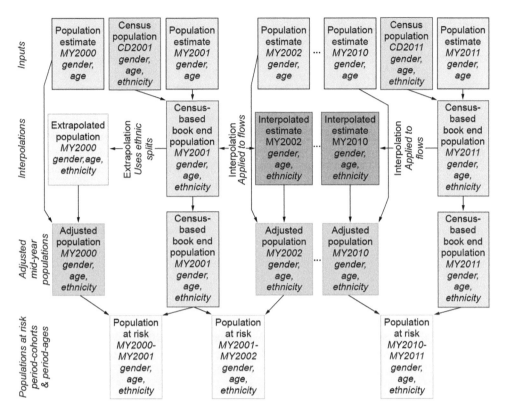

Figure 21.3 Scheme for estimating ethnic-specific populations and PAR, mid-year 2001 to mid-year 2011

Notes: MY = mid-year (30 June/1 July), CD = Census Date (29 April 2001).

The equation is shown using a nested 'loop' where the outer loop, y, is the age at MY2010–MY2011 and the inner loop, x, is the age being estimated from the new-born cohort to the year before the target age in MY2010–MY2011, i.e. (y-1). In order to interpolate between births in each year and the CBBE population for an age which is linked by cohort survival to births, we need to implement the fertility component estimation for the maternal age range 15 to 49 (discussed in Section 21.6).

An additional computation is needed for the mid-2000 population where we extrapolate the ethnic composition of the 2001 CBBE population backwards using ethnic shares of each LAD, gender and age population group, adjusting to ensure that shares are non-negative and sum to 100 per cent as follows:

$$P^{i,y}_{g,x,\star} = \sum^{e} P^{i,y}_{g,x,e} \tag{21.3}$$

$$S^{i,y}_{g,x,e} = 100 \times [P^{i,y}_{g,x,e} / P^{i,y}_{g,x,\star}] \tag{21.4}$$

$$S^{i,2000}_{g,x-1,e} = S^{i,2001}_{g,x,e} - \left(\frac{1}{10}\right) \times \left[S^{i,2011}_{g,x+10,e} - S^{i,2001}_{g,x,e}\right] \tag{21.5}$$

$$S_{g,x-1,e}^{i,2000'} = S_{g,x-1,e}^{i,2000} \times \left[100 \bigg/ \sum^{e} S_{g,x-1,e}^{i,2000} \right] \qquad (21.6)$$

$$P_{g,x-1,e}^{i,2000} = P_{g,x-1,\star}^{i,2000} \times [S_{g,x-1,e}^{i,2000'} / 100] \qquad (21.7)$$

where S is percentage share, $\star$ indicates summation over the e subscript it replaces and the prime indicates a revised estimate. Equation (21.3) sums the interpolated LAD mid-year populations over ethnicity, (21.4) computes the percentage shares, (21.5) carries out the extrapolation, (21.6) makes sure the shares add up to 100 and (21.7) generates the extrapolated populations.

We handle the oldest cohorts using the same equations as for the standard cohorts. Prior to these computations we must estimate the populations by single year of age over 100, which are not available at LAD scale. We use populations estimated by ONS (2015b) of the very old (including centenarians), decomposed by single years of age from 90 to 104 with 105+ as the final age group for 2002 to 2014 to decompose LAD populations aged 100+:

$$P_{g,x,e}^{i} = P_{g,100+,e}^{i} \times [P_{g,x}^{UK} / P_{g,100+}^{UK}], \quad x = 100, \dots, 104, 105+ \qquad (21.8)$$

We assume that all persons aged 105+ are 105 and that no one is 106+.

From the sequence of estimated LAD mid-year populations by gender, age and ethnicity, we computed PAR to use with component rates described in section 21.6. The PAR are required for mid-year to mid-year intervals and are computed as averages of pairs of mid-year population estimates, as shown in Figure 21.3.

The PAR for the standard period-cohorts (and the oldest period-cohorts) are computed as averages of two successive mid-year populations, as follows:

$$PAR_{g,x_{pc},e}^{i,\{y,y+1\}} = \tfrac{1}{2} \times \left[P_{g,x,e}^{i,y} + P_{g,x+1,e}^{i,y+1} \right] \qquad (21.9)$$

where the pair of times, $\{y, y+1\}$, refers to a mid-year to mid-year interval and the subscript x_{pc} refers to a period-cohort with start age x. The PARs for the fertility model that estimates births by ethnicity uses a period-age PAR.

21.6 The estimation of the ethnic components of change

The methods used to estimate the components of ethnic population change between censuses are now outlined.

Fertility estimates between CBBEs

Figure 21.4 shows the sequence of computations for fertility. The CBBE fertility rate estimates for calendar years 2001 and 2011 were based on a method developed by Norman et al. (2014) for 2001 and also used for calendar year 2011. From aggregate census tables of the population by gender, age and ethnicity were extracted infants aged 0 and mothers aged 15–44 by ethnicity to compute child–woman ratios for LADs. Births data by age of mother were used with mid-year female populations by age to compute age-specific and total fertility rates (no ethnicity).

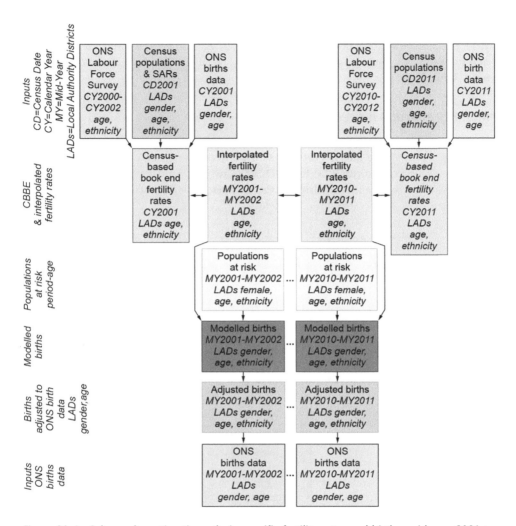

Figure 21.4 Scheme for estimating ethnic-specific fertility rates and births, mid-year 2001 to mid-year 2011

Census data from the Samples of Anonymised Records and from several years of the Labour Force Survey were used to compute ethnic- and age-specific fertility rates for England as a whole. These proxy estimates were then combined to produce CBBE estimates (pink box in Figure 21.4) of ethnic- and age-specific fertility rates for LADs (Norman et al., 2014, Table 4). Fertility rates for mid-year intervals between the 2001 and 2011 CBBE estimates (light orange box in Figure 21.4) were derived through linear interpolation, adapting the population interpolation method shown in equation (21.1), allowing for the switch from CBBE calendar years to mid-year to mid-year intervals:

$$f_{x,e}^{i,\{y,y+1\}} = f_{x,e}^{i,2001} + \left(y - \tfrac{1}{2}\right) \times \frac{\left[f_{x,e}^{i,2011} - f_{x,e}^{i,2001}\right]}{10} \, \forall \, y = 1,10 \tag{21.10}$$

285

Then the interpolated rates were multiplied by PAR defined by equation 21.9 (green boxes) to yield modelled births (blue boxes). Modelled births by age and ethnicity were then adjusted to sum to ONS births data for LADs provided in the official reconciled populations and components (ONS, 2013).

Mortality estimates between CBBEs

Figure 21.5 sets out the computations needed to estimate deaths by period-cohort and ethnicity for mid-year to mid-year intervals between CBBEs. Rees et al. (2009) estimated ethnic mortality rates and life expectancies (LEs) for LADs using two methods: the first used knowledge of long-term illness by ethnicity available in the censuses, while the second used knowledge of

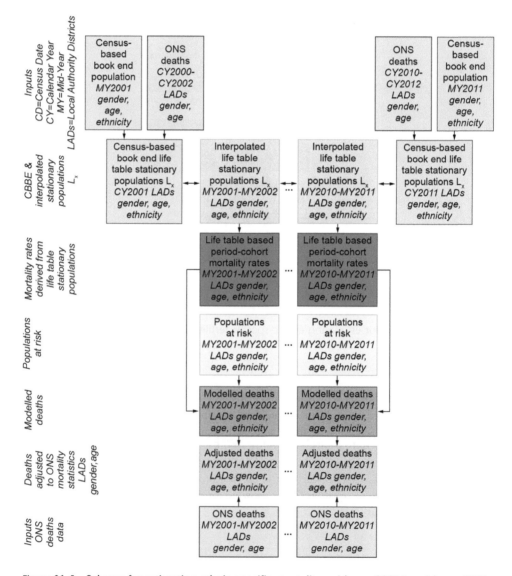

Figure 21.5 Scheme for estimating ethnic-specific mortality, mid-year 2001 to mid-year 2011

the geographical distribution of ethnic groups. In this analysis, the Geographical Distribution Method (GDM) was used, which involved computing a national estimate of ethnic mortality by using a weighted sum of local authority mortalities, the weights being based on the geographical distribution of each ethnic group. The national estimates were used in each LAD but adjusted to match the overall mortality rates. A fully satisfactory estimate of ethnic mortality must await incorporation of ethnicity into death registration.

The CBBE estimates refer to the three calendar years that bracket the census, following standard ONS practice, and combine CBBE estimates for MY2001 and MY2011 of the geographical distribution of ethnic groups with the deaths data for years 2000–02 and 2010–12. Between CBBE estimates, mortality rates are interpolated for mid-year to mid-year intervals (Figure 21.5). The interpolated life table variable was the stationary population, L_x, for mid-year to mid-year intervals $y = 0, 9$ or MY2001–MY2002 to MY2010–MY2011:

$$L_{g,x,e}^{i,\{y,y+1\}} = L_{g,x,e}^{i,2001} + \left(\frac{y + 0.5}{10}\right) \times \left[L_{g,x,e}^{i,2011} - L_{g,x,e}^{i,2001}\right] \tag{21.11}$$

From these interpolated stationary populations, we computed the mortality rates needed for forming the assumptions for our ethnic group projections. The mortality rates, m, are occurrence-exposure rates for period-cohorts which are applied in the projection model to PAR averaged between successive mid-years. They are computed as follows for period-cohorts indexed by the x subscript of the mortality rate age $0–1, 1–2, \ldots, 98–99$ and age $99–100$:

$$m_{g,x,e}^{i} = \left[L_{g,x,e}^{i} - L_{g,x+1,e}^{i}\right] / \frac{1}{2}\left[L_{g,x,e}^{i} + L_{g,x+1,e}^{i}\right] \tag{21.12}$$

For the final period-cohort, aged 100+ to 101+, the equation is modified to:

$$m_{g,100+,e}^{i} = \left[L_{g,100+,e}^{i} - L_{g,101+,e}^{i}\right] / \frac{1}{2}\left[L_{g,100+,e}^{i} + L_{g,101+,e}^{i}\right] \tag{21.13}$$

For the first period-cohort, the equation uses the life table radix, l_0, fixed at 100,000 for all population groups and an empirical factor is computed from infant mortality data for England and Wales, 2000–14 (Rees, 2015), which estimates the fraction of a year spent alive by babies born in those years as 0.225. The mortality rates for the period-cohort linking the new-born to age 0 are estimated from:

$$m_{g,b,e}^{i} = \left[l_0 - L_{g,0,e}^{i}\right] / \left[0.225l_0 + 0.775L_{g,0,e}^{i}\right] \tag{21.14}$$

Internal migration estimates between CBBEs

The scheme for estimating internal migrations between CBBEs is set out in Figure 21.6. Lomax (2013) constructed a dataset which is a complete origin by destination by age array of migration flows that harmonises definitions between home countries and estimates missing elements such as flows between LADs in different home countries (bottom row of Figure 21.6). The technique employed to create the dataset involved use of 2001 Census migration flows. This provided a complete UK origin–destination dataset for the year prior to the 2001 Census as initial estimates of flows between LADs for intermediate years from 2000–01 to 2010–11. The census flows were adjusted using iterative proportional fitting methods (Lomax and Norman, 2016) to inter-census marginal flow data from National Health Service (NHS) datasets recording changes of patient addresses.

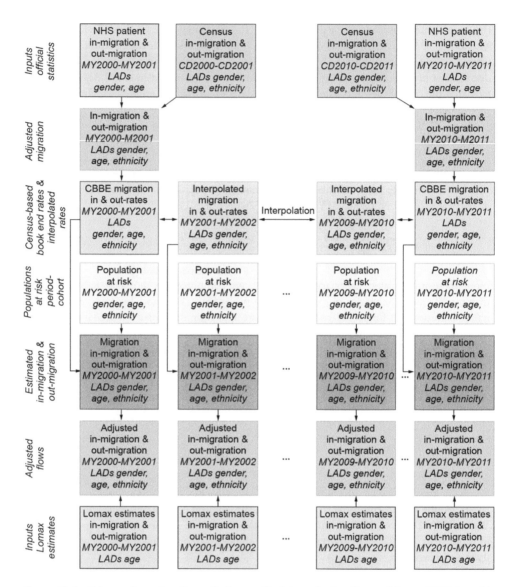

Figure 21.6 Scheme for estimating ethnic-specific internal migration, mid-year 2001 to mid-year 2011

The estimations have been extended to use census migration data for 2011 as well as 2001 and have been linked to available census cross-tabulations of LAD to LAD flows by ethnicity, making assumptions when a full classification by ethnicity could not be released because of disclosure rules (Figure 21.6, rows 1 and 2). For use in experimental projections (Rees et al., 2015) which used the bi-regional projection model, the flow arrays were aggregated to sets of LAD inflows and outflows from/to the rest of the UK. These flows were used to estimate in- and out-migration for the CBBE intervals MY2000 to MY2001 and MY2010 to MY2011 and divided by the appropriate PAR to yield in-migration and out-migration rates. Thereafter, in- and out-migration rates for intermediate mid-year to mid-year periods were interpolated using the methods set out in Section 21.5.

The interpolated rates were employed along with PAR to produce estimates of internal migration inflows and outflows (Figure 21.6, row 5) and a final adjustment was made to align the ethnic-specific LAD migration inflows and outflows with estimates for all groups. The inflows and outflows in this case were not extracted from the ONS (2013) reconciled populations and components dataset but from the updated estimates made by Lomax, which dealt with cross-border flows more accurately. The final adjustment built in the known variation in the volume and directions of internal migration during the decade, associated with the economic conditions in each year – the boom years of mid-2000 to mid-2007 followed by the severe recession of mid-2007 to mid-2010, followed by a slow recovery (Lomax et al., 2013, 2014).

International migration estimates between CBBEs

Figure 21.7 presents the scheme for estimating international migration for LADs. For this component we estimate the flows of immigrants and emigrants rather than rates. The

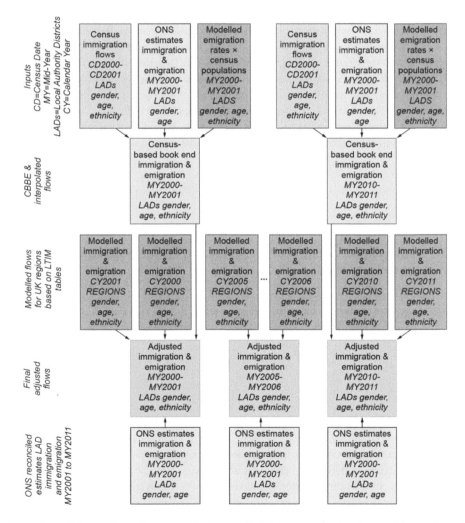

Figure 21.7 Scheme for estimating ethnic-specific international migration, mid-year 2001 to mid-year 2011

reason for estimating flows rather than rates is that, in the planned projections, we have left open the form in which international migration is incorporated. A challenge in estimating international migration both for CBBE years and in between was assigning an ethnic composition to both national and international flows. To help with this we made use of the official International Passenger Survey (IPS) and Long Term International Migration (LTIM).

Census migration tables provide immigration flow data for the single years prior to the 2001 and 2011 Censuses, classified by ethnicity. No equivalent information is available for emigration, so we modelled these flows by applying ONS emigration rates for LAD populations (not distinguishing ethnicity) to the census population classified by ethnicity. ONS emigration rates are themselves a modelled estimate dependent on the LAD population, immigration in previous years and various attributes. The CBBE immigration and emigration flows were then interpolated using equations equivalent to those for the population (Section 21.5).

IPS and LTIM data for 2000 to 2011 point to considerable variation in the country of origin of immigrants with a major increase in immigrants from eastern Europe occurring after eight countries joined the EU in May 2004 and rising numbers of immigrants from southern Europe as a result of the global financial crisis starting in 2008. Using data from the IPS/LTIM data series, we estimated a time series of immigration and emigration flows for calendar years 2000 to 2014, using new definitions of country of birth groups introduced in 2014. Using 2001 and 2011 Census tables of country of birth by ethnicity we computed the probability that a person born in a country group would be a member of one of the 12 ethnic groups. Multiplication of the country of birth time series by these probabilities yielded immigration and emigration totals by ethnicity for the nine regions of England and for Wales, Scotland and Northern Ireland. The LAD scale immigration and emigration flows were controlled to these regional compositions by ethnicity to produce adjusted international migration flows.

21.7   Demographic accounts for reconciling ethnic population and components

After the components of change between CBBEs have been estimated, it is necessary to reconcile the demographic component flows with start and end populations in each time interval. To carry out the reconciliation, we constructed demographic accounts. These are tables which assemble the component and population data in a format for checking and adjustment. If the end populations are inconsistent with the start populations and component flows, we adjust the component flows using an iterative proportional fitting routine.

There are many ways in which an arithmetic balance could be achieved. We make judgements, varying by age, about how the closure error across components should be distributed, based on knowledge of the data sources and estimation methods. We assume the bulk of the closure error derives from international migration and that emigration is far more uncertain than immigration. Errors in internal migration play an important part; we assign equal weights to both in- and out-migration as the data sources and estimation methods are common to the two flows. Finally, we regard the births and deaths estimates as most reliable but judge that errors in deaths estimates increase at older ages.

21.8   Discussion and conclusions

In this chapter we have described methods used to estimate ethnic population change in UK local areas. It will be possible to look at annual trends and fluctuations in the growth of minority ethnic groups and at any declines in the combined White British and Irish majority group at the geographical scale of LAD. The new estimates proved crucial for forming projection assumptions for components. In this process the availability of census datasets which reliably measure the population distribution and age–sex structure of ethnic group populations was vital. Ethnicity is not recorded on a systematic basis in any of the systems used to gather component change data (Norman et al., 2010). The registration of births and deaths does not require the identification of the ethnicity of the new-born or of the deceased. Where such information is collected, it is on a voluntary basis or through a third party. There are difficulties in determining the ethnicity of migrants. The NHS registration system is used as a proxy for migration but patients do not have to declare their ethnic identity. The IPS only asks questions about country of previous or next residence or country of birth or citizenship. The APS, despite its name, is still focused on gathering labour market statistics and fails to provide reliable estimates of population ethnicity outside the big cities because the sample size is limited. These deficiencies in capturing people's ethnicity in administrative records or surveys led us to place reliance on the results of two successive censuses which both record ethnicity.

To produce population and component estimates by ethnicity for LADs we employed an age and time framework that connects people, born into the same cohort, between censuses. However, the available sources yield demographic statistics for different time points and time intervals. A first step was to define, therefore, CBBE time points for population and CBBE time intervals for components. Two CBBE time intervals were defined: the first, employed for births and deaths, was based on calendar years around 2001 and 2011; the second, for internal and international migration, used mid-year to mid-year intervals prior to the censuses. Using CBBE populations and components as anchor points we were able to estimate intervening populations and components. We reviewed possible proxy variables that might indicate how much trends in ethnic populations departed from the linear over the ten years between CBBEs but rejected all candidates. We employed linear interpolation between CBBEs. In the case of international migration we felt a linear interpolation did not fit the story of immigration waves during the decade and developed new estimates of the ethnic composition of immigration and emigration flows based on IPS/LTIM tables. Previous work had shown how important international migration was for local ethnic population change.

The methods described here were implemented through a program written in R. The equations served as the design for writing the program; the program served as a check on the consistency of the equations. Because many readers might find these equations a block to understanding rather than an aid, we have used flow charts to show how different inputs feed into the procedures used for estimating the populations and change components. We concluded our explanation of the methodology by describing the demographic accounting frameworks into which the population estimates and components fit and pointing to the need for a final round of adjustment based on intelligent adjustment when the accounts did not precisely balance.

Without two reliable UK censuses in 2001 and 2011, none of this work would have been possible. National statistical agencies, pressured by national governments seeking to limit public expenditure, are investigating how to save on the expense of censuses through use of administrative data, surveys and 'big data'. These sources can provide valuable univariate population attribute information (Ajebon and Norman, 2015) but not cross-tabulations, especially with an ethnic group dimension. Cutting out censuses removes the most reliable knowledge about our population and prejudices the reliability of future prognoses. Censuses tell us what kind of people live where at regular time intervals and, using census information, we will be able to make better forecasts of how the population will change in the future.

Acknowledgements

The research reported in this chapter was supported by Economic and Social Research Council (ESRC) award ES/L013878/1 for the project Evaluation, Revision and Extension of Ethnic Population Projections – NewETHPOP and by the University of Leeds Higher Education Innovation Fund. The datasets used in the analysis were sourced from the ONS, National Records Scotland (NRS) and the Northern Ireland Statistics & Research Agency (NISRA) under the terms of the UK's Open Government Licence and are Crown copyright.

References

Ajebon, M. and Norman, P. (2015) Beyond the census: A spatial analysis of health and deprivation in England. *GeoJournal*, DOI: 10.1007/s10708-015-9624-8.

Bryant, J.R. and Graham, P.J. (2013) Bayesian demographic accounts: Subnational population estimation using multiple data sources. *Bayesian Analysis*, 8(3): 591–622.

Clark, S. (2015) Modelling the impacts of demographic ageing on the demand for health care services. PhD Dissertation, School of Geography, University of Leeds, Leeds. Available at: http://etheses.whiterose.ac.uk/11676/.

Finney, N. and Simpson, L. (2008) Population dynamics: The roles of natural change and migration in producing the ethnic mosaic. *CCSR Working Paper 2008–21*, Cathie Marsh Centre for Census and Survey Research, University of Manchester. Available at: http://hummedia.manchester.ac.uk/institutes/cmist/archive-publications/working-papers/2008/2008-21-population-dynamics.pdf.

GLA (2015) 2014 round ethnic group population projections. Greater London Authority, London Datastore. Available at: data.london.gov.uk/dataset/2014-round-ethnic-group-population-projections.

Large, P. and Ghosh, K. (2006a) A methodology for estimating the population by ethnic group for areas within England. *Population Trends*, 123: 21–39.

Large, P. and Ghosh, K. (2006b) Estimates of the population by ethnic group for areas within England. *Population Trends*, 124: 8–25.

Lomax, N. (2013) Internal and cross-border migration in the United Kingdom: Harmonising, estimating and analysing a decade of flow data. PhD Dissertation, School of Geography, University of Leeds, Leeds. Available at: http://etheses.whiterose.ac.uk/5839/.

Lomax, N. and Norman, P. (2016) Estimating population attribute values in a table: 'Get me started in' Iterative Proportional Fitting (IPF). *Professional Geographer*, 68(3): 451–461.

Lomax, N., Norman, P., Rees, P. and Stillwell, J. (2013) Subnational migration in the United Kingdom: Producing a consistent time series using a combination of available data and estimates. *Journal of Population Research*, 30(3): 265–288.

Lomax, N., Stillwell, J., Norman, P. and Rees, P. (2014) Internal migration in the United Kingdom: Analysis of an estimated inter-district time series, 2001–2011. *Applied Spatial Analysis and Policy*, 7(1): 25–45.

Norman, P., Rees, P., Wohland, P. and Boden, P. (2010) Ethnic group populations: the components for projection, demographic rates and trends. In Stillwell, J. and van Ham, M. (eds.) *Ethnicity and Integration*. Springer, Dordrecht, pp. 289–315.

Norman, P., Rees, P. and Wohland, P. (2014) The use of a new indirect method to estimate ethnic-group fertility rates for subnational projections for England. *Population Studies*, 68(1): 43–64.

ONS (2011) Population Estimates by Ethnic Group, 2002–2009. Statistical Bulletin, Office for National Statistics. Available at: ons.gov.uk/ons/rel/peeg/population-estimates-by-ethnic-group--experimental-/current-estimates/population-density--change-and-concentration-in-great-britain.pdf.

ONS (2013) Population Estimates for England and Wales, Mid-2002 to Mid-2010 Revised (Subnational). Statistical Bulletin. Available at: www.ons.gov.uk/ons/dcp171778_307723.pdf.

ONS (2015a) ONS Census Transformation Programme: Administrative Data Update. Available at: www. ons.gov.uk/ons/guide-method/census/2021-census/progress-and-development/research-projects/ beyond-2011-research-and-design/research-outputs/administrative-data-update.pdf.

ONS (2015b) Estimates of the Very Old (including Centenarians), England and Wales, and United Kingdom, 2002 to 2014. Statistical Bulletin, Office for National Statistics. Available at: www.ons.gov.uk/ons/ dcp171778_418037.pdf.

ONS (2016) Annual Population Survey. NOMIS, Official Labour Market Statistics. Available at: www. nomisweb.co.uk/.

Rees, P. (2015) Factor to use in computing period-cohort mortality rates for the new born, based on ONS 2001–2014 infant mortality data. Spreadsheet available from the author, School of Geography, University of Leeds, Leeds.

Rees, P., Wohland, P. and Norman, P. (2009) The estimation of mortality for ethnic groups within the United Kingdom. Social Science and Medicine, 69(11): 1592–1607.

Rees, P., Wohland, P. and Norman, P. (2016) The United Kingdom's multi-ethnic future: How fast is it arriving? In Lombard, J., Stern, E. and Clarke, G.P. (eds.) Applied Spatial Modelling and Planning. Routledge, London, pp. 157–171.

Rees, P., Wohland, P., Norman, P. and Boden, P. (2011) A local analysis of ethnic group population trends and projections for the UK. Journal of Population Research, 28: 149–183.

Rees, P., Wohland, P., Norman, P. and Boden, P. (2012) Ethnic population projections for the UK, 2001–2051. Journal of Population Research, 29: 45–89.

Rees, P., Wohland, P., Clark, S., Lomax, N. and Norman, P. (2015) Ethnic population projections for the UK and local areas, 2011–2061: New results for the fourth demographic transition. Paper presented at the British Society for Population Studies, Annual Conference, Leeds, 7–9 September. Available at: www.ethpop.org/Presentations/NewETHPOP/BSPS2015/BSPS%20Ethnic%20Population% 20Projections%20Phil%20Rees.pdf.

Wohland, P. (2016) SN 852508 NEWETHPOP-Ethnic population projections for UK local areas, 2011–2061. UK Data Service, University of Essex. Available at: https://discover.ukdataservice.ac.uk/ catalogue/?sn=852508&type=Data%20catalogue.

Wohland, P., Rees, P., Nazroo, J. and Jagger, C. (2015) Inequalities in healthy life expectancy between ethnic groups in England and Wales in 2001. Ethnicity and Health, 20(4): 341–353.

Wohland, P., Rees, P., Norman, P., Boden, P. and Jasinska, M. (2010) Ethnic population projections for the UK and local areas, 2001–2051. Working Paper 10/02, School of Geography, University of Leeds, Leeds. Available at: www.geog.leeds.ac.uk/fileadmin/documents/research/csap/working_papers_new/2010-02.pdf.

The prevalence of informal care and its association with health

Longitudinal research using census data for England and Wales

*Maria Evandrou, Jane Falkingham, James Robards
and Athina Vlachantoni*

22.1 Introduction

The provision of informal care towards family members, friends or neighbours has received increasing attention over the last two decades by academics and policy makers alike as a result of a combination of demographic trends and policy changes. On the one hand, the increasing number of people surviving to old age directly affects the demand for informal care provision, while cuts in local authority budgets are resulting in a greater reliance on the informal sector for the provision of social care for older people (Commission on Funding of Care and Support, 2011). At the same time, other demographic changes, such as increasing life expectancy for both men and women at older ages, mean that older fit individuals are better able to provide support to their spouses, other members of their family and beyond (Grundy and Tomassini, 2010; Norman and Purdam, 2013; Robards et al., 2012).

The relationship between informal care provision and the health of the carer is complex, and our understanding of such relationships can differ depending on the methodology used (Vlachantoni et al., 2013). Cross-sectional analysis of the association between informal caregiving and the carer's health (and by extension, mortality) shows mixed results and is dependent upon the particular characteristics of the person providing the care and the person receiving it. For instance, O'Reilly et al. (2008) found that, generally, carers are less likely than non-carers to report a limiting long-term illness (with the exception of men providing more than 50 hours per week); in contrast, Doran et al. (2003), who analysed the 2001 UK Census, found the opposite; and finally Ross et al. (2008), using data from the English Longitudinal Study of Ageing (ELSA), found no differences in health among carers and non-carers aged 50 and over.

In contrast to cross-sectional studies, longitudinal analyses allow us to understand the direction of causality between care provision and health; however, even this kind of analysis presents challenges in terms of unravelling the interaction of caregiving with other roles undertaken by individuals over their lifecourse. O'Reilly et al. (2008) examined individuals' caring status in the 2001 Northern Ireland Census and their mortality four years later, and showed that although overall caregivers presented lower mortality rates than non-caregivers, nevertheless the mortality risk increased with the intensity of care provided. This highlights the importance of considering the intensity of care provision (usually measured in hours per week), and changes to such intensity over time. Other longitudinal research from the UK and beyond has found that caregivers show a lower mortality risk than non-carers, or that caregiving has no negative impact on one's mortality (Rahrig Jenkins et al., 2009; Fredman et al., 2010). Finally, longitudinal research has also highlighted the complexity of disentangling individuals' caring roles from other roles over their lifecourse, for instance the parental and paid worker roles (Glaser et al., 2005).

Against this background, this chapter presents analyses of the Office for National Statistics Longitudinal Study (ONS LS), a 1 per cent extract of 2011 Census records matched to responses from the same individuals at the 2001 and earlier censuses. The chapter first outlines the trends in the provision of informal care in the UK between 2001 and 2011 in Section 22.2, along with information on the key characteristics of carers. The evidence in this section is contextualised against existing research on the characteristics of informal carers. The remainder of the chapter then investigates the relationship between informal care provision and the carer's health from different angles using a longitudinal approach. Section 22.3 examines the relationship between different trajectories of informal care provision between 2001 and 2011 and self-reported health status in 2011, while Section 22.4 investigates the impact of health on the propensity to provide informal care in 2011, among those providing care in 2001. The final section discusses the policy implications of current patterns of informal care provision for the future design of policies aimed at supporting informal carers and care recipients.

22.2   Patterns of informal care provision between 2001 and 2011

A question about provision of unpaid (informal care) was first included in the 2001 UK Census and repeated in 2011. The 2001 Census indicated that approximately 5.8 million individuals provided informal care to other persons (Doran et al., 2003), and among them, over 1 million were aged 65 and over. The intensity of informal care provision was also highlighted, showing that more than one-fifth of all informal carers were caring for at least 50 hours per week and among those, more than one-quarter reported 'not good' general health (Doran et al., 2003). The provision of high-intensity care was most prevalent across the latter part of the lifecourse, with almost half of all carers aged 85 years and over providing 50 hours of care or more per week (Evandrou, 2005). Indeed, research has shown that although informal care provision is more common in mid-life, the intensity of caregiving increases later in life, particularly within spousal relationships (Dahlberg et al., 2007; Robards et al., 2015).

Analysis of the 2011 Census indicated that in England and Wales, the prevalence of informal care had increased from 2001, particularly for more intense caring, that is, 20 or more hours per week (ONS, 2013a). A geographical pattern was also evident from the analysis of the 2011 Census, with the South West region of the country showing higher proportions of the population providing informal care, compared to London. This is likely due to a younger age structure of the population in London compared to an older age structure in the South West. Gender differences have been found to permeate patterns of informal care provision, often interacting with individuals' marital status (Young and Grundy, 2008). The role of gender in care provision continued to be significant in 2011, with women constituting 58 per cent of all carers, although the gender gap narrows in later life (ONS, 2013b).

Table 22.1 compares the prevalence of informal care in the ONS LS with the aggregate census figures for England and Wales in 2001 and 2011. Overall, 10.8 per cent of ONS LS members reported providing some level of care in 2001 and 10.6 per cent in 2011. There is a small

Table 22.1 Number and percentage of informal carers in aggregate census data and the ONS LS by caring intensity, 2001 and 2011

Caring level	Informal caring at 2001 Census				Informal caring at 2011 Census			
	ONS LS		Census		ONS LS		Census	
	N	%	N	%	N	%	N	%
No care provided	457,662	89.2	46,824,111	90.0	521,681	89.4	50,275,666	89.7
1–19 hours per week	37,567	7.3	3,555,822	6.8	38,796	6.6	3,665,072	6.5
20–49 hours per week	6,074	1.2	573,647	1.1	8,428	1.4	775,189	1.4
50+ hours per week	11,663	2.3	1,088,336	2.1	14,738	2.5	1,359,985	2.4
Total carers	*55,304*	*10.8*	*5,217,805*	*10.0*	*61,962*	*10.6*	*5,800,246*	*10.3*
Total	538,190	100	52,041,916	100	583,643	100	56,075,912	100

Sources: Evandrou et al. (2015, Table 1). Aggregate England and Wales informal caring percentages are from ONS (2013a) and are adjusted. Authors' own analysis of ONS LS unadjusted data. Excluding edited carer cases from 2001 Census.

Table 22.2 Number and percentage of informal carers in the ONS LS by caring intensity and sex, 2001 and 2011

Caring level	Informal caring at 2001 Census					Informal caring at 2011 Census				
	N		%			N		%		
	Male	Female	Male	Female	All	Male	Female	Male	Female	All
No care provided	225,318	232,344	90.5	88.0	89.2	257,607	264,074	90.8	88.0	89.4
1–19 hours per week	16,592	20,975	6.7	7.9	7.3	16,692	22,104	5.9	7.4	6.6
20–49 hours per week	2,417	3,657	1.0	1.4	1.2	3,554	4,874	1.3	1.6	1.4
50+ hours per week	4,595	7,068	1.8	2.7	2.3	5,798	8,940	2.0	3.0	2.5
Total carers	*23,604*	*31,700*	*9.5*	*12.0*	*10.8*	*26,044*	*35,918*	*9.2*	*12.0*	*10.6*
Total	248,922	264,044	100	100	100	283,651	299,992	100	100	100

Sources: Evandrou et al. (2015, Table 2). Authors' own analysis of ONS LS. Excluding edited carer cases from 2001 Census.

difference in the proportion of non-carers between the ONS LS sample and the 2001 Census (89.2 per cent compared to 90 per cent) and individuals providing between one and 19 hours of care per week (7.3 per cent compared to 6.8 per cent). However, the difference is less marked in 2011 (6.6 per cent compared to 6.5 per cent). Looking at the patterns of change across the decade, reassuringly the ONS LS reflects the same pattern of change in the profile of care seen in the aggregate census results; between 2001 and 2011, it can be seen that there is an increase among carers providing more than 50 hours of care per week in both the ONS LS sample (2.3 to 2.5 per cent) and the full census (2.1 to 2.4 per cent). In some cases, analyses are restricted to those aged 25–74 years at the census or edited cases have been removed to avoid issues arising from 2001 Census data processing edit rules (Buxton and Smith, 2010). Having benchmarked the ONS LS data, the remainder of this chapter presents results from the ONS LS only.

Looking at the gender differences in informal care provision, Table 22.2 indicates that 12 per cent of women and 9.5 per cent of men reported providing care in 2001 and this did not change substantially by 2011. However, between 2001 and 2011, there were changes in the intensity of care provided, with the percentage of men and women providing 1–19 hours of informal care per week decreasing (from 6.7 to 5.9 per cent for men and 7.9 to 7.4 per cent for women) and the percentage of individuals providing medium (20–49 hours) and high-intensity care (50 hours or more) per week increasing.

Focusing on the factors associated with the provision of caring (of any intensity) in 2001 and in 2011, cross-sectional multivariate analysis shows significant stability over time in the characteristics of informal carers (Table 22.3). In terms of demographic characteristics, men were less likely to provide care than women, and individuals aged between 45 and 54 were more likely to provide care compared to other age groups. Elsewhere it is shown that Pakistani and Bangladeshi individuals were found to be more likely to provide care compared to White British individuals at both time points (Evandrou et al., 2015). Socio-economic factors were also found to be important predictors of providing care, for example social renters (i.e. renting from a local authority or housing association) were more likely to provide care than owner-occupiers, while individuals

Table 22.3 Binary logistic regression of any level informal caring, 2001 and 2011 Censuses

	Model 1 – Caring at 2001					Model 2 – Caring at 2011				
	N	%	OR	Sig.	95% CI	N	%	OR	Sig.	95% CI
Sex										
Female (ref.)	196,858	51.2	1.00			217,226	51.3	1.00		
Male	187,487	48.8	0.80	0.000	0.79 0.82	206,169	48.7	0.76	0.000	0.74 0.77
Age										
55–64 (ref.)	58,783	15.3	1.00			70,336	16.6	1.00		
16–19	23,623	6.1	0.23	0.000	0.21 0.25	25,870	6.1	0.24	0.000	0.22 0.26
20–34	101,625	26.4	0.31	0.000	0.29 0.32	111,094	26.2	0.32	0.000	0.31 0.33
35–44	80,325	20.9	0.57	0.000	0.55 0.59	81,874	19.3	0.53	0.000	0.51 0.54
45–54	73,018	19.0	0.98	0.269	0.95 1.01	82,188	19.4	0.92	0.000	0.89 0.95
65–74	46,971	12.2	0.66	0.000	0.63 0.68	52,033	12.3	0.71	0.000	0.68 0.73
Health, 2001										
Fairly good (ref.)	94,242	24.5	1.00							
Good	253,222	65.9	0.74	0.000	0.73 0.76					
Not good	36,881	9.6	0.85	0.000	0.82 0.88					
Health, 2011										
Fair (ref.)						56,100	13.3	1.00		
Very good						179,014	42.3	0.79	0.000	0.77 0.82
Good						164,761	38.9	0.97	0.058	0.94 1.00
Bad						18,487	4.4	0.74	0.000	0.70 0.77
Very bad						5,033	1.2	0.54	0.000	0.49 0.59
LLTI										
Yes, limited a lot/little (ref.)	67,952	17.7	1.00			69,491	16.4	1.00		
No LLTI or health problem	316,393	82.3	0.86	0.000	0.83 0.89	353,904	83.6	0.68	0.000	0.66 0.71

Sources: Adjusted from Evandrou et al. (2015, Table 13). Authors' own analysis of ONS LS. Excluding edited carer cases from 2001 Census.

Note: The model also controlled for marital status, ethnic group, housing tenure, economic activity, highest educational qualification, car access and Government Office Region.

looking after the home and those working part-time were more likely to provide care than those working full-time (Evandrou et al., 2015).

The comparison of self-reported health as a factor associated with care provision between 2001 and 2011 is complicated by changes in the response options between these time points. In both years, individuals who reported 'fairly good'/'fair' health were the most likely to be providing care compared to all other categories. In 2001, individuals reporting 'good' health were slightly less likely to be caring than those reporting 'not good' health, whereas in 2011, those reporting 'very bad' or 'bad' health were less likely to be caring than those reporting 'very good' health. Compared to individuals with a long-term limiting illness (LLTI), those who reported no such an illness were more likely to be providing informal care in 2001 and 2011, even after controlling for age; this is in line with other research findings examining the relationship between health status and caring using cross-sectional data. Below, this relationship is further explored by exploiting the longitudinal nature of the ONS LS.

22.3   The association between caring trajectories and health

Understanding the relationship between health status and caring is complicated by the fact that there may be a selection effect of individuals into the caring role who are sufficiently fit to provide care. Moreover, the relationship may be mediated over time by the intensity and duration of care provided, and a change in health status may result in some individuals ceasing to care. Thus, health influences care and, in turn, caring influences health.

In the next two sections we: (i) explore the relationship between different trajectories of care between 2001 and 2011 on health outcomes in 2011; and (ii) investigate the role of health as a predictor of providing care in 2011, controlling on being a carer in 2001 (and thus controlling for initial selection effects). The analysis therefore sheds new light on both sides of the 'care–health' relationship.

Table 22.4 uses linked data from 2001 and 2011 to examine trajectories of caring among those who were aged 25–75 in 2001 and who were still present in England and Wales at the 2011 Census (aged 35–85). The typology distinguishes between carers and non-carers; and among carers, the groups are further distinguished between light carers (1–19 hours of care provided per week) and heavy carers (20 hours or more per week). The majority of individuals were not providing care at either 2001 or 2011 (74 per cent), while just over 11 per cent were not caring in 2001 but were caring in 2011 ('starters'); almost 10 per cent were caring in 2001 but not caring in 2011 ('stoppers'); and finally just over 5 per cent were caring in both 2001 and 2011 ('repeat carers').

Focusing on the prevalence of poor health ('bad' or 'very bad') in the different sub-groups, the table indicates that individuals who had shifted from no care to providing heavy care, or the opposite, showed the highest percentages of reporting poor health in 2011 (13 and 17 per cent respectively), possibly reflecting the demanding nature of heavy caring for the former group, and the difficulty of sustaining heavy care provision over a prolonged period for the latter group.

Table 22.5 uses the typology above in order to show the likelihood of reporting poor health in 2011. It indicates that, after controlling for a range of demographic, health-related, area-related and socio-economic characteristics, individuals who were not caring in 2001 but had started caring in 2011, and those who were caring at both time points, were less likely to report poor health than non-carers (odds ratios (OR) of 0.64 and 0.56 respectively). By contrast, individuals who were caring in 2001 but were not caring in 2011 were approximately 9 per cent more likely than non-carers to report poor health in 2011, which may suggest a negative impact on health of the carer's role. Once the intensity of the caring role is taken into account (model 2), Table 22.5 shows that individuals who had a heavy caring role in 2001 and provided no care in 2011 were

Table 22.4 Caring trajectories among ONS LS members aged 25–75 years (2001) resident at both 2001 and 2011 Censuses; and percentage of each care group reporting poor health

Carers' trajectories	%	N	% of care group reporting poor health in 2011
(a) Not caring at 2001 and 2011	73.6	198,753	8.2
(b) Not caring at 2001; caring at 2011	11.4	30,682	11.4
Non-carer to light carer	7.6	20,559	4.3
Non-carer to heavy carer	3.7	10,123	12.6
(c) Caring at 2001; not caring at 2011	9.6	25,939	9.6
Light carer to non-carer	7.2	19,338	7.8
Heavy carer to non-carer	2.4	6,601	17.1
(d) Caring at 2001 and 2011	5.4	14,680	5.4
Persistent light carer	2.5	6,747	3.5
Carer, increasing intensity	0.9	2,466	9.5
Carer, decreasing intensity	0.5	1,305	8.5
Persistent heavy carer	1.5	4,162	14.3
Total	100.0	270,054	8.3

Source: Vlachantoni et al. (2016). Available under the terms of the Creative Commons Attribution License (CC BY). https://creativecommons.org/licenses/by/4.0/ DOI : http://dx.doi.org/10.1016/j.ssmph.2016.05.009. Authors' own analysis of ONS LS.

Note: Light carer is defined as those providing 1–19 hours of care a week; heavy carer is defined as those providing 20 or more hours of care a week.

Table 22.5 Regression odds ratios for reporting poor health at 2011 among ONS LS members aged 35–85 years (2011) resident at 2001 and 2011

Carer status	Model 1				Model 2			
	OR	P (sig)	95% CI		OR	P (sig)	95% CI	
(a) Not caring at 2001; not caring at 2011	1							
Non-carer					1			
(b) Not caring at 2001; caring at 2011	0.64	0.000	0.60	0.67				
Non-carer to light carer					0.58	0.000	0.54	0.63
Non-carer to heavy carer					0.69	0.000	0.64	0.74
(c) Caring at 2001; not caring at 2011	1.09	0.001	1.04	1.15				
Light carer to non-carer					1.03	0.390	0.97	1.09
Heavy carer to non-carer					1.22	0.000	1.13	1.33
(d) Caring at 2001 and 2011	0.56	0.000	0.52	0.60				
Persistent light carer					0.45	0.000	0.50	0.68
Carer, increasing intensity					0.58	0.000	0.50	0.68
Carer, decreasing intensity					0.64	0.000	0.51	0.80
Persistent heavy carer					0.63	0.000	0.56	0.70

Source: Vlachantoni et al. (2016). Available under the terms of the Creative Commons Attribution License (CC BY), https://creativecommons.org/licenses/by/4.0/ DOI : http://dx.doi.org/10.1016/j.ssmph.2016.05.009. Authors' own analysis of ONS LS.

Note: Controls for health at baseline (2001 Census), LLTI at baseline (2001), sex, age (2011), ethnic group (2011), change in marital status (2001–11), housing tenure (2011), highest educational qualification (2011), car access (2011) and household LLTI (2001 and 2011).

about 22 per cent more likely than non-carers to report poor health, whereas there was no significant difference for individuals who moved from a light caring role to providing no care in 2011 (OR 1.03), highlighting the potential negative impact on health of heavy caring. Regardless of the intensity of care provision, individuals who either started providing care in 2011, or who provided care in both time points, were less likely to report poor health in 2011 than non-carers.

Overall, individuals who provided light care in both 2001 and 2011 were the least likely to report poor health, while those who had moved from heavy to light care between 2001 and 2011, and those who had taken up light caring in 2011 from no caring in 2001, showed lower odds of reporting poor health compared to non-carers. Such findings indicate the possibility that individuals with relatively good health are selected into the caring role, and that a certain level of good health is compatible with, or indeed required for, the provision of informal care to other individuals. When the analysis was run separately for men and women, broadly similar results were found with two exceptions: first, men who moved from no caring to light caring were slightly more likely than women making the same transition to report poor health; and, second, men who shifted from light to heavy caring were also more likely than women making such a transition to report poor health (Vlachantoni et al., 2016).

22.4 Among carers in 2001, the role of health as a predictor of caring again in 2011

In order to further understand the role of health status on an individual's risk of providing informal care, this final section presents the results of a model focusing on the sub-sample of individuals who were providing care in 2001, with care provision in 2011 as the outcome variable (Table 22.6). A range of demographic characteristics were significant. Being male (compared to female) was associated with a lower likelihood of providing care again in 2011, while individuals aged between 35 and 44 years in 2001 were the most likely to also be caring ten years later. Marital status was also an important predictor of repeated caring between 2001 and 2011, with

Table 22.6 Among informal carers in 2001, binary logistic regression to predict provision of informal care (of any intensity) in 2011

	N	%	Odds Ratio	Sig.	95% CI		
Sex							
Female (ref.)	25,416	59.3	1.00				
Male	17,413	40.7	0.90	0.000	0.86	to	0.94
Age, 2011							
45–54 (ref.)	8,832	20.6	1.00				
26–34	2,025	4.7	0.47	0.000	0.41	to	0.53
35–44	4,553	10.6	0.73	0.000	0.67	to	0.79
55–64	12,581	29.4	0.93	0.028	0.87	to	0.99
65–74	10,142	23.7	0.71	0.000	0.65	to	0.77
75–84	4,696	11.0	0.62	0.000	0.56	to	0.69
Caring level, 2001							
50+ hours per week (ref.)	8,033	18.8	1.00				
1–19 hours per week	30,268	70.7	0.41	0.000	0.39	to	0.44
20–49 hours per week	4,528	10.6	0.65	0.000	0.60	to	0.70

	N	%	Odds Ratio	Sig.	95% CI		
Marital status, 2011							
Married or in a registered same-sex civil partnership (ref.)	27,867	65.1	1.00				
Never married and never registered a same-sex civil partnership	5,199	12.1	0.73	0.000	0.68	to	0.79
Separated, but still legally married or in a same-sex civil partnership	1,021	2.4	0.65	0.000	0.57	to	0.75
Divorced or formerly in a same-sex civil partnership now legally dissolved	4,381	10.2	0.65	0.000	0.60	to	0.70
Widowed or surviving partner from a same-sex civil partnership	4,361	10.2	0.24	0.000	0.22	to	0.27
Health, 2011							
Fair (ref.)	9,991	23.3	1.00				
Very good	10,709	25.0	0.87	0.000	0.81	to	0.94
Good	18,223	42.5	0.93	0.017	0.87	to	0.99
Bad	3,141	7.3	0.69	0.000	0.63	to	0.76
Very bad	765	1.8	0.49	0.000	0.41	to	0.59
LLTI, 2011							
Yes, limited a lot / little (ref.)	13,445	31.4	1.00				
No LLTI	29,384	68.6	0.74	0.000	0.69	to	0.79

Source: Authors' own analysis of ONS LS. Excluding edited carer cases from 2001 Census.

Note: The model also controlled for highest educational qualification, economic activity, ethnic group and housing tenure.

those who were married being more likely to be also caring in 2011, reflecting both the provision of spousal care and also care for parents and parents-in-law (being married increases the risk of having a surviving parent-in-law). Finally, the model shows that carers providing between one and 19 hours or between 20 and 49 hours of care per week in 2001 were less likely to also be caring in 2011 compared to those providing 50 hours or more of care per week in 2001.

Turning to the role of health status, the model shows that among those who were providing care in 2001 (and thus controlling for any initial healthy carer effect), those reporting their health to be 'fair' in 2001 were most likely to also be caring in 2011. Those reporting 'bad' or 'very bad' health were, perhaps not surprisingly, the least likely to be caring, but interestingly those whose health was 'very good' or 'good' were also less likely to be caring again in 2011 compared to those in fair health. Individuals who reported a LLTI in 2001 were more likely to be caring again in 2011 compared to those without such illness. Such findings reflect a complex relationship between health status and the activity of caring, indicating that poor general health is incompatible with a sustained caring role.

22.5 Discussion of findings and conclusion

Our understanding of patterns of informal care provision has come a long way since the introduction of the relevant question in the 2001 UK Census, and linked data in the form of the ONS LS has allowed for the exploration of changes in such patterns over time. The analysis shown in this chapter reaffirms the complexity of the relationship between informal care provision and the carer's health, particularly when examined over time.

In summary, the cross-sectional analysis showed that individuals reporting fair health were the most likely to be providing informal care in both 2001 and 2011, but at the same time those who reported a LLTI were more likely to be caring than those who did not report such an illness. Taking a longitudinal perspective, examining the association of health on the likelihood of providing repeat care, after controlling for a range of demographic, socio-economic and area-related factors, it was found that, among carers in 2001, reporting fair health and a LLTI were both associated with higher odds of caring (again) in 2011. Looking at the other side of the equation, examining the provision of care in both 2001 and 2011, and accounting for the intensity of care provision, showed that individuals who were providing care in 2001 but had stopped providing care ten years later were more likely than non-carers to report poor health, particularly those who had been involved in heavy care provision. By contrast, those who had either taken up care provision in 2011, or had reported caring at both time points, showed the lowest odds of reporting poor health in 2011, compared to non-carers.

Taken together, these findings indicate that, holding a range of other factors constant, the provision of intensive (i.e. over 20 hours a week) informal care repeatedly or over an extended period of time may be associated with poor health outcomes, and that future analysis in this area requires more information reflecting the type of caring activity being provided. If further development of the caring questions in the 2021 Census were to be planned, inclusion of additional questions on the direction and type of care would be extremely useful. From a policy perspective, the findings presented in this chapter emphasise the challenge faced by all informal carers, particularly those providing heavy caring at more than one time point. Interventions which are aimed at supporting carers, such as periods of respite care or support with the combination of paid work and informal care provision, are important contributions which can alleviate part of the challenges faced by many carers.

When interpreting these key findings, it is important to take into account certain limitations of the research. First, as the analysis uses data from the 2001 and 2011 Censuses, it is not possible to ascertain the duration of the caring role between these two time points. As a result, any causation implied in the findings relates to repeated rather than continuous caring between 2001 and 2011. The second limitation is that information about the care recipient is not available, which means that our understanding of the quality of the relationship between the carer and the person cared for is essentially limited, although the importance of this dimension has been noted in the literature (Keene and Prokos, 2008). Equally important to take into account, but also missing from this dataset, is information on the extent to which carers rely on formal sources of support, for example that provided by the state following assessment or provided by the private sector. Such information is critical as it provides us with an understanding of the demand for informal care provision, particularly for tasks which are more personal than instrumental in nature (Vlachantoni et al., 2015). A final limitation relates to the multi-faceted nature of the concept of health, including more nuanced physical and mental dimensions, which may require more complex indicators than self-reported general health and the report of a limiting long-standing illness. Such a detailed exploration of the carer's health status is beyond the scope of the dataset discussed in this chapter, but has been undertaken elsewhere (see for example, Kenny et al., 2014).

At the same time, evidence on the relationship between informal caregiving and health needs to be contextualised in the broader framework of individuals' complex lives, as additional roles and changes in such roles may also affect such a relationship. Existing research, including that shown in this chapter, has evidenced the importance of other demographic and socio-economic characteristics for understanding whether individuals start/stop/continue to provide care and, if so, how much care. As informal care provision is expected to become an increasingly common activity in the lives of future cohorts, the examination of the short- and long-term effects of care provision on the carer's health will remain a key issue on policy and research agendas alike.

Acknowledgements

The authors wish to acknowledge the support of colleagues in the Engineering and Physical Sciences Research Council (EPSRC) Care Life Cycle (CLC) project (grant number EP/ H021698/1) and the Economic and Social Research Council (ESRC) Centre for Population Change (CPC) (grant numbers RES-625-28-0001 and ES/K007394/1) at the University of Southampton.

The permission of the Office for National Statistics to use the Longitudinal Study is gratefully acknowledged. The authors alone are responsible for the interpretation of the data.

This work contains statistical data from ONS which is Crown copyright. The use of the ONS statistical data in this work does not imply the endorsement of the ONS in relation to the interpretation or analysis of the statistical data. This work uses research datasets which may not exactly reproduce National Statistics aggregates.

We are grateful for the assistance of user support officers Julian Buxton, Lorraine Ireland, Shayla Leib, Kevin Lynch, Nicola Rogers, James Warren and the ONS LS Development Team for the extract from the dataset provided, their guidance on the dataset, and clearance of outputs.

The permission of Dr Paul Norman, School of Geography, University of Leeds, to use the 2011 Carstairs Index of Deprivation he created is gratefully acknowledged. The Carstairs Index has previously been used in conjunction with the ONS LS (Boyle et al., 2004; Norman et al., 2005; Norman and Boyle, 2014).

References

Boyle, P., Norman, P. and Rees, P. (2004) Changing places: Do changes in the relative deprivation of areas influence limiting long-term illness and mortality among non-migrant people living in non-deprived households? *Social Science & Medicine*, 58: 2459–2471.

Buxton, J. and Smith, N. (2010) Self-rated health and care-giving in the 2001 Census: Implications of non-response for analyses using the ONS Longitudinal Study. *CeLSIUS Technical Paper*.

Commission on Funding of Care and Support (2011) Fairer Care Funding. Report of Commission on Funding of Care and Support. Available at: http://www.thirdsectorsolutions.net/assets/files/Fairer-Care-Funding-Report%20Dilnot%20July%202011.pdf.

Dahlberg, L., Demack, S. and Bambra, C. (2007) Age and gender of informal carers: A population-based study in the UK. *Health and Social Care in the Community*, 15(5): 439–445.

Doran, T., Drever F. and Whitehead, M. (2003) Health of young and elderly informal carers: Analysis of UK census data. *British Medical Journal*, 327(7428): 1388.

Evandrou, M. (2005) Health and social care. In Soule, A., Baab, P., Evandrou, M., Balchin, S. and Zealey, L. (eds.) *Focus on Older People (Focus On Series)*. Office for National Statistics and Department for Work and Pensions, Palgrave Macmillan, Basingstoke, pp. 51–66.

Evandrou, M., Falkingham, J., Robards, J. and Vlachantoni, A. (2015) Who cares? Continuity and change in the prevalence of caring and characteristics of informal carers, in England and Wales 2001–2011. *CPC Working Paper 68*, ESRC Centre for Population Change, UK.

Fredman, L., Cauley J.A., Hochberg, M., Ensrud K.E. and Doros, G. (2010) Mortality associated with caregiving, general stress, and caregiving-related stress in elderly women: Results of caregiver-study of osteoporotic fractures. *Journal of the American Geriatric Society*, 58(5): 937–43.

Glaser, K., Evandrou, M. and Tomassini, C. (2005) The health consequences of multiple roles at older ages in the UK. *Health and Social Care in the Community*, 13(5): 470–477.

Grundy, E. and Tomassini, C. (2010) Marital history, health and mortality among older men and women in England and Wales. *BMC Public Health*, 10: 554.

Keene, J.R. and Prokos, A.H. (2008) Widowhood and the end of spousal care-giving: Relief or wear and tear? *Ageing & Society*, 28(4): 551–570.

Kenny, P., King, M.T. and Hall, J. (2014) The physical functioning and mental health of informal carers: Evidence of care-giving impacts from an Australian population-based cohort. *Health and Social Care in the Community*, 22(6): 646–659.

Norman, P. and Boyle, P. (2014) Are health inequalities between differently deprived areas evident at different ages? A longitudinal study of census records in England and Wales, 1991–2001. *Health & Place*, 26: 88–93.

Norman, P. and Purdam, K. (2013) Unpaid caring within and outside the carer's home in England and Wales. *Population, Space and Place*, 19(1): 15–31.

Norman, P., Boyle, P. and Rees, P. (2005) Selective migration, health and deprivation: A longitudinal analysis. *Social Science & Medicine*, 60 (12): 2755–2771.

ONS (2013a) 2011 Census Analysis: Unpaid care in England and Wales, 2011 and comparison with 2001. Available at: www.ons.gov.uk/ons/dcp171766_300039.pdf.

ONS (2013b) The gender gap in unpaid care provision: Is there an impact on health and economic position? Available at: www.ons.gov.uk/ons/dcp171776_310295.pdf.

O'Reilly, D., Connolly, S., Rosato, M. and Patterson C. (2008) Is caring associated with an increased risk of mortality? A longitudinal study. *Social Science & Medicine*, 67(8): 1282–1290.

Rahrig Jenkins, K., Kabeto, M.U. and Langa, K.M. (2009) Does caring for your spouse harm one's health? Evidence from a United States nationally-representative sample of older adults. *Ageing & Society*, 29(2): 277–293.

Robards, J., Evandrou, M., Falkingham, J. and Vlachantoni, A. (2012) Marital status, health and mortality. *Maturitas*, 73(4): 295–299.

Robards, J., Vlachantoni, A., Evandrou, M. and Falkingham, J. (2015) Informal caring in England and Wales – stability and transition between 2001 and 2011. *Advances in Life Course Research*, 24: 21–33.

Ross, A., Lloyd, J., Weinhardt, M. and Cheshire, H. (2008) *Living and Caring? An Investigation of the Experiences of Older Carers.* A Report of Research carried out by the National Centre for Social Research for the ILC-UK, London.

Vlachantoni, A., Evandrou, M., Falkingham, J. and Robards, J. (2013) Informal care, health and mortality. *Maturitas*, 74(2): 114–118.

Vlachantoni, A., Robards, J., Falkingham, J. and Evandrou, M. (2016) Trajectories of informal care and health. *Social Science & Medicine: Population Health*, 2: 495–501.

Vlachantoni, A., Shaw, R., Evandrou, M. and Falkingham, J. (2015) The determinants of receiving social care in later life in England. *Ageing & Society*, 35(2): 321–345.

Young, H. and Grundy, E. (2008) Longitudinal perspectives on caregiving, employment history and marital status in midlife in England and Wales. *Health & Social Care in the Community*, 16(4), 388–399.

Using 2011 Census data to estimate future elderly health care demand

Stephen Clark, Mark Birkin, Alison Heppenstall and Philip Rees

23.1 Introduction

Many western societies are predicting an important shift in the composition of their populations. The clear historical and anticipated future trend is for the elderly population to increase, both in numbers and as a proportion of the total population (Rechel et al., 2009; European Commission, 2014). For England, the latest 2014-based principal projection by the Office for National Statistics (ONS, 2015) shows the English population aged 50 and older increasing from 19.4 million in 2014 to 24.8 million in 2034, a gain of 5.4 million. This is in the context of a more modest increase of 2.1 million in the 49 and younger population.

The question arises as to what this ageing phenomenon will mean for society at large (Rutherford, 2011; House of Lords, 2013). Whilst longer life expectancies are to be celebrated, the ageing of the population is, at best, seen as a challenge (Christensen et al., 2009) and, at worst, a threat (Laurence, 2002). But, in terms of provision of services for an ageing population, health is an area where the impact of an ageing population will be most keenly felt (Wanless, 2004; Craig and Mindell, 2005). Here, the composition of this projected population will be important and questions arise around its economic, social and cultural composition. For example, the ethnic composition of the ageing population may change over time, with a trend towards greater diversity; the lifestyle history of the population will evolve, with downward trends in smoking and rising obesity; and the employment history will continue to transition away from manual occupations such as mining, steel and textiles towards retail and service occupations. Since health care planning and delivery are carried out at the local level, it is also important to gain an understanding of the changes in the geographic distribution of this demand.

The outline of the chapter is as follows. Section 23.2 reviews the determinants of health and the need for care. Section 23.3 describes the information on health provided by the 2011 Census, which, in general, is available for local and small geographical areas. Section 23.4 summarises the spatial microsimulation methodology used to create a synthetic population with census attributes and detailed health characteristics derived from a longitudinal survey (see Chapter 19). Section 23.5 explains how the synthetic population was projected into the future and presents results for local authority districts in England in 2031. The final section provides a summary of outcomes of the research.

23.2 Determinants of health care

The health care demands associated with an elderly population and the determinants affecting this demand are intensively studied. Some research examines the prevalence of specific morbidities (Seshamani and Gray, 2004); other work looks at either general health or the presence of limiting activities (Lubitz et al., 2003). Many studies, particularly from North America, model health care costs as a proxy for ill health (Denton et al., 2002). The general findings are that as people get older they are more prone to develop morbidities or be in generally worse health. This means that, all other things being equal, an older population will tend to have worse health (Alemayehu and Warner, 2004). However, chronological age itself may not be the actual driver of health status. Some studies have reported that it is the remaining years of life that are important, with an individual's health deteriorating in the final months or years before death (Zweifel et al., 1999). Others take the argument further and suggest that it is not age or the remaining years of life that are important but the presence of a disabling condition (de Meijer et al., 2011).

Gender and ethnic differences influence health status as well. Females live longer than males but these extra years are not necessarily spent in good health. In regard to ethnicity, there are some morbidity conditions that are more prevalent within certain ethnic groups; for example, diabetes is more prevalent in the South Asian population (HSCIC, 2005). A person's socio-economic status, measured using income, wealth, education or employment indicators, can also influence health Clark, 2015, p. 41.

A range of data exists to study the health outcomes for an aged population. Population censuses provide extensive coverage of the characteristics of the population and provide rich detail on the many determinants often cited as important for health outcomes. However, the actual information collected in the census on health is general and rarely touches on specific morbidities. In addition, the time interval between censuses can be long. To try and overcome these issues, governments often commission national sample surveys that can either be general in nature or focused on a particular public policy issue, such as health. In an era where administrative systems are increasingly being implemented and co-ordinated, scope has arisen to use such data for researching health status.

The literature identifies two main approaches used to model elderly health care demand, either statistical modelling or simulation. Statistical modelling is the approach most often used (see Brailsford et al., 2009, for a review of such methodologies). The advantage of this approach is that it is grounded in statistical theory, which allows for various interpretative and testing regimes to be followed. The disadvantage is that statistical modelling is rigid in both its outcome and the reliance on the assumptions that underlie the modelling technique. Just as there are many statistical techniques which can be used, there is also a wide variety of simulation methods. These methods attempt to replicate the composition or behaviour of a population, either real or hypothetical. They commonly use the technique of Monte Carlo simulation to replicate a decision-making process within the simulation system (Brailsford, 2007). The advantage of such an approach is that the simulation can be built using information on the processes being simulated or by simple rules informed from an understanding of the dynamics of health care. However, the drawback is that it is sometimes difficult to disentangle these dynamics, particularly when unexpected or little-understood interactions occur.

23.3 The 2011 Census and health

Questions on health are a fairly recent addition to the UK censuses, starting in the 1991 Census with a question on whether the individual considered that he or she had a long-term illness which limited their activities. A question on self-assessed general health was added in the 2001 Census, using a three-point scale of "Good, Fairly good or Not good". Also in 2001, a question was asked about the amount of time devoted to caring for family members, neighbours or friends.

Prior to the 2011 Census, the ONS conducted a consultation exercise on the health questions in the census (ONS, 2010). There was strong support for retaining the health questions in the census and for continuity so that long-term trends could be assessed. There was, however, scope for some changes. The question on the presence of a limiting long-term illness (LLTI) was expanded from a binary "yes or no" response to a "no, limited a little or limited a lot" response. The general health question was expanded from a three-point to a five-point scale. The question on the amount of caring provided remained unchanged from 2001.

The UK censuses are therefore able to capture the variations in self-rated health in a population, at geographic scales that vary from small areas to countries. What the censuses do not provide are any details on which specific conditions or morbidities cause individuals to assess their health as less than good. For general health planning and the allocation of resources, these general measures may be sufficient, but more detailed local health planning information on which morbidities are present are of greater value. This information could then influence the relative allocation of resources to various sectors of the health system, for example to local pharmacies, general practitioners, hospitals or specialist treatment centres. Whilst surveys are available that begin to capture this detail of information, they are not comparable to the census in being able to provide the geographically specific information. What is therefore required is a way to combine the rich geographic details of the census with the information-rich detail contained in these surveys.

23.4   Spatial microsimulation

A technique that is often used to achieve this goal is spatial microsimulation. The technique attempts to reconstruct a population of individuals for a specific area from a sample population. Individuals are chosen, with replacement, from this sample population, based on a comparison of their individual characteristics and the known aggregate characteristics of the population of the area. Thus, if an area contains a count of 200 individuals who are male, aged 55 to 59 and of Chinese ethnicity, the task is to repeatedly select individuals from the sample population to meet this constraint. The spatial microsimulation task is to estimate a set of area-specific weights to apply to the sample population so that, when the sample population is aggregated using these weights, they reproduce the aggregate constraint counts. Spatial microsimulation has been used widely in the field of health (Brown, 2011). This technique is able to produce synthesised local populations which may then be used in further modelling exercises. Applications in the area of health planning have included morbidity prevalence estimates (Clark et al., 2014; Shulman et al., 2015), the prevalence of obesity (Edwards et al., 2011), people who smoke (Smith et al., 2011) and care needs (Lymer et al., 2009).

Immediately, the value of the census tabulations becomes apparent in that they are able to provide the constraint count tables. These counts are very accurate estimates for small areas and are flexible in the range of multi-dimensional cross-tabulations that are possible. Hermes and Poulsen (2012) recommend using such multi-dimensional tables as constraints in spatial microsimulations and the Detailed Characteristics and Local Characteristics tables from the 2011 Census are ideally suited to this purpose.

Attention then turns to a source for the individual microdata to be sampled. The microdata outputs from the 2011 Census are not suitable since they do not add any extra information. There are, however, some government surveys that may be of value. These include the Health Survey for England (HSfE) (Joint Health Surveys Unit, 2012) and the English Longitudinal Study of Ageing (ELSA) (Institute for Fiscal Studies, 2015). The HSfE is able to provide a detailed picture of the health of the English population and also trends in various health-related activities, for example smoking, drinking, gambling and physical activity. Since the HSfE is a general health survey, it is not able to give particular prominence to sub-sections of the population, although there are occasional one-off survey boosts to the survey to highlight, for example, the health of ethnic minorities, young people or the elderly. Of particular concern to a study of the elderly population is that the HSfE does not include in its sample residents in communal establishments such as residential or nursing homes. Outputs from the 2011 Census show that, particularly for the very old, such residents are a substantial part of the elderly population.

The ELSA is a survey that is particularly geared to gaining an insight into the lives of the elderly population of England. The purpose of the survey is to examine the life histories of the ageing population of England in order to better understand the impact of both ageing and the passage of time on health. ELSA data are not limited to physical and psychological health outcomes but include a wide range of objective and subjective measures about the respondents and their households, including work and pensions, income and assets, housing and social participation. ELSA participants are aged 50 or older and at each biennial survey wave attempts are made to contact the same individuals. The ELSA asks the same range of questions at each wave and questions about morbidities associated with an elderly population are given particular prominence. The feature of the ELSA that makes it particularly suitable as a sample population is that it surveys individuals who have moved into a communal residential setting.

Candidate constraints

Anderson (2007) sets out four criteria for the selection of suitable constraints for a spatial microsimulation:

1 compatibility of definition in the constraint and sample populations;
2 availability for the unit of analysis (in his case, households, but here, individuals);
3 reasonable predictive power at the small area level; and
4 good predictive power at the unit of analysis level.

Here, a joint consideration of criteria 2 and 3 is used to identify a long list of candidate constraint variables, which is then refined in light of the first and fourth criteria. Examination of the literature suggests that there are differences in health outcomes by age, with ill health becoming more common at older ages. Gender can also influence health outcomes. It is likely that the presence of a morbidity condition will cause a person to assess his/her own health as either "not good" or as "limiting their activities". Individuals from certain ethnic groups have differing health profiles for some morbidity conditions. Someone who is cohabiting or in a relationship has better health outcomes compared to someone who has always lived alone or who is separated or widowed. The socio-economic status of the individual needs also to be taken account of. This can be measured directly using the individual's socio-economic status classification (NS-SeC) (Rose et al., 2005) or indirectly using information on the tenure of the household in which they live (as a proxy for wealth), the level of highest qualification (as a proxy for both wealth and income) or through the level of vehicle ownership (as a proxy for both income and mobility). These variables therefore emerge as candidate constraints under criteria 2 and 3. A detailed examination of the 2011 Census and the ELSA Wave 5 variables demonstrates that variables with similar definitions in both datasets can be identified. This satisfies Anderson's first criterion, leaving the final criterion, the performance of the variable in predicting the outcome of interest, to be considered.

Anderson tackled the fourth criterion using an incremental series of nested binary logistic regressions with a succession of candidate explanatory variables. Here a similar approach is used by estimating three hazard models, a form of logistic regression (Singer and Willet, 2003), one each for the presence of cardio vascular disease (CVD), diabetes or high blood sugar (DHBS) and respiratory illness (RI), and testing the suitability of each constraint variable. In order of importance, the hazard models identified the presence of a LLTI, age, gender and ethnicity as important socio-demographic influences on the incidence of the three morbidities (ethnicity was particularly influential on diabetes). Of the candidate socio-economic variables, living arrangements, car ownership and NS-SeC were influential for the morbidities. The influences of tenure, education qualification and the amount of caregiving were found to be poor. The results of the hazard modelling suggest that the 2011 Census tables shown in Table 23.1 could be used as possible constraints.

Tables DC1117 (age and gender), DC2101 (ethnicity) and DC6114 (NS-SeC) are based on the usual resident population, which is the population under study. Tables LC3101 (disability),

Table 23.1 Census 2011 tables which are candidates to serve as constraints for the microsimulation model

Table	Age bands	Categories
DC1117 (Age structure)	Single years of age	Not applicable
LC3101 (Disability)*	50–54 55–59	Limited, not limited
DC2101 (Ethnicity)	60–64 65–69 70–74 75–79 80–84 85 and older	English/Welsh/Scottish/Northern Irish/British, Irish, Gypsy or Irish Traveller, Other White, White and Black Caribbean, White and Black African, White and Asian, Other Mixed, Indian, Pakistani, Bangladeshi, Chinese, Other Asian, African Caribbean, Other Black, Arab, Any other ethnic group
DC1108 (Living arrangements)*		Married couple, Cohabiting, Single, Married (not a couple), Separated, Divorced, Widowed
DC4109 (Vehicle ownership or use)*		No vehicle, One vehicle, Two or more vehicles
DC6114 (NS-SeC)	50–64 65 and older	1.1 Large employers and higher managerial and administrative occupations, 1.2 Higher professional occupations, 2. Lower managerial, administrative and professional occupations, 3. Intermediate occupations, 4. Small employers and own account workers, 5. Lower supervisory and technical occupations, 6. Semi-routine occupations, 7. Routine occupations, 8. Never worked and long-term unemployed, L14.1 Never worked, L14.2 Long-term unemployed, L15 Full-time students
DC3402 (Disability)+	50–64 65–74 75–84 85 and older	Limited a lot, Limited a little, Not limited, Staff member, Family member

Notes:
* The population base is residents in households.
+ The population base is residents in institutional establishments.

DC1108 (living arrangements) and DC4109 (vehicle ownership) are based on those resident in households, which is a subset of the usually resident population. The remaining table DC3402 (disability) is based on those who are resident in communal establishments. It is desirable to have a consistent population in all the constraint tables; here the residential population of the area. (This is possible with 2011 Census counts since they have not been subject to post-tabular disclosure control; see Stillwell and Duke-Williams (2007) for its impact on 2001 Census outputs.) This consistent population is achieved by identifying a 'residual' population in each local authority district (LAD) in the household and communal establishment residential populations about whom nothing is known other than their 'otherness' in regard to the substantive population of the table. Thus, if the level of household vehicle ownership is known for 1,000 residents in households in the LAD and the usual residential population of the LAD is 1,100, then a residual category of 'communal' is created in table DC4109 with a count of 100 about whom no vehicle ownership is known. Similarly, if there are 100 communal residents in the LAD whose disability status is known in table DC3402, a category of 'household' is created in this table with a count of 1,000. Since these residual categories now exist in the tables, it is necessary to provide individuals in the sample population with these residual characteristics who can be sampled. So, for example, this is achieved by associating the level of vehicle ownership for a sample individual who is resident in a communal establishment as 'residual' (i.e. unknown).

The *Flexible Modelling Framework* (FMF) software package is used to derive the LAD-specific spatial weights (Harland, 2013). This implementation of spatial microsimulation uses the combinatorial optimisation approach (Voas and Williamson, 2000). In this approach, an objective function is defined that measures how well the weighted aggregate counts in the sample population agree with the constraint counts in each constraint table. Commonly, the Total Absolute Error (TAE), which measures the absolute value of the difference between the weighted aggregate counts and the census tabulated counts, is chosen as both the objective function and a measure of goodness-of-fit. Rather than use all seven constraint tables in the weight estimation process, only five are used, with vehicle ownership and NS-SeC held back for validation purposes.

To derive the spatial weights for a population of just over 18 million people, the FMF takes 12 hours on a quad core i5-2300 PC with 4 GB of RAM running *Windows* 7. The outputs of the FMF are contained in two files: a sample weights file that contains pairs of zone identifiers (the LAD codes) and sample member identifiers (ELSA participant codes) and a statistical fit file that reports the TAE at the end of each annealing stage for each LAD. The average final TAE value is just 39 and the largest is 272 for Birmingham. This means that just 272 of the 292,565 individuals aged 50 and older living in Birmingham (0.093 per cent) have been mis-allocated to one of the constraints across all five constraint tables.

Prevalence estimates for 2011

As well as the constraint variables that correspond with 2011 Census variables, the ELSA also contains information about the individual that has no equivalent in the census, such as the morbidity status of the individual. Using the LAD-specific spatial weights estimated by the FMF and knowledge of which individuals in the sampling population have a morbidity condition at Wave 5, it is possible to estimate the number of individuals in each LAD with a morbidity condition and also the prevalence rate. Table 23.2 reports the ten LADs with the highest estimated prevalence rates and the ten LADs with the lowest, along with their 2001 ONS area classification type (ONS, 2003).

For those LADs with high CVD prevalence rates, there is a range of LAD types. Some LADs will have a population which is generally in poorer health, the causes of which could be employment history or lifestyle factors (e.g. Tendring, a depressed coastal district). Another explanation could be the ethnic makeup of the area, with some ethnic groups tending to have a higher

Table 23.2 Highest and lowest LADs for prevalence of CVD, DHBS, RI or CVD and RI combined, England, 2011

Ten highest LADs

Rank	Cardiovascular disease (CVD)	%	Diabetes or high blood sugar (DHBS)	%
1	Leicester (CI)	18.9	Newham (LC2)	17.3
2	Tendring (CC)	18.1	Brent (LC2)	17.1
3	Sandwell (CI)	18.0	Leicester (CI)	16.4
4	Tower Hamlets (LC)	18.0	Tower Hamlets (LC)	16.4
5	Eastbourne (RC)	18.0	Harrow (LS)	15.6
6	Wolverhampton (CI)	17.9	Slough (LS)	15.5
7	Rother (CC)	17.9	Hackney (LC2)	15.5
8	Christchurch (CC)	17.9	Ealing (LS)	15.4
9	Thanet (CC)	17.8	Hounslow (LS)	15.3
10	Wyre (CC)	17.6	Redbridge (LS)	15.2

Rank	Respiratory illness (RI)	%	CVD and DHBS combined	%
1	Leicester (CI)	22.2	Newham (LC)	7.3
2	Tower Hamlets (LC)	22.0	Leicester (CI)	7.2
3	Newham (LC2)	21.3	Tower Hamlets (LC)	7.1
4	Sandwell (CI)	20.9	Brent (LC)	6.6
5	Wolverhampton (CI)	20.9	Harrow (LS)	6.2
6	Blackburn with Darwen (CI)	20.8	Hounslow (LS)	6.1
7	Knowsley (IH)	20.7	Ealing (LS)	6.1
8	Hounslow (LS)	20.7	Slough (LS)	6.1
9	Birmingham (CI)	20.6	Redbridge (LS)	5.9
10	Slough (LS)	20.6	Manchester (CI)	5.6

Ten lowest LADs

Rank	Cardiovascular disease (CVD)	%	Diabetes or high blood sugar (DHBS)	%
1	South Northamptonshire (PST)	13.6	Elmbridge (PSE)	10.5
2	Daventry (PST)	13.6	Stroud (PST)	10.5
3	Basingstoke and Deane (PSE)	13.5	Bracknell Forest (PSE)	10.5
4	Milton Keynes (NGT)	13.4	South Northamptonshire (PST)	10.5
5	Lambeth (LC2)	13.4	Wokingham (PSE)	10.5
6	Southwark (LC2)	13.3	Uttlesford (PSE)	10.4
7	Hart (PSE)	13.3	Chiltern (PSE)	10.4
8	West Berkshire (PSE)	13.2	Mid Sussex (PSE)	10.4
9	Wokingham (PSE)	13.2	East Hertfordshire (PSE)	10.4
10	Bracknell Forest (PSE)	13.1	West Berkshire (PSE)	10.3

(Continued)

Table 23.2 (Continued)

Rank	Respiratory illness (RI)	%	CVD and DHBS combined	%
1	Surrey Heath (PSE)	17.1	South Hams (CC)	3.3
2	South Northamptonshire (PST)	17.1	South Cambridgeshire (PSE)	3.3
3	Waverley (PSE)	17.1	Bracknell Forest (PSE)	3.3
4	Mid Sussex (PSE)	17.0	Guildford (PSE)	3.3
5	Guildford (PSE)	17.0	South Northamptonshire (PST)	3.3
6	Kensington and Chelsea (LC)	16.9	Basingstoke and Deane (PSE)	3.3
7	Hart (PSE)	16.9	East Hertfordshire (PSE)	3.2
8	Rutland (PST)	16.8	West Berkshire (PSE)	3.2
9	Tandridge (PSE)	16.8	Surrey Heath (PSE)	3.2
10	Southwark (LC2)	16.8	Hart (PSE)	3.1

Notes: CI = Centres with Industry; CC = Coastal and Countryside; LC = London Centre; RC = Regional Centres; PST = Prospering Smaller Towns; PSE = Prospering Southern England; NGT = New and Growing Towns; LC2 = London Cosmopolitan; LS = London Suburbs; IH = Industrial Hinterlands; LAD = Local Authority District (the lowest tier of local government in England).

prevalence of the conditions that contribute to CVD (e.g. Tower Hamlets with a large Bangladeshi population). A final explanation could be related to a generally aged population living in the area, with the incidences of CVD accumulating in an older age structure (e.g. Eastbourne). For DHBS, the most important defining characteristic is the ethnic composition of the LAD, with all the top ten LADs having a high proportion of their population from the Black and minority ethnic (BME) community, in particular from the South Asian groups, which tend to have higher rates of diabetes than the general population. With RI, there appears to be both an ethnic and deprivation dimension to those LADs in the top ten, indicating that this morbidity affects individuals differently depending on their ethnicity or deprivation. The LADs with the highest prevalence for the comorbidity of CVD and DHBS appear to be those most impacted by DHBS prevalence, with very similar LADs featuring in the top ten for both DHBS and CVD and DHBS.

The spatial pattern for the ten LADs with the lowest prevalence rates appears clearer than for the ten highest. The vast majority of LADs are prosperous authorities located in southern England, but there are also some London boroughs reporting low prevalence rates. The very diverse and cosmopolitan nature of these boroughs can lead to divergence in the expected outcomes for some morbidity conditions. A case in point is the London borough of Southwark, which has the lowest estimated prevalence rate for RI and a low prevalence for CVD. Southwark is an authority with a large black African and Caribbean population (in the 2011 Census, these ethnic groups make up 17 per cent of the 50 and older population). The HSCIC report (2005) shows that illnesses of the heart and circulatory and respiratory systems are lower for these ethnic groups than for the general population. In other cosmopolitan London boroughs with a very different ethnic mix (say a larger south Asian population), a different prevalence profile is relevant and this is borne out by an inspection of the top ten prevalence rates in Table 23.2.

23.5 Prediction framework

The framework for producing predictions of these prevalence rates is shown in Figure 23.1. The top half of this figure, the spatial microsimulation to estimate a base 2011 population of individuals in each English LAD, has already been described.

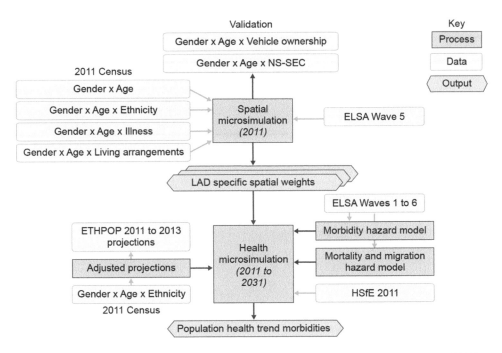

Figure 23.1 The framework for predicting future numbers of people with particular morbidities

Source: Adapted from Clark (2015, Figure 5.1, p. 96).

Future ethnic composition

In projecting the morbidity status of the population into the future, it is necessary to take account of the demographic composition of the population. The composition of the 50 and older population in each LAD will be influenced by those turning 50 years of age each year, those who die and those who migrate in and out of the LAD. In this framework these population dynamics are captured by use of an external population projection. The ONS provides population projections that estimate the size of the population in each LAD by gender and single year of age but include no other characteristic (ONS, 2014). However, there is available another set of population projections that provide an additional ethnic breakdown of these populations (Rees et al., 2011, 2012, 2013). These ETHPOP projections are based on information collected in the 2001 Census and subsequent vital statistics on births, deaths and migration up to 2008, but crucially the projections preceded the 2011 Census. When 2011 Census counts were published, a comparison of the ETHPOP projected population for 2011 with the 2011 Census population indicated a need for revision of the projections (Rees et al., 2016). For this study, the ETHPOP projections were updated using a methodology based on that described in Rees and Clark (2014). The steps were:

1   Re-base projections to 2011 Census counts.
2   Utilise subsequent ETHPOP cohort projections.
3   Incorporate a projection adjustment factor based on 2001 to 2011 performance.
4   Constrain the population by gender and age to ONS 2012 sub-national mid-year population projections.

Stephen Clark, Mark Birkin, Alison Heppenstall and Philip Rees

Table 23.3 Yearly adjustment for the White population of Plymouth

White	2001	2011	Growth	Yearly adjustment
Census	110,314	120,221	$\dfrac{120,221}{110,314} = 1.090$	$\left[\dfrac{1.090}{1.158}\right]^{\frac{1}{10}} = 0.9940$
ETHPOP	110,506	128,000	$\dfrac{128,000}{110,506} = 1.158$	

Sources: 2001 Census data from table S101; 2011 Census data from table DC2101; ETHPOP data from www.ethpop.org.

The annual projection adjustment factor is calculated from:

$$\text{Yearly adjustment} = \left[\frac{\dfrac{2011\ \text{Census}}{2001\ \text{Census}}}{\dfrac{2011\ \text{ETHPOP}}{2001\ \text{ETHPOP}}}\right]^{\frac{1}{10}} \tag{23.1}$$

where '2001 Census' is the ethnic-specific count of the older population in the LAD from the 2001 Census (Table S101); '2011 Census' is the ethnic-specific count of the older population in the LAD from the 2011 Census (Table DC2101); '2001 ETHPOP' is the mean of the TREND-EF and UPTAP-ER 2001 projections of the older population in the LAD from ETHPOP; and '2011 ETHPOP' is the mean of the TREND-EF and UPTAP-ER 2011 projections of the older population in the LAD from ETHPOP.

This annual adjustment factor captures how the ETHPOP projection needs to be adjusted to reproduce the 2011 outcome (see Table 23.3 for an example). Where the counts are small numbers (which is possible for LADs with small ethnic minority populations), the range of this adjustment is controlled to lie within an empirically derived credible interval calculated on the annual projection adjustments used in other similar LADs.

Dynamic microsimulation

Given an estimated 2011-based population for each LAD and an indication of the future composition of its population in terms of gender, age and ethnicity, the next stage is to evolve this population over the medium term, here to 2031. This is achieved through the following processes:

1   Age the population by two years. The dynamic microsimulation works in steps of two years (the time interval between ELSA waves).
2   Replenish the population at younger ages. As the population ages, the lower age range in each LAD will increase. This necessitates the creation of a population of 'replenishers' at ages 50 and 51 at each time step.
3   Update the morbidity status of individuals. This is achieved using hazard models estimated using ELSA data to predict if an individual will develop a morbidity based on their gender, age, ethnicity, smoking status, presence of a comorbidity, place of residence and time. This probability of occurrence is converted into an actual occurrence using a Monte Carlo approach.
4   The structure of the population is adapted to conform to the revised ETHPOP projections for the LAD.

Table 23.4 The numbers and percentages of people aged 50 and older with CVD, DHBS, RI and CVD and DHBS combined, England, 2011 to 2031

Year	Population aged 50 and older	Cardiovascular disease (CVD)		Diabetes or high blood sugar (DHBS)		Respiratory illness (RI)		CVD and DHBS combined	
	Number (thousands)	Number (thousands)	%	Number (thousands)	%	Number (thousands)	%	Number (thousands)	%
2011	18,230	2,874	15.8	2,190	12.0	3,409	18.7	749	4.1
2013	18,965	2,804	14.8	2,425	12.8	3,396	17.9	747	3.9
2015	19,690	2,738	13.9	2,657	13.5	3,470	17.6	751	3.8
2017	20,400	2,635	12.9	2,834	13.9	3,496	17.1	733	3.6
2019	21,083	2,532	12.0	2,980	14.1	3,490	16.6	708	3.4
2021	21,733	2,451	11.3	3,129	14.4	3,562	16.4	696	3.2
2023	22,305	2,292	10.3	3,192	14.3	3,477	15.6	649	2.9
2025	22,755	2,207	9.7	3,300	14.5	3,482	15.3	631	2.8
2027	23,129	2,076	9.0	3,359	14.5	3,439	14.9	593	2.6
2029	23,540	1,944	8.3	3,386	14.4	3,367	14.3	553	2.3
2031	24,029	1,859	7.7	3,436	14.3	3,376	14.1	534	2.2

The result of this dynamic microsimulation in terms of the prevalence counts and rates for the English population is shown in Table 23.4. For CVD, the count of those with the morbidity falls by 1 million over the 20-year time period and consequently the prevalence rate halves. This result is driven by changes in lifestyle, occupation and medical advances, for example the use of statins to lower cholesterol (Law et al., 2003). The picture for DHBS is the reverse. The count increases by 1.2 million and the rate increases by 2 per cent. The main drivers behind this are poor diet, lack of exercise (which impacts on obesity) and an ageing of the (now) middle-aged south Asian population into our 50 year and older age range. With respiratory illness, the count remains largely static over time but because of a larger population at risk, the rate decreases. The reason for this is almost certainly the reduced prevalence of smoking in the population in future years but may also be influenced by changes in occupations. The number with the comorbidity of CVD and DHBS falls by a modest 200,000 and the prevalence rate reduces by half, a consequence of the fall in the prevalence of CVD.

The national trends shown in Table 23.4 are simply aggregates of the results for each English LAD. The results for each English LAD in 2031 are shown in Figure 23.2. This figure represents the prevalence counts as circles, with the area of the circle being proportional to the number of people estimated to have the morbidity, whilst the shading indicates the prevalence rate for the LAD. One distinct pattern in the DHBS map is high prevalence rates in LADs located in West Yorkshire, Greater Manchester, the West Midlands and London plus Derby, Nottingham and Leicester. These are all authorities with large BME communities in 2011 and the size of 50 and older populations in these communities is projected to increase by 2031.

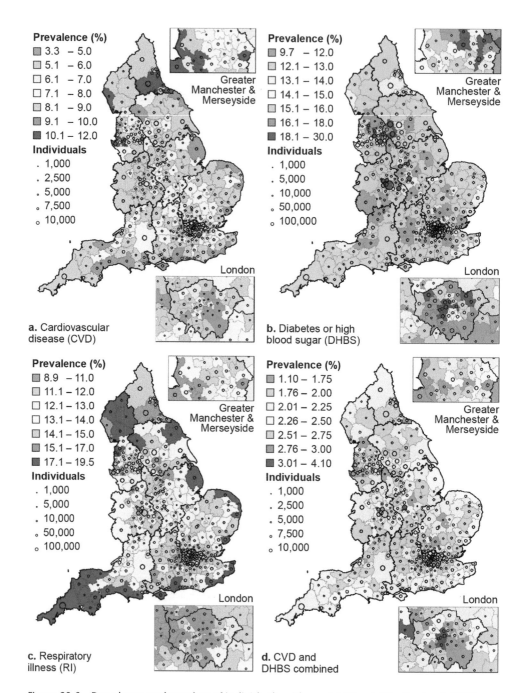

Figure 23.2 Prevalence and number of individuals with morbidities, LADs, England, 2031

Source: Adapted from Clark (2015, Figure 9.3, p. 228).

23.6 Conclusions

Access to the 2011 Census tabulations has provided a vital component in the dynamic simulation used to project to 2031 the prevalence counts and rates for three important morbidities. The census tables provide a very good estimate of the important co-variates associated with these morbidities at a useful geographic scale. However, they are not sufficient. A further dataset, the ELSA, is needed to provide additional details on specific health outcomes, in this instance the case study morbidities. The census counts from 2001 and 2011 are also useful to provide a trend in the changing size and composition of the elderly population, particularly in terms of its ethnicity.

Looking at the outcomes of this study, the predicted downward trend in the morbidity for CVD agrees with recent experience Clark, 2015, p. 39. The hotspots for DHBS prevalence are located in areas where the ethnic composition of the population is such that high prevalence rates would be expected. Changes in smoking behaviour and exposure to toxins in the workplace are stabilising numbers with respiratory illness leading to a fall in the prevalence rates.

This framework for analysis and prediction is transferable to other countries that have access to the same types of data, that is, census or administrative data to provide the detail about the whole population and a sample survey to provide the detail for the domain of interest, such as health.

Acknowledgements

The funding for this study has been provided by the Economic and Social Science Research Council as part of the National Centre for Research Methods initiative. ELSA was developed by a team of researchers based at the National Centre for Social Research, University College London and the Institute for Fiscal Studies. The data were collected by the National Centre for Social Research. The funding is provided by the National Institute of Aging in the United States, and a consortium of UK government departments co-ordinated by the ONS. The data were made available through the UK Data Archive. The developers and funders of ELSA and the Archive do not bear any responsibility for the analyses or interpretations presented here. All census materials are Crown copyright and reproduced with the permission of the Controller of HMSO.

References

Alemayehu, B. and Warner, K.E. (2004) The lifetime distribution of health care costs. *Health Service Research*, 39(3): 627–642.

Anderson, B. (2007) Creating small area income estimates for England: Spatial microsimulation modelling. *Chimera Working Paper Number 2007–07*, Chimera, University of Essex. Available at: http://opendepot.org/166/1/CWP-2007-07-Income-Deprivation-England.pdf.

Brailsford, S.C. (2007) Tutorial: Advances and challenges in healthcare simulation modelling. In Biller, S.G., Hsieh, B., Shortle, M.H., Tew, J.D. and Barton, R.R. (eds.) *Proceedings of the 2007 Winter Simulation Conference*, Curran Associates Inc., Washington DC. Available at: http://ieeexplore.ieee.org/stamp/stamp.jsp?arnumber=4419754.

Brailsford, S.C., Harper, P.R., Patel, B. and Pitt, M. (2009) An analysis of the academic literature on simulation and modelling in health care. *Journal of Simulation*, 3: 130–140.

Brown, L. (2011) Editorial: Special issue on 'Health and Microsimulation'. *International Journal of Microsimulation*, 4(3): 1–2.

Christensen, K., Doblhammer, G., Rau, R. and Vaupel, J.W. (2009) Ageing populations: The challenges ahead. *The Lancet*, 374: 1196–1208.

Clark, S.D. (2015) Modelling the impacts of demographic ageing on the demand for health care services. PhD Thesis, School of Geography, University of Leeds, Leeds. Available at: http://etheses.whiterose.ac.uk/11676/.

Clark, S.D., Birkin, M. and Heppenstall, A.J. (2014) Sub-regional estimates of morbidities in the English elderly population. *Health and Place*, 27: 176–185.

Craig, R. and Mindell, J. (2005) Health Survey for England 2005. The health of older people: Summary of key findings. Available at: https://catalogue.ic.nhs.uk/publications/public-health/surveys/heal-surv-heal-old-peo-eng-2005/heal-surv-heal-old-peo-eng-2005-rep-v5.pdf.

de Meijer, C., Koopmanschapa, M., d'Uvac, T.B. and van Doorslaera, E. (2011) Determinants of long-term care spending: Age, time to death or disability? *Journal of Health Economics*, 30(2): 425–438.

Denton, F.T., Gafni, A. and Spencer, B.G. (2002). Exploring the effects of population change on the costs of physician services. *Journal of Health Economics*, 21: 781–803.

Edwards, K.L., Clarke, G.P., Thomas, J. and Forman, D. (2011) Internal and external validation of spatial microsimulation models: Small area estimates of adult obesity. *Applied Spatial Analysis and Policy*, 4(4): 281–300.

European Commission (2014) The 2015 Ageing Report. Underlying Assumptions and Projection Methodologies. *European Economy* 8. Available at: http://ec.europa.eu/economy_finance/publications/european_economy/2014/pdf/ee8_en.pdf.

Harland, K. (2013). Microsimulation model user guide: Flexible modelling framework. *National Centre for Research Methods Working Paper 06/13*. Available at: http://eprints.ncrm.ac.uk/3177/2/microsimulation_model.pdf.

Hermes, K. and Poulsen, M. (2012) A review of current methods to generate synthetic spatial microdata using reweighting and future directions. *Computers, Environment and Urban Systems*, 36: 281–290.

House of Lords (2013) Ready for Ageing? House of Lords Select Committee Report. Available at: www.parliament.uk/business/committees/committees-a-z/lords-select/public-services-committee/report-ready-for-ageing/.

HSCIC (2005) Health Survey for England 2004: The Health of Minority Ethnic Groups – Headline Tables. Health and Social Care Information Centre. Available at: http://content.digital.nhs.uk/catalogue/PUB01209/heal-surv-hea-eth-min-hea-tab-eng-2004-rep.pdf.

Institute for Fiscal Studies (2015) ELSA Main Reports. Available at: www.elsa-project.ac.uk/publications.

Joint Health Surveys Unit (2012) Health Survey for England – Trend Tables. Available at: https://catalogue.ic.nhs.uk/publications/public-health/surveys/heal-survey-eng-2011-tren-tabl/HSE2011-Trend-commentary.pdf.

Laurence, J. (2002) Why an ageing population is the greatest threat to society. *The Independent*. 10 April. Available at: www.independent.co.uk/news/uk/home-news/why-an-ageing-population-is-the-greatest-threat-to-society-5361944.html.

Law, M.R., Wald, N.J. and Rudnicka, A.R. (2003) Quantifying effect of statins on low density lipoprotein cholesterol, ischaemic heart disease, and stroke: Systematic review and meta-analysis. *British Medical Journal*, 326: 1423–1429.

Lubitz, J., Cai, L., Kramarow, E. and Lentzner, H. (2003) Health, life expectancy and health care spending among the elderly. *New England Journal of Medicine*, 349(11): 1048–1055.

Lymer, S., Brown, L., Harding, A. and Yap, M. (2009). Predicting the need for aged care services at the small area level: The CAREMOD spatial microsimulation model. *International Journal of Microsimulation*, 2(2): 27–42.

McCulloch, A. (2011) Health Service funding. Available at: https://web.archive.org/web/20140830000806/http://www.significancemagazine.org/details/webexclusive/1385441/Health-Service-funding.html.

McCulloch, A. (2012) Standardising death. Available at: https://web.archive.org/web/20140830000930/http://www.significancemagazine.org/details/webexclusive/1501135/Standardising-Death.html.

ONS (2003) National Statistics 2001 Area Classification. Available at: www.ons.gov.uk/ons/guide-method/geography/products/area-classifications/ns-area-classifications/index/index.html.

ONS (2010) Final Recommended Questions for the 2011 Census in England and Wales Health. Available at: www.ons.gov.uk/ons/guide-method/census/2011/the-2011-census/2011-census-questionnaire-content/question-and-content-recommendations/final-recommended-questions-2011---health.pdf.

ONS (2014) Methodology: 2012-based Sub-national Population Projection. Available at: www.ons.gov.uk/ons/rel/snpp/sub-national-population-projections/2012-based-projections/rpt-snpp-2012-based-methodology-report.html.

ONS (2015) National Population Projections. 2014-based Statistical Bulletin. Available at: www.ons.gov.uk/ons/rel/npp/national-population-projections/2014-based-projections/stb-npp-2014-based-projections.html.

Rechel, B., Doyle, Y., Grundy, E. and McKee, M. (2009) How can health systems respond to population ageing? *Health Systems and Policy Analysis*. World Health Organization, on behalf of the European Observatory on Health Systems and Policies. Available at: http://www.euro.who.int/__data/assets/pdf_file/0004/64966/E92560.pdf.

Rees, P. and Clark, S. (2014) The Projection of Ethnic Group Populations Aged 18 and Over for Westminster Parliamentary Constituencies in Great Britain for Election Years 2015, 2020, 2025, 2030 and 2035. A Report to the Policy Exchange, London.

Rees, P., Wohland, P. and Norman, P. (2013) The demographic drivers of future ethnic group populations for UK local areas 2001–2051. *Geographical Journal*, 179(1): 44–60.

Rees, P., Wohland, P. and Norman, P. (2016) The United Kingdom's multi-ethnic future: How fast is it arriving? In Lombard, J., Stern, E. and Clarke, G.P. (eds.) *Applied Spatial Modelling and Planning*. Routledge, London, pp. 157–171.

Rees, P., Wohland, P., Norman, P. and Boden, P. (2011). A local analysis of ethnic group population trends and projections for the UK. *Journal of Population Research*, 28(2–3): 149–183.

Rees, P., Wohland, P., Norman, P. and Boden, P. (2012) Ethnic population projections for the UK, 2001–2051. *Journal of Population Research*, 29(1): 45–89.

Rose, D., Pevalin, D.J. and O'Reilly, K. (2005) The National Statistics Socio-economic Classification: Origins, Development and Use. Institute for Social and Economic Research, University of Essex. Available at: www.ons.gov.uk/ons/guide-method/classifications/archived-standard-classifications/soc-and-sec-archive/the-national-statistics-socio-economic-classification--origins--development-and-use.pdf.

Rutherford, T. (2011) Population ageing: Statistics. House of Commons Library Standard Note: SN/SG/3228. Available at: www.parliament.uk/briefing-papers/SN03228.pdf.

Seshamani, M. and Gray, A. (2004). Ageing and health-care expenditure: The red herring argument revisited. *Health Economics*, Vol 13: 303–314.

Shulman, H., Birkin, M. and Clarke, G.P. (2015) A comparison of small-area hospitalisation rates, estimated morbidity and hospital access. *Health & Place*, 36: 134–144.

Singer, J.D. and Willett, J.B. (2003) *Applied Longitudinal Data Analysis*. Oxford University Press, Oxford.

Smith, D., Pearce, J.R. and Harland, K. (2011). Can a deterministic spatial microsimulation model provide reliable small-area estimates of health behaviours? An example of smoking prevalence in New Zealand. *Health & Place*, 17: 618–624.

Stillwell, J. and Duke-Williams, O. (2007) Understanding the 2001 UK census migration and commuting data: The effect of small cell adjustment and problems of comparison with 1991. *Journal of the Royal Statistical Society: Series A*, 170(2): 425–445.

Voas, D. and Williamson, P. (2000) An evaluation of the combinatorial optimisation approach to the creation of synthetic microdata. *International Journal of Population Geography*, 6(5): 349–366.

Wanless, D. (2004) Securing our Future Health: Taking a Long-Term View. Final Report. HM Treasury, London. Available at: http://www.yearofcare.co.uk/sites/default/files/images/Wanless.pdf.

Zweifel, P., Felder, S. and Meiers, M. (1999) Ageing of population and health care expenditure: A red herring? *Health Economics*, 8: 485–496.

Using census microdata to explore the inter-relationships between ethnicity, health, socio-economic factors and internal migration

Fran Darlington-Pollock, Paul Norman and Dimitris Ballas

24.1 Introduction

The UK census introduced two new questions in 1991, asking respondents to identify their ethnic group and confirm the presence or absence of limiting long-term illness (LLTI). Although census microdata have since been extensively used to examine social and spatial inequalities in ethnicity and in health, examples of which feature in this book (Chapters 21–23), relatively less work examines ethnic inequalities in health. In the context of increasing ethnic diversity, the pathways by which ethnic inequalities in health are maintained, widened or even narrowed must be scrutinised. In this chapter, we present analyses for 1991 to 2011 using the Samples of Anonymised Records (SARs) (we refer to the 2011 individual safeguarded sample as SARs) and the Office for National Statistics Longitudinal Study (ONS LS). These analyses evaluate one possible explanation for changing ethnic health gradients in the population: selective sorting, a process that helps us interpret the complex inter-relationships between ethnicity, health, socio-economic factors and internal migration. These inter-relationships will exist for international migration, but here we concentrate on internal, sub-national migration.

Applied to processes of residential mobility (across any sub-national geographic scale) or social mobility, the selection hypothesis holds that those in better health or with more health-enabling behaviours are more likely to move to (or remain in) more advantaged circumstances. This can be defined by area type (e.g. deprivation) or measures of social status such as social class (Bartley and Plewis, 1997, 2007; Norman et al., 2005; Boyle et al., 2009; Exeter et al., 2011). Conversely, those in poorer health or with fewer health-enabling behaviours are more likely to move to (or remain in) more disadvantaged circumstances. Movement between areas or area types and social statuses may therefore be influenced by health-related behaviours or physical and mental health, with groups of different health status *selected* into different circumstances.

In the context of residential mobility or migration, area deprivation is particularly important as change of address may also be associated with deprivation change (Norman et al., 2005). However, as individuals can experience deprivation change without moving house through area regeneration or decline, deprivation mobility also reflects an important independent dimension of selective sorting (see Boyle et al., 2004). 'Deprivation mobility' is used to identify any situation whereby an individual experiences a change in the level of deprivation of their residential area between two specified time points. This may be through migration or residential mobility – these groups are referred to as 'movers' – or due to the changing circumstances of the area relative to other areas in England – these groups are referred to as 'stayers'.

While residential mobility or migration have been explored alongside deprivation mobility, few studies also simultaneously examine social mobility (Boyle et al., 2009 is a notable exception). Where opportunities for either social mobility, residential mobility or deprivation mobility vary, not only by health but also between ethnic groups, this may differently influence health gradients in the population.

First, we use cross-sectional data from the SARs to illustrate ethnic differences in rates and types of migration, and how these change over time. We then examine the inter-relationships between health, social class and migration in the context of ethnic inequalities in health, using binary logistic regression models to calculate modelled probabilities of LLTI by ethnic group according to migrant status and social class. Variations in the probabilities of LLTI are illustrative of the extent and magnitude of ethnic inequalities in health over time. Then, we use longitudinal data which track individuals over time from the ONS LS to examine how transitions between area types or social classes for different population sub-groups influence health gradients. To examine how transitions influence health gradients, we compare the health of different groups transitioning into and out of the extremes of the deprivation spectrum or social hierarchy. Following previous work (Brimblecombe et al., 1999, 2000; Connolly et al., 2007; Norman and Boyle, 2014), we then adopt a 'put people back approach' to compare health gradients for the population by their original social class or area type with the health gradients that arise after the population experiences social mobility, residential mobility or deprivation mobility. The difference between the original health gradients and those after transitioning to new states allows an appraisal of the effect of these changes.

24.2   Health, migration and ethnicity: disentangling relationships and understanding inequality

The concept of 'selective sorting' holds that differently healthy groups will be *selected* and *sorted* into different life circumstances, whether defined by social class or area type, through social mobility or deprivation mobility (whether by migration or area type change). In this section, we focus on migration, thereby emphasising the inter-relationships between migration and social mobility and the likelihood that a migration event may be accompanied by some form of social mobility (Fielding, 1992; Ewens, 2005).

Migration is inherently selective by age, sex, housing tenure, socio-economic position and educational attainment (Plane, 1993; Boyle et al., 1998; Champion and Ford, 1998; Norman et al., 2005; Brown et al., 2012). Health is another distinguishing characteristic of migrants, varying with age and differently determining choice (not) to migrate (see Darlington et al., 2015a, for a review of migration and health). Younger migrants are more likely to be in better health than non-migrants, whereas older migrants are more likely to be in poorer health (e.g. Bentham, 1988; Boyle et al., 2002; Larson et al., 2004; Norman et al., 2005). Thus, while younger migrants may move in good health for employment or education opportunities, older migrants are more likely to move in poor health, seeking formal or informal care. Health and age not only influence choice (not) to migrate, but also destination (Norman and Boyle, 2014). Although health can also be analysed as a consequence of migration, this stretches beyond the scope of this chapter (see Darlington et al., 2015a). Nevertheless, if we assume living in more or less deprived circumstances is harmful or beneficial to health, it logically follows that moving between states of (dis)advantage will influence health outcomes.

The selectivity of migration is as applicable to ethnic minority groups as to the majority population. Research finds that the patterning of migration by housing tenure, educational attainment and socio-economic status is generally comparable between ethnic groups (e.g. Simpson and Finney, 2009; Raymer and Giulietti, 2009; Catney and Simpson, 2010). We therefore anticipate that migration will similarly vary across the lifecourse in line with certain life events or socio-demographic attributes by ethnic group (Finney, 2011). However, as Finney and Simpson (2008, p. 64) point out, "even if the same determinants of migration are recognised for each ethnic group, variation in group migration rates will be observed because of compositional effects". The authors find that differences in the composition of ethnic groups largely explain differences in migration patterns. Migrant characteristics not only vary by stage in the lifecourse but also relate to distance moved. Ethnic differences in distance moved have been determined, with Champion (1996) finding that 55 per cent of migrants from minority ethnic groups (MEGs) moved less than 5 km, compared with 47 per cent of Whites. Champion also found that of all the ethnic groups, Black groups moved over the shortest distances. Similar findings have been reported more recently by Finney and Simpson (2008).

As the influence of migrant characteristics appears to hold across ethnic groups, any process of selective sorting may operate similarly across ethnic groups. However, the extent of the influence will vary according to the composition of each ethnic group. Most importantly, the more advantaged the group, the higher the likely rates of migration (Catney and Simpson, 2010). This may differently influence health gradients if certain groups are less likely to move away from more disadvantaged circumstances. If we are to evaluate whether selective sorting (i) operates and (ii) differently influences (changing) health gradients in the overall population, or indeed (changing) health gradients by ethnic group, we must attempt to disentangle the complexities of the health–migration relationship. This will also shed light on the nature of ethnic inequalities in health in contemporary society. This section will therefore present analyses using the SARs, with the intent of delineating the (varying) health–migration relationship for different ethnic groups.

Data and methods

The SARs sample for this analysis is restricted to household residents in England aged between 16 and 74. International migrants, aged 0–15 and 75+, and residents in communal establishments such as care homes or prisons are therefore excluded. Excluding these sub-groups is common practice in the extant literature on selection effects, migration and health (e.g. Norman et al., 2005). The sample is restricted by age because of incomplete socio-economic data for the excluded ages. The SARs offer users a high degree of flexibility in the choice of variables and categories that can be explored, and the degree of statistical control in the modelling process (Gould and Jones, 1996; Norman and Boyle, 2010).

Migration here refers to any change of address across *any* geographic scale within England. Migrants are thus identified according to the SARs one-year migration variable, comparing sample members' current address with their address one year prior to the census. We only include sample members who change address within England, with moves between England and Wales, Scotland or Northern Ireland excluded on the same basis as international migrants. As migrants are identified by a one-year migration variable, this only excludes recent international migrants. In this chapter, we refer to all SAR members who change address as 'movers' and those who do not as 'stayers'. While our definition of 'movers' is a catch-all, we also differentiate movers by distance moved. Mover type distinguishes short (0–14 km), mid (15–149 km) and long-distance (150+ km) according to natural breaks in the distance moved by all movers in 1990–91, 2000–01 and 2010–11. These categories are comparable (if not identical) to previous work distinguishing migrants by distance moved (e.g. Boyle et al., 2002; Finney and Simpson, 2008).

Health is defined as the presence or absence of LLTI. While there are some changes to the question wording or response options over time, these are not expected to substantially bias results. In 2011, respondents distinguish between 'limited a little' and 'limited a lot', and we combine these responses in line with ONS outputs (e.g. ONS, 2014). In this chapter, we report on five ethnic groups which are consistently and (relatively) meaningfully defined over time in the SARs: White (including White Other who could not be consistently distinguished); Black Caribbean; Black African; Indian; and Pakistani and Bangladeshi (combined due to small numbers). Although broad, these groups allow for quite detailed analysis of ethnic differences in society, a benefit of the SARs and the large sample sizes available. Previous research on selective sorting and social mobility has used the Registrar General's schema of social class (e.g. Bartley and Plewis, 1997; Boyle et al., 2009), as do we. The 2001 and 2011 Socio-economic Classification (NS-SeC) information is converted to social class with all SARs members not assigned to a social class defined as 'unclassifiable'. To increase sample size, the top and bottom two social classes are combined: I (Professional) and II (Managerial and Technical) combined; IIIN (Skilled non-manual); IIIM (Skilled manual); IV (Partly skilled) and V (Unskilled) combined; U (Unclassifiable). Other variables included in the analysis as controls include tenure, educational attainment, country of birth and Government Office Region (GOR).

To illustrate how rates of migration vary between ethnic groups and over time, we first summarise rates of migration by ethnic group in 1991, 2001 and 2011, and differentiate migrants by distance moved. Binary logistic regression is then used to model the odds of LLTI, adjusting for an interaction between mover distance status and tenure, for different population subsets by age (see Dale et al., 2000, for a comprehensive discussion of binary logistic regression). Our models adjust for sex, ethnicity, country of birth, social class, educational attainment, tenure, mover status, GOR, and the interaction between tenure and mover status. By modelling health for different population subsets, accounting for the contingency of the health–migration relationship with age and the interaction with tenure (see Boyle et al., 2002), we illustrate how selective sorting may operate differently between ethnic groups due to ethnic variations in the nature of the health–migration relationship. As tenure also influences the nature of the relationship between migration and health, we also account for this in the presented results. All independent variables are known determinants of health (e.g. Kunst et al., 2005; Chandola et al., 2007; Mackenbach, 2012) or migration. We present modelled probabilities of LLTI for different age groups by ethnic group, mover status and social class in 2011. Although data are available for 1991 and 2001, the patterns are largely consistent and therefore not presented. Where deviations do occur, these will be highlighted, particularly where they reflect a change in the health–migration relationship.

Results

Table 24.1 summarises differences in the rates of migration within ethnic groups in 1991, 2001 and 2011, further differentiating migrants within ethnic groups by migrant type. The population became more mobile between 1991 and 2011, although there are marked variations between ethnic groups. Black Caribbeans, and Pakistanis and Bangladeshis are consistently the least mobile contrasting with high rates of migration amongst Black Africans. While Indians are less mobile than Whites in 1991 (7.4 per cent compared to 9.0 per cent), this reverses by 2011 (11.9 per cent compared to 10.6 per cent). Although Black Caribbeans are consistently the

Table 24.1 Proportion of movers and stayers and movers by mover type by ethnic group, 1991, 2001 and 2011

Percentage of population (%)	White	Black Caribbean	Black African	Indian	Pakistani and Bangladeshi	Total population[†]
1991 Stayer	91.0	92.4	83.5	92.6	92.2	91.0
Mover	9.0	7.6	16.5	7.4	7.8	9.0
Of movers						
Short distance	71.3	85.8	84.4	70.4	79.2	71.6
Mid distance	19.4	9.4	10.9	20.7	13.9	19.2
Long distance	9.3	4.8	4.7	9.0	6.9	9.2
2001 Stayer	89.1	90.5	83.0	90.3	89.9	89.0
Mover	10.9	9.5	17.0	9.7	10.1	11.0
Of movers						
Short distance	71.2	84.8	78.2	70.0	81.5	71.6
Mid distance	19.7	11.9	15.7	20.1	11.7	19.4
Long distance	9.2	3.4	6.1	9.9	6.8	9.0
2011 Stayer	89.4	90.5	81.9	88.1	90.2	88.9
Mover	10.6	9.5	18.1	11.9	9.8	11.1
Of movers						
Short distance	71.8	79.1	76.3	70.4	79.3	72.4
Mid distance	21.1	16.7	17.1	21.6	14.8	20.5
Long distance	7.2	4.2	6.6	8.0	5.9	7.1

[†] Includes Chinese, Mixed and Other

Source: Samples of Anonymised Records.

least likely to move, this group experiences the greatest growth in rates of migration between 1991 and 2011.

Table 24.1 also breaks movers down by mover distance type. Black Caribbeans, Black Africans and, to some extent, Pakistanis and Bangladeshis have the lowest proportion of migrants moving over long distances and the highest moving over short distances. Higher proportions of White and Indian migrants move over greater distances.

Differences between ethnic groups in the nature of migration events, such as distance moved, might be attributable to differences in composition and health status. For example, although not presented in this chapter, analysis of SARs data indicates that short-distance moves are consistently more likely amongst those with LLTI than those without, whereas mid- and long-distance moves are less likely for those with LLTI. Mid- and long-distance moves are also less likely amongst lower social classes. As certain MEGs are disproportionately concentrated in lower social classes and often in poorer health (e.g. Mindell et al., 2014; Darlington et al., 2015b), opportunities for migration and the nature of the move vary between ethnic groups.

Figure 24.1 presents modelled probabilities of LLTI for different age groups by ethnicity, mover status and social class in 2011. Despite the SARs being a large sample, results should be cautiously interpreted due to small sample sizes in some population sub-groups. Although the original SARs sample was restricted to ages 16–74, this sample is extended to include ages 75 and over, notwithstanding incomplete socio-economic data. The expansion of the sample is to illustrate a change in the age-selectivity of the health–migration relationship between 2001 and 2011. In 1991 and 2001, migrants aged 65–74 have higher probabilities of LLTI than their stable counterparts, indicative of the poorer health of older migrants and the likely motivations for their

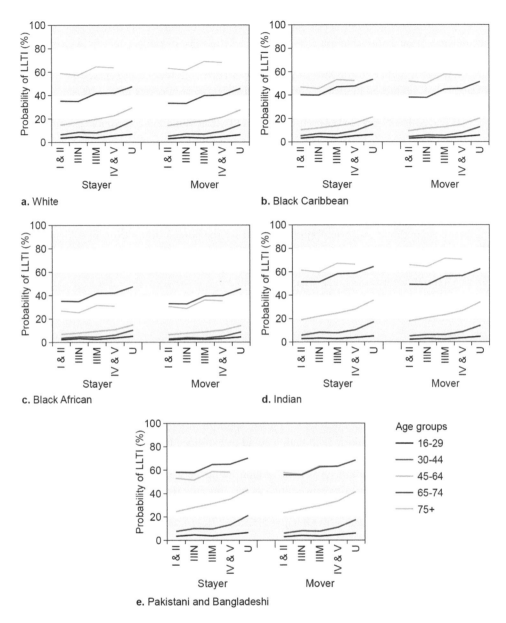

Figure 24.1 Modelled probability of LLTI by age, ethnicity, mover status and social class, 2011

Source: Samples of Anonymised Records.

Note: Probability of LLTI for Unclassifiable in 2011 not shown due to incomplete socio-economic data.

move. Salomon et al. (2012) find that global healthy life expectancy has increased between 1990 and 2010. Population ageing and wider improvements in healthy life-expectancy appear to have delayed moves in older ages precipitated by poorer health. Expanding the sample to those aged 75 and over in 2011 illustrates this change.

For minority groups, the difference in probability of LLTI between ages 65–74 and 75+ is notably smaller than for the White majority, particularly for Pakistanis and Bangladeshis.

Nevertheless, stayers aged 75+ consistently have lower probabilities of LLTI than movers, although the differences are marginal in some cases. Conversely, movers in younger ages always have lower probabilities of LLTI than stayers. While the social patterning to health is consistent within ethnic groups whereby higher probabilities of LLTI are associated with lower social classes, the influence of social class and indeed migrant status on health varies between ethnic groups by age such that overall ethnic inequalities vary notably between age groups.

MEGs aged 16–29 have lower probabilities of LLTI than Whites in comparable circumstances and of the same mover status. Indians have the best health of those aged 16–29 as indicated by the lowest probabilities of LLTI (between 2.3 per cent and 6.6 per cent). For all social classes and mover statuses aged 30–44 years, Indians, Black Caribbeans and Black Africans report better health than their White peers. For Indians, this reflects a marked improvement from 1991 and 2001. Conversely, increasing age brings poorer health for Pakistanis and Bangladeshis relative to Whites in comparable circumstances. As the population ages, this heralds greater inequalities for the South Asian groups, who are found to experience the poorest health from ages 45 and over. Thus, despite comparable socio-economic circumstances, minority groups are often in poorer health when in older age than the White majority, despite their advantaged health in early adulthood (although, as cross-sectional data, it is possible that those aged 45 and over in 2011 may have also been in poor health when aged 16–29).

Discussion

Marked variations in the propensity to migrate by ethnic group are observed in 1991, 2001 and 2011. South Asian groups are notably less mobile than Black Africans. Further, despite the younger age structure of MEGs compared to the White majority, this does not necessarily result in increased mobility for these groups overall despite the fact that migration is most likely in younger ages. This is perhaps indicative of fewer opportunities for migration amongst minority groups than the White majority, although it could be explained by differences in the association between migration and lifecourse events for different ethnic groups (Finney, 2011). Differences in the rates of migration and types of migration events between ethnic groups may also be explained by compositional differences, such as the concentration in different social classes, or differences in health. Where MEGs experience disadvantage, as often observed for Black Caribbeans, Pakistanis and Bangladeshis (Jivraj and Khan, 2015), opportunities for, and distances of, migration will vary.

The contrasting socio-economic and spatial experiences of different ethnic groups will have different implications for any process of selective sorting. To illustrate this, consider the detrimental and limiting relationship between poor health, disadvantage and the likelihood of migrating. As poor health and disadvantage both limit opportunities for migration, any process of selective sorting for such groups will vary from that of more advantaged, healthier groups. Despite this hypothesised difference, while the health–migration relationship appears to vary by age, this does not necessarily vary by ethnic group. Migration consistently 'selects' healthy young migrants while older migrants are more likely to be in poorer health. The point at which 'young' becomes 'old', however, has changed over time; given that the statutory retirement age is increasing, this may continue to change. The tipping point between younger, healthy migrants and older, unhealthy migrants may contribute to the 'kink' in the LLTI curve observed in other work (Marshall and Norman, 2013).

However, the extent of the influence of socio-economic status (and region, although not shown) on the health–migration relationship does appear to vary between ethnic groups: certain minority groups are less able to reap the health benefits of more advantaged circumstances than others. Further, as ethnic inequalities in health widen with increasing age, an ageing and increasingly ethnically diverse population may be faced with persisting and widening gaps in health that must be addressed in policy. Variations in the relationship between health and migration, and wider variations in the socio-economic composition of each ethnic group will be likely to influence the operation of selective sorting between social classes and area types. Thus, the influence on (changing) health gradients may vary and deserves investigation.

24.3 Transitions between area types and social classes: can selective sorting influence changing health gradients?

A wide body of literature investigates selective sorting and health gradients, whether in terms of social mobility (e.g. Bartley and Plewis, 2007) or migration and deprivation change (e.g. Boyle et al., 2004; Norman et al., 2005; Exeter et al., 2011). Essentially, these studies ask whether transitions of differently healthy groups between areas or social classes maintain, widen or constrain health gradients. To establish whether and how transitioning groups influence (changing) health gradients, we must compare health between transitioning groups (see Boyle et al., 2009; Norman and Boyle, 2014).

For selective sorting to widen health gradients, the health of those entering the most advantaged circumstances must be better than the health of those leaving. Similarly, the health of those entering the least advantaged circumstances must be poorer than the health of those leaving. For selective sorting to maintain health gradients, the health of the downwardly mobile must be better than the health of the upwardly mobile for either those transitioning around the most advantaged circumstances, or for those transitioning around the least advantaged. If the health of the downwardly mobile is consistently better than the health of the upwardly mobile, it is possible that selective sorting narrows rather than widens or maintains existing health gradients.

Data and methods

To examine transitions over time, we turn to the ONS LS rather than the cross-sectional SARs. The LS, introduced in Chapter 9, is a 1 per cent sample of the population of England and Wales, linking the decennial census to life event information (such as births or deaths) and cancer registrations. We select a closed sample of LS members resident in England who were present at the 2001 and 2011 Censuses. All LS members with incomplete ethnicity or deprivation data are excluded, as are international migrants, residents in communal establishments and those in poor health in 2001. As a closed sample, we can assume the cohort is relatively healthy since all persons survive the ten-year period.

The variables utilised vary slightly from those used in the SARs analysis, with additional variables derived to identify transitions between area types and social classes. Social class, health and the ethnic groups are as defined in the previous SARs work. However, this chapter will focus on the experiences of the total population (of which Whites are the majority) compared to the experiences of Indians, Pakistanis and Bangladeshis (sample sizes for Black Caribbeans and Black Africans are too small for meaningful analysis). Movers are identified at source by a ten-year rather than one-year migration variable, comparing LS members' address at 2011 with their address at 2001 (though address-specific information is not available for analysis). International migrants and moves between England and Scotland, Wales or Northern Ireland are excluded on the same basis as the previous SARs analysis. Area types are defined by deprivation, measured according to the Carstairs Index (Morris and Carstairs, 1991).

Table 24.2 Transition categories

Transitions between area types (Deprivation mobility)	Transitions between social classes (Social mobility)
Q1 – Q1	I and II – I and II
Q2–Q4 into Q1	IIIN–IIIM into I and II
Q1 into Q2–Q4	I and II into IIIN–IIIM
Q2–Q4 – Q2–Q4	IIIN–IIIM – IIIN–IIIM
Q5 into Q1–Q4	IV and V into I–IIIM
Q1–Q4 into Q5	I–IIIM into IV and V
Q5 – Q5	IV and V – IV and V

The Carstairs Index provides a summary of deprivation in each area type relative to other areas at that time and cannot here be used to describe absolute change in deprivation over time (see Chapter 30). Scores are categorised into quintiles whereby Q1 is the least deprived and Q5 is the most deprived.

Following previous research in this area (Norman et al., 2005; Boyle et al., 2009; Exeter et al., 2011), seven transition categories are defined for deprivation change (transitions between area types) and social mobility (transitions between social classes) in Table 24.2. As individuals can change area types without migration, following area decline or regeneration, we distinguish between movers and stayers in this analysis. Transitions are therefore identified for movers and stayers who may transition between area types due to migration *or* deprivation change, but who may also transition between social classes.

Two methods can be used to identify whether selective sorting may contribute to (changing) health gradients. In this section, we present examples of each method. First, we compare the health of groups transitioning between area types and social classes, calculating Standardised Illness Ratios (SIRs) by transition category for movers and stayers by ethnic group. SIRs of more than 100 indicate greater than expected levels of poor health (LLTI) whereas those of less than 100 indicate lower than expected levels. Expected rates are calculated using a standard population of all LS members present in the closed cohort. This method reveals whether migration attenuates the health–deprivation or health–social class relationship, and is illustrative of the inter-dependency between migration, deprivation mobility (area type change) and social mobility.

Second, we summarise the overall influence of the transitions on health–deprivation and health–social class gradients. We compare health gradients in the population by social class or area type in 2011 with the health gradient that would have occurred if the population did not redistribute between 2001 and 2011. In other words, we 'put people back' into their class or area type of origin while maintaining their health status at destination. Health gradients are calculated using SIRs by deprivation quintile and social class. The Relative Index of Inequality (RII) is then calculated to summarise the extent of health inequality across the population by social class or area type at destination and origin (see Mackenbach and Kunst, 1997, for details of the calculation and interpretation of the RII). If the RII is higher by destination, transitions may widen health gradients, whereas if the RII is lower at the destination, transitions may narrow or constrain health gradients. This method does not distinguish between 'movers' and 'stayers'. Similar approaches are effectively employed in the literature (e.g. Brimblecombe et al., 1999, 2000; Connolly et al., 2007; Norman and Boyle, 2014).

Evaluating selective sorting: comparing transitions and summarising gradients

Table 24.3 summarises the SIRs for movers and stayers by transition category for the total population, Indians and Pakistanis and Bangladeshis. For clarity, transitions equating to increasing disadvantage are shown in italic whereas those equating to decreasing disadvantage are underlined. SIRs which are significantly different to the national morbidity level are starred. Distinguishing movers and stayers highlights the influence of migration on (changing) health gradients, while calculating SIRs by transition category for movers and stayers combined reveals the overall influence of selective sorting on health gradients. It is worth reiterating that the transitions between area types for movers arise because these groups have changed address, whereas transitions between area types for stayers arise because the areas in which these groups reside have changed. For groups transitioning between area types or social classes, movers are more likely to be in better health than stayers when transitioning around the least deprived areas or the top social classes. Conversely, movers transitioning around the most deprived areas or bottom social classes are more likely to be in poorer health than stayers. This is illustrative of the complex health–migration relationship, and the extent to which health differently influences types of moves.

For the total population, transitions between area types contribute to widening health gradients, evident in the successive increases in the SIRs for the total population when moving from the top (Q1: Q1) to bottom (Q5: Q5) transition category. However, the SIRs of movers and stayers reveal that much of this is driven by the transitions of movers between areas rather than stayers whose area type changes. A similar pattern is apparent for Pakistani and Bangladeshi movers and stayers combined, although it is worth noting that the confidence intervals

Table 24.3 Standardised illness ratios by ethnic group, mover status and transition category

	Total population			Indian			Pakistani and Bangladeshi		
	Mover	Stayer	Total	Mover	Stayer	Total	Mover	Stayer	Total
Q1:Q1	53.7*	65.4*	62.8*	30.1*	56.4*	51.7*	57.8	136.9	118.0
Q2–Q4: Q1	63.6*	73.7*	67.8*	59.6*	60.5*	59.9*	71.1	79.7	73.0
Q1: Q2–Q4	68.5*	70.4*	69.3*	51.7*	84.2	66.3	63.8	89.5	74.5
Q2–Q4: Q2–Q4	83.6*	84.1*	84.0*	65.6*	83.0*	78.9*	87.0	132.5*	118.4*
Q5: Q1–Q4	97.1	106.5*	100.5	99.1	148.0*	113.9	140.5*	153.6*	144.1*
Q1–Q4: Q5	113.6*	92.2*	103.4	89.4	111.2	104.0	132.2	169.5*	150.1*
Q5: Q5	131.0*	114.4*	118.0*	118.8	120.9*	120.5*	162.6*	168.8*	167.0*
I and II: I and II	64.3*	71.0*	68.8*	40.7*	71.3*	61.4*	108.1	114.5	105.6
IIIN–IIIM: I and II	78.4*	84.3*	82.0*	74.5	92.4	87.2	95.7	131.3	130.9
I and II: IIIN–IIIM	82.9*	84.7*	84.1*	67.3	108.6	95.5	84.7	144.5	101.0
IIIN–IIIM: IIIN–IIIM	100.4	93.9*	95.7*	111.5	123.6*	120.7*	153.5*	174.2*	180.1*
IV and V: I–IIIM	120.2*	113.0*	115.4*	159.0	117.6	129.7	199.3	212.0*	163.8*
I–IIIM: IV and V	136.7*	118.6*	124.3*	217.0*	172.0*	183.5*	215.6	191.9	180.2*
IV and V: IV and V	158.9*	127.6*	134.6*	260.9*	190.1*	202.6*	209.1*	219.9*	211.1*

Source: ONS Longitudinal Study.

Note: Statistically significant results are starred. Underlining denotes decreasing disadvantage; italic denotes increasing disadvantage.

are wider for these minority groups who are less likely to move (see Section 24.2) and highly concentrated in the most deprived areas (Jivraj and Khan, 2015). Conversely, the overall contribution of transitions between area types for Indians appears to maintain rather than widen existing health gradients. For example, Indian movers transitioning out of Q1 have better health than those entering, while those entering Q5 have better health than those leaving. For the total population, whether movers or stayers, transitions between social classes appear to contribute to widening health gradients. For Indians, although the collective influence of the social class transitions of stayers and movers may contribute to widening health gradients, the health gradient amongst Indians who change address is effectively maintained rather than widened by transitions between social classes. The patterning of the SIRs by transition category for Pakistanis and Bangladeshis suggests that social mobility maintains existing health gradients. Differences in health between the transitioning groups are notably smaller for these ethnic groups than observed for either the total population or Indians.

While the results presented in Table 24.3 are illustrative of the influence of transitions at the extremes of the deprivation spectrum or social class structure, this does not account for the majority of the population who transition between (or churn within) the middle deprivation quintiles (Q2–Q4) or the middle social classes (IIIN–IIIM). The RII, however, accounts for differences across all quintiles and social classes by comparing the health of the population at destination area type or social class with the health of the population put back to origin area type or social class (see Section 24.3). As such, it may be more revealing as to the influence of selective sorting for minority groups, particularly when including those not assigned a social class (Unclassifiable). Table 24.4 summarises the RII for the total population, Indians and Pakistanis and Bangladeshis. The RII for 'Transitions' summarises the deprivation or social class health gradients in 2011, allowing the population to transition between social classes or area types in the 2001–11 period.

The RII for 'No transitions' calculates the health gradient if the population retain their 2011 health status but are put back to their area type or social class of origin. For all population groups, transitions result in a higher RII value than when the population is put back into their origin circumstance (no transitions). In other words, transitions between area types or social classes appear to widen health gradients for both the overall population and the Indians and Pakistanis and Bangladeshis.

Discussion

Differences in the health of movers compared to stayers at either end of the deprivation or social class spectrum illustrate the complex relationship between health and migration. We suggest that moves for more deprived groups or lower social classes are more likely to relate to poor health

Table 24.4 Assessing the contribution of transitions between area types or social classes to the health gradients: comparing the RII between ethnic groups

RII	Health–deprivation gradient		Health–social class gradient	
	Transitions	No transitions	Transitions	No transitions
Total population	2.23	2.10	2.48	1.80
Indian	2.54	2.37	3.84	1.68
Pakistani and Bangladeshi	1.73	1.63	2.17	1.41

Source: ONS Longitudinal Study.

and be over shorter distances. This contrasts with the greater distances we anticipate are made by movers in better health, less deprived areas and higher social classes. This would account for the health differences between moves at either end of the hierarchies.

Comparing the health of different population groups transitioning between areas and social classes suggests that selective sorting can contribute to widening health gradients both in the overall population and by ethnic group. Where differences emerge amongst the minority groups, this may reflect differences in the propensity to migrate or (lack of) opportunities to move away from more deprived areas. However, this may also reflect limitations in the methods used to assess the influence on health gradients such as the focus on transitions between the extremes of the deprivation quintiles or social class structures or the exclusion of the population not assigned to a class. This 'Unclassifiable' group comprises a substantial proportion of MEGs and their exclusion may therefore distort results. By using alternative measures and including the Unclassifiable groups, we can understand more about how selective sorting may differently influence (changing) ethnic health gradients.

When evaluating the influence of selective sorting on health gradients using a measure that accounts for differences across all deprivation and social class categories, a slightly different picture emerges to that revealed in Table 24.3. The RII suggests that movement within and between the middle deprivation quintiles may be particularly important for changing minority ethnic health gradients, with transitions between all area types and social classes (including the Unclassifiable groups) appearing to consistently contribute to widening health gradients. Indeed, including the Unclassifiable group suggests that social mobility markedly increases health gradients for the MEGs. Future work should examine the nature of mobility into and out of this Unclassifiable group for MEGs (and self-employment as captured in the NS-SEC).

24.4 Conclusion

Selective sorting is one possible mechanism by which we can evaluate whether ethnic inequalities in health may be changing over time. Moreover, by using cross-sectional and longitudinal census microdata we can explore the extent of ethnic inequalities in health and the context within which selective sorting may vary between ethnic groups. The results presented in this chapter have shown that while the inter-relationships between migration, health and socio-economic circumstances hold across ethnic groups, there are important variations contributing to widening health inequalities. For example, health inequalities between ethnic groups vary markedly with age with younger minority groups in better health than the White majority. However, older MEGs are more likely to be in poorer health. These analyses also indicate that selective sorting can contribute to widening health gradients in the overall population and by ethnic group; understanding why and how health inequalities manifest is the first vital step to addressing persisting inequalities.

Given the results of these analyses, it is possible to identify a number of future avenues for research. First, differences in the patterning of health between ethnic groups by age deserve consideration (building on Norman and Boyle, 2014). In particular, it is important to establish whether and why the health of MEGs appears to deteriorate faster with increasing age than observed in the White majority. This may reflect the influence of accumulated disadvantage harming the health of minority groups over the lifecourse. Conversely, this may reflect cohort effects masked in cross-sectional data; the health of minority groups aged 30 and over in 2011 may have also been poorer than the White majority when aged 16–29, while those minority groups currently (in 2011) in better health than the White majority at ages 16–29 may continue to experience better health. Second, in the context of selective sorting, transitions between area types and social classes in the middle of the spectrum deserve more attention, as does the question of whether propensity (not) to migrate may differently influence the relationship between selective sorting and health–deprivation gradients for different ethnic groups.

Third, it is possible that the measure of social class is not appropriate to capture the breadth of experiences within MEGs. Indeed, it has been argued elsewhere that multi-dimensional measures of socio-economic status are more suited for analyses of ethnically differentiated experiences of social mobility than the uni-dimensional measures used in this analysis (see Harding, 2003).

Notwithstanding the limitations and the directions for future research they highlight, these do not undermine the conclusions drawn from this analysis. The methods and measures employed are widely and effectively used in comparable literature, and combining approaches, as demonstrated in this chapter, enhances the conclusions drawn. While more can be done, such as using an alternative measure of socio-economic status or examining in more detail why different ethnic groups are differently likely to migrate or experience social mobility, these analyses do begin to disentangle the relationships between ethnicity, socio-economic status, health and migration.

References

Bartley, M. and Plewis, I. (1997) Does health-selective mobility account for socio-economic differences in health? Evidence from England and Wales, 1971 to 1991. *Journal of Health and Social Behaviour*, 38: 376–386.

Bartley, M. and Plewis, I. (2007) Increasing social mobility: An effective policy to reduce health inequalities. *Journal of the Royal Statistical Society: Series A*, 170 (2): 469–481.

Bentham, G. (1988) Migration and morbidity: Implications for geographical studies of disease. *Social Science & Medicine*, 26(1): 49–54.

Boyle, P., Halfacree, K. and Robinson, V. (1998) *Exploring Contemporary Migration*. Longman, Harlow.

Boyle, P., Norman, P. and Popham, F. (2009) Social mobility: Evidence that it can widen health inequalities. *Social Science & Medicine*, 68: 1835–1842.

Boyle, P., Norman, P. and Rees, P. (2002) Does migration exaggerate the relationship between deprivation and limiting long-term illness? A Scottish analysis. *Social Science & Medicine*, 55: 21–31.

Boyle, P., Norman, P. and Rees, P. (2004) Changing places: Do changes in the relative deprivation of areas influence limiting long-term illness and mortality among non-migrant people living in non-deprived households? *Social Science & Medicine*, 58: 2459–2471.

Brimblecombe, N., Dorling, D. and Shaw, M. (1999) Mortality and migration in Britain: First results from the British Household Panel Survey. *Social Science & Medicine*, 49: 981–988.

Brimblecombe, N., Dorling, D. and Shaw, M. (2000) Migration and geographical inequalities in health in Britain. *Social Science & Medicine*, 50: 861–878.

Brown, D., O'Reilly, D., Gayle, V., Macintyre, S., Benzeval, M. and Leyland, A.H. (2012) Socio-demographic and health characteristics of individuals left behind in deprived and declining areas in Scotland. *Health & Place*, 18: 440–444.

Catney, G. and Simpson, L. (2010) Settlement area migration in England and Wales: Assessing evidence for a social gradient. *Transactions of the Institute of British Geographers*, 35(4): 571–584.

Champion, A. (1996) Internal migration and ethnicity in Britain. In Ratcliffe, P. (ed.) *Ethnicity in the 1991 Census. Volume 3. Social Geography and Ethnicity in Britain: Geographical Spread, Spatial Concentration and Internal Migration*. The Stationery Office, London, pp. 135–193.

Champion, A. and Ford, T. (1998) *The Social Selectivity of Migration Flows Affecting Britain's Larger Conurbations: An Analysis of Regional Migration Tables of the 1981 and 1991 Census*. University of Newcastle, Newcastle.

Chandola, T., Ferrie, J., Sacker, A. and Marmot, M. (2007) Social inequalities in self reported health in early old age: Follow-up of prospective cohort study. *British Medical Journal*, 334: 990.

Connolly, S., O'Reilly, D. and Rosato, M. (2007) Increasing inequalities in health: Is it an artefact caused by the selective movement of people? *Social Science & Medicine*, 64: 2008–2015.

Dale, A., Fieldhouse, E. and Holdsworth, C. (2000) *Analyzing Census Microdata*. Arnold, London.

Darlington, F., Norman, P., and Gould, M. (2015a) Health and internal migration. In Smith, D.P., Finney, N., Halfacree, K. and Walford, N. (eds.) *Internal Migration, Geographical Perspectives and Processes*. Ashgate, Surrey, pp. 113–128.

Darlington, F., Norman, P., Ballas, D. and Exeter, D. (2015b) Exploring ethnic inequalities in health: Evidence from the Health Survey for England, 1998–2011. *Diversity and Equality in Health and Care*, 12(2): 54–65.

Ewens, D. (2005) Moving home and changing school. Widening the understanding of pupil mobility. *DMAG 2005-32*, Greater London Authority, Data Management and Analysis Group, London.

Exeter, D. J., Boyle, P. and Norman, P. (2011) Deprivation (im)mobility and cause-specific premature mortality in Scotland. *Social Science & Medicine*, 72: 389–397.

Fielding, A. J. (1992) Migration and social mobility: South East England as an escalator region. *Regional Studies*, 26(1): 1–15.

Finney, N. (2011) Understanding ethnic differences in the migration of young adults within Britain from a lifecourse perspective. *Transactions of the Institute of British Geographers*, 36(3): 455–470.

Finney, N. and Simpson, L. (2008) Internal migration and ethnic groups: Evidence for Britain from the 2001 Census. *Population, Space and Place*, 14(2): 63–83.

Gould, M. and Jones, K. (1996) Analyzing perceived limiting long-term illness using U.K. microdata. *Social Science & Medicine*, 42 (6): 857–869.

Harding, S. (2003) Social mobility and self-reported limiting long-term illness among West Indian and South Asian migrants living in England and Wales. *Social Science & Medicine*, 56: 335–361.

Jivraj, S. and Khan, O. (2015) How likely are people from minority ethnic groups to live in deprived neighbourhoods? In Jivraj, S. and Simpson, L. (eds.) *Ethnic Identity and Inequalities in Britain : The Dynamics of Diversity*. Policy Press, Bristol, pp. 199–214.

Kunst, A.E., Bos, V., Lahelma, E., Bartley, M., Lissau, I., Regidor, E., Mielck, A., Cardano, M., Dalstra, J.A.A., Geurts, J.J.M., Helmert, U., Lennartsson, C., Ramm, J., Spadea, T., Stronegger, W.J. and Mackenbach, J.P. (2005) Trends in socio-economic inequalities in self-assessed health in 10 European countries. *International Journal of Epidemiology*, 34: 295–305.

Larson, A., Bell, M. and Young, A.F. (2004) Clarifying the relationships between health and residential mobility. *Social Science & Medicine*, 59: 2149–2160.

Mackenbach, J.P. (2012) The persistence of health inequalities in modern welfare states: The explanation of a paradox. *Social Science & Medicine*, 75: 761–769.

Mackenbach, J.P. and Kunst, A.E. (1997) Measuring the magnitude of socio-economic inequalities in health: An overview of available measures illustrated with two examples from Europe. *Social Science & Medicine*, 44(6): 757–771.

Marshall, A. and Norman, P. (2013) Geographies of the impact of retirement on health in the United Kingdom. *Health & Place*, 30: 1–12.

Mindell, J., Knott, C.S., Ng Fat, L.S., Roth, M.A., Manor, O., Soskolne, V. and Daoud, N. (2014) Explanatory factors for health inequalities across different ethnic and gender groups: Data from a national survey in England. *Journal of Epidemiology and Community Health*, 68(12): 1133–1144.

Morris, R. and Carstairs, V. (1991) Which deprivation? A comparison of selected deprivation indexes. *Journal of Public Health*, 13(4): 318–326

Norman, P. and Boyle, P. (2010) Using migration microdata from the samples of anonymised records and the longitudinal studies. In Stillwell, J., Duke-Williams, O. and Dennett, A. (eds.) *Technologies for Migration and Commuting Analysis: Spatial Interaction Data Applications*. IGI Global, Hershey, pp. 133–151.

Norman, P. and Boyle, P. (2014) Are health inequalities between differently deprived areas evident at different ages? A longitudinal study of census records in England and Wales, 1991–2001. *Health & Place*, 26: 88–93.

Norman, P., Boyle, P. and Rees, P. (2005) Selective migration, health and deprivation: A longitudinal analysis. *Social Science & Medicine*, 60(12): 2755–2771.

ONS (2014) How do people rate their general health? An analysis of general health by disability and deprivation. Available at: www.ons.gov.uk/ons/dcp171776_353238.pdf.

Plane, D.A. (1993) Demographic influences on migration. *Regional Studies*, 27(4): 375–383.

Raymer, J. and Giulietti, C. (2009) Ethnic migration between area groups in England and Wales. *Area*, 41(4): 435–451.

Salomon, J.A., Wang, H., Freeman, M.K., Vos, T., Flaxman, A.D., Lopez, A.D. and Murray, C.J.L. (2012) Healthy life expectancy for 18 countries, 1990–2010: A systematic analysis of the Global Burden of Disease Study 2010. *The Lancet*, 380(9859): 2144–2162.

Simpson, L. and Finney, N. (2009) Spatial patterns of internal migration: Evidence for ethnic groups in Britain. *Population, Space and Place*, 15(1): 37–56.

25

Changes in social inequality, 2001–11

Danny Dorling

25.1 Introduction

Social inequality has many meanings and many key aspects of social inequality cannot be studied through a population census that does not ask people about their income or wealth. However, the censuses do ask questions that make it possible to draw some general conclusions as to the direction of travel within UK society. In this chapter, aided by the cartograms that Bethan Thomas has drawn, I use census data to demonstrate that social and geographical polarisation continued to grow in the 2000s. The chapter reveals that the 'South' of England continued to prosper far more than the 'North', Wales, Scotland or Northern Ireland. This represents a continuation and possibly an acceleration of earlier trends.

The maps of local authorities in Figures 25.1a–f provide a key to all the cartograms used in this chapter. Each local authority district is drawn here with its area proportional to its population. The cartograms have been drawn to ensure that every local authority district remains contiguous with its actual neighbours. The colours used in these maps are chosen to differentiate between the areas. The boundaries of major cities have a slightly thicker black outline. Unitary authorities in England are coloured purple. The English district councils within each county are distinctly shaded.

25.2 Population density and potential

The first form of social inequality to consider is simply the density at which people live in the UK. This varies by local authority district from under 0.1 people per hectare (in Highland, Scotland) to over 138 people per hectare in 2011 (in Islington, London). Most people in the UK live at densities greater than five people per hectare as shown in Figure 25.2a. Only those living in the districts coloured red lived at a lower population density than this. By 2011, the population density of the UK was 2.6 people per hectare, an area of 10,000m². In 2001, the UK population density had been 2.4 people per hectare. The overall population density in the UK is not high; it is lower than in some other European countries and many of our cities have a lower population density than many cities in other parts of Europe. However, between the last two UK censuses, the total population rose by 7.4 per cent and so the overall density of the population rose by exactly the same amount. This amounts to an extra person for every five hectares in just ten years.

Figure 25.1 Cartograms of districts of the UK

c. Local authority districts comprising the 'West of England' and Wales

d. Local authority districts comprising the 'South East and East of England'

Figure 25.1 (Continued)

e. Local authority districts of Northern Ireland   f. Main towns/cities of the United Kingdom

Figure 25.1 (Continued)

Figure 25.2b shows density change between 2001 and 2011 and makes it clear that most people in the UK have not experienced much rise. Some 68 per cent saw a fall or no noticeable rise in population density where they lived. In contrast, 5 per cent of the population in 2011 lived in areas that had seen an additional 10 people or more resident per hectare since 2001. That is an extra one person for every 1,000 m² in their localities. A large detached house with a large garden covers an area of that size, so it is as if one more person has arrived in such a house in each place. Of course, in the densely populated parts of the UK, there are very few such houses and many terraced houses or blocks of flats and so, in reality, it is simply an extra person arriving, net, in just a single one of the many properties that cover a space of 1,000 m² in these districts.

All of the districts of very rapid population growth were in London. They are the areas coloured blue in Figure 25.2b and most of them are in Inner London. The sole exception is Kensington and Chelsea where the population actually fell even as many more rooms were added to properties there through extensive basement extensions and similar building work. Outside London in the few places where there were population declines these were greatest in South Tyneside (-0.7 people per hectare); Sefton (-0.6); Knowsley (-0.5); Sunderland (-0.4); and Barrow-in-Furness (-0.3); and Burnley, East Dunbartonshire, Redcar and Cleveland, West Dunbartonshire and Inverclyde (all -0.2 people per hectare).

The censuses allow us to measure the change in the number of rooms in a district as well as people (Tunstall, 2015). Social inequality has increased even when simply measured by the increase in population where there already were the most people in most of the UK. Already

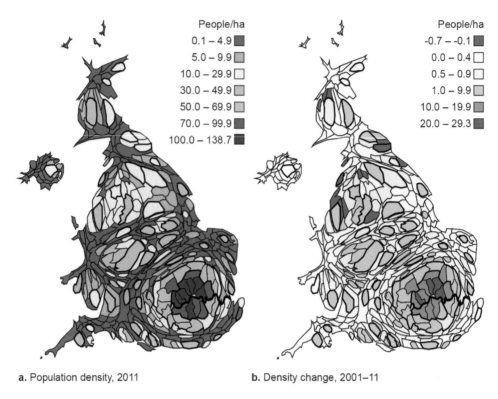

People/ha
0.1 – 4.9 ■
5.0 – 9.9 ▢
10.0 – 29.9 ☐
30.0 – 49.9 ▢
50.0 – 69.9 ▢
70.0 – 99.9 ▢
100.0 – 138.7 ■

People/ha
-0.7 – -0.1 ■
0.0 – 0.4 ☐
0.5 – 0.9 ☐
1.0 – 9.9 ▢
10.0 – 19.9 ▢
20.0 – 29.3 ■

a. Population density, 2011

b. Density change, 2001–11

Figure 25.2 Population density in 2011 and change, 2001–11

overcrowded areas became, in general, more overcrowded. Further analysis of the census reveals that what has tended to happen in recent decades is that already overcrowded areas became more overcrowded within those properties in which people were already most crammed; they saw the greatest rise in densities. Elsewhere, even within London boroughs, other houses acquired more rooms and/or fewer people to occupy those rooms, so that overall, the living space was shared less and less efficiently.

Population potential is a measure of how near someone is to everyone else. It can be calculated for a single state such as the UK, or for any point on the planet taking into account everyone else on the planet. It is measured as the sum of all people in every other district when each group of people is divided by how far away they are from the centre of their district, with distance here measured in metres. The population of the district for which potential is being calculated is not included, but every other district in the UK is. A low population potential is a measure of the remoteness of that district. The place with the highest population potential in the UK in 2011 was the City of London (Figure 25.3a) with a surrounding pressure of 1,292 people per metre away from its centre, 149 people per metre higher than the measure recorded in 2001 (Figure 25.3b). Population potential has risen the most where it was highest to begin with. That is, because population growth was concentrated in London in the 2000s, it is almost exclusively within London that people have experienced the arrival of large numbers of additional people, far more than have left. Migration is the key reason for this rise.

It would have been possible for population density to have become more polarised and population potential to have become less polarised had the northern cities of England, and Cardiff,

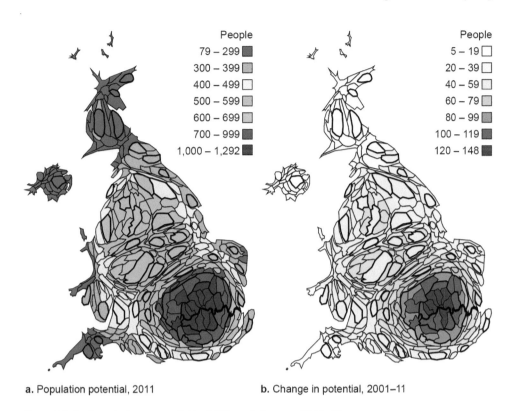

People

79 – 299 ■
300 – 399 ▨
400 – 499 ☐
500 – 599 ▨
600 – 699 ▨
700 – 999 ■
1,000 – 1,292 ■

People

5 – 19 ☐
20 – 39 ☐
40 – 59 ▨
60 – 79 ▨
80 – 99 ■
100 – 119 ■
120 – 148 ■

a. Population potential, 2011   **b.** Change in potential, 2001–11

Figure 25.3 Population potential in 2011 and change, 2001–11

Belfast, Edinburgh and Glasgow grown more quickly than London; then, the population potential across the UK would have become more evenly distributed despite the populations of cities still growing more quickly than rural areas. The opposite happened. Planning controls meant that it was mainly in the cities that any growth occurred, but that growth was far higher within London than elsewhere. Thus, if you wish to place a business within a certain number of miles of a very large market, the nearer you place it to London, the larger that market will be and this became even more the case in 2011 as compared to 2001. Social inequality in terms of remoteness and centrality also rose between the 2001 and 2011 Censuses.

25.3 Area classifications

Following the release of the 2011 Census, the Office for National Statistics (ONS) produced its own classification of areas based upon the data collected. This is shown in the first of the pair of maps in Figure 25.4. There are 77 local authority districts labelled by the ONS as 'Mining Heritage and Manufacturing' centres. On average in these places, just 0.3 per cent of the population were affluent enough to qualify for inhabitants to pay inheritance tax upon their deaths in 2011; that proportion rose to 0.4 per cent in 2012 and 0.5 per cent in 2013 (Dorling and Thomas, 2016). Slightly better off are the 53 areas labelled 'Scottish and Northern Irish Countryside', in which the respective proportions were 0.6 per cent in 2011, 0.7 per cent in 2012 and 0.9 per cent in 2013; less than one person in every 100 had been dying with enough wealth to qualify to pay inheritance tax. This is an example of how it is possible to combine

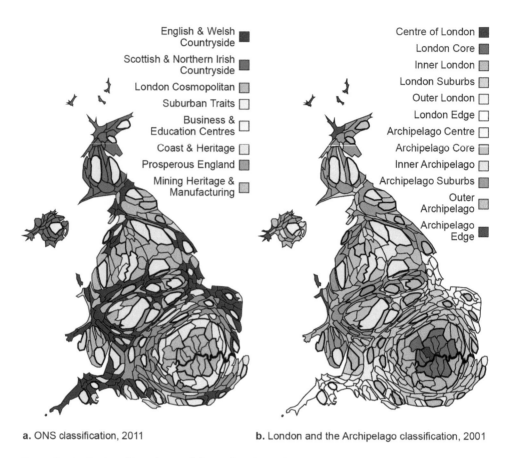

a. ONS classification, 2011   **b.** London and the Archipelago classification, 2001

Figure 25.4   ONS and London and the Archipelago classifications of local authority districts

census and tax data to show just how few people are very rich within the ONS 2011 Census classification areas.

The second area classification in Figure 25.4 is one that Bethan Thomas and I first produced shortly after 2001 following the release of data from that census. In that classification, six groups of places in the south of England are *London Areas* and the rest of the UK is divided into six other areas, mainly because of their differing levels of remoteness and all described as parts of an *Archipelago*. Thus, in the rest of this chapter when the discussion is about the London Areas what is meant is what you might think of as southern England. The reason for using the term 'London' is that London has come to so dominate areas in the south of England. It is distance from London that tends to determine, *above any other factor anyone has been able to measure,* whether an area in the south of England appears to perform poorly or well economically. House prices depend most on distance from London, salaries, average levels of qualifications and excuses for giving ever greater remuneration become most far-fetched the further into the centre of London you travel.

In contrast, once outside the six 'London' areas, the UK becomes split up into an archipelago of economic islands of mild prosperity – the successful (and less successful) university cities of the periphery – the old centres of industry. And then, the further you are from any of those mildly prosperous archipelago islands, the worse many economic conditions appear to become. House prices fall, salaries and wages are lower, and more people exist on benefits, average qualifications are lower and the vilification of people for not being able to do or get a job (and calling them scroungers) grows. What Owen Jones termed "the demonization of the working class" in his 2012 book, *Chavs*, is mainly a demonisation of the working class of the remoter parts of the North, Wales, Scotland and Northern Ireland.

The division into 12 regions shown above in the second cartogram in Figure 25.4 was produced because, when Bethan Thomas and I finished our analysis of the 2001 Census, we came to the conclusion that the 2001 Census had revealed the UK to be dividing very quickly and very abruptly. It is a good basis for the study of changes since 2001 as it was defined in 2001 and so, unlike the ONS 2011 classification, it was not partly defined by what you might also want to use it to measure.

The first six of 12 regions all had London in their titles: *London Centre*, *London Core*, *Inner London*, *London Suburbs*, *Outer London* and *London Edge*. None of the last three of these regions actually included any area that was or is formally part of London, but London was included in their titles because distance to London had such a dominant effect on their fortunes in the years between 1991 and 2001. Thus, the entire south of England was labelled *the London Areas*, stretching as far north as Lincolnshire and as far west as Cornwall.

Census (and other) migration data can be studied to reveal that Londoners (disproportionately) retire to Lincolnshire and holiday in Cornwall. The most affluent had their second homes in such places. The poorest of southerners might one day end their days in the cheaper parts of the great metropolis' outer edges. The index of deprivation that was published in September 2015 revealed poverty rates rising around the periphery of the south of England, especially along the east coast. This was presumably because families with very little were being pushed out of areas that had become too expensive in the south or could not move to places like east London anymore to try their luck. East London had gentrified.

For several decades there has been a distinct lack of much migration out of the south of England across the outer border of the London Areas – in a northerly direction across what had been the old North–South divide. Occasionally a reporter from a southern newspaper would venture out of their southern comfort zone and do a little tour of the North. Politicians mostly only headed out of these areas when they needed votes. The affluent youth of the south might venture north for a few years to be university students. However, in recent years it has been Southampton, Reading, Bristol and Exeter universities that have grown most in student numbers. For decade after decade, far more northerners moved south than southerners ever crossed north and west. And then, since 2001, migration from abroad also began to become more and more concentrated in London and the South.

The remaining six regions of the UK that Bethan Thomas and I identified following the release of the 2001 Census, we termed the *Archipelago*. These were those places centred on a string of urban centres all well over an hour and a half commuting time from London by the fastest form of travel. The largest of these is centred on what is now called the *Northern Powerhouse* of Manchester, connected by old railway lines through to Leeds and Sheffield. Less well connected again are the islands of Cardiff, Birmingham, Liverpool, Newcastle, Belfast, Glasgow and Edinburgh. These are all highpoints in the sea of peripheral places. Around many of these places is what we called the *Archipelago Core* – economically lower-lying land but still part of this string of islands. Then there is the *Inner Archipelago* that you can see defined in Figure 25.4b – places a little less well connected again; then the large mass of *Archipelago Suburbs* – places from which you might commute to a peripheral centre; and then there is the beginnings of the *Outer Archipelago*; and finally the most isolated places of all, the *Archipelago Edge*. These are areas that were most often left behind in the past, and that can appear to be very behind the times now.

London and the Archipelago was how Bethan and I summed up the new human geography that the 2001 Census had revealed. It did not matter where a place was in the South, or what people did there. Irrespective of all that, in the 1990s it would prosper just because it was in the South. Living in the new expanded London Areas many people appeared to become more prosperous but life was less easy, more overcrowded and more stressful. And then it became too expensive to buy property, and the South became home to 'generation rent'.

25.4 Rising social inequality

The most basic demographic change is in the numbers of people living in each part of the UK and this has been highlighted in the cartograms above. By 2011, it was clear that the greatest population increases since 2001 had been in the Centre of London (+20.3 per cent), followed by London Core (+16.7 per cent), and what we called Inner London (+11.7 per cent); all these places having between two and almost five times faster population increases than those seen in both the Core and the Suburbs of the Archipelago (+4.5 per cent). The outer regions of the Archipelago and the Edge did not see such slower population growth, with the Edge experiencing the national average rise of 7.5 per cent, the same change as in what we termed the 'new' Outer London (the area just outside of the Greater London Authority). It is possible that the start of a counterurbanisation is taking place with very remote areas seeing a slight revival in their populations. This is only visible on a graph (Figure 25.5). The Archipelago Centre may well buck the trend because universities there were continuing to expand, but even given that expansion in student numbers, the rise in population was small compared to that experienced within the three most central London regions. Table 25.1 shows the number of people living in each region and how that changed between the censuses.

The population change that has occurred in each of our 12 regions can be broken down in many ways. It is possible to look at the change in people working in each part of the country by this regional classification, the qualifications they have, the kinds of jobs they have, where they come from and what kinds of families they form, if any. Increasing numbers of young people in

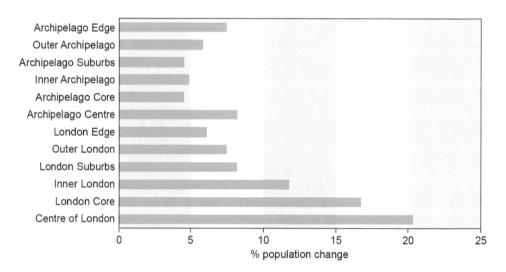

Figure 25.5 Population change (%) by region in the UK, 2001–11

Table 25.1 Population by region, 2001 and 2011, absolute and relative change

	Region	2001	2011	Increase	% increase
1	Centre of London	188,000	227,000	39,000	20.3
2	London Core	2,841,000	3,316,000	475,000	16.7
3	Inner London	6,973,000	7,792,000	819,000	11.7
4	London Suburbs	9,902,000	10,711,000	809,000	8.2
5	Outer London	4,603,000	4,946,000	343,000	7.5
6	London Edge	3,760,000	3,989,000	229,000	6.1
7	Archipelago Centre	6,903,000	7,468,000	565,000	8.2
8	Archipelago Core	7,778,000	8,129,000	351,000	4.5
9	Inner Archipelago	2,346,000	2,460,000	114,000	4.9
10	Archipelago Suburbs	11,217,000	11,725,000	508,000	4.5
11	Outer Archipelago	1,722,000	1,822,000	100,000	5.8
12	Archipelago Edge	556,000	598,000	42,000	7.5
	UK	58,789,000	63,182,000	4,393,000	7.5

the London Areas have to delay forming families because the cost of housing there has risen so much and we can begin to see the implications of that in all kinds of other trends.

There are advantages to viewing the UK as split between two groups of six regions. Within this new classification, areas are viewed differently from the stereotypical idea of where they are and hence how the population within them might normally be expected to behave given the traditional regional typology. Take, for example, the district of Nuneaton in Warwickshire. This area is normally stereotyped as a textile and manufacturing centre in the Midlands of England, but those are industries that largely disappeared from the area decades ago. Nuneaton is now largely a dormitory town for surrounding areas to which people commute for office work. It is still home to some electronics and other light industries, but not that many. Politically, from the general elections of 1935 through to 1983, Nuneaton was a safe Labour seat; but in both 2010 and 2015, its voters returned a Conservative MP, doubling his majority in the most recent election. It was seen as one of the key potential 'swing seats' of the 2015 general election, one that Labour should have won if it was to do better overall. However, is it really that kind of an area any longer?

According to the regional classification used here, Nuneaton has been part of what Bethan and I have called 'London Suburbs' since at least 2001. Thought of that way, there is nothing special about Nuneaton. People behave in that district much more like people in the other areas we have labelled as London Suburbs, partly because, increasingly, they are similar people and these are becoming more similar places:

> Nuneaton has not one but two shopping centres in its centre. If it's not market day, then both can look a little forlorn. The main shopping area, which includes the Abbeygate centre, boasts a number of betting shops, moneylenders, and no fewer than 22 charity shops. Inside the Myton Hospices shop, Marian Ferris reckoned coach parties from as far away as Bath came to savour the unique consumer experience of Nuneaton's abundance of charity shops. There were plans, she said, for Marks and Spencer to return to the town. "That will lift spirits. And they say Primark is coming here too. That will really lift Nuneaton." (Anthony, 2015)

In September 2015, the deputy governor of the Bank of England linked increased immigration, which is part of the reason for the rapid rises in people in ethnic minority groups in some

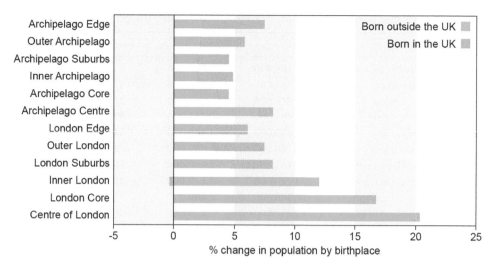

Figure 25.6   Population change (%) by region and birthplace in the UK, 2001–11

parts of London (and the centre of the Archipelago), to the growth in low wage groups there. He said: "Perhaps easier immigration has made low-skilled labour easier to come by . . ." (Milliken, 2015). Census analysis reveals that a majority of the London population growth that has occurred in recent years is due to an increase in people living in the south of England who were not born in the UK. In 2001, there were 4.9 million people living in the UK who were not born in the UK. That number grew to 8 million by 2011. Of the net increase of 3.1 million non-UK born people living in the UK by 2011, 2 million was of a rise in the non-UK born population within the six London Areas. That contrasts with a 0.7 million increase in the UK born population of the London Areas over that same decade.

Only at the London Edge and in what we here call Outer London (which – remember – is outside traditional London) has the bulk of the increase in population been of people born in the UK (Figure 25.6). Even in the London Suburbs, a majority of population increase has been due to a greater increase in the non-UK born population. In Nuneaton, the UK born population rose by 2,630 between the censuses (from 114,127 to 116,757); the non-UK born population is still very small but rose by 3,480 (from 5,015 to 8,495).

Further out from the London Suburbs, if you look at the demographics you will find that the rising UK born population are mostly of elderly people moving away from London, but managing to still remain in the South, if only on its edge. It is true that even on the edge of the Archipelago there has been an increase in non-UK born that is considerable. But proportionally it is eight times smaller than the increase in non-UK born in the centre of London. The statistics in the graph shown in Figure 25.6 are of the increase in population of each group as a proportion of the total population of both groups combined (in others words, all people) in 2001. The only area to see a reduction in either group is Inner London, where the UK born population fell fractionally.

25.5   A concentration of bankers and landlords

To determine where the 'haves' might be, look first at the finance and insurance sector (Figure 25.7). Between 2001 and 2011, it was in the centre of London that those jobs became more common. To be exact it was in the very heart of London that the proportion of people holding

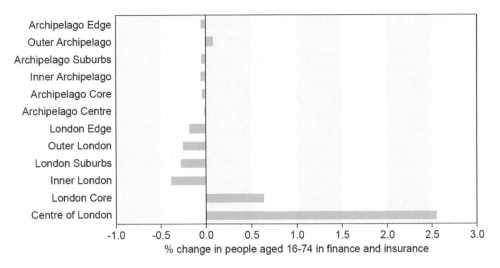

Figure 25.7 Change in share (%) of the entire population working in finance and insurance by region in the UK, 2001–11

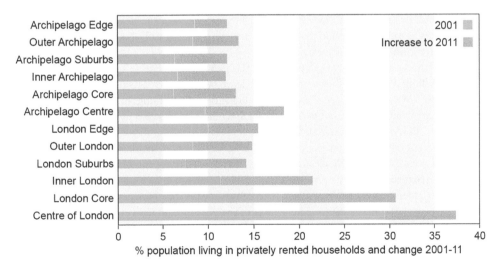

Figure 25.8 Change in share (%) of the entire population renting privately by region in the UK, 2001–11

those jobs was increasingly likely to be sleeping on census day, Sunday 27 March 2011. The rise of 2.6 per cent and 0.6 per cent in the heart of the capital is what now underpins the huge differences between average salaries and wages seen in these parts of London and those enjoyed or just survived on elsewhere. The census shows us the changes that underlie so much more.

Take, for example, those who have to pay rent to private landlords (Figure 25.8). The number of people who had a private landlord almost doubled in the inter-census decade in what we define as Inner London to become more than a fifth of all residents by 2011; and there was a 70 per cent increase in London's Core (a 12.6 per cent 2001–11 increase on the 18.1 per cent

345

2001 share). In the Core, an additional eighth of the population were renting just ten years on from 2001. In contrast, the rise in private renting on the Archipelago Edge was the slowest in the UK, but still a rise of over 40 per cent.

By just concentrating on the increase in private renting alone, the very different patterns and new diverging trends within London and the Archipelago become clear. And these changes are reinforcing all the other changes that more detailed census analysis can reveal. Some people are becoming very wealthy in the South, partly because a small group of the wealthy can now collect rent from a much larger group who have no choice but to pay it, rent which is on average a far greater monthly amount and greater proportion of their incomes than was the case in 2001.

25.6 Conclusion

Just a tiny number of examples have been given in this chapter of how the census can be used to reveal that social inequality in the UK is growing, especially as shown through the changing fortunes of people living in different areas. It is vital to realise that this does not mean that all is becoming better in the South. Rents are rising faster than incomes in the South. Traffic jams become longer as more people commute from more densely packed areas and as they have to change jobs more often and look further afield for work, especially better-paid work. Government minsters tell the average and the poor to decide whether they can still afford to live in London, and, by implication, leave London if they cannot 'take the heat': "Housing Minister Brandon Lewis today said Londoners had to make a 'judgement call' about whether they could afford to live in the capital" (Crerar, 2015).

In 2001, we could not have predicted that things would change so quickly that a minister would be telling people that they would have to leave our largest city, their city, if they could not afford to stay. We could not see the economic crash of 2008 coming, or the subsequent referendum in Scotland over whether to leave the UK when – had just 5.5 per cent of people voted differently – the Scots would have left the UK.

Whether all the constituent parts of the UK stay attached in the next ten years is far from certain. In June 2016, the UK electorate voted to leave the European Union, while a majority of voters in Scotland chose the remain option. This difference may result in a new referendum on Scottish independence in due course. The census reveals that Northern Ireland has suffered greatly economically in the 2000s. The centres of the Archipelago have experienced deeper cuts to services than they or anywhere else in the UK has ever suffered before. The bailout of the banks has saved the wealth of the centre of London, but at what cost? So much can happen in a decade, but it takes a census to show us the net effect.

If we do not look back at past censuses and compare them with the last one, it becomes easy to think it was always like this. Social inequalities of all kinds are growing within the UK. The censuses can only reveal some of these because so few questions are asked in the census. Most importantly, income and wealth have never been ascertained (income is asked in the US census). However, some things are not polarising as quickly as others and are even becoming less polarised. The final graph in this chapter shows one of these trends which is not of increased segregation. After decade upon decade of a rising concentration of graduates in London, the rate at which that concentration is growing is declining. London still has the most graduates but, partly due to saturation in its centre and partly because such a concentration of very affluent people requires more poorly-paid 'servants' to look after them, and perhaps partly because of social progress elsewhere (Dorling, 2010), the census measure of the concentration of graduates is finally no longer accelerating, as Figure 25.9 indicates.

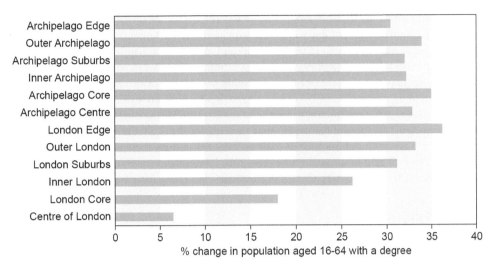

Figure 25.9 Relative increase in people aged 16–64 holding a university degree by region, percentage rise, 2001–11

Figure 25.9 shows that a third of all the graduates working or living across almost the entire Archipelago in 2011 were not there in 2001. The same is true of London Edge, Outer London and the London Suburbs; but in Inner London, the rise in graduates during the 2000s was of just a quarter more than its 2001 total, in London Core it is less than a fifth more, and in the Centre of London, graduate growth was less than one extra graduate arriving by 2011 for every 15 already there in 2001.

Sometimes censuses foretell the future. In 1991, the only area to have a majority of working age people not in a marriage was London. By 2011, what had been common in London was common everywhere. In 2001, the centre of London was being taken over by finance, increasing at a rate that appeared unsustainable, and so it proved to be, although in the very heart of the capital there are still more bankers in 2011 than there were in 2001 (Figure 25.7). For the first time in parts of London, it became evident by 2011 that there were absolute falls in the UK born population (Figure 25.6) – in what we call 'Inner London'. However, since 2011, the only traditional region to see a rise in the UK born population was London, and that was due to the births of children to immigrants in the capital. Other data, not shown here, suggest that most of those newly born children will be living in a household paying rent to a private landlord.

At current rates of increase, by 2031, a majority of people living in Inner London will have a private landlord (Figure 25.8) and by then there will be over 70 million people living in the UK (Table 25.1). Most of that increase – again, if current trends continue – will be in the capital, which is not impossible as London is currently the least densely populated mega-city on the planet (compare Figure 25.2 to data for other very large cities). If current growth in population potential is part of a longer-term trend (Figure 25.3) then that is making old typologies of areas redundant (Figure 25.4). Social inequality and polarisation will continue to rise, the winners and the dreamers will move to the centre from all over the world, and the losers will drop out to the periphery. All these predictions could be as wrong as they might now appear sensible, but they do tell us to watch current trends very carefully to see if the pace of change either accelerates or slows and even reverses.

All it would take to secure a rapid reversal would be another financial disaster, or for the Thames Barrier to fail and London to flood, or confidence in the London housing market to fall, or for the population to vote to leave the European Union (which happened in June 2016), or the Labour Party to become a normal left-wing European party again and promote equality in practice rather than just rhetoric, or one of 101 other possible scenarios, and what now looks like a long-term trend may well cease overnight. So much is possible. However, in the future, we may never really know what might have happened in the 2011 to 2021 decade. Should the economic situation worsen, Members of Parliament may not vote for the 2021 Census Act when it is laid before the House of Commons in 2019. We may not get to draw these pictures again for some time if no census is held in the near future. The 2010 Government tried to cancel the 2011 Census and failed. Of course, a 2021 Census is promised – but events, dear boys (and girls), events.

Acknowledgements

I am extremely grateful to Bethan Thomas for drawing all the maps used in this chapter and for all her help and support over the two decades we have worked together. John Stillwell and David Dorling very kindly commented on earlier drafts of this chapter. John also introduced me to migration data and David taught me how to program a computer just under four decades ago.

References

Anthony, A. (2015) With only five days to go, why can't Britain make up its mind? *The Guardian*, 3 May. Available at www.theguardian.com/politics/2015/may/03/nuneaton--election-2015-marginal-seat-where-voters-like-change-but-not-that-much.

Crerar, P. (2015) Where people live in London is a 'judgement call' says housing minister Brandon Lewis. *The Evening Standard*, 1 December. Available at: www.standard.co.uk/news/london/living-in-london-is-a-judgement-call-says-housing-minister-brandon-lewis-a3127041.html.

Dorling, D. (2010) Expert view: one of Labour's great successes. *The Guardian*, 28 January. Available at: www.dannydorling.org/?page_id=576.

Dorling, D. and Thomas, B. (2016) *People and Places: A 21st Century Atlas of the UK*. Policy Press, Bristol.

Jones, O. (2012) *Chavs: The Demonization of the Working Class*. Verso, London.

Milliken, D. (2015) Low-skilled migration, under-investment may have hurt UK wages – BoE. Reuters, 23 September. Available at: http://uk.reuters.com/article/2015/09/23/uk-britian-boe-broadbent-idUKKCN0RN1W520150923.

Tunstall, B. (2015) Relative housing space inequality in England and Wales, and its recent rapid resurgence. *International Journal of Housing Policy*, 15(2): 105–126.

The spatial-temporal exploration of health and housing tenure transitions using the Northern Ireland Longitudinal Study

Myles Gould and Ian Shuttleworth

26.1   Introduction

This chapter seeks to demonstrate the nature and value of using the Northern Ireland Longitudinal Study (NILS) as a source of census microdata for longitudinal analysis of health. We do this through our involvement in a 'beta testing' project which aimed to assess the linkage of the 2011 Census to the NILS data spine. It also makes use of the Northern Ireland (NI) Census Area Statistics (CAS). The substantive example considers how individuals make transitions between 2001 and 2011 with respect to their health status, housing tenure (position in housing market) and residential movement (characterised in two different ways, as explained in due course). In other words, this is an application that simultaneously considers individual well-being (and its geographical clustering), spatial mobilities, and also the relative importance of variations between people and places using longitudinal microdata as well as standard CAS. We begin by considering the literature that informs the research application and, at the end of Section 26.2, we describe the organisation of the remainder of the chapter.

26.2   Health, place and movement

The substantive application is informed by three different strands of literature. The first is a more general literature on the nature of health inequalities and specifically that on socio-demographic and geographical variations in self-reported illness (the focus of this chapter), observed morbidity and recorded mortality (Curtis, 2004; Gatrell and Elliott, 2009). Moreover, there have been in the UK numerous official inquiries that have reviewed persistent evidence of socio-economic and geographical health inequalities (Marmot, 2010; UK Parliament House of Commons Health Committee, 2009; Gordon et al., 1999; Townsend et al., 1992). There is documented evidence, using aggregated and disaggregated census data, that has shown differences in health status by sex, age, social class/occupational group and ethnicity, as well as geographical variations at fine spatial scales within cities and urban districts, at coarser regional and sub-regional scales, and also between rural and urban communities (Congdon, 1995; Gould and Jones, 1996; Norman and Boyle, 2014; Senior et al., 2000; Shouls et al., 1996; Sloggett and Joshi, 1994).

The second strand of literature reflects the much wider and enduring debate both in social science and public health about compositional (e.g. individual) versus contextual (e.g. neighbourhood) explanations for geographical variations in health outcomes (Macintyre et al., 1993; Duncan et al., 1998; Curtis and Jones, 1998). Sloggett and Joshi (1994), for example, have argued for compositional explanations for mortality using the England and Wales Longitudinal Study (LS); whereas Smith and Easterlow (2005) argue that contextual explanations are overstated and present some compositional qualitative research findings of health discrimination and entrapment in the housing system. Others have provided evidence of contextual variations, the classic example being the 'Alameda County Study' in the USA (Hann et al., 1987), whilst Gould and Jones (1996) have demonstrated residual variation in self-reported illness after taking account of the social and demographic composition of somewhat coarse SAR areas (combinations of contiguous local authorities) using multi-level analysis of the 1991 Census microdata. Jones and Duncan (1995) demonstrate the importance of cross-level interactions between individual (i.e. compositional) and ecological (i.e. contextual) variables and we deploy this multi-level methodology in the analysis presented later in the chapter.

Arising from this wider debate has been a plea for place-sensitive research that identifies the 'actual' area characteristics associated with remaining contextual variations after accounting for the demographic and socio-economic composition of these places. Macintyre et al. (1993, 2002) provide a useful classification of five different types of area characteristics/effects implicated in health inequalities, with the later paper making an explicit distinction between effects associated with 'infrastructure and opportunity structures' (i.e. physical features of the environment shared by all residents, availability of (un)healthy environments and services provided to support people) and 'collective social functioning' (i.e. socio-cultural features of a neighbourhood and its reputation). Jones and Duncan (1995) provide a similar typology of place effects but additionally note the potential importance of processes of *selective spatial mobility*, which provides the third strand of literature informing this chapter.

A growing number of studies have considered the inter-connections between migration, social mobility, selection effects and gradients in health outcomes but without consensus, although many studies note that younger migrants tender to be healthier and more mobile than older counterparts (Bartley and Plewis, 2007; Bentham, 1988; Boyle, 2004; Boyle et al., 2009; Connolly et al., 2007; Norman et al., 2005; Norman and Boyle, 2014). Chapter 24 in this volume by Darlington-Pollock and Norman considers selective sorting and the nexus of ethnicity, health, socio-economic factors and internal migration. A fuller consideration of the geography of health and migration can be found in Boyle and Norman (2009) and Darlington et al. (2015). Smith and Easterlow (2005) also provide a valuable consideration of selective health entrapment, placement and displacement in risky/healthy spaces mediated through the housing system, demonstrating that different people's trajectories are bound up with their health histories.

Address changes (e.g. internal migration) can lead to changes in housing tenure type (e.g. moves from private renting to owner occupation) but can also be the result of moves within housing sectors (e.g. between owner-occupied houses). This, coupled with moves between more and less socially deprived places, means that there are many possible housing and social transitions that an individual can experience through time, all or any of which could have implications for their health as they are exposed to, or removed from, potential hazards such as poverty, pollution, lack of recreational space, social exclusion, loneliness, and so on (Kawachi and Berkman, 2003). This is the inspiration for the empirical case study of the transitions seen in NI which we present later, but the theoretical implications are not considered here more fully due to reasons of limited space. It should also be noted that this NI case study is related to a particular devolved, socio-historical and political context during a very interesting period of accelerated change and transition from sectarianism to peaceful coexistence (Shuttleworth et al., 2014).

Section 26.3 contains a brief brushstroke consideration of both the wider NI data landscape and more specifically the characteristics of the NILS relevant to the substantive application under consideration (the NILS has already been described in Chapter 9). In Section 26.4, we report an aggregate-level analysis of spatial clustering of limiting long-term illness and housing tenure using Anselin's (1995) Local Indicators of Spatial Association (LISA) as useful background to describing the geographical structure of NI, and the contexts which influence the chances for individuals to make health transitions. In Section 26.5, we present a longitudinal analysis of changing health and housing tenure using both descriptive statistics and some multi-level statistical modelling of the NILS microdata. The chapter concludes with some general observations about both the research potential of the NILS to tackle other research issues and also its future position in the emerging new data landscape in NI and the UK as a whole.

26.3   Longitudinal NILS microdata and Census Area Statistics

The NILS is one of the family of UK LSs. Its individual-level microdata have similar characteristics to the Census Samples of Anonymised Records (SARs) and Small Area Microdata (SAM) (Dale et al., 2000), but are generally superior for a number of reasons. As has been highlighted in Chapter 9, the NILS data spine is an approximately 28 per cent sample of the NI population, containing members with 104 of 365 birthdates drawn from the NI health card registration system which is analogous to the National Health Service Central Register (NHSCR) in England and Wales. These birth dates are unknown so as to preserve confidentiality and data security; NILS members are therefore anonymous.

These health service data provide basic demographic information on age and gender but a further strength is that they also give address information which is updated when someone reports a move to their GP or another health professional, thus allowing analysis of residential movement. The health card registration records are also linked to other data: the 1981, 1991, 2001 and 2011 Censuses and also vital statistics (although the latter are not considered in this chapter). Other researchers have used linked data on property characteristics provided by the NI Land and Property Services (LPS); and other opportunities exist to link to other administrative data using health card registration numbers (as a consequence of different devolved legislation in NI).

With such large amounts of information available, research access to the NILS is carefully controlled in much the same way as the other LSs. Only approved projects by recognised 'safe' researchers are allowed. There is an extra hurdle of medical relevance because of legal considerations, as access to the core NILS data – based on health card data – is open via a legal pathway that allows it to be shared for health-research purposes. The data can only be accessed in a secure environment and research outputs are only released after careful vetting and confirmation that they are non-disclosive. Other aspects of the NILS 'user journey', including initial training, supervised data preparation and analysis in a controlled environment, are discussed elsewhere (Chapter 9).

The NILS' sampling fraction exceeds that of the other UK LSs and is the starting point of the NILS' extra analytical possibilities. It means that the sample is of sufficient size to consider people and place, investigating individuals and their spatial contexts, using the geography of super output areas (SOAs), of which there are 890 in NI. These are aggregated from output areas (OAs, see Chapter 6) and are designed to be both spatially compact and socially homogenous and meet a population size threshold. In NI, this is approximately 2,000 people on average which means typically around 500 NILS members per SOA. This is sufficient for descriptive mapping approaches besides statistical analyses that categorise SOAs into larger aggregated classes. Moreover, and uniquely, this fine grained geographical detail can be explicitly analysed and handled (e.g. in the English and Welsh LS, SOAs are indicated by type but are not locationally identifiable during analysis). There is considerable flexibility in choice of variable categories/tabulations – because the data are analysed in a secure data environment, not forgetting it is now possible to compare individuals' responses for up to four censuses in NI. For the purposes of the current chapter it also enables multi-level analysis to be carried out, where the number of higher-level SOA units is an important consideration (Duncan et al., 1998).

Multi-level modelling has many analytical advantages; it is a way to partition variance between people and places, and to examine how, for example, the relationships between health outcomes and personal characteristics, such as living in owner-occupied housing, vary between different places. This chapter therefore explores what analytical and conceptual benefits can be gained from linking individual-level microdata such as that in the NILS to standard CAS at SOA level; and how combining individual and contextual levels of analysis and exploring cross-level interactions (simultaneously between individual and area-level covariates) can give extra insights. The other great strength of the NILS arises from the way data from different censuses have been linked to a common spine. This permits more detailed understanding and analysis than standard microdata products such as the SARs/SAM. The great strength of the NILS – and indeed the other LSs in the UK – is that they are longitudinal and can be used to examine transitions over time in ways which are impossible using cross-sectional microdata like the SAR. In the example we develop in this chapter, it is possible to investigate individual transitions in health and housing tenure status (here between 2001 and 2011), as well as relating these to whether individuals have changed address or not, together with the opportunity for setting this all in the social context in which SOAs started out at the time of the 2001 Census and where they ended up in 2011.

26.4   Aggregate analysis

Aggregated 2011 CAS mapped at SOA level in NI illustrate the spatial pattern of responses to the question about limiting long-term illness (LLTI), thus providing an indication of the spatial context for the analysis of NILS members (Figure 26.1). Not all these individuals will have the same opportunities available and existing geographies of inequality are not only a measure of context but also shape the opportunities open to individuals regarding spatial mobility (or immobility). For example, someone resident in an area with high social disadvantage, surrounded by other SOAs of high social disadvantage, will have far fewer chances to become upwardly socially mobile if moving locally (as most people do) than a similar person living in a disadvantaged SOA which is surrounded by less disadvantaged places. Understanding the geographical structure of the population is, therefore, key to understanding how life chances are structured, and what spatial mobility and immobility mean in making transitions up and down the health and housing ladders. To paraphrase Marx, people make their own histories but they do not make them in geographies of their own choosing.

Spatial autocorrelation indicators are one way to explore how populations are structured and how they vary geographically. Those shown in Figures 26.1 and 26.2 make use of Anselin's LISA and are implemented using the *GeoDa*™ software (Anselin, 1995, 2003). These are a local formulation of Moran's *I*, involving a decomposition of a global indicator based on each area's contribution and indicating whether a location differs from the mean value, and, when transformed using z-scores, can be used to assess whether an area is statistically different from its neighbours (Anselin, 1995). LISA scores are signed such that positive values indicate similarly co-located areas, whilst negative values represent dissimilarity. In Figure 26.1, SOAs shaded white show a random patterning of health status, whilst areas shaded bright red are areas of high rates of illness surrounded by SOAs with similar high illness rates (high–high), and deep blue distinguishes areas of low rates of illness bordered by similarly low rates of illness (low–low). In contrast, the pink areas represent places with high rates of illness surrounded by areas with low rates (high–low), whilst the pale blue areas represent SOAs with low rates of illness surrounded by high rates (low–high). In the case of high rates of LLTI (Figure 26.1) there is generally little spatial clustering of SOAs particularly in western and central parts and around the coast; but there are some pockets of SOAs with high rates surrounded by other high rate areas in Strabane, and Derry/Londonderry. There is, however, extensive clustering in east and south Belfast. With respect to areas of low rates of LLTI, there is clustering in some parts of the districts of Antrim, Down, Lisburn, Banbridge, Craigavon, Limavady and Coleraine. There are virtually no SOAs classified as low surrounded by high, nor high surrounded by low.

Table 26.1 provides a summary of the global and LISA cluster typology means for the different census variables mapped to give indications of how the classifications are scaled. So, for example,

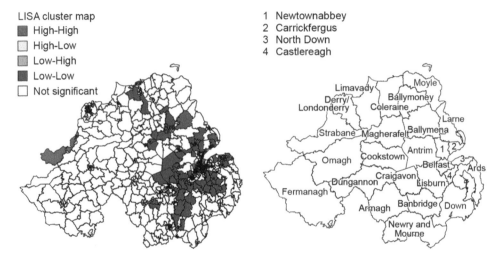

Figure 26.1 LISA mapping of the percentage of people with very LLTI in 2011

Source: Northern Ireland Statistics and Research Agency (NISRA), 2011 NI Census Area Statistics and Digitised Boundary Data.

Table 26.1 Mean SOA area rates for different census variables classified by different area cluster typologies and overall global averages

Cluster type (2011)	Mean		
	Very limited and limited long-term illness (%)	Owner occupied LISA (%)	Social renting LISA (%)
Random	11.8	69.6	13.0
High–high	19.2	83.8	40.8
Low–low	7.8	37.4	4.5
Low–high	10.7	57.8	8.9
High–low	14.4	74.0	20.9
Global	11.9	68.2	14.7

Source: NISRA, 2011 NI Census Area Statistics.

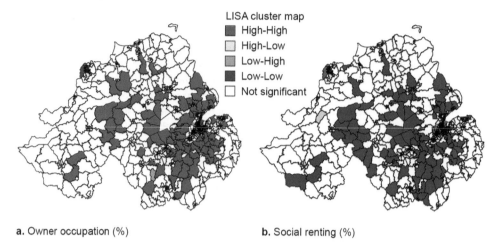

a. Owner occupation (%) **b.** Social renting (%)

Figure 26.2 LISA mapping of a. percentage of owner occupiers in 2011, and b. percentage of social renters in 2011

Source: NISRA, 2011 NI Census Area Statistics and Digitised Boundary Data.

SOAs classified as high surrounded by high on average have nearly a fifth (19.2 per cent) of people reporting LLTI, which is considerably higher than that for the global average, and also for SOAs which are randomly distributed (both of which are close to 12 per cent), and much higher than areas classified low–low.

The maps in Figure 26.2 show the spatial structure and clustering in housing tenure for owner occupiers and social renters. Areas of high levels of owner occupation are often similarly surrounded by other areas with high rates; and also areas of low rates of social renting are often surrounded by other SOAs with low rates. Spatial clustering of low rates of owner occupation and high rates of social renting are found in Derry/Londonderry and also in east and south Belfast (Figures 26.2a and 26.2b). However, the patterns for Belfast are more complicated than for LLTI shown in Figure 26.1, with some juxtaposition of SOAs with high rates of owner occupation being surrounded by low rates (Figures 26.2a), and also some areas with low rates of social renting being surrounded by high rates (Figure 26.2b). This reflects more spatial differentiation of housing tenure, a finding which accords with work reported elsewhere on the segregation of community background and socio-economic deprivation – where clustering of the latter was less marked than the former (Shuttleworth et al., 2013, 2014).

Figure 26.2 also shows that there are lots of SOAs located away from urban centres which display no spatial clustering of areas with low or high rates of owner occupation or social renting. We now turn our attention to consideration of individual-level health and tenure transitions.

26.5 Analysis of NILS data

Table 26.2 shows the health transitions between 2001 and 2011 for individual NILS members included in our analysis with both absolute cell counts and row percentages displayed. The vast majority of individuals (166,101) reported no LLTI at either time points, with approximately 81 per cent of individuals who reported no illness in 2001 also doing so in 2011, whilst approximately one in five people who started with no illness in 2001 transitioned to illness in 2011. Approximately 77 per cent of the 55,929 people who were ill at the 2001 starting point remained

Table 26.2 Limiting long-term illness transitions, 2001–11

		Limiting illness 2011		
		No	Yes	Total
Limiting illness 2001	No	166,101	39,403	205,504
		80.8%	19.2%	100%
	Yes	13,051	42,878	55,929
		23.3%	76.7%	100%
	Total	179,152	82,281	261,433
		68.5%	31.5%	100%

Source: NILS.

Table 26.3 Percentage of NILS members remaining in LLTI 2001–11, classified by different area cluster typologies

Cluster type (2011)	% Remaining in LLTI, 2001–11, by SOA area classification			
	Very limited and long-term illness LISA (%)	Owner occupied LISA (%)	Private renting LISA (%)	Social renting LISA (%)
Random	15.8	15.9	16.5	15.7
High– high	26.4	12.3	17.6	26.9
Low–low	10.9	25.2	14.3	12.7
Low–high	16.2	17.9	26.9	16.6
High–low	16.6	17.8	15.3	16.2

Source: NILS and NISRA, 2011 NI Census Area Statistics.

so in 2011, whilst around 23 per cent transitioned to no LLTI in 2011. There is evidence of 36 per cent of people (95,332) reporting illness at either one or the other of the two time points, with some 16 per cent of NILS members (42,878) reporting illness at both time points. These figures and proportions for LLTI are higher than the responses for the other census question on self-reported general health (not presented here). There is, however, a caveat with the analysis presented, as responses to the 2011 LLTI question have been pragmatically recoded/combined (into two categories to facilitate comparison with the 2001 question which was binary in nature). Moreover, there are some subtle differences in the wording of the question between 2001 and 2011, although they remain 'broadly comparable' (NISRA, 2012). These rates, based on the NILS sample, correspond closely with those recorded in aggregate key statistics and particularly when comparing the 'closed' starting point sample with the full 2001 Census cross-section.

Table 26.3 classifies NILS members who remain in LLTI between 2001 and 2011 using the LLTI question and the LISA area classification described in the previous section. The area-level typology was brought into the secure data analysis environment (with NISRA approval) and then attached to study members using SOA codes. Comparisons are made for four different area-variable classifications for the 2011 end point: LLTI affecting daily activities and three housing tenures. Whilst NILS members who remain in LLTI live in all five categories of the LLTI area typology, not unsurprisingly the largest proportion of people (26.4 per cent) live in areas of high levels of self-reported LLTI surrounded by other areas with high rates, whilst the lowest proportion of individuals remaining in LLTI live in areas of low LLTI surrounded by other SOAs with low rates of LLTI. For owner occupation, the largest proportion of NILS members remaining ill is found for those living in areas of low owner occupation surrounded by other areas of low owner occupation. For the social renting category, the largest proportion of remaining ill individuals is found for high areas surrounded by other high areas. The pattern is slightly different for private renting with the biggest proportion (26.9 per cent) of NILS members remaining ill being found in areas of low rates of private renting surrounded by high rates. Overall, these results are reassuring in confirming that individuals who remain in LLTI between 2001 and 2011 are more likely to live in spatially clustered areas of social disadvantage, high levels of rented accommodation and self-reported illness. We now consider this further for the NILS using the different approach of multi-level modelling to simultaneously analyse individual and area characteristics, and the interaction of the two (Jones and Duncan, 1995).

Multi-level analysis

Attention now turns to the simultaneous multi-level modelling of changes in health status and individual and area characteristics. Due to limitations of space, we give a flavour of what is possible by focusing on two illustrative models: (i) transitioning from no LLTI in 2001 to LLTI in 2011 – that is, modelling individuals related to the first row of Table 26.2; (ii) staying in LLTI but including LISA indicators (discussed in the previous section) – but this time related to all individuals included in Table 26.2. In the former case 201,314, and in the latter case 252,784, individuals are actually statistically modelled, which is slightly smaller than the numbers shown in Table 26.2 and reflects non-response (missing/edited) observations for variables included as individual-level covariates.

Model 1 includes a number of individual characteristics: age, sex, 2001 housing tenure, occupational status, education and community background (Catholic, Protestant and other religion); as well as a typology summarising whether an individual has changed housing tenure and/or moved address during the period 2001–11. It should be noted that it is possible for individuals to move address but not change SOA; and also it is theoretically possible, albeit unlikely, for someone to change tenure but not change address by either re-mortgaging or buying from a landlord. The model also includes first-order interactions between 2001 (starting) tenure and the changed tenure/moved address (between 2001 and 2011) typology.

Big overall effects are found for housing tenure (not presented here because of space limitations), with those in social and private rented accommodation in 2001 generally having the largest probability of transitioning to ill-health between 2001 and 2011 when compared to those in owner occupation who have not changed tenure since 2001. These probabilities are slightly increased for both social and private renters who have moved but not changed tenure (there is no difference for owner occupiers). Indeed, considering the interaction between 2001 starting tenure and the changed tenure–changed/not changed address terms reveals more nuanced findings. The effects of changing tenure and moving/not moving address also increase the probability of transitioning to ill-health for individuals who started in owner occupation in 2001 and have subsequently 'moved down' the housing market rather than staying put both spatially and as owner occupiers. For individuals who started as social and private renters in 2001, there are reductions in the probabilities of transitioning to ill-health if they changed tenure; similarly for those who have changed address. The biggest effect is for social renters who change tenure and not address and, moreover, the probability is the same for those in private renting. In all these cases, these individuals are likely to have moved up the housing market over the ten-year period and may have experienced improvements in their well-being. Increasing age and being retired also increase the probability of transitioning to LLTI; whilst having an educational qualification of any level reduces the probability of transitioning, with the biggest reduction for those with degrees relative to those with no qualifications. The fixed-effect parameters are statistically significant with the exception of the contrast categories for students, the economically inactive and those with no community background (religion) – in each case these all represent small numbers of NILS members. The main effect for not changed tenure, but moved address is not statistically significant (but the higher-level interaction terms with social and private rented tenure are) and is interpreted as there being no difference in health transitions for owner occupiers who do not change tenure (the base comparison) but who do or do not move.

The statistical model does not include LISA indicators but does include an SOA continuous scale index of multiple deprivation instead and its linear *cross-level interaction* with a four-way typology of individuals who have changed address and/or housing tenure (by multiplying the variables together and also including them as parameters in the model). The model parameters associated with these interaction terms are statistically significant, and the meaning of these results is most readily appreciated when presented graphically, as in Figure 26.3. The slopes shown in the figure are the product of multi-level regression of the 201,314 cases (not shown), using the particular model parameters (mentioned above) with the model prediction and graphing tools contained in the *MLwiN* software package (Rasbash et al., 2013). The horizontal axis shows the range of deprivation scores and the value of 17.8 represents the median stereotypical SOA deprivation level in 2001. The vertical axis shows the predicted probability of transitioning into ill-health based on the multi-level regression. The first thing to note is that all four lines (representing different combinations of individuals changing tenure and address and based on model predictions for people who are of average age and have all other base category characteristics[1]) rise as SOA-level deprivation increases – with steepest lines found for individuals 'not changed tenure, not changed address'; and 'not changed tenure, changed address'. In more affluent areas, the probability of transitioning to illness is relatively low and similar to those who have 'not changed tenure, not changed address'; 'not changed tenure, changed address'; and 'changed tenure,

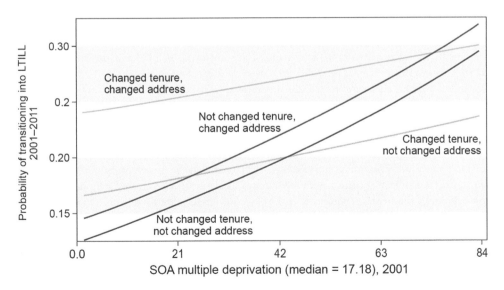

Figure 26.3 Cross-level interaction of individual changed tenure/address and multiple deprivation area effects on transitioning from no limiting illness to illness (Model 1)

Source: Based on authors analysis of NILS.

Table 26.4 Multi-level models of staying in LLTI, 2001–11, that include LISA area-level cluster typologies

Cluster type (2001)	Nature of area-level fixed effects	
	(i) Limiting illness LISA	*(ii) Owner occupied LISA*
Random	Base category	Base category
High–high	Large positive, significant	Large positive, significant
Low–low	Large negative, significant	Large negative, significant
Low–high	Not significant	Not significant
High–low	Not significant	Not significant

Source: Summary of authors' analysis of NILS and NISRA, 2011 NI CAS.

not changed address'. For those who have 'changed tenure, changed address', there is a considerable increased probability of transitioning to ill-health. At the other end of the distribution, for places with very high levels of multiple social deprivation, the differences between types of people are more marked for those 'not changed tenure, changed address' and having higher probabilities of transitioning to ill-health; now followed by 'changed tenure, changed address'; and 'not changed tenure, not changed address'. The 'changed tenure, not changed address' category has the lowest probability of transitioning to ill-health at high levels of deprivation.

Inspection of maps of SOA-level random terms (not presented here) indicates that there is some statistically significant remaining residual geographical variation in the probability of transition from no illness to illness. We suggest that this is a consequence of complex historical/political geographies of community background and local-level housing market structure (Shuttleworth et al., 2013). That said, we argue that geographical variations in health transitions are not purely compositional, but that context does matter to a degree – here, measured in terms of multiple deprivation (but we have explored other variables too) and their interactions with individual tenure/address changes at two census points. Moreover, individuals starting tenure in 2001 and their interaction with 2001–11 tenure and address change do matter for transitions to ill-health.

Finally, we turn to two models of probabilities of staying in LLTI but this time including LISA area-level (summarised in Table 26.4) and individual-level predictors. The main effects for individual characteristics are broadly as they were previously. This time we include the LISA cluster scores for: (i) LLTI affecting daily activities; and (ii) owner occupation as main area effects. Because it is a nominal classification, the LISA indicators are not readily modelled and portrayed as a cross-level interaction; but this specification is parsimonious and has reasonable model fit. The summarised results in Table 26.4 (second column) indicate that SOAs with the high rates of LLTI surrounded by similar high rates in 2011 predicted the largest probability of individuals staying in LLTI; whilst SOAs with low rates of LLTI surrounded by low areas have decreased probabilities of individuals remaining in LLTI. Similar results are found for owner occupation (Table 26.4, third column) and provide further evidence of structural clustering of health transitions (cf. Table 26.3) – here, staying in LLTI.

26.6 Conclusions and the future

The analytical and conceptual benefits that can be gained from linking and combining individual-level NILS microdata CAS have been shown to give extra/nuanced insights. We have demonstrated using the NILS data that there were transitions in self-reported long-term illness between 2001 and 2011 – and that there are changes in both directions. Big overall individual-level effects for housing tenure were found, with those in social and private rented accommodation in 2001 generally having the largest probability of transitioning to ill-health between 2001 and 2011 when compared to those in owner occupation who have not changed tenure since 2001. The mapping of LISA scores for CAS data showed that there are distinctive and complex geographies in NI for self-reported illness and housing tenure. Whilst many areas did not significantly vary from overall average rates, other areas did. Multi-level analysis provided evidence of some remaining residual geographical variation in the probability of health transitions. We argue that context does matter to a degree – here, measured in terms of multiple deprivation – and its interaction with individual tenure/address changes at two census points.

One criticism that can be levelled at the analysis presented here is that it does not fully capture the richness of multiple inter-censal residential moves that are particularly important for people living in the private rented sector, nor does it include specific data about the quality and value of housing stock in which these people live. Regarding the first point, we note that whilst some use of data from health card registrations was made to derive indicators of change of address between 2001 and 2011, more use could have been made of information on the multiple number/frequency of moves made between censuses. This might facilitate getting purchase on the transitory nature of private renters and whether this group are at higher risk of falling or remaining in poor health compared to those in other tenures, as well as better understanding their '(re)placement' (i.e. geographies). However, a caveat must be added that certain demographic groups (e.g. young males) can delay registering change of doctors and/or addresses, resulting in some under-reporting of residential moves in the NILS (Barr and Shuttleworth, 2012).

We noted earlier in the chapter that NILS projects can be linked to the NI LPS, but this was not possible in our case as the work was undertaken as a beta testing project. This is something that could be done in the future. Additional data on housing stock (e.g. year built, habitable square meterage) and capital value could be gathered to derive better-defined proxies of household wealth and answer more specific research questions about the role of individual and contextual characteristics on changing health status. Moreover, data contained in the census/NILS on the number of rooms and central heating could also be utilised (and were not included in our analyses for issues of parsimony).

The analytical power of the NILS, and indeed the other LSs, is set to increase when data from other censuses, for example that planned for 2021, are added. This will give the ONS English and Welsh LS a run of 50 years of data from 1971 with six censuses and the NILS 40 years of data (from 1981) with five censuses. This means that the NILS and the other LSs are very powerful data resources to examine long-term social change across populations and neighbourhoods and their value will increase over time as more data are added. Moreover, the close link between the NILS and the Administrative Data Research Centre for Northern Ireland (ADRC-NI) offers exciting new prospects for the ethical and appropriate linkage of other official data. These ADRC-NI projects involve limited-life, one-off data linkage, extraction and analysis. They have the benefits of using 100 per cent sampling and the potential, in theory, to link a number of secondary data sources, albeit with complex logistical and legal constraints.

The NILS is perhaps logistically easier to use and access (albeit with access controls), making use of existing ongoing collection of linked census with other data (vital statistics, health card registrations and property valuation data). Putting it another way, the NILS provides a now well-tested data resource (at the time of writing over a hundred NILS projects have been initiated), a large set of sampled records, and a relatively confined (although extensive) set of data variables, many of which are repeatedly measured at different censuses (albeit with subtle changes in question wording and coding – as in the case of self-reported illness). Indeed, the analysis we have described in this chapter should be seen as the start of a much longer project, illustrating what can already be done, and setting out a route for what might be done in the future with emerging data.

Acknowledgements

The help provided by the staff of the Northern Ireland Longitudinal Study and the NILS Research Support Unit (RSU) is acknowledged. The NILS is funded by the Health and Social Care Research and Development Division of the Public Health Agency (HSC R&D Division) and NISRA. The NILS-RSU is funded by the Economic and Social Research Council and the Northern Ireland Government. We are very grateful to John Stillwell for very useful comments on the initial draft of this chapter. The authors alone are responsible for the interpretation of the data and any views or opinions presented are solely those of the authors and do not necessarily represent those of NISRA/NILS.

Note

1   That is a 42 year old female Protestant who is employed, lives in owner occupation and has no qualifications.

References

Anselin, L. (1995) Local indicators of spatial association – LISA. *Geographical Analysis*, 27: 93–115.
Anselin, L. (2003) *GeoDa™ 0.9 User's Guide*. Spatial Analysis Laboratory, Department of Agricultural and Consumer Economics, University of Illinois, Illinois.
Barr, P.J. and Shuttleworth, I. (2012) Reporting address changes by migrants: The accuracy and timeliness of reports via health card registers. *Health & Place*, 18: 595–604.
Bartley, M. and Plewis, I. (2007) Increasing social mobility: An effective policy to reduce health inequalities. *Journal of the Royal Statistical Society: Series A*, 170: 469–481.
Bentham, G. (1988) Migration and morbidity: Implications for geographical studies of disease. *Social Science and Medicine*, 26(1): 49–54.
Boyle, P. (2004) Population geography: Migration and inequalities in mortality and morbidity. *Progress in Human Geography*, 28: 767–776.
Boyle, P. and Norman, P. (2009) Migration and health. In Brown, T., McLafferty, S. and Moon, G. (eds.) *A Companion to Health and Medical Geography*. Wiley-Blackwell, Chichester, pp. 346–374.
Boyle, P., Norman, P. and Popham, F. (2009) Social mobility: Evidence that it can widen health inequalities. *Social Science and Medicine*, 68: 1835–1842.
Congdon, P. (1995) The impact of area context on long term illness and premature mortality: An illustration of multi-level analysis. *Regional Studies*, 29: 327–344.
Connolly, S., O'Reilly, D. and Rosato, M. (2007) Increasing inequalities in health: Is it an artefact caused by the selective movement of people? *Social Science and Medicine*, 64: 2008–2015.
Curtis, S. (2004) *Health Inequality: Geographical Perspectives*. Sage, London.
Curtis, S. and Jones, I.R. (1998) Is there a place for geography in the analysis of health inequality? *Sociology of Health and Illness*, 20(5): 645–672.
Dale, A., Holdsworth, C. and Fieldhouse, E. (2000) *The Analysis of Census Microdata*. Edward Arnold, London.
Darlington, F., Norman, P. and Gould, M. (2015) Migration and heath. In Smith, D., Finney, N., Halfacree, K. and Walford N. (eds.) *Internal Migration: Geographical Perspectives and Processes*. Ashgate, Farnham, pp. 113–128.
Duncan, C., Jones, K. and Moon, G. (1998) Context, composition and heterogeneity. *Social Science and Medicine*, 46: 97–117.
Gatrell, A.C. and Elliott, S.J. (2009) *Geographies of Health*. Second edition. Wiley-Blackwell, Chichester.
UK Parliament House of Commons Health Committee (2009) *Health Inequalities: Vol. 1 Report, Together with Formal Minutes*. The Stationery Office, London.

Gordon, D., Shaw, M., Dorling, D. and Davey Smith, G. (eds.) (1999) *Inequalities in Health: The Evidence (The Evidence Presented to the Independent Inquiry into Inequalities in Health, Chaired by Sir Donald Acheson)*. The Policy Press, Bristol.

Gould, M.I. and Jones, K. (1996) Analysing perceived limiting long-term illness using UK census microdata. *Social Science and Medicine*, 42(6): 857–869.

Hann, M., Kaplan, G.A. and Camacho, T. (1987) Poverty and health: Prospective evidence from the Alameda County Study. *American Journal of Epidemiology*, 125: 989–998.

Jones, K. and Duncan, C. (1995) Individuals and their ecologies: Analysing the geography of chronic illness within a multilevel modelling framework. *Health and Place*, 1: 27–40.

Kawachi, I. and Berkman, L.F. (eds.) (2003) *Neighborhoods and Health*. Oxford University Press, Oxford.

Macintyre, S., Ellaway, A. and Cummins, S. (2002) Place effects on health: How can we conceptualise, operationalise and measure them? *Social Science and Medicine*, 55: 125–139.

Macintyre, S., Maciver, S. and Soomans, A. (1993) Area, class and health: Should we be focusing on places or people? *Journal of Social Policy*, 22: 213–234.

Marmot, M.G. (2010) *Fair Society, Healthy Lives: The Marmot Review (Strategic Review of Health Inequalities in England post-2010)*. Institute of Health Equity, University College London, London.

NISRA (2012) *Comparability of the Census Questionnaire in Northern Ireland Between 2001 and 2011*. Northern Ireland Statistics and Research Agency, Belfast.

Norman, P. and Boyle, P. (2014) Are health inequalities between differently deprived areas evident at different ages? A longitudinal study of census records in England and Wales, 1991–2001. *Health and Place*, 26: 88–93.

Norman, P., Boyle, P. and Rees, P. (2005) Selective migration, health and deprivation: A longitudinal analysis. *Social Science and Medicine*, 60: 2755–2771.

Rasbash, J., Browne, W., Healy, M., Cameron, B. and Charlton, C. (2013) *MLwiN Version 2.29*. Centre for Multilevel Modelling, University of Bristol, Bristol.

Senior, M., Williams, H. and Higgs, G. (2000) Urban–rural mortality differentials: Controlling for material deprivation. *Social Science and Medicine*, 51: 289–305.

Shouls, S., Congdon, P. and Curtis, S. (1996) Modelling inequality in reported long term illness: Combining individual and area characteristics. *Journal of Epidemiology and Community Health*, 50: 366–376.

Shuttleworth, I., Barr, P. and Gould, M. (2013) Does internal migration in Northern Ireland increase religious and social segregation? Perspectives from the Northern Ireland Longitudinal Study (NILS) 2001–2007. *Population, Space and Place*, 19(1): 72–86.

Shuttleworth, I., Gould, M. and Barr, P. (2014) Perspectives on social segregation and migration: Spatial scale, mixing and places. In Shuttleworth, I., Lloyd, C. and Wong, D. (eds.) *Social Segregation: Concepts, Processes and Outcomes*. Policy Press, Bristol, pp. 197–220.

Sloggett, A. and Joshi, H. (1994) Higher mortality in deprived areas: Community or personal disadvantage? *British Medical Journal*, 309(6967): 1470–1474.

Smith, S.J. and Easterlow, D. (2005) The strange geography of health inequalities. *Transactions of the Institute of British Geographers*, 30: 173–190.

Townsend, P., Davidson, N. and Whitehead, M. (eds.) (1992) *Inequalities in Health: The Black Report, The Health Divide*. Second edition. Penguin, Middlesex.

Changing intensities and spatial patterns of internal migration in the UK

John Stillwell, Nik Lomax and Stephane Chatagnier

27.1 Introduction

This chapter makes use of 2011 Census origin–destination migration data together with data from the 2001 Census to compare the intensities and spatial patterns of migration at local authority district (LAD) scale in the UK. We set out to answer three research questions. First, can we make consistent comparisons of origin–destination migration data in the UK between 2001 and 2011? Before any analysis can be undertaken, issues of inconsistency must be addressed and adjustments to the data tables undertaken. Second, how have all age and age group migration intensities and impacts changed in the UK in the 2000s? Indicators measuring migration intensity and impact are used to answer this question. Third, are there scale and zonation effects associated with these migration indicators and have they changed over time? This analysis requires use of specialist software to compute migration indicators at different scales and with alternative zone configurations.

The following section briefly outlines previous studies which have used the UK census to measure migration patterns and provides an overview of the migration indicators that we have chosen to use for making comparisons. Section 27.3 identifies the problems with inconsistency between the 2001 and 2011 Census origin–destination migration data and describes the methods used to make these data comparable. Section 27.4 provides an assessment of changes in national migration intensity and impact before local variations and changes are considered in Section 27.5. An assessment of the scale and zonation effects on migration intensity and impact is made in Section 27.6, and some concluding remarks are offered in the final section.

27.2 Studies and indicators of migration in the UK

Internal migration in the UK is an important component of population redistribution at the sub-national scale and one which has captured the attention of social scientists since the nineteenth century. Although its temporal infrequency is a serious shortcoming, the decennial population census is the most reliable and comprehensive dataset available for examining the intensity, composition and geographical pattern of this human behaviour. As such, it has been used in a number of studies looking at UK internal migration. In the last decade, Champion (2005) provides detailed insights into the migration propensities of different population sub-groups using the 2001 Census, whilst Dennett and Stillwell (2010a) have produced an area classification based on 2001 flows as a means of summarising migration patterns in the UK at the LAD scale. Fielding (2012) used census data at the regional scale as evidence for a set of theoretical perspectives which underpin migration behaviour. Census-based time-series studies of migration in the UK are less common, largely due to difficulties in obtaining consistent data at an appropriate spatial scale over an extended period. Unlike Australia, where the census is taken every five years and there is a five-year migration question that provides continuous temporal coverage of migration from one census to the next, the one-year migration question in the UK census means that only a snapshot of migration every ten years is available for comparison. Recent work by Champion and Shuttleworth (2016a), however, has used data on individuals from the Longitudinal Study (LS) in England and Wales whose attributes have been linked across five censuses between 1971 and 2011. The results based on these longitudinal microdata suggest that aggregate mobility between censuses (i.e. those who lived at a different address ten years apart) has been fairly stable over longer distances (moves over 10 km) but has experienced steady decline over shorter distances (moves of less than 10 km) over the last 40 years.

Given the importance of understanding how internal migration propensities and patterns change over time and the difficulties with one-year aggregate and origin–destination migration data, time-series analysis of migration flows is reliant on patient registration data from the National Health Service Central Register (NHSCR) or from the Patient Register Data System (PRDS). Stillwell et al. (1992) report on fluctuations in national internal migration rates between 1975 and 1989 using NHSCR data, attributing much of the variation to the economic cycle and using census data to validate the estimates derived from administrative data. More recently, a long-run annual time series (1976–2011) of moves between and within 80 relatively large spatial zones (harmonised health areas) in England and Wales has been used by Champion and Shuttleworth (2016b) and Lomax and Stillwell (2017) to establish that there is little evidence of a long-term decline in longer-distance migration over time comparable to that which Cooke (2013) reports for inter-county migration intensities in the USA.

Undertaking analysis of change over time at a more spatially detailed scale in the UK requires some estimation work and necessarily a shorter timeframe, due largely to data availability. Lomax et al. (2013) offer a methodology for combining annual register data from the four home nations, calibrated against census data, to generate consistent estimates of migration between LADs for the period 2002 to 2011. Aggregate results from this dataset are reported in Lomax et al. (2014) and an extended time series which assesses changing propensities by age between 2001–02 and 2012–13 is presented in Lomax and Stillwell (2017). Both studies report that there has been an overall decline in migration rates between LADs in the 12 years since the 2001 Census, with the most marked declines evident at the young (0–14) and older (60–75+) ages. What is clear from these studies is that, sooner or later, census data are required for calibration or to fill in missing information. It is also apparent that a direct comparison between migration flows recorded by the 2001 and 2011 Censuses has yet to be carried out and this provides the primary rationale for this chapter.

Measuring and assessing the characteristics of internal migration evident in any dataset requires a strategy for summarising large volumes of data, given that migration data are multi-dimensional: every migrant has an origin, a destination and a number of demographic attributes (even if only age and sex are available). National or system-wide indicators can be used to reveal what is happening in the migration system as a whole. Bell et al. (2002) present indicators for a number of domains such as the intensity, distance, connectivity and impact of migration and these have been used to compare migration within Australia and the UK (e.g. Stillwell et al., 2000, 2001), and more recently, to make cross-national comparisons across a much larger range of countries (Bell et al., 2015; Stillwell et al., 2016; Rees et al., 2016).

In this chapter, we use three indicators of migration. The first of these is the crude migration intensity (*CMI*) defined for any age group at any spatial scale as:

$$CMI = 100 \left(\sum_{i \neq j} Mij / \sum_i Pi \right) \tag{27.1}$$

where $\sum_{i \neq j} Mij$ in our case is the total migration between LADs in the UK and $\sum_i Pi$ is the national population. Following Bell et al. (2015), the *CMI* is distinguished from the aggregate *CMI* (or *ACMI*), the migration rate that includes flows of migrants of all ages within as well as between LADs. The second indicator is the net migration rate (*ANMR*) defined as:

$$ANMR = 100 \ (0.5) \sum_i | D_i - O_i |) / \sum_i P_i \tag{27.2}$$

where D_i are the inflows to LAD *i* and O_i are the outflows from LAD *i*. This indicator is an overall measure of migration impact on population redistribution, telling us the number of persons who changed region of residence per 100 persons resident in the country. When the same numerator is divided by the total flows in the system, a third indicator is derived which is known as the migration effectiveness index (*MEI*) at LAD scale:

$$MEI = 100 (0.5) \sum_i | D_i - O_i |) / \left(\sum_{i \neq j} M_{ij} \right) \tag{27.3}.$$

The *MEI* can be interpreted as the number of persons who changed LAD of residence per 100 migrants who entered or departed. Low positive or negative values are observed when migration streams and counter-streams are well balanced, whilst high values indicate that there is significant asymmetry in the internal migration system, with some LADs effectively gaining or losing population at the expense of other LADs.

These three indicators of intensity and impact are connected since the overall impact of migration on population redistribution (*ANMR*) is the product of the migration intensity (*CMI*) and the spatial imbalance of flows (*MEI*):

$$ANMR = (CMI * MEI) / 100 \tag{27.4}$$

Thus, for migration between 404 UK districts in 2010–11:

$$ANMR = (100(2794882 / 63182178)) \ (50(272250 / 2794882)) \ /100 \tag{27.5}$$
$$ANMR = 4.43 * 4.87 / 100 = 0.215 \tag{27.6}.$$

Each of these national indicators can be computed for national aggregate (all age) migrants or for specific age groups. Local indicators can be similarly defined for each LAD *i* as CMI_i, NMR_i and MEI_i. In this chapter we explore how each of the national and local indicators changes in the UK between 2001 and 2011, but need first to consider what datasets the two censuses provide and what estimation or adjustments are required to make the data comparable.

27.3 Census data and estimation methods

Origin–destination migration data collected by the three national statistical offices in the UK on census day are collated by the Office for National Statistics (ONS) into a single set of Special Migration Statistics (SMS) tables for the UK. These data are deposited with the UK Data

Service-Census Support and the *WICID* interface offers the option to construct tables based upon user defined specifications or, in the case of 2011 data, to download the tables in their entirety. The 2001 Census tables were released at three levels of geography: LAD, ward and output area (OA), available by age and sex as well as a number of other variables, including economic activity and tenure type. The level of detail available diminishes as the geography becomes finer: 24 age groups at LAD scale compared with three age groups at OA scale, for example. For the 2011 release, the geographic levels on offer remained broadly consistent (albeit with some changes to the boundaries) but the number of variables available increased substantially, largely due to a change in disclosure control methodologies (see Chapter 7 of this book).

The data available from each census and the adjustments made to ensure consistency largely dictate what data can be used for the comparison presented later. For the purpose of this study, we have limited comparison to 11 age groups (plus an all-age group) and, in order to keep the volume of analysis at a manageable level, grouped together males and females to report results for total persons in each age group. We use the LAD-level data from both censuses (i.e. the coarsest geography of the SMS) as this provides the most detailed age breakdown and requires the smallest amount of adjustment to ensure consistency. Migrants within each LAD are available, so overall levels of migration which occur in the UK at all spatial scales can be identified although intra-LAD flows are not included in our scale analysis. Aside from disclosure control, other issues when making comparisons between the 2001 and 2011 Census data are changes in geographical boundaries, changes in definitions and issues around misreporting of previous address in the 2001 Census. In order to produce a consistent dataset to compare 2001 with 2011 flows, these issues need to be addressed.

Disclosure control is the mechanism used to prevent the release of data which could potentially be used to identify an individual reported in the census. Whilst this is not an issue confined to the migration data tables, it is particularly pertinent, as there are many cells which contain small numbers in the SMS, especially in the sparse interaction matrices disaggregated by age. The 2001 disclosure control method was termed the 'small cell adjustment method' (SCAM) where cells containing a value of 1 or 2 were adjusted to either 3 or 0 probabilistically, although data for flows terminating in Scotland were not affected by the SCAM methodology as General Register Office (GRO) Scotland decided not to adopt the SCAM procedure. The impacts of this method on the sparsely populated interaction matrices which constitute the SMS in 2001 have been assessed by Stillwell and Duke-Williams (2007) and (fortunately) this method was abandoned for 2011 in favour of 'record swapping' (discussed in detail in Chapter 7 of this book). The effect on the SMS is that the available data in 2011 are less heavily adjusted than they were in 2001. We have not made any adjustments for the effects of SCAM in 2001, partly because any method could potentially skew the data and partly because, in aggregate, the SCAM adjustment returned a total number of migrants at LAD level close to what was actually reported in the raw data.

Boundary adjustments between 2001 and 2011 saw the number of LADs in England and Wales decline from 376 to 346; these boundary adjustments were aggregations of local authorities, including the amalgamation of the six non-metropolitan districts of Cornwall to produce a single unitary authority in 2009. This makes the task of adjusting the 2001 Census data to match the boundaries available for the 2011 data fairly straightforward. First, moves between the LADs in 2001 which make up the larger LADs in 2011 are converted from inter-area moves to intra-area moves (i.e. moves within LADs rather than moves between LADs); and, second, moves into and out of the larger LADs are aggregated so that their interactions with all other areas take into account the moves reported for the former constituent LADs. The resulting adjusted dataset therefore retains all moves reported in the 2001 Census but reallocates some inter-LAD moves as intra-LAD moves.

Issues with boundary inconsistency in Northern Ireland are less straightforward. In 2001, migration was reported for the 18 Parliamentary Constituencies (PCs) which overlapped with the 26 Local Government Districts (LGDs) used in 2011 (and indeed used for all other data table releases of the 2001 Census). As these PC areas cannot be cleanly aggregated to the LGDs, data at the LGD scale were constructed from the ward-level tables (the second tier of geography). The knock-on effect of this adjustment is the limited number of age groups available for the more detailed geography, which is the reason why we report the results for 11 age groups in this chapter, rather than the 24 groups available in the 2001 Census tables at LAD level. Similar to the adjustments made for England and Wales, no moves are lost (i.e. the total number of UK moves is retained) but there is a reallocation of inter-district and intra-district migration, as defined by the LGD boundaries. The 32 Council Areas (CAs) in Scotland are consistent in 2001 and 2011 so no adjustment is needed. After all the boundary adjustments are made, we are left with a consistent geographical system of 404 UK 'LADs' in 2001 and 2011. One final note on geography: whilst there are in total 406 UK LADs in 2011, the SMS report flows for 404 LADs since the City of London has been combined with Westminster and the Isles of Scilly merged with Cornwall to overcome the problem of small numbers reported in the former of each pair.

Migration is reported in the census using the response given by the household reference person to a question asking: 'What was your usual address one year ago?' One possible response in 2001 was 'no usual address', intended to identify only 'a child born after 29 April 2000', who would not have been in existence one year before the census date (ONS, 2014). However, this question caused substantial confusion, with people ticking the 'no usual address' response for themselves or for other members of their household who were not aged under 1 year old at the time of the census. In total, ONS report 467,036 individuals identified as having 'no usual address one year ago' in the 2001 Census SMS. Of these, an estimated 463,605 (99.27 per cent) were aged 1 year or over at the time of the 2001 Census, so should have been included with some origin stated, either within or outside the UK. This confusion, and the resulting over-count of people who had no known origin, meant that the 'no usual address' response was removed for the 2011 Census (ONS, 2012). Of course it is possible that some people did not have a usual address one year prior to the census (for example, someone who was homeless on 29 April 2000), but it is impossible to accurately distinguish these responses. Therefore, the following method assumes that all respondents had a usual address somewhere.

This disparity requires us to allocate the 'missing' migrants to allow for comparison between consistent datasets in the two census years. In essence the method used to reallocate these migrants requires that the origin unknown flows are first split into the three possible origins for each LAD: a move from within the same area (an intra-LAD move); a move from somewhere else in the UK (an inter-LAD move); or a move from outside of the UK (an into-LAD move). The age and sex structure of those with 'no usual address' is reported in table MG101 of the 2001 SMS for each LAD, so the allocation to destinations is a straightforward apportionment based upon the percentage of total inflow to that district from each of the three possible origins for each age and sex group. Allocation of the origins for inter-LAD migrants requires a similar approach, whereby each age and sex group is assigned an origin LAD based on the proportion of the total flow in that group for each pairwise interaction.

One further issue relating to both censuses is the count of infants aged under 1, since the migration question captures only those alive at the start and end of the 12-month period. The ONS makes an estimate of these flows based on the number of female migrants of child-bearing age so they are included in our analysis; in 2011, estimates of those aged 0 years are included in only one of the SMS tables (MM01BUK). Having addressed the issues of geographical inconsistency and the allocation of those migrants who stated 'no usual address' for the year before the 2001 census, we are now able to assess changes in national migration intensity and impact between 2001 and 2011. We present results for 11 age groups where moves are defined as between or within 404 LADs for the UK.

27.4   Changes in national migration intensity and impact

The adjusted flow estimates suggest that total migration in the UK increased from 6.63 million to 6.9 million between the censuses but the overall intensity of migration dropped from 11.3 persons per 100 of the population (Table 27.1) to just under 11 per cent between the two 12-month periods. The rate of inter-LAD migration decreased marginally from 4.5 to 4.4 per cent and the rate of mobility within LADs dropped from 6.8 to 6.5 per cent of the population although this shorter-distance residential mobility accounted for a slightly larger share of total migration in 2010–11.

Figure 27.1 contains graphs that illustrate the variation in the migration indicators for each of the 11 age groups into which the aggregate migration flow is divided, together with changes

Table 27.1 Migration in the UK, 2000–01 and 2010–11

Migration flow	2000–01			2010–11		
	Flow	Share (%)	Rate (%)	Flow	Share (%)	Rate (%)
Inter-LAD	2,655,584	40.0	4.51	2,794,882	40.5	4.42
Intra-LAD	3,982,289	60.0	6.76	4,106,665	59.5	6.50
All migration	6,637,873	100.00	11.27	6,901,547	100.00	10.92

Source: Authors' estimates based on data from 2001 and 2011 Censuses.

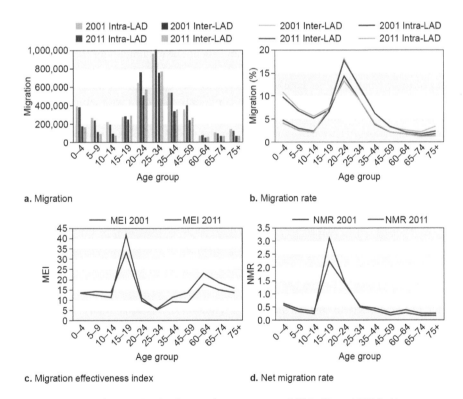

a. Migration   b. Migration rate

c. Migration effectiveness index   d. Net migration rate

Figure 27.1   National migration indicators by age group, 2000–01 and 2010–11

Source: Authors' estimates derived from data from the 2001 and 2011 Census SMS.

between 2000–01 and 2010–11. The age groups are relatively broad and of uneven interval but they do allow us to identify characteristics of migration at different stages during the lifecourse. Migrants in the first three age groups, all of which are five years in length, exhibit a decline in mobility from when they are infants to when they are children of secondary school age, a decline that is paralleled by those of their parental group aged in their 30s and early 40s. However, the migration rates for children decline between the two periods and both the migration effectiveness and the net migration rate scores are low.

The 15–19 age group includes those leaving home to go to higher or further education or who join the labour market and whose migration intensity rises as a consequence; this is the only age group in which, for 2010–11, the number of migrants between LADs was greater than those moving within LADs; and the inter-LAD migration rate increased by over 10 per cent. Both graphs e and f in Figure 27.1 suggest that it is migrants in this age group that have the most impact on the redistribution of the respective population, that is, these migrants have the most asymmetric patterns of movement between LADs. Moreover, flows of migrants in their late teens have undergone a significant increase in effectiveness between the two periods. Migrants in the next two age groups are likely to be migrating for job reasons although marriage and cohabitation may also be important explanatory factors. Figure 27.1b indicates that whilst migrants in the 20–24 age group have the highest system-wide migration intensity of all ages, their migration effectiveness shown in Figure 27.1c is lower than that of children, although their national net migration rate does not fall to the same extent.

Figure 27.1a shows that most migrants fall into the 25–34 age group. Whilst familiar motivations for this group are associated with changes in job or the need for a larger home to accommodate children, the net migration indicators suggest that many migration streams in this age category are cancelled out by movements in the opposite direction and there is virtually no change between 2001 and 2011. Rates of migration continue to diminish for the migrants within and between LADs until the final group of those aged 75 and over, when migration into institutions or to be closer to other family members becomes an important motivation. Migration effectiveness peaks at age 60–64, an age group commonly associated with retirement. Although net migration rates for those aged 45–74 fall marginally between the two census periods, a decline in migration effectiveness is apparent for the 45–59 and 60–64 year olds in particular, despite the fact that the number of 60–64 year olds migrating between LADs increased by 30 per cent. Migration is becoming less important for redistributing the population in these older ages.

27.5 Comparison of local migration impact

Hitherto, we have not looked at the spatial patterns of migration flows within the UK. In this section, we have chosen to use a 'local' migration effectiveness indicator defined as:

$$MEI_i = 100 \ (D_i - O_i) / (D_i + O_i) \tag{27.5}$$

for each LAD i to illustrate how the impact of migration varies geographically. The proportional symbol maps in Figure 27.2 show the variation of *MEI* for migrants of all ages at district level in the two census periods. The size of each circle in the two maps is proportional to the *MEI* value with red circles indicating net migration gain and blue indicating net migration loss. Generally speaking, the pattern of aggregate net migration in 2000–01 in England and Wales is characterised by net losses from metropolitan LADs and net gains to non-metropolitan LADs as conceptualised by the process of counterurbanisation (Fielding, 1982; Champion, 1989) and reported in Dennett and Stillwell (2010b). This pattern of urban loss and rural gain is not repeated in

a. 2000-01   b. 2010-11

Figure 27.2 Migration effectiveness by LAD, all ages, 2000–01 and 2010–11

Source: Authors' estimates derived from data from the 2001 and 2011 Census SMS.

Scotland and Northern Ireland, where many rural LADs experienced depopulation and therefore net losses of migrants in 2000–01. The effectiveness of net migration losses in more rural LADs in Northern Ireland appears to have increased by 2010–11 and many LADs in the East and West Midlands have witnessed a shift from positive to negative migration effectiveness, whilst several LADs in the South East have moved in the other direction and the effectiveness of migration gains to most LADs along the south coast has diminished.

All age *MEI* values are, of course, the composite scores resulting from different patterns of movement by those in age groups experiencing different lifecourse events. In this instance, we have selected two age groups to demonstrate this age variation in migration effectiveness and the changes that have occurred between the two census periods. The two age groups are those for which largest changes at the national level have been observed (Figure 27.1c). The first is the 15–19 age group whose national *MEI* increased and Figure 27.3 presents the local effectiveness indices for each LAD ranked according to their value in 2001 (solid line). Only 85 out of 404 LADs recorded positive *MEI* scores in 2001 and the large majority of these were university towns and cities with Cambridge and Oxford being at the top of the ranking with *MEI* vales of 43 and

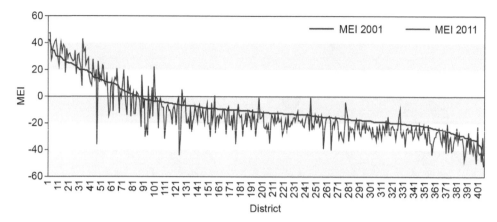

Figure 27.3 UK LADs ranked by their *MEI* for 15–19 year olds in 2000–01

Source: Authors' estimates derived from data from the 2001 and 2011 Census SMS.

41 respectively in 2001. Dungannon in Northern Ireland sits at the other end of the ranking with a score of -52 in 2001 and several of the LADs in Northern Ireland are among the LADs with highest negative *MEI* in both periods. Not only does the graph indicate the asymmetry of flows in this age group but it also illustrates the tendency for positive *MEI* values to be more positive in 2011 and negative *MEI* values to be more negative; for values above zero, the dotted line is above the solid line whereas for values below zero, the dotted line is below the solid line. There are several exceptions to this trend, among which the most prominent is Harrogate in North Yorkshire, ranked forty-sixth in 2001 with a positive *MEI* of 14 but having a negative *MEI* value of 36 in 2011. Harrogate's net balance switched from a gain of 528 in 2001 to a loss of 1,071 in 2011 due primarily to a decline in inflows but also to an increase in outflows.

The second group includes those aged between 45 and 74, whose national migration effectiveness indices for the three composite age groups all declined between 2001 and 2011. The geographical patterns of *MEI* shown in both maps in Figure 27.4 reflect that of counterurbanisation, with net losses from the large cities and gains further down the urban–rural hierarchy, but the key difference in 2011 is the reduced size of the circles representing both loss and gain that confirms a diminution of the counterurbanisation as reported by Lomax et al. (2014) using time-series data derived from administrative sources.

27.6 Effects of scale and zonation on migration intensity and impact

Most empirical studies of internal migration make use of data that are released by the national statistics office for the country concerned at a limited number of spatial scales. In the UK, the ONS released migration flow data from the 2001 and 2011 Censuses at three spatial scales and responsibility lies with the researcher to select whichever scale is deemed most appropriate for the analysis to be undertaken, or to perform some aggregation to generate flows for a higher-level geography. In this chapter we have used LADs as our chosen geography since this scale is convenient for understanding variations across the UK in intensities and patterns based on matrices that do not suffer the problems of sparsity observed at ward or output area levels. However, in making this choice, we ignore the issue of the sensitivity of migration indicators to spatial scale, referred to more generally as the Modifiable Areal Unit Problem (MAUP) (Openshaw, 1984). Recent software has been developed as part of the IMAGE (Internal Migration Across the GlobE)

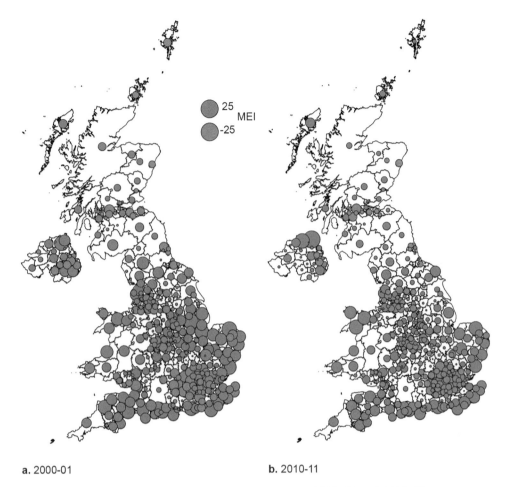

a. 2000-01   **b.** 2010-11

Figure 27.4   Migration effectiveness by LAD, ages 45–74, 2000–01 and 2010–11

Source: Authors' estimates derived from data from the 2001 and 2011 Census SMS.

project[1] which enables both the scale effect (the variation in a migration indicator value due to the number of zones) and the aggregation or zonation effect (the variation in an indicator value due to the zone boundary configuration at any one scale) to be investigated.

The *IMAGE Studio* (Stillwell et al., 2014) requires a shapefile containing the polygons of a set of basic spatial units (BSUs), the 404 LADs in our case, together with a 404 by 404 migration flow matrix and a set of populations at risk for the 404 BSUs. The *IMAGE Studio* contains two different aggregation algorithms for generating '*n*' contiguous aggregate statistical regions (ASRs) from the 404 BSUs. The first of these, the initial random aggregation (IRA) algorithm (Openshaw, 1977), provides a high degree of randomisation to ensure that the resulting aggregations are different during the iterations. The second is the IRA-wave algorithm, a hybrid version of the original IRA algorithm with strong influences from the mechanics of the breadth-first search (BFS) algorithm, which begins by selecting *n* BSUs randomly and assigning each one to an empty ASR. Using an iterative process, each BSU is allocated to one of the *n* ASRs based on contiguity. The advantages of the IRA-wave algorithm are the swiftness for producing a large number of initial aggregations and the relatively well-shaped ASRs that are generated in comparison to the more irregular shapes of the IRA algorithm. The *IMAGE Studio* supports both algorithms for experimentation on different degrees of randomness but also, importantly, allows the user to perform multiple aggregations of the BSUs at any spatial scale. This means that the user can specify the scale step and the number of configurations of ASRs at each scale. In our case, with 404 BSUs, by selecting multiple aggregation with a scale step of 10 and 50 configurations, the *IMAGE Studio* will aggregate the BSU data to 10, 20, 30, 40 . . . 380, 390, 400 ASRs, producing 50 different configurations for each of these scales. It will then aggregate the original BSU to BSU migration flow matrix and the vector of BSU populations for each set of ASRs that have been computed and these aggregated flows and populations can then be used to produce a range of migration indicators that include the *CMI*, *ANMR* and the *MEI* at each spatial scale for each configuration of ASRs. The *IMAGE*

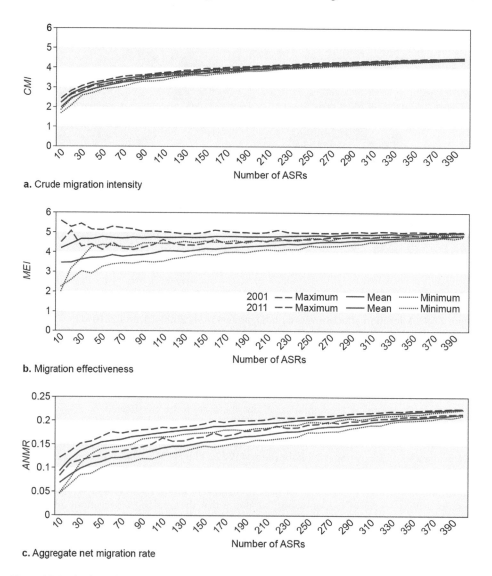

a. Crude migration intensity

b. Migration effectiveness

c. Aggregate net migration rate

Figure 27.5 Scale and zonation effects on aggregate migration indicators, 2000–01 and 2010–11

Source: Authors' estimates derived from data from the 2001 and 2011 Census SMS.

Studio has been used to calculate these indicators for both all-age and age-group migration for both 2000–01 and 2010–11 periods. In Figure 27.5, the solid lines in the graphs show the mean values of the 50 configurations for each indicator, whereas the hashed lines show the minimum and maximum values respectively. The shape of each line provides a visual representation of the scale effect, as the number of ASRs increases on the horizontal axis from 10 to 400, whilst the range around the mean gives an indication of the zonation effect at each scale. The *CMI* in 2001 varies across the scale range, falling from 4.5 migrants per 1,000 population with 400 ASRs to 2.2 per 1,000 between 10 ASRs; the scale effect becomes progressively larger as the number of ASRs gets smaller, particularly when below 100 zones. The mean value of the *CMI* at 400 ASRs in 2011 is 4.42, reflecting the decreased propensity to migrate between LADs in the more recent period, but the scale effect is similar. Whereas there is clear evidence of the effect of scale on the *CMI* for both periods, the zonation effect is much less pronounced in both cases. The graphs for the *MEI*, on the other hand, suggest much less scale effect, particularly in 2001 when the *MEI* shows greater stability across scale except for when the number of ASRs gets relatively small; the *MEI* for 2011 is lower than that for 2001 and reduces as the number of ASRs gets smaller, suggesting that net migration becomes less important with larger zones. In both cases, the zonation effect is more pronounced than with the *CMI*, increasing as the number of zones gets smaller. The third variable, the *ANMR*, reflects both the lower values of both *CMI* and *MEI* components in 2011 compared with 2001 and follows the shape of the *CMI* over scale, reducing as the number of ASRs gets smaller.

The two graphs in Figure 27.6 exemplify how the mean national *MEI* values for each age group change according to the number of ASRs in 2000–01 and 2010–11. In both periods, the *MEI* is constant over scale or reduces marginally in 10 out of the 11 age groups, but the anomaly is that of the 15–19 age group, whose trajectory experiences a strong scale effect; migration effectiveness for this group reduces as the number of zones get smaller and the size of the zones gets larger. Moreover, the scale effect for this age group increases between the two periods. One further noteworthy feature of these graphs is that cross-overs occur, indicating that at different scales, migration in certain age groups will be more effective in terms of redistribution than others. For example, migration of 15–19 year olds is most effective at scales involving 400 ASRs

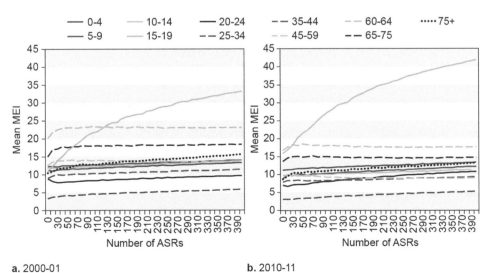

Figure 27.6 Migration effectiveness by scale, age groups, 2000–01 and 2010–11

Source: Authors' estimates derived from data from the 2001 and 2011 Census SMS.

down to 130 ASRs, but then migrants aged 0–14 have the highest *MEI* scores when ASRs become fewer and increase in size.

27.7 Conclusions

A number of conclusions can be drawn from the work reported in this chapter. First, although a wealth of migration flow data is available from consecutive censuses in the UK, careful preparation and adjustment of data is required in order to facilitate robust comparisons of intensities and impacts between 2011 and 2001. In some respects, the issues of boundary and definitional change are less problematic than those apparent when comparing 2001 with 1991 data (Stillwell and Duke-Williams, 2007). Second, our estimates of migration between and within LADs at the national level suggest that there has been a small reduction in the intensity of migration both within and between LADs. Third, whilst the variation in the intensity of migration by age group is well known, the analysis of migration effectiveness suggests some quite profound variations and impacts on population redistribution for certain age groups. The net migration pattern of 15–19 year olds with gains to LADs with higher education establishments and losses from other LADs contrasts markedly with the overall spatial pattern of metropolitan gains and non-metropolitan losses and the impact of these flows, as measured by the *MEI*, has increased between 2001 and 2011. In contrast, the impact of migrants in the older working and retirement age groups has diminished as propensities to leave the metropolitan areas have waned, leading to a less distinctive process of counterurbanisation. Moreover, there is evidence of increasing rural-to-urban migration during the 2000s from annual administrative data on inter-district flows in the UK (Stillwell et al., 2015; Lomax and Stillwell, 2017).

The analysis of the scale and zonation effects of migration within the UK suggests that whilst the *CMI* and the *ANMR* are sensitive to scale, the *MEI* tends to be constant across scale for all aggregate and age-specific migration apart from the 15–19 age group, whose migration intensity and pattern characteristics are unique. Our finding that the *MEI* for aggregate internal migration flows tends to be stable over scale conforms with evidence from other countries around the world reported in Rees et al. (2016). However, a zonation effect is more apparent with *MEI* than the other two indicators for the UK, particularly as the number of zones falls to less than 100. This implies that the zone shape plays an increasingly important role in determining migration effectiveness as the zones increase in size. Finally, scale is important as far as certain indicators of migration are concerned when making comparisons between 2001 and 2011; we have observed how the effectiveness of late teenage migration predominates at scales above 130 zones but declines at scales with fewer zones. Further research using other variables such as migration distance and spatial connectivity is required to extend our understanding of changes in internal migration in the UK and to make use of the extensive datasets available from the census to understand the effects of scale and zonation on migration within as well as between LADs.

Acknowledgement

The first author is grateful for Economic and Social Research Council (ESRC) funding for research grant ES/J02337X/1: UK Data Service-Census Support.

Note

1 www.gpem.uq.edu.au/qcpr-image.

References

Bell, M., Blake, M., Boyle, P., Duke-Williams, O., Rees, P., Stillwell, J. and Hugo, G. (2002) Cross-national comparison of internal migration: Issues and measures. *Journal of the Royal Statistical Society: Series A (Statistics in Society)*, 165(3): 435–464.

Bell, M., Charles-Edwards, E., Kupiszewska, D., Kupiszewski, M., Stillwell, J. and Zhu, Y. (2015) Internal migration and development: Comparing migration intensities around the world. *Population and Development Review*, 41(1): 33–58.

Champion, A.G. (ed.) (1989) *Counterurbanisation: The Changing Pace and Nature of Population Deconcentration.* Routledge, New York.

Champion, A.G. (2005) Population movement within the UK. In Chappell, R. (ed.) *Focus on People and Migration.* Macmillan, Basingstoke, pp. 92–114.

Champion, A.G. and Shuttleworth, I. (2016a) Is longer-distance migration slowing? An analysis of the annual record for England and Wales. *Population, Space and Place.* DOI: 10.1002/psp.2024.

Champion, A.G. and Shuttleworth, I. (2016b) Are people moving address less? An analysis of migration within England and Wales, 1971–2011, by distance of move. *Population, Space and Place.* DOI: 10.1002/psp.2026.

Cooke, T.J. (2013) Internal migration in decline. *The Professional Geographer*, 65(4): 664–675.

Dennett, A. and Stillwell, J. (2010a) Internal migration in Britain, 2000–01, examined through an area classification framework. *Population Space and Place*, 16(6): 517–538.

Dennett, A. and Stillwell, J. (2010b) Internal migration patterns by age and sex at the start of the 21st century. In Stillwell, J., Duke-Williams, O. and Dennett, A. (eds.) *Technologies for Migration and Commuting Analysis; Spatial Interaction Data Applications.* IGI Global, Hershey, pp. 153–174.

Fielding, A. (1982) Counterurbanisation in Western Europe. *Progress in Planning*, 17(1): 1–52.

Fielding, A. (2012) *Migration in Britain: Paradoxes of the Present, Prospects for the Future.* Edward Elgar, Cheltenham.

Lomax, N. and Stillwell, J. (2017) Temporal change in internal migration in the United Kingdom. In Champion, A., Cooke, T. and Shuttleworth, I. (eds.) *Internal Migration in the Developed World: Are We Becoming Less Mobile?* Routledge, London.

Lomax, N., Norman, P., Rees, P. and Stillwell, J. (2013) Subnational migration in the United Kingdom: Producing a consistent time series using a combination of available data and estimates. *Journal of Population Research*, 30: 265–288.

Lomax, N., Stillwell, J., Norman, P. and Rees, P. (2014) Internal migration in the United Kingdom: Analysis of an estimated inter-district time series, 2001–2011. *Applied Spatial Analysis and Policy*, 7(1): 25–45.

ONS (2012) 2011–2001 Census in England and Wales Questionnaire Comparability. Available at: https://www.ons.gov.uk/census/2011census/2011censusdata/2011censususerguide/comparabilityovertime.

ONS (2014) 2011 Census Analysis: Internal and international migration of older residents (aged 65 and over) in England and Wales in the year prior to the 2011 Census. Available at: www.ons.gov.uk/ons/dcp171776_377047.pdf.

Openshaw, S. (1977) Algorithm 3: A procedure to generate pseudo-random aggregations of N spatial units into M spatial units, where M is less than N. *Environment and Planning A*, 9: 1423–1428.

Openshaw, S. (1984) *The Modifiable Areal Unit Problem.* Concepts and Techniques in Modern Geography, 38, GeoBooks, Norwich.

Rees, P., Bell, M., Kupiszewski, M., Kupiszewska, D., Bernard, A., Charles-Edwards, E. and Stillwell, J. (2016) The impact of internal migration on population redistribution: An international comparison. *Population, Space and Place.* DOI: 10.1002/psp.2036.

Stillwell, J. and Duke-Williams, O. (2007) Understanding the 2001 UK Census migration and commuting data: The effect of small cell adjustment and problems of comparison with 1991. *Journal of the Royal Statistical Society: Series A (Statistics in Society)*, 170(2): 425–445.

Stillwell, J., Lomax, N. and Sander, N. (2015) Monitoring and visualising sub-national migration trends in the United Kingdom. In Geertman, S., Ferreira, J., Godspeed, R. and Stillwell, J. (eds.) *Planning Support Systems and Smart Cities.* Springer, Dordrecht, pp. 427–445.

Stillwell, J., Rees, P. and Boden, P. (1992) Internal migration trends: An overview. In Stillwell, J., Rees, P. and Boden, P. (eds.) *Migration Processes and Patterns. Volume 2; Population Redistribution in the United Kingdom.* Belhaven Press, London, pp. 28–55.

Stillwell, J., Daras, K., Bell, M. and Lomax, N. (2014) The IMAGE Studio: A tool for internal migration analysis and modelling. *Applied Spatial Analysis and Policy*, 7(1): 5–23.

Stillwell, J., Bell, M., Blake, M., Duke-Williams, O. and Rees, P. (2000) Net migration and migration effectiveness: A comparison between Australia and the United Kingdom, 1976–96. Part 1: Total migration patterns. *Journal of Population Research*, 17(1):17–38.

Stillwell, J., Bell, M., Blake, M., Duke-Williams, O. and Rees, P. (2001) Net migration and migration effectiveness: A comparison between Australia and the United Kingdom, 1976–96. Part 2: Age-related migration patterns. *Journal of Population Research*, 18(1):19–39.

Stillwell, J., Bell, M., Ueffing, P., Daras, K., Charles-Edwards, E., Kupiszewski, M. and Kupiszewska, D. (2016) Internal migration around the world: Comparing distance travelled and its frictional effect. *Environment and Planning A*. DOI: 10.1177/0308518X16643728.

Commuting intensities and patterns in England and Wales, 2001–11

Thomas Murphy, John Stillwell and Lisa Buckner

28.1 Introduction

In 2011, nearly two-thirds of the population of 41.1 million individuals aged between 16 and 74 in England and Wales either worked at home or commuted to work using some mode of travel. On the basis of personal experience, commuting to and from work is an activity that most people endure and few enjoy; it can make significant demands on our financial resources as well as on our time and, if problems arise that delay our arrival or departure, it can be a very stressful and counterproductive experience. According to the National Travel Survey (NTS) (Department for Transport, 2015), whilst travel to work represents 16 per cent of all the journeys made in England, the distance travelled to work is by far the longest of all the types of journeys that we make, representing 20 per cent of total distance travelled.

Whilst the NTS provides a wealth of information about travel trends at the national level, the census provides by far the most authoritative, comprehensive and spatially detailed data on the daily mass movement of people between their locations of usual residence and their workplaces. In this chapter, our aims are, first, to provide some indication of the magnitude, complexion and geographical pattern of commuting in England and Wales based on data extracted from the 2011 Census undertaken by the Office for National Statistics (ONS), and, second, to outline where changes have occurred since the previous census in 2001. In doing so, we identify variations in commuting rates based on different socio-demographic variables and on the alternative travel modes that commuters use to make their journeys. We attempt to offer some possible explanations for these variations whilst recognising that the first decade of the twentieth century saw significant societal and attitudinal changes as well as developments in infrastructure and policy at different levels. Amongst the wider changes with impacts on commuting behaviour are the relatively high levels of population growth and immigration (ONS, 2012a), the continuing tertiarisation of the economy (ONS, 2013a), increased female participation in the labour force (Scheiner and Kasper, 2003; ONS, 2013b), increasing fuel costs (BBC, 2013a) and environmental awareness in line with international trends (Lorenzoni et al., 2007), urban regeneration (Seo, 2002; Ogden and Hall, 2000) and further political devolution (BBC, 2015a, 2015b). Whilst the impact of any one of these factors may be felt nationally, their singular or combined effect may result in varying geographical impacts, particularly where key policy or infrastructure initiatives, such as the introduction of the congestion charge in London or the opening of the new metro line in Manchester, have also occurred.

The chapter is organised as follows. In the next section, we introduce the aggregate and microdata used subsequently and introduce the system of districts employed for examining spatial variations and changes between 2001 and 2011. Section 28.3 considers commuting in England and Wales at the national level and identifies variations in commuting propensities between population sub-groups disaggregated by a number of socio-demographic variables. It also examines the popularity of different modes of travel to work and how these have changed over the decade. In Section 28.4, we distinguish commuters from homeworkers and present the results of fitting binary logistic regression models to microdata from both the 2001 and 2011 Censuses in order to capture variations in commuting intensities at the national level for a range of characteristics that take confounding variables into account. Geographical variations in commuting and homeworking by former Government Office Region are also introduced in the modelling in this section before spatial patterns of commuting rates are examined at local authority district (LAD) scale in Section 28.5. The ONS 2011 classification of districts into supergroups is used to summarise these variations and then some conclusions are presented in the final section.

28.2   Census data

Commuting data collected by ONS in England and Wales come in a variety of forms, both cross-sectional and longitudinal, and in aggregate counts or individual cases, extracted from the questions in the 2001 and 2011 Censuses that asked: (i) "In your main job, what is the address of your workplace?" and (ii) "How do you usually travel to work?". Using the responses to these and other questions, ONS has produced estimated univariate measures of commuting behaviour such as aggregate commuter flows into, within and out of an area, distance travelled and origin–destination flow data. Multivariate data are produced from cross-tabulations of the responses to these questions with responses to other census questions to produce counts of commuters by age, sex, ethnic group, family status, socio-economic group, occupation, employment status, industry and mode of transport. All of the commuting variables are available through the aggregate and interaction commuting datasets as published outputs. In this chapter, our primary sources of data are the aggregate statistics and the individual records contained in the 2001 Census Samples of Anonymised Records (SARs) and the 2011 Census microdata.

Aggregate commuting data from the 2001 Census are published as tables in the Key Statistics (KS), Standard Tables (ST) and Census Area Statistics (CAS) at different geographical levels, whereas 2011 Census tables with aggregate commuting data are available from the Quick Statistics (QS), Local Characteristics (LC) and Detailed Characteristics (DC) as explained in Chapter 5 and accessed either through the *Nomis* or *InFuse* online interfaces. In this chapter, two spatial scales are used: first, the national level, which includes the whole of England and Wales; and, second, a set of 348 LADs with consistent boundaries between 2001 and 2011 that includes 22 Welsh principal areas, 32 London boroughs (and the City of London), 36 metropolitan districts, 56 unitary authorities and 201 non-metropolitan districts and takes into account the English local government reorganisation in 2009, which involved changes to LAD boundaries due to the abolition of certain non-metropolitan areas in the North East, North West, West Midlands, East of England and the South West regions. We have chosen LADs instead of functional regions such as Travel To Work Areas (TTWAs) for a number of reasons. First, TTWAs covering the whole of the UK numbered 228 in 2011 (ONS, 2015) compared with 243 in 2001 (Coombes et al., 2005) and their boundaries are not consistent over time, making comparison between the two census dates difficult. Second, the choice of LADs allows more detailed geographical variation across the country to be observed since they are more numerous than TTWAs. Third, we are not using origin–destination flow data in this chapter since our focus is not on distance travelled, directional movements or methods of flow visualisation such as that reported by Nielsen and Hovgesen (2008) or more recently by Rae (2011) and exemplified in Chapter 15 of this book by O'Brien and Cheshire.

The other key sources of census data used in this chapter are the 2001 Individual Licensed SARs consisting of 1,843,525 cases and the 2011 Census Microdata Individual Safeguarded Sample (Regional) containing 2,848,149 cases. A major benefit of using these microdata files is the opportunity they provide to cross-tabulate the variables available as required; in this chapter, a selection of socio-demographic characteristics including sex, age and ethnicity are used, together with a region variable, in order to assess the probability that an individual is a commuter or a homeworker when controlling for the other variables included.

Whilst it is important to understand the changing volume of commuters in 2001 and 2011, comparison between the two censuses is more consistent when flows are standardised by populations at risk (PAR) of commuting, but there is debate about the denominator that should be used in the calculation of commuting rates. When computing a national commuting rate, for example, some might argue that the total population might be appropriate; others would prefer to identify those of workforce age, such as those aged 16–65 or 16–74 or just those aged over 16, but this calls into question whether or not there are optimal cut-off points for working age at both the young and old end of the age spectrum. The national school leaving age of 16 makes this age reliable for those beginning their working lives though many delay the commencement of work by taking up secondary and higher education opportunities. In Scotland, unlike England, Wales and Northern Ireland, Special Travel Statistics (STS) are available that report on the journey to study as well as the journey to work. Further alternative denominators might be the economically active population of working age (say 16–74), which includes those who are unemployed, or the working age population, which would include full-time students. Since the number of commuters is very closely correlated with the number employed, we have chosen to define national and district aggregate commuting rates by dividing the number of commuters by the population aged 16–74 of the area concerned and this principle has been applied to compute rates disaggregated by the various characteristics used.

There are two small additional measurement or definitional issues in making comparisons between 2001 and 2011. The first is that male and female commuters are reported for ages 16–74 in the 2001 Census and 16+ in the 2011 Census and consequently an adjustment has been made to estimate a consistent 16–74 count for men and women in 2011. Second, having produced counts of those working at or from home in 2001, the ONS initially released data on those working at home in 2011, thereby rendering the homeworker counts incomparable with those of 2001. However, a new table was released subsequently that provided counts of homeworkers on a similar basis in both years which thus enabled the estimation of those working at home and those working from home in 2011, a distinction that is unavailable in 2001. Finally, it is worth remembering that the commuting data for both 2001 and 2011 contain errors relating to the mode of travel; in particular, the counts of those travelling by bicycle and on foot contain numbers of individuals making daily journeys of impossible length, for example between Manchester and Birmingham. These errors are likely to occur because respondents have misread the census questionnaire and ticked the wrong mode of travel box or they have mistakenly entered the wrong workplace address. These errors are ignored in the analysis which follows.

28.3   Commuting at the national level

The magnitude and composition of the population aged 16–74 commuting to work in England and Wales is changing. Aggregate statistics from the 2001 and 2011 Censuses[1] indicate that the number

Table 28.1 Commuting aged 16–74 in England and Wales by sex and age, 2001 and 2011

Categories	Number of commuters (millions)		Usually resident population (millions)		Commuting rate (%)		Change
	2001	2011	2001	2011	2001	2011	2001–11
England and Wales	23.63	26.53	37.61	41.13	62.8	64.5	1.7
Male	12.79	14.03*	18.50	20.39	69.1	68.8	–0.3
Female	10.84	12.49*	19.10	20.74	56.7	60.2	3.5
Aged 16–24	3.25	3.38	5.68	6.66	57.3	50.7	–6.6
Aged 25–64	20.01	22.21	27.56	29.62	72.6	75.0	2.4
Aged 65–74	0.37	0.93	4.37	4.85	8.4	19.2	10.8

* Estimates based on commuters aged 16+.

Source: Derived from the 2001 I-SAR and 2011 Census aggregate data.

of commuters increased by 2.9 million persons over the decade, a rise of 12.3 per cent to 26.5 million in 2011 (Table 28.1). Around 57 per cent of this growth was female, despite the fact that female commuters comprised only 47 per cent of all commuters in 2011, but perhaps the most significant change over the period was the increase in the number of commuters in the oldest of the three age groups for which consistent data are available between the two censuses. The number of commuters aged 65–74 increased by 154 per cent whereas the usually resident population in this age group increased by 11.1 per cent. As a consequence, the commuting rate for this group rose from 8.4 per cent in 2001 to 19.2 per cent in 2011 or 10.8 percentage points compared with the increase of 1.7 percentage points for those aged 16–74. The main anomaly in these national statistics is the change in commuting for the youngest age group, whose volume of commuters increased but whose commuting rate decreased by 6.6 percentage points or 11.5 per cent over the decade. Whilst the female commuting rate increased by 3.5 percentage points, the rate for males decreased by 0.3 percentage points, although these changes are based on estimates of males and females aged 16–74 in 2011.

Further information about variations in commuting at the national level can be derived from the Individual Sample of Anonymised Records (I-SAR) for 2001 and census microdata on individuals for 2011. In Table 28.2, national data are presented for three further variables: ethnicity, limiting long-term illness (LLTI) and dependent children. The microdata indicate that commuting rates for ethnic minorities were all lower than the White majority in 2001, with lowest rates for Bangladeshis (33.1 per cent) and Pakistanis (36.5 per cent), although these groups saw the largest increases between the two points in time. Ethnic inequalities in the labour market over the last two decades documented by Kapadia et al. (2015) confirm very low rates of economic activity amongst Bangladeshi and Pakistani women in particular. In these minority ethnic groups, families tend to be large and women are the homemakers. Commuting rates of Indians increased to the same level as the White population in 2011 and it was only the Chinese whose commuting rates remained much the same at both census dates. As might be expected, those with LLTI have much lower rates of commuting than those without, but contrary to expectation, those with dependent children, whose commuting rate was much lower than those without children in 2001, have higher rates of commuting than their counterparts in 2011; the increase in commuting for those with children has risen by 25 per cent compared with a fall in commuting rates for those without children.

Thus, in summary, whilst the increases seen in the absolute numbers of commuters in England and Wales between 2001 and 2011 reflect demographic growth (ONS, 2012a), the aggregate

Table 28.2 Commuting aged 16–74 in England and Wales by ethnicity, LLTI and dependent children, 2001 and 2011

Variables	Categories	Number of commuters		Population 16–74		Commuting rate		Change
		2001	2011	2001	2011	2001 (%)	2011 (%)	2001–11
Sample, England and Wales		703,407	1,330,320	1,163,463	2,065,163	60.5	64.4	4.0
Ethnic group	White	656,413	1,172,529	1,067,755	1,787,288	61.5	65.6	4.1
	Indian	13,572	36,756	23,902	55,824	56.8	65.8	9.1
	Pakistani	5,207	17,578	14,259	36,877	36.5	47.7	11.1
	Bangladeshi	1,772	6,679	5,349	14,316	33.1	46.7	13.5
	Other Asian	2,888	19,370	5,611	32,471	51.5	59.7	8.2
	Black	13,280	38,365	25,540	65,686	52.0	58.4	6.4
	Chinese	2,750	8,714	5,674	17,944	48.5	48.6	0.1
	Other	7,525	30,329	15,373	54,757	48.9	55.4	6.4
LLTI	LLTI	48,513	94,013	234,789	330,780	20.7	28.4	7.8
	No LLTI	654,894	1,236,307	1,085,663	1,734,383	60.3	71.3	11.0
Dependent children	No	320,811	585,196	465,957	907,481	68.8	64.5	-4.4
	Yes	269,430	497,549	593,939	708,547	45.4	70.2	24.9

Source: Derived from the 2001 I-SAR and 2011 Census microdata.

commuting rates shown in Table 28.1 suggest that a higher proportion of the overall population at risk was commuting. One reason for this is the increased female participation in the workforce (Scheiner and Kasper, 2003), but also important, despite Government concern over the number of people leaving work before the state pension age (Department for Work and Pensions, 2014), is the increasing proportion of those working to an older age and retiring later (ONS, 2012b). Increases are apparent across the ethnicity spectrum.

Both the 2001 and 2011 Censuses across the UK contained a question about the usual mode of travel to work, indicating that respondents should tick the box for the mode used for the longest part, by distance, of their usual commute. Comparable results are produced for 11 modes of travel, including a category for those who work mainly at or from home which, in 2011, accounted for 10.3 per cent of all commutes (Figure 28.1) The number of homeworkers increased by 553,465 to over 2.7 million in 2011, a figure eclipsed only by the number of car drivers, whose increase over the period was nearly 1.4 million (top graph in Figure 28.1). However, whilst homeworking increased as a proportion of all commuters from 9.1 per cent to 10.3 per cent, or 287,183 people (bottom graph in Figure 28.1), the proportion of car drivers actually fell by from 55.2 per cent to 54.5 per cent, or 202,793 individuals. The proportion of all commuters travelling as passengers in cars and vans declined by the largest amount (1.3 per cent) of all modes and the percentage driving to work also fell by 0.8 per cent, whilst train passengers increased their share by 0.9 per cent and those travelling by underground, metro, light rail and tram also increased their share by 0.7 per cent. These last two modes of travel experienced an absolute increase of commuters, as did those travelling by bus, minibus and coach and on foot, but, like car drivers, these three categories claimed lower shares of total commuters in 2011 than they did in 2001.

By showing the absolute numbers of commuters derived from both censuses and the numbers in the changing proportions of total commuting, Figure 28.1 provides both positive and negative

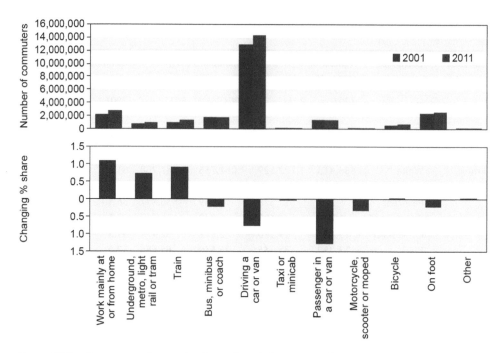

Figure 28.1 Commuters by mode of travel and changes in share, England and Wales, 2001 and 2011

Source: Derived from the 2001 and 2011 Census aggregate data.

messages for policy makers. Car drivers make up a lower share of all commuters to work, despite having increased substantially in absolute terms; car passengers have decreased in absolute and relative terms; commuters by train, underground, bicycle, taxi and other means have increased their numbers and shares, whereas bus travellers and those on foot have increased in number but their shares of all commuters have fallen. We will return to explore some of the spatial changes by mode of travel later in the chapter.

28.4 Modelling commuting and homeworking

It is a moot point whether or not homeworkers should be classified as commuters. In this section, we have decided to separate them from those commuting by one or other of the various modes of travel and present the results of fitting binary logistic regression (BLR) models to the 2001 and 2011 microdata in order to establish that the variations in commuting and homeworking rates identified in the last section persist when we account for the influence of other variables. In addition, we introduce a geographic variable to capture the variation in behaviour by region of usual residence.

The categorical response or dependent variable in the two BLR models is either an individual who commutes to work (1) or not (0), or someone who works at or from home (1) or not (0), given a set of predictor or independent variables, which include those introduced in Tables 28.1 and 28.2. Thus, BLR quantifies the relative prevalence of predefined behavioural patterns (of commuting or homeworking) for the selection of socio-demographic and geographic variable categories in relation to a set of corresponding reference categories. In other words, in BLR, the dependent variable is determined by which variable category an individual is in, for example whether

Table 28.3 Odds ratios for commuters and homeworkers (aged 16–74) by selected characteristics, England and Wales, 2001 and 2011

Characteristics		Commuting		Homeworking	
Variables	Categories	OR (Exp(β))		OR (Exp(β))	
		2001	2011	2001	2011
Sex	Male	1.000	1.000	1.000	1.000
	Female	0.429*	0.572*	0.795*	0.764*
Age group	16–24	1.000	1.000	1.000	1.000
	25–44	1.444*	1.874*	1.968*	2.018*
	45–64	0.778*	1.194*	3.285*	3.311*
	65–74	0.025*	0.070*	8.951*	8.176*
Ethnic group	White	1.000	1.000	1.000	1.000
	Indian	0.802*	0.884*	1.366*	0.938*
	Pakistani	0.245*	0.325*	1.196*	1.040
	Bangladeshi	0.184*	0.274*	2.430*	1.149*
	Other Asian	0.503*	0.584*	1.454*	1.099*
	Black	0.595*	0.586*	0.872*	0.789*
	Chinese	0.647*	0.678*	2.219*	1.690*
	Other	0.493*	0.534*	1.242*	0.988
LLTI	LLTI	1.000	1.000	1.000	1.000
	No LLTI	8.790*	7.220*	0.810*	0.850*
Dependent children	No Dependent Children	1.000	1.000	1.000	1.000
	Dependent Children	0.792*	0.921*	1.069*	1.177*
Occupation	Professional and Managerial			1.000	1.000
	Non-Professional and Non-Managerial			1.069*	0.938*
Region of usual residence	North East	1.000	1.000	1.000	1.000
	North West	1.188*	1.136*	1.085*	1.122*
	Yorkshire and Humber	1.264*	1.119*	1.109*	1.195*
	East Midlands	1.341*	1.221*	1.176*	1.266*
	West Midlands	1.279*	1.155*	1.165*	1.222*
	East of England	1.467*	1.308*	1.233*	1.333*
	South East	1.545*	1.346*	1.304*	1.499*
	South West	1.381*	1.295*	1.478*	1.613*
	London	1.271*	1.252*	1.148*	1.268*
	Wales	1.074*	1.067*	1.267*	1.326*
Constant		0.574*	0.493*	0.043*	0.045*

* refers to OR significance at 95 per cent confidence level.

Source: Columns 2–4 derived from the 2001 I-SAR and 2011 microdata.

they are male or female. The columns of Table 28.3 contain the odds ratios (ORs) for commuting and homeworking rates respectively for 2001 and 2011 with asterisks indicating significance at a 95 per cent confidence level. The OR is represented by the term Exp(β), computed by raising the base of the natural log to the power β, where β is the slope of the logistic regression equation, and indicates for each non-reference category the relative likelihood that an outcome will occur compared to the corresponding reference category.

The ORs in Table 28.3 provide evidence that the socio-demographic and geographic variations identified above remain when controlling for the different variables. Women are less likely to be commuters or homeworkers when compared to men. Those aged 25–44 in both years and those aged 45–64 in 2011 are more likely to be commuters than those aged 16–24, but there is an increasing likelihood of homeworking as age increases when compared with age 16–24. All ethnic minority populations have lower odds of commuting than Whites, especially Bangladeshis who, together with Chinese, have a higher likelihood than Whites of being homeworkers. The Black ethnic minority group is the only group that has a lower OR for homeworking than the White group in both periods, though Indians also have a lower OR in 2011. As one might expect, the ORs for commuters with no LLTI are much greater than their counterparts with LLTI but the difference is reversed for homeworkers, although the differential is much less. Those with dependent children are less likely to be commuters and more likely to be homeworkers than those without dependent children. The occupation variable for commuting is redundant because only commuters (not the rest of the population in the sample) are categorised as either professional and managerial or non-professional and non-managerial; for homeworkers, the OR for non-professional and non-managerial is slightly above the reference category in 2001 and slightly below it in 2011. Finally, the North East appears to be the region with the lowest chance of being a commuter or a homeworker; the relative odds of being a commuter are highest for those living in the South East whereas the relative odds of homeworking are highest in the South West.

The BLRs show that the value of the constant for commuting decreased from 0.574 in 2001 to 0.493 in 2011, whilst the constant for homeworking increased slightly from 0.043 to 0.045. This means that whilst the overall likelihood of commuting decreased over the decade, the overall likelihood of homeworking increased when controlling for the variables in the two models.

The inclusion of a geographic variable in the modelling of both commuters and homeworkers indicates that when we take into consideration the other socio-demographic factors, there is significant variation between regions in both commuting and homeworking propensities. The rates of commuting in 2011 vary from 54.4 per cent of the population aged 16–74 in the North East to 60.7 per cent in the South East, highlighting the North–South divide in the UK (Martin, 1988), with economic activity rates generally being lower in northern England and Wales than in southern England (Anyadike-Danes, 2004), although comparison between the two census years indicates rates declining in the South East and East of England and increasing fastest in the North East. The same regional divide is apparent for homeworking but, in this case, the variation has increased between 2001 and 2011. We now turn our attention to exploring these spatial inequalities in more detail.

28.5 Spatial patterns of aggregate commuting and homeworking

In this section of the chapter, we maintain the division between those commuting to work and those working at or from home and compare the LAD-level patterns of commuting rates (Figure 28.2a), where commuters are represented as a percentage of the population aged 16–74, with homeworkers expressed as a percentage of total commuters aged 16–74 (Figure 28.2b). As one might expect, the spatial patterns evident in the two maps are rather different. In the former case, the geography reflects the core–periphery distribution observed at regional level, with highest rates for LADs in the South East such as Crawley, Rushmoor, Wandsworth and Bracknell Forest, all of which have rates over 66 per cent of their respective populations at risk and all of which are within London's hinterland. On the other hand, there are some relatively large and peripheral LADs, including Ceredigion, West Somerset and East Lindsey, where commuting involves less than half their usually resident populations aged 16–74. Nottingham and Rother LADs also have

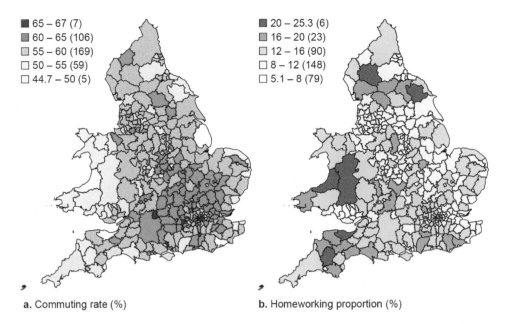

Legend (left map):
- ■ 65 – 67 (7)
- ■ 60 – 65 (106)
- ☐ 55 – 60 (169)
- ☐ 50 – 55 (59)
- ☐ 44.7 – 50 (5)

Legend (right map):
- ■ 20 – 25.3 (6)
- ■ 16 – 20 (23)
- ☐ 12 – 16 (90)
- ☐ 8 – 12 (148)
- ☐ 5.1 – 8 (79)

a. Commuting rate (%)

b. Homeworking proportion (%)

Figure 28.2 Commuting rates and homeworking proportions in 2011, by local authority district in England and Wales

Source: Derived from 2011 Census aggregate data.

commuting rates below 50 per cent. These low commuting rates may be due to high levels of economic inactivity or the presence of a large number of students in these LADs. In between these extremes, the rates are generally higher in more central parts of the country and lower in coastal and in more rural areas.

There is considerable spatial variation around the national mean of 10.3 per cent in the proportion of all commuters who work from or at home and the pattern is, in some respects, the inverse of the commuting rate shown in Figure 28.2a. The homeworking proportion is highest in rural West Somerset, where it reaches over 25 per cent, and is over 20 per cent in West Devon, Eden and Ryedale and in parts of Wales (Powys and Ceredigion). Rural areas have high homeworking rates because farmers work from home and the LADs in highest class are the extreme rural areas in which farming occupations are most important in relative terms. The pattern of homeworking differs from that of commuting in that the lowest rates of homeworking are not found in the South East and central England but in some of the urban LADs in the north of England such Hull, Knowsley and South Tyneside, as well as Blaenau Gwent in South Wales, all of which have homeworking rates of between 5 per cent and 6 per cent. The pattern more generally reflects urban–rural contrasts with higher rates in rural areas of Yorkshire and the Humber, the North East and the North West, Wales and the South West. The South West is the region with the highest homeworking rate of all regions at 12.5 per cent and London's rate at 9.5 per cent is below the national average.

Although the national commuting rate increased between 2001 and 2011, the pattern of percentage point change in commuting rates (Figure 28.3a) shows that 153 LADs, particularly in central and southern England, had lower commuting rates in 2011 than they did in 2001 and the greatest falls (over 3 per cent) were in Hart, Daventry, Test Valley, Wyre Forest and Malvern Hills. On the other hand, the largest increases in commuting rate were experienced in the London

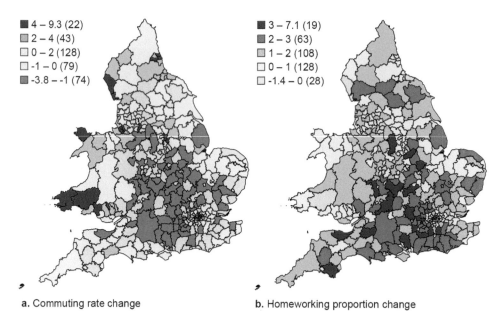

■ 4 – 9.3 (22)	■ 3 – 7.1 (19)
▨ 2 – 4 (43)	▨ 2 – 3 (63)
□ 0 – 2 (128)	▨ 1 – 2 (108)
□ -1 – 0 (79)	□ 0 – 1 (128)
▨ -3.8 – -1 (74)	□ -1.4 – 0 (28)

a. Commuting rate change   b. Homeworking proportion change

Figure 28.3 Changes in commuting and homeworking rates, 2001–11, by local authority district in England and Wales

Source: Derived from the 2001 and 2011 aggregate data.

boroughs of Tower Hamlets, Hackney and Newham, in the older industrial centres including Liverpool and Manchester in the North West, Gateshead and South Tyneside in the North East and Merthyr Tydfil in South Wales, as well as in some of the rural peripheral LADs in Wales and northern England.

In contrast to commuting change, many areas of central and southern England experienced increases in homeworking, most noticeably in West Somerset and Cotswold, whereas only 28 LADs scattered throughout the country saw their homeworking shares decrease. The negative relationship between the two change variables is illustrated in Figure 28.4 by the downward slope of the regression line, although the R^2 of 0.27 suggests that the correlation between the two variables is not strong.

Table 28.4 presents an alternative method of summarising the changes in commuting and homeworking rates by making use of the 2001 Census LAD classification for Great Britain[2] into the seven supergroups relevant for England and Wales, ranked on the basis of the changes in commuting rates between 2001 and 2011. The two supergroups containing the largest proportions of commuters and homeworkers appear at either end of the league table. LADs in 'London Cosmopolitan' had the largest increase in commuting rate whilst those in 'Prosperous England' experienced a decline overall in commuting rate but the largest increase in homeworking. 'Business and Education Centres', which include the major provincial towns and cities, and 'Mining Heritage and Manufacturing' LADs, some of the country's old industrial areas, have seen relatively large increases in commuting but relatively low increases in homeworking. LADs in the 'Coast and Heritage' and 'English and Welsh Countryside' supergroups have seen the largest increases in homeworking with relatively low rise in commuting, and LADs classified as 'Suburban Traits' have low increases in both commuting and homeworking.

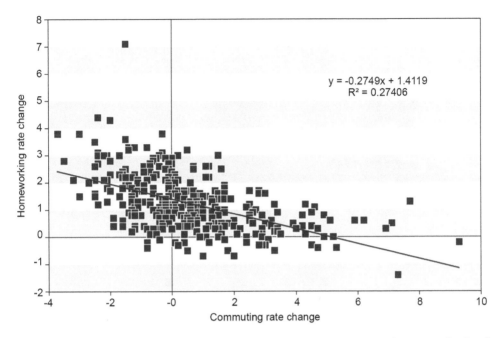

Figure 28.4 Relationship between changing commuting and homeworking rates by local authority district, 2001–11

Source: Derived from the 2001 and 2011 aggregate data.

Table 28.4 LAD commuting and homeworking rates for 2011 Census supergroups

2011 supergroup	Commuters (%)	Home-workers (%)	Commuting rate, 2011	Home-working rate, 2011	Change in commuting	Change in home-working
London Cosmopolitan	20.79	26.81	59.12	9.53	2.93	0.90
Business and Education Centres	8.08	7.55	55.15	8.03	1.76	0.69
Mining Heritage and Manufacturing	14.22	12.51	57.17	8.34	1.43	0.53
Coast and Heritage	12.14	9.51	56.01	10.88	0.72	1.24
Suburban Traits	6.11	6.44	60.03	8.85	0.32	0.68
English and Welsh Countryside	15.13	17.47	56.81	13.21	0.30	1.67
Prosperous England	23.52	19.71	61.22	12.51	-1.09	2.05

Source: Derived from the 2001 and 2011 aggregate data.

Figure 28.5 shows the proportions of commuters (excluding homeworkers) who travelled to work in 2011 using a selection of six different modes. Given the different percentages of individuals in each category, the class intervals of each map are unique but the distributions are mapped as quantiles, with the same shading colours used for each map and with roughly one-fifth of all the LADs in each quantile. Travel by car or van (Figure 28.5a) is by far the most popular means of transport in much of the country but not within London, where shares fall to below 10 per

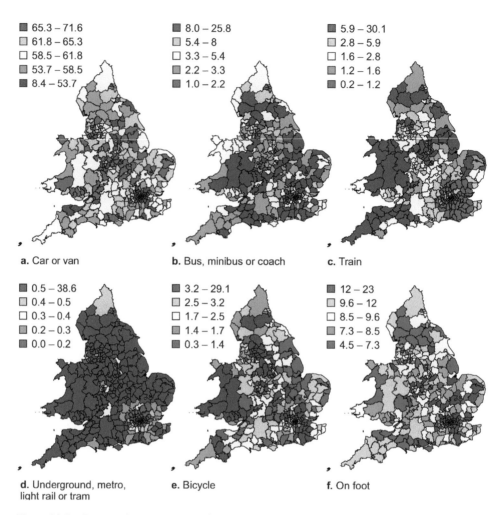

Figure 28.5 Commuting proportions by mode of transport for districts, 2011

Source: Derived from the 2011 aggregate data.

cent in City of London, Westminster, Camden and Islington in comparison with Midland LADs such as South Derbyshire, South Staffordshire and North West Leicestershire where seven out of ten people travel by car or van. Whilst the provincial cities tend to have relatively low commuting rates, many of the more rural areas of the country have relatively high shares of car drivers. In contrast, London and the provincial metropolitan areas have the highest shares of those using buses, minibuses or coaches (Figure 28.5b), whereas the more rural LADs have the lowest shares in the category, although the range of values is much narrower than for cars and vans. Thus, for example, a quarter of commuters use the bus in the London boroughs of Hackney and Southwark, whereas only 1 per cent use this mode in LADs such as Powys, North Dorset, Rutland and Cotswold. On average, 55 per cent used cars or vans to get to work in 2011 whereas 7 per cent used buses. A further 5 per cent travelled to work on the train but the pattern of train commuters (Figure 28.5c) is highly skewed towards Greater London and the South East, with London boroughs of Bromley and Lewisham having shares of 30 per cent and 27.5 per cent respectively, but almost 90 per cent of LADs in England and Wales have less than 10 per cent of their commuters using the train and over half having less than 2 per cent commuting by train. Inevitably, the pattern of preference for commuting by rail is linked to the availability of railway lines and stations and the dominance of London reflects this spatial variation in provision of infrastructure.

Even more spatially polarised is the pattern of commuting shares in 2011 by underground, light rail and tram (Figure 28.5d), where preference is inevitably confined to those districts containing transport systems of this type (London, Newcastle, Manchester, Liverpool, Sheffield, Birmingham), or some of their adjacent areas, given the availability of track and of park and ride facilities. In many of the boroughs of Inner London, this mode of transport commands at least a third of all commuters and 3.8 per cent of all commuters use this means of transport. Of the two 'green' options, those who walk to work represent nearly 10 per cent of all commuters whereas nearly 3 per cent use their bicycles. The maps in Figures 28.5e and 28.5f illustrate the two distributions and it is perhaps not unsurprising that Cambridge is the district where cycling is most popular as a means of getting to work, with 30 per cent of commuters choosing this mode. Figure 28.5e shows relatively high levels of cycling across many parts of eastern England where flat topography may well have an incentivising effect on preference. The nation's walking capital is Norwich, also in the East of England, where almost 23 per cent of commuters prefer to commute on foot, although walking to work is also popular in Exeter and Scarborough, where shares are over 20 per cent.

Changes that have taken place in modal split between 2001 and 2011 are illustrated by the maps in Figure 28.6 which, in this case, use the same class intervals and shading colours for capturing variation in the change in the percentage shares and are therefore directly comparable. A north–south division is evident in the changing share of car or van drivers, with increases in many parts of northern England and Wales, particularly in South Wales, where Blaenau Gwent, Rhondda Cynon Taf, Caerphilly and Neath Port Talbot all have increases above 5 percentage points. In contrast, car or van driving has become much less popular in London and the South East with falls of over 8 per cent in this mode's share of all commuters in the London boroughs of Hackney, Southwark, Richmond upon Thames and Lewisham, as well as in Brighton and Hove on the south coast. The increases in many parts of the 'north' may be due to increases in the cost of public transport, especially train travel (BBC, 2013b, 2015c), but may also be due to the reduction or withdrawal of bus services in many rural areas due to budget constraints and government attempts to reduce subsidies (Gray et al., 2006). Outside London, the large majority of LADs have a reduced share of people travelling to work by bus, minibus or coach (Figure 28.6b), most noticeable in some of the LADs in the large conurbations like Sheffield, Rotherham, Darlington and Leeds, whereas bus travel in London has increased in all boroughs apart from Hackney.

The changes in both car and bus travel in London are very likely to reflect the impact of the imposition of a cordon around central London and the levying of a congestion charge of £5 per vehicle in 2003, increasing to £10 in 2011, thereby encouraging commuters to seek alternative means of travel other than the car. Transport for London reports that, since the congestion charge was introduced, more than £1.2 billion of revenue has been reinvested in transport, including £960 million on improvements to the bus network and £260 million on roads and bridges, road safety, local transport and borough plans, sustainable transport and the environment.[3] These policy and investment changes are also responsible, at least in part, for the increase in the share of commuters in London by train (Figure 28.6c), underground (Figure 28.6d) and bicycle (Figure 28.6e). The proportion of cyclists in Hackney increased by 7.6 per cent and in Islington by 4.4 per cent in contrast to most of the rest of the country, where a declining share was typical. Walking to work, on the other hand, did not increase its modal share in London boroughs over

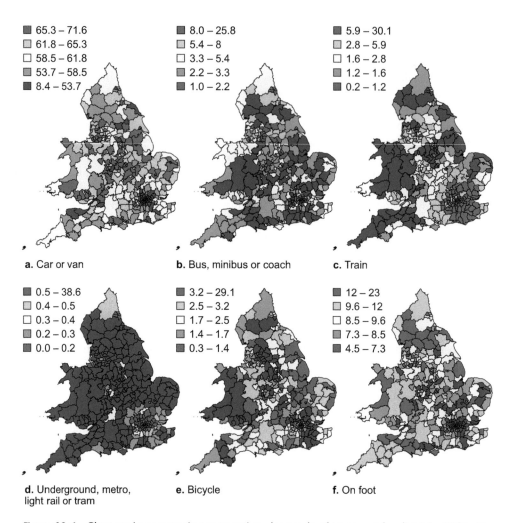

Figure 28.6 Changes in commuting proportions by mode of transport for districts, 2001–11
Source: Derived from the 2001 and 2011 aggregate data.

the period as it did in several other large cities like Bristol, York, Manchester, Liverpool, Newcastle and Sheffield and in big towns like Exeter, Lincoln, Southampton, Ipswich and Oxford.

28.6 Conclusions

This chapter set out to examine variations in commuting propensities and patterns and how they have changed, using aggregate and microdata from the 2001 and 2011 Censuses. The analyses assist in addressing the need for evidence-based policy recommendations. Such policy at the local level is increasingly important due to the devolution of some economic and transport policy to city regions, with the recent announcement of the devolution of a number of policy areas to Greater Manchester (BBC, 2015a) and to other combined local authorities (BBC, 2015b) and the implementation of the 'Northern Powerhouse' initiative by the UK Government (BBC, 2015d). These changes may in turn lead to a greater divergence of commuting patterns across the country in future as local and regional authorities become increasingly able to pursue policies suited to local needs and demands.

The chapter has shown that there have been increases in the number of commuters and commuting rates in England and Wales and every macro region, driven by increases in female participation and high levels of immigration as increasing numbers of women and immigrants are now part of the commuting population as well as the population at risk. Whilst these increases continue a trend observed by Nielsen and Hovgesen (2008) of increased commuting throughout much of England and Wales between 1991 and 2001, the census data also suggest that there has been convergence in commuting rates between males and females and between 'older' (65–74) and 'middle-aged' (25–64) commuters. However, at the same time there has been divergence between 'younger' and 'middle-aged' commuters, with a substantial (and perhaps worrying) fall in the commuting rate of the 16–24 age group.

At the LAD level, much of England and Wales has experienced substantial increases in commuting rates. As these trends show no sign of going into reverse, and as the total population is expected to continue to increase (ONS, 2011), those authorities responsible for the provision of transport networks should be considering substantial improvements and additions to existing transport networks. As the era of 'Predict and Provide' in relation to road transport has now passed and has fallen out of favour with policy makers, it is likely that any substantial improvements in capacity to meet demand will have to be made to the existing underground, metro, light rail, tram, train and bus networks. Probably the most well-known proposal is that for High Speed 2 (HS2), the planned high-speed railway linking London with the city centres of Birmingham (Phase 1, which is to begin in 2017 and complete in 2026) and then on to Leeds and Manchester (Phase 2) for completion by 2033. Focusing on improving, extending and building new public transport networks would be likely to have positive outcomes in terms of ridership.

The chapter has also shown us that the number of homeworkers and the prevalence of homeworking both increased between 2001 and 2011, that homeworking is more common amongst males, those aged 65–74, the Chinese ethnic group, those with a LLTI, those with dependent children and those in non-professional and non-managerial occupations. Spatially, homeworking is more common in rural areas and in central and southern England. The increased prevalence of homeworking potentially presents policy makers with both opportunities and problems. The rise of homeworking is a potential opportunity, as it creates an opportunity to tackle traffic congestion, and the associated economic, social and environmental problems, through the promotion of homeworking practices. However, the increased prevalence of homeworking is a potential problem as it may negatively affect the ridership and, therefore, financing of existing and new public transport networks.

Finally, the chapter has indicated that there has been a general increase in the percentage of people commuting to work by underground, metro, light rail and tram, train and on foot. In contrast, there has been a general decrease in the percentage of those commuting to work by bus, minibus or coach and as a passenger in a car. In addition to these national changes in modal split, the chapter has shown that continual infrastructure developments and changing preferences also lead to district variations in modal split due to spatial variations in transport networks and commuting preferences. Increasing fuel costs and environmental awareness may have led to many commuters choosing to travel to work by public transport rather than driving in order to reduce their commuting costs and the environmental impact of their commuting behaviour, and urban regeneration may have led to an increase in walking and cycling to work as increasing numbers of people live in city centres and inner-city areas close to their places of work.

However, many of the explanations for changes in the modal split are fundamentally linked to changes in urban structure, such as urban regeneration, suburbanisation and urban–rural migration. Therefore, if the relevant national, regional and local authorities seek to encourage the use of public transport it would be advisable for them to consider prioritising certain types of development, urban or otherwise, over others. For instance, the building of a new suburb or industrial estate far from a city centre, with little or no access to existing public transport networks, is unlikely to facilitate the use of public transport by those living or working there. Likewise, if authorities are aiming to increase walking and cycling to work it would be unwise for the same authorities to allow further residential or commercial developments on the rural–urban fringe; they should instead be increasing the density of urban development by prioritising high-density mixed-use developments in city centres and inner-city areas where walking or cycling to work is more likely.

We conclude that census data are of huge value in understanding patterns and trends in commuting at the LAD level, as documented in this chapter, but they also provide evidence of changes within our cities and large towns and how local labour market areas are evolving in the face of enormous pressures for continued residential and workplace development.

Acknowledgements

This work has been supported by PhD funding from the University of Leeds and research grant funding from the Economic and Social Research Council (ESRC) grant ES/J02337X/1: UK Data Service-Census Support).

Notes

1 Extracted using *InFuse* (UK Data Service-Census Support).
2 Details of ONS 2011 classifications are available at: www.ons.gov.uk/ons/guide-method/geography/products/area-classifications/ns-area-classifications/ns-2011-area-classifications/index.html.
3 https://tfl.gov.uk/modes/driving/congestion-charge/changes-to-the-congestion-charge.

References

Anyadike-Danes, M. (2004) The real North–South divide? Regional gradients in UK male non-employment. *Regional Studies*, 38(1): 85–95.
BBC (2013a) Petrol price 'may go up 4p' as retailers urge review. BBC News. Available at: www.bbc.co.uk/news/business-21208524.
BBC (2013b) Rail commuters hit by 4.2% average fare rise. BBC News. Available at: www.bbc.co.uk/news/uk-20881684.
BBC (2015a) Budget 2015: New powers announced for Greater Manchester. BBC News. Available at: www.bbc.co.uk/news/uk-england-manchester-33448965.
BBC (2015b) Elected mayors for north-east of England as devolution deal announced. BBC News. Available at: www.bbc.co.uk/news/uk-england-34609507.
BBC (2015c) Rail fares: Minister Patrick McLoughlin defends rises. BBC News. Available at: www.bbc.co.uk/news/uk-30652098.
BBC (2015d) Budget 2015: George Osborne sets out 'northern powerhouse' vision. BBC News. Available at: www.bbc.co.uk/news/uk-england-31942291.
Coombes, M., Raybould, S. and Wymer, C. (2005) *Travel to Work Areas and the 2001 Census: Initial Research*. Office for National Statistics, London.
Department for Transport (2015) National Travel Survey, England 2014. Statistical release, 2 September. Available at: www.gov.uk/government/uploads/system/uploads/attachment_data/file/457752/nts2014-01.pdf.
Department for Work and Pensions (2014) *Fuller Working Lives – A Framework for Action*. DWP, London. Available at: www.gov.uk/government/publications/fuller-working-lives-a-framework-for-action.

Gray, D., Shaw, J. and Farrington, J. (2006) Community transport, social capital and social exclusion in rural areas. *Area*, 38(1): 89–98.

Kapadia, D., Nazroo, J. and Clark, K. (2015) Have ethnic inequalities in the labour market persisted? In Jivraj, S. and Simpson, L. (eds.) *Ethnic Identity and Inequalities in Britain; The Dynamics of Diversity*. Policy Press, Bristol, pp. 161–179.

Lorenzoni, I., Nicholson-Cole, S. and Whitmarsh, L. (2007) Barriers perceived to engaging with climate change among the UK public and their policy implications. *Global Environmental Change*, 17(3–4): 445–459.

Martin, R. (1988) The political economy of Britain's North–South divide. *Transactions of the Institute of British Geographers*, 13(4): 389–418.

Nielsen, T.A.S. and Hovgesen, H.H. (2008) Exploratory mapping of commuter flows in England and Wales. *Journal of Transport Geography*, 16(2): 90–99.

Ogden, P.E. and Hall, R. (2000) Households, reurbanisation and the rise of living alone in the principal French cities, 1975–1990. *Urban Studies*, 37(2): 367–390.

ONS (2011) Summary: UK population projected to reach 70 million by mid-2027. Available at: www.ons.gov.uk/ons/dcp171780_240701.pdf.

ONS (2012a) 2011 Census: Population estimates for the United Kingdom. 27 March. Available at: www.ons.gov.uk/ons/dcp171778_292378.pdf.

ONS (2012b) Average age of retirement rises as people work longer. Available at: www.ons.gov.uk/ons/dcp29904_256641.pdf.

ONS (2013a) Index of services. April. Available at: www.ons.gov.uk/ons/dcp171778_315661.pdf.

ONS (2013b) Full report – women in the labour market. Available at: www.ons.gov.uk/ons/dcp171776_328352.pdf.

ONS (2015) *Overview of 2011 Travel To Work Areas*. ONS, Titchfield. Available at: www.ons.gov.uk/ons/guide-method/geography/beginner-s-guide/other/travel-to-work-areas/index.html.

Rae, A. (2011) Flow-data analysis with geographical information systems: A visual approach. *Environment and Planning B*, 38(5): 776–794.

Scheiner, J. and Kasper, B. (2003) Lifestyles, choice of housing location and daily mobility: The lifestyle approach in the context of spatial mobility and planning. *International Social Science Journal*, 55(176): 319–332.

Seo, J.K. (2002) Re-urbanisation in regeneration areas of Manchester and Glasgow: New residents and the problems of sustainability. *Cities*, 19(2): 113–121.

Analysing relative decline in cities with the British census

Mike Coombes and Tony Champion

29.1 Introduction

City decline is a potentially relevant policy issue in the UK because it has been identified in other early industrialising countries (e.g. Turok and Mykhnenko, 2007). British urban policy now emphasises 'agglomeration economies' in cities providing growth potential, but a longer perspective recalls the severe decline seen in some major British cities not so long ago (Martin et al., 2014). Research has shown city decline to be multi-dimensional with mutually reinforcing factors including de-industrialisation and population loss leading to increased poverty (e.g. Power et al., 2010).

Measurement of city decline can simply involve the absolute reduction in an indicator such as population or employment, but this does not address the widespread concern with the *relative decline* of cities which under-perform when compared to others. Given the understanding of city decline as multi-dimensional, there were a range of different possible emphases for the analyses below; in practice, the focus is placed upon aspects of relative decline likely to increase the risk of poverty, because that was the context for the research project drawn upon here (Pike et al., 2016).

The empirical work reported in this chapter highlights relative decline over a single decade between the 2001 and 2011 Censuses. Highlighting poverty risk puts the spotlight on cities with job shortfalls and this rules out the complete limitation of the analyses to census dates and thereby to census data because the last inter-censal decade blurs together a period of growth up to 2008 with the subsequent recession and the following period of public sector austerity policies. Yet clearly the census will be a key source, not only because of its reliability but also due to its back catalogue offering measures of longer-term dynamics, while its broad topic coverage supports an analysis of decline which is multi-faceted.

The next section tackles the crucial initial question which concerns the objects of study in these analyses of the cities of the UK, because it is necessary to be clear what definition of cities is adopted, and also why it was adopted. The third section examines demographic change, not only because of the focus on population loss in many international city decline studies, but also to understand the supply side of the labour market whose trends shape city poverty risk levels. The fourth section of the chapter summarises the analyses which synthesise key indicators within an index of relative decline, leading to some modelling of the index results. The final section of the chapter reflects on the role played by census data on stocks and flows – from both the most recent and some earlier censuses – in these analyses of relative decline among the cities of the UK.

29.2   City labour market areas

Assessing the scale and nature of decline in UK cities requires a decision on the areas to be analysed: which urban areas will be considered 'cities' for this analysis. The most widely used definition distinguishes cities from other urban areas by reference to urban population size, as with the Primary Urban Areas (PUAs) researched in *State of the English Cities* (Parkinson et al., 2006), and later paralleled by equivalent cities elsewhere in the UK (as in Champion, 2014). The 2001 Census was used to identify PUAs as urban areas with 125,000 residents or more, but for the analyses in this chapter a 100,000 threshold brings in a further ten cities to create 74 'PUA+s' that are deemed to be cities. The cities are grouped in some of the analyses that follow by either their size range or broad region of the UK. Figure 29.1 shows the spread of the cities across the UK and the size band to which each city belongs.

The definition of PUA+s is rooted in a functional conception of cities because it sets urban areas within their wider local labour markets. It is the labour market which shapes local levels of job opportunity and poverty risk. Official labour market areas in the UK are termed Travel-To-Work Areas (TTWAs), which are defined by analysing census commuting data (Coombes, 2010). These areas were fundamental in defining the PUA+s, but this highlights a problem for work relying on census flow datasets. Such datasets are always among the very last to be made available by the Office for National Statistics (ONS), so updating functional city definitions cannot be done until long after other census data has become available. The original PUAs were defined to support early analyses of stock data from the 2001 Census, so the definitions had to use the available TTWAs, which were still based on commuting behaviour as in 1991. This time lag was entirely due to the late release of flow data.

29.3   Population dynamics over decades

Population censuses are invaluable in providing the long-term context for city decline in Britain. This part of the chapter begins by tracing the evolution of city decline since 1901, then presents components of population change for the decade 2001–11. Long-term analyses face the problem of reporting areas – usually administrative areas – that change; here, data up to 1981 relies on 'apportioned' data from Mounsey (1982). For the three subsequent decades, the official mid-year estimates (MYE) are used: these estimates rely on census results, but seek to compensate for known census deficiencies such as under-enumeration. The 2011 MYEs also had to be apportioned because of the administrative area changes in 2009 in parts of England.

By dividing the period 1901–2011 into four periods, population decline appears to affect a steadily increasing number of cities in Britain. Just three cities had fewer residents in 1921 than in 1901, with the number declining rising to five in the 1921–51 period and eight over the next three decades. For the final 30-year period, the MYEs used here (which count students at term-time addresses from 1981 to 2011), show that there were ten declining cities. Yet with a decade-by-decade analysis decline can be seen to have passed its peak. This confirms recent work on 'urban renaissance' in the UK (Champion, 2015). Analysis by single decades (apart from the 1931–51 20-year period necessitated by no census being taken in 1941) shows city decline peaking in the 1970s when as many as 23 – almost one in three – cities experienced population shrinkage. The number declining fell to 22 in the 1980s and then 19 in the 1990s. Since then a

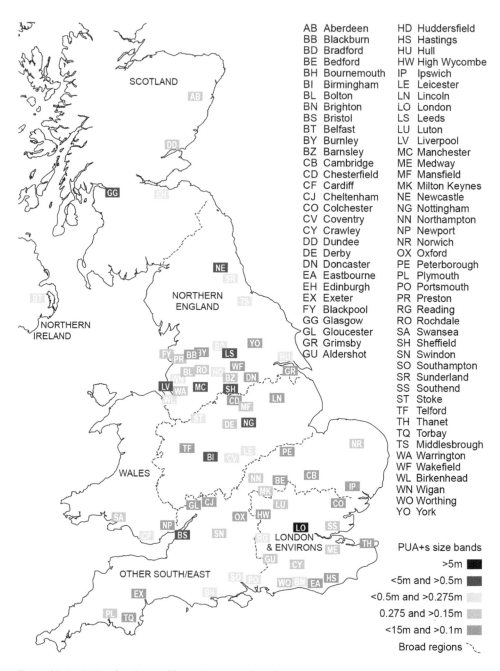

AB	Aberdeen	HD	Huddersfield
BB	Blackburn	HS	Hastings
BD	Bradford	HU	Hull
BE	Bedford	HW	High Wycombe
BH	Bournemouth	IP	Ipswich
BI	Birmingham	LE	Leicester
BL	Bolton	LN	Lincoln
BN	Brighton	LO	London
BS	Bristol	LS	Leeds
BT	Belfast	LU	Luton
BY	Burnley	LV	Liverpool
BZ	Barnsley	MC	Manchester
CB	Cambridge	ME	Medway
CD	Chesterfield	MF	Mansfield
CF	Cardiff	MK	Milton Keynes
CJ	Cheltenham	NE	Newcastle
CO	Colchester	NG	Nottingham
CV	Coventry	NN	Northampton
CY	Crawley	NP	Newport
DD	Dundee	NR	Norwich
DE	Derby	OX	Oxford
DN	Doncaster	PE	Peterborough
EA	Eastbourne	PL	Plymouth
EH	Edinburgh	PO	Portsmouth
EX	Exeter	PR	Preston
FY	Blackpool	RG	Reading
GG	Glasgow	RO	Rochdale
GL	Gloucester	SA	Swansea
GR	Grimsby	SH	Sheffield
GU	Aldershot	SN	Swindon
		SO	Southampton
		SR	Sunderland
		SS	Southend
		ST	Stoke
		TF	Telford
		TH	Thanet
		TQ	Torbay
		TS	Middlesbrough
		WA	Warrington
		WF	Wakefield
		WL	Birkenhead
		WN	Wigan
		WO	Worthing
		YO	York

PUA+s size bands

>5m

<5m and >0.5m

<0.5m and >0.275m

0.275 and >0.15m

<15m and >0.1m

Broad regions

Figure 29.1 PUA+s by size and broad regional location

Source: Adapted from Pike et al. (2016, Map 1, p. 8).

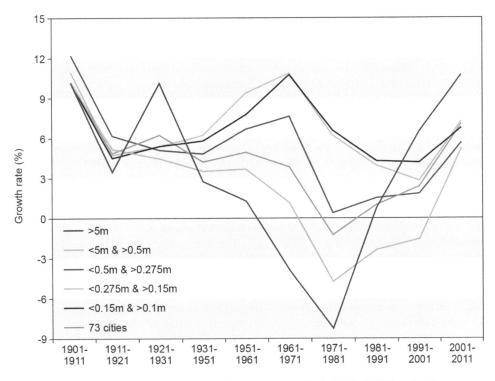

Figure 29.2   Population growth rates over inter-censal periods 1901–2011 for city size groups
Source: Based on mid-year population estimates available from ONS. Crown copyright data.

big change took place: only two cities had smaller populations in 2011 than in 2001, the same number as had seen decline in the first decade here (1901–11).

These trends can be readily seen in Figure 29.2 which shows the population growth rates across the inter-censal periods from 1901 for each of the five city size groups shown in Figure 29.1 (except for Belfast). Looking first at all 73 cities together, the average annual rate of growth halved between the first and second decades, then had stabilised at about 0.5 per cent before dropping into overall loss in the 1970s. Since then, it has climbed steadily, and the rate for 2001–11 was the highest since 1901–11. These first and last decades are also similar in the smallness of the variation between the city size groups, apart from London's rate being higher than for the other four in the latest decade. The city size group differences started widening in 1931–51 and reached their maximum extent in the 1970s, which became known as the 'counterurbanisation decade' due to a clear negative relationship then between growth rate and city size. London's dramatic recovery since then, returning to its 1901–11 rate for 2001–11, has led a broader urban renaissance, though the next largest city group did not see its recovery take off to a similar degree until 2001–11.

There are two main reasons for this patterning of population trends. One is variation over time in the country's overall demographic dynamism, with decadal growth plummeting from 5 per cent in the 1960s to under 1 per cent in the 1970s, but then rising again to 2.9 per cent in the 1990s and reaching 7.1 per cent in the decade to 2011, the highest rate for a century. This 'high tide' has lifted the growth rate of almost all cities. When taken together the 74 cities have sometimes exceeded and sometimes lagged the national growth rate, while individual cities have strongly differing trends. In aggregate the cities had grown at only half the UK rate in the

1970s, but this gap had narrowed in the 1990s and by 2001–11 the cities were growing slightly faster than the national average. As might well be expected, cities of northern England feature heavily among the lists of declining cities: Burnley and Sunderland were the only declining cities in 2001–11; Liverpool, Dundee, Glasgow, Birkenhead and Sunderland had the highest loss rate in 1991–2001, Glasgow, Liverpool, Dundee and Sheffield in 1981–91.

During the twentieth century, the differences between cities in their growth rates led to some substantial changes in British city size rankings. At the top of the scale, a big gap persistently separates London from all other cities, but over the century Manchester took second place from Birmingham. Of the other large cities Bristol rose four places to seventh. Most of the biggest upward shifts were by cities from the middle or lower ranks in 1901 (e.g. Southend climbed fully 45 places over the century). Burnley was the city declining fastest on this measure, ending the century 31 places lower than in 1901. This decline was nearly matched by Dundee and also Blackburn, with falls of 29 and 28 places respectively. Data on the cities' decline or growth by individual decades offers little evidence of their fortunes being reversed, suggesting a strong degree of path dependency (and consequently this indicator of change in city size ranking has been included in the analyses below of relative decline among the 74 cities over recent years).

Differential city performance over the decade 2001–11 can be better understood by examining the major components of demographic change. The main sources of UK data on components of population change are non-census, especially records of births, deaths and changes of address, although the 'other changes' element takes account of differences between the census-based estimates for 2011 and the population numbers rolled forward from the 2001 Census.

Table 29.1 shows the results for the 74 cities in aggregate and by the five size groups in Figure 29.1. The bottom line shows that international migration is clearly the prime driving force in the growth of urban Britain in 2001–11, adding 0.47 per cent a year to city population, although natural change is not much less influential at 0.35 per cent. Partially offsetting these growth drivers is the 0.16 per cent a year net loss of people from these cities to the rest of the UK, continuing the counterurbanisation relationship evident in the table's data for the size groups (i.e. the smaller the cities, the more positive are their within-UK migration rates). The opposite urbanisation relationship holds for the international migration component.

Of these two opposing migration forces, it is the within-UK component's counterurbanisation that is the more powerful in redistributing population between the city size groups.

Table 29.1 Components of population change 2001–11 for city size groups, percentage per year (compound rate)

City size groups (millions)	Overall change	Natural change	Within-UK migration	International migration	Other changes	Migration and other
>5	1.07	0.79	-0.74	1.01	0.02	0.28
<5 and >0.5	0.52	0.24	-0.16	0.39	0.05	0.28
<0.5 and >0.275	0.56	0.24	-0.02	0.31	0.03	0.32
<0.275 and >0.15	0.72	0.26	0.14	0.29	0.03	0.46
<0.15 and >0.1	0.68	0.19	0.17	0.27	0.05	0.49
All 74 cities	0.70	0.35	-0.16	0.47	0.03	0.35

Source: Mid-year population estimates available from ONS. Crown copyright data.

Note: 'Other changes' chiefly comprises undocumented migration estimated by comparing 2011 Census results with population estimates rolled forward from the 2001 Census.

Table 29.1 shows that although the net international migration rate does fall with city size, the difference between the 1.01 per cent of the largest city (London) and the smallest city size group's 0.27 per cent is only 0.74 per cent points and this is a smaller gap than the 0.91 percentage points difference between the internal migration rates of London and the smallest size group (-0.74 per cent as against +0.17 per cent). Thus the two components in combination fuel more growth in smaller cities than in large, and this net effect is reinforced by the undocumented migration driving the values in the final column of Table 29.1, where the overall migration effect is lowest for the two largest city size groups at 0.28 per cent a year, rising to 0.49 per cent for the smallest city size.

Table 29.1 also reveals the population dynamics of London (the sole 5 million+ city): when compared to the other city size groups it has very high rates not only of net international migration and natural change but also of net within-UK migration loss. Data for individual cities confirms the major role played by net international migration in the growth rate of cities: just two of the 74 (Birkenhead and Torbay) had a negative overseas migration balance in 2001–11. At the other extreme, this component gave London, Peterborough and Luton more than a 10 per cent boost to their populations and, partially due to the inflow of young migrants, the same three cities were also among those with the highest natural change rates, while also seeing the highest rates of loss from within-UK migration over the decade. These dynamic cities' labour supply growth sets them apart from conventional city decline, although without equal growth in jobs they can face increased poverty risk. In the case of London, the strong net gain of young adults and net out-migration of older people reflects the 'escalator' pattern identified by Fielding (1992) whereby young people move to the UK's largest concentration of jobs to more rapidly advance their careers. Later work on linked census data showed that many young migrants to London are highly skilled, constituting a within-UK 'brain drain' (Champion et al., 2007).

29.4 Index of relative decline

The selectivity of migration flows sees less prosperous cities experiencing net losses of the more high-skilled workforce that is a key to sustained economic growth (OECD, 2009). Yet in some cities, in some periods, growth from migration and/or high fertility rates may outstrip economic growth so that poverty risk rises there too. These diverse city trends mean that any one indicator will fail to identify the type of relative decline that has occurred in some cities. The solution is to develop an index of relative decline that synthesises the information from several diagnostic indicators. As with the Index of Multiple Deprivation (McLennan et al., 2011), the index 'scores' then represent a multi-dimensional analysis of each place relative to its comparators. The indicators are rendered as rankings, with the final score for each city produced by adding its rank score on each indicator. The range of issues represented by the index militates against the census being the only data source required.

Robust measures of employment rates – a key indicator of poverty risk for the European Commission (Melo et al., 2016) – rely upon data from the census. Census datasets are also used in measuring other key factors in relative decline such as internal migration. The basic trend in job availability is crucial too, with an essential separation of pre-recession years from the later period. Hence the following indicators were included in the index, with census data used for the first of these:

- percentage point change in the level of employment rate 2001–11;
- percentage change in full-time-equivalent jobs at local workplaces 1998–2008; and also
- percentage change in full-time-equivalent jobs at local workplaces 2009–12.

Table 29.2 City scores on the index of relative decline, by broad region

Broad region	Score on the index of relative decline			
	High	Moderate	Low	% High
London and environs	0	6	8	0.0
Other South/East	0	11	7	0.0
Midlands	2	6	2	20.0
Northern England	14	10	0	58.3
Scotland/Wales/N. Ireland	4	1	3	50.0
All UK PUA+s	20	34	20	27.0

Source: Authors' own calculations.

Turning to population change as the most readily recognised measure of decline, it was argued above that the relevant demographic dynamics are not limited to the most recent decade but extend to relative change over a much longer period. There is also the key process of net migration of young adults, itself potentially linked to the proportion of the economically active age group with higher skills. As a result, four census-based indicators are added to the index:

- percentage change in total population 2001–11;
- change in city size rank position 1901–2001;
- estimated rate of net in-migration 2001–11 of those aged 15–19 at the start of the decade; and
- percentage point change in the share of those aged 16–64 with a degree or higher qualification 2001–11.

The main focus for the research is on cities with high scores on the relative decline index and so interest centres on the 'top 20' and the contrast between them and the 20 cities at the other end of the ranking (which leaves 34 mid-range PUA+s). Table 29.2 distributes these three categories of cities across the broad UK regions shown in Figure 29.1. There is a dramatic north–south contrast evident, with not one of the 32 cities from the southern regions of England among the 'top 20' on the relative decline index, whereas northern England includes no city in the 'bottom 20' (i.e. cities with the least evidence of decline). The two other broad regions of the UK – England's Midlands, and the three devolved territories combined, include cities with both high and low index values. Among the list of 'top 20' cities on this relative decline index is at least one city from each of the four countries of the UK.

29.5   Factors behind relative decline

The index is a new measure, but the north–south contrast in its values is very familiar. Exploring the factors associated with relative decline calls for an econometric model which draws on relevant literature (e.g. Duranton and Puga, 2013). The model assesses various structural differences between cities at, or before, the start of the 2001–11 period that are central to the relative decline index. Table 29.3 lists the model variables and shows their sources, thereby identifying those using data from a census.

The emphasis upon employment rates is here given a longer-term perspective by the model including 1981 employment rates. One of the leading indicators in much economic modelling is

Table 29.3 Independent variables for models of the relative decline index

Variable	Source
1931 mining/manufacturing	Population Census (derived from www.visionofbritain.org.uk)
1981 employment rates	Population Census
% public services	Business Register and Employment Survey
% EAA with degrees	Population Census
Workforce size (logged)	Population Census
% out-commuting	Population Census
Firms/workforce ratio	VAT Registrations; Population Census
1961 over-shadowing	Population Census, Census of Distribution (as in Carruthers, 1967)
GVA/head	Regional Gross Value Added (Income Approach)
1991–2001 net in-migration	Mid-year Population Estimates
London train time	Shortest time departing before 0900 (as in Parkinson et al., 2006)

Note: If a date is not included in a variable name, the measure relates to 2001.

GVA/head which is seen as the best available local productivity measure. Another factor widely cited (e.g. Martin et al., 2014) is the skill level of people in the economically active ages (EAA): percentage EAA with degrees is the variable in the models here. The entrepreneurship issue was covered by a firms/workforce ratio whose denominator uses census workplace data, while the role of the public sector is approximated by a percentage of public services measure using employment sector data. A very much longer-term influence is tested by the 1931 mining/manufacturing variable which measures historic dependence upon key sectors subsequently affected by de-industrialisation.

A geographical influence that has attracted much attention recently is agglomeration (OECD, 2009), with the key measure workforce size (logged here, as is usual). London has a very dominant role in the UK economy and so a London train time measure is used to reflect relative peripherality from the capital. Dominant cities at the regional scale attract higher-order services that are increasingly concentrated in fewer locations, so cities near to a larger neighbour have lost amenities in a process already seen by Carruthers (1967), whose measure is the basis for the 1961 over-shadowing measure (the one binary variable in the models). Interaction between cities is increasing as an aspect of globalisation (Martinez-Fernandez et al., 2012), and here the percentage out-commuting variable reflects one form of integration with surrounding areas. The final potential predictor variable is the estimated rate of 1991–2001 net in-migration of the EAA.

The regression results place the greatest emphasis on the influence of the variable on high-level skills, and this chimes with recent research (e.g. Martin et al., 2014). In marked contrast, two other variables identified by the modelling are the rather unfamiliar ones of London train time and 1961 over-shadowing. The variable representing agglomeration effects, workforce size, only highlights the labour market size of the city itself, whereas these variables in the model highlight the national and regional context of cities. One modelling variant also found the 1931 mining/manufacturing variable to have a significant influence, reflecting the long-term industrial legacy of many relatively declining cities in the UK.

29.6 Conclusions

The analyses reported here illustrate the essential contribution of census data to both the measurement and the greater understanding of the different trajectories of cities in the UK. Datasets on population stocks and flows shed some light on the processes of change shaping levels of poverty risk in cities. In fact few, if any, cities are seeing absolute population loss, but growth is slower in most northern cities, while net migration by young adults in particular centres on flows to the capital. Migration flows tend to be towards places with more job opportunities and although this potentially reduces labour market imbalances, in practice regional imbalances tend to be sustained due to prosperous places attracting the more highly skilled people from other cities.

Census datasets also played a part in the models here exploring the results of the index of relative decline. These analyses suggest that the risk of a UK city experiencing relative decline recently was lower if the city had:

- more highly qualified people in its working age group;
- no over-shadowing larger city nearby;
- faster access to London; as well as (less certainly)
- little history of dependence for work on mining or manufacturing.

The issue of over-shadowing relates to contrasting rates of growth in neighbouring cities, which is reflected in commuting trends and stronger inter-dependencies between adjacent cities. Growth of net in-commuting increases the pressures on locally available jobs, and those requiring higher skills – usually better paid jobs – are the most likely to attract longer-distance commuters further from the city (Champion et al., 2009).

This process can be seen in changing commuting patterns, as with neighbouring cities Preston and Blackburn (Lancashire) and also Leeds and Bradford (West Yorkshire). Hall et al. (2001) showed that while Leeds and Preston became service centres when better-paid jobs replaced those in old industries, their slightly smaller neighbours struggled to make the transition from their old industrial bases. The 1921 Census gives the earliest commuting data, revealing a flow from Leeds to Bradford of 1,635 which was larger than that in the opposite direction (847). Similarly the flow from service centre Preston to manufacturing-based Blackburn outnumbered that in the opposite direction. Subsequently the decline in the manufacturing centres meant that even by 1951 flows from Leeds to Bradford and from Preston to Blackburn had fallen but those from Bradford and Blackburn to their more successful neighbours had nearly doubled. The 2011 Census then shows flows to strongly growing Leeds and Preston from their over-shadowed neighbour cities becoming roughly twice as high as flows in the opposite direction, massively reversing the position seen nearly a century earlier.

The analyses of relative city decline linked to the risk of poverty have found that flow and stock data derived from the population census are necessary but not sufficient for a full picture of trends and their drivers. One aspect of decline for which the census is invaluable is the longer-term context to the present. Examples here have included tracing the trajectory of city populations through the twentieth century, and comparing the relative attractiveness of cities to commuters from as far back as 1921. Another is that the census, either directly or indirectly (as in enabling the revision of annual population estimates for each decade), provided the majority of the variables used in these analyses: five of the seven indicators used in the relative decline index, as well as eight of the 11 independent variables in models of the index scores.

At the same time, census datasets had to be supplemented by other data to capture the multi-faceted nature of city decline. This was most notably the case in separating out job growth rates for before and after the 2008–09 recession and also allowing the inclusion of an income-related variable when the UK census, unlike censuses in some other countries, does not collect data on this critical concern for any study of poverty risk. Other data sources also had to be used to break population change down into its demographic components, because this requires a continuous record of births, deaths and migration which the census does not provide.

Looking finally to the future, it is encouraging that a decision has been made to collect similar census data in 2021, although after this the proposed generation of 'census-like' information from administrative and survey sources raises concerns over flow data in particular, because there is no evidence that robust small area commuting and migration datasets can be obtained from any source other than a census.

Acknowledgements

The research drawn upon in this chapter (Pike et al., 2016) was commissioned in 2014 by the Joseph Rowntree Foundation, whose funding is acknowledged. The study was completed by the authors in conjunction with colleagues Andy Pike, Danny MacKinnon, David Bradley, Liz Robson and also Colin Wymer (who drew Figure 29.1), together with Andy Cumbers (Glasgow University). Census datasets are Crown copyright: the numerous datasets from Censuses 1901–2011 have been accessed from several different sources.

References

Carruthers, W. (1967) Major shopping centres in England and Wales, 1961. *Regional Studies*, 1: 65–81.

Champion, T. (2014) Population in cities: The numbers. *Foresight Future of Cities Working Paper*, Government Office for Science, London.

Champion, T. (2015) Urban population – can recovery last? *Town & Country Planning*, 84: 338–343.

Champion, T., Coombes, M. and Brown, D. (2009) Migration and longer-distance commuting in rural England. *Regional Studies*, 43: 1245–1259.

Champion, T., Coombes, M., Raybould, S. and Wymer, C. (2007) *Migration and Socioeconomic Change: A 2001 Census Analysis of Britain's Larger Cities*. Policy Press, Bristol.

Coombes, M. (2010) Defining labour market areas by analysing commuting data: Innovative methods in the 2007 review of Travel-to-Work Areas. In Stillwell, J., Duke-Williams, O. and Dennett, A. (eds.) *Technologies for Migration and Commuting Analysis: Spatial Interaction Data Applications*. IGI Global, Hershey, PA (USA), pp. 227–241.

Duranton, G. and Puga, D. (2013) The growth of cities. *DP9590, Centre for Economic Policy Research*, London School for Economics, London.

Fielding, A. (1992) Migration and social mobility: South East England as an escalator region. *Regional Studies*, 26: 1–15.

Hall, P., Marshall, S. and Lowe, M. (2001) The changing urban hierarchy in England and Wales. *Regional Studies*, 35: 775–806.

Martin, R., Gardiner, B. and Tyler, P. (2014) The evolving economic performance of UK cities: City growth patterns 1981–2011. *Foresight Future of Cities Working Paper*, Government Office for Science, London.

Martinez-Fernandez, C., Audirac, I., Fol, S. and Cunningham-Sabot, E. (2012) Shrinking cities: Urban challenges of globalization. *International Journal of Urban and Regional Research*, 36: 213–225.

McLennan, D., Barnes, H., Noble, M., Davies, J., Garrett, E. and Dibben, C. (2011) *The English Indices of Deprivation*. Communities and Local Government, London.

Melo, P., Copus, A. and Coombes, M. (2016) Modelling small area at-risk-of-poverty rates for the UK using the World Bank methodology and the EU-SILC. *Applied Spatial Analysis and Policy*, 9(1): 97–117.

Mounsey, H. (1982) The cartography of time-changing phenomena: The animated map. PhD Thesis, Durham University, Durham. Available at: http://etheses.dur.ac.uk/10276/.

OECD (2009) *How Regions Grow: Trends and Analysis*. OECD Publishing, Paris.

Parkinson, M., Champion, T., Evans, R., Simmie, J., Turok, I., Cookston, M., Katz, B., Park, A., Berube, A., Coombes, M., Dorling, D., Glass, N., Hutchins, M., Kearns, A., Martin, R. and Wood, P. (2006) *State of the English Cities*. Office of the Deputy Prime Minister, London.

Pike, A., MacKinnon, D., Coombes, M., Champion, T., Bradley, D., Cumbers, A., Robson, L. and Wymer, C. (2016) *Uneven Growth: Tackling Declining Cities*. Joseph Rowntree Foundation, York.

Power, A., Plöger, J. and Winkler, A. (2010) *Phoenix Cities: The Fall and Rise of Great Industrial Cities*. Policy Press, Bristol.

Turok, I. and Mykhnenko, V. (2007) The trajectories of European cities, 1960–2005. *Cities*, 24: 165–182.

The changing geography of deprivation in Great Britain

Exploiting small area census data, 1971 to 2011

Paul Norman and Fran Darlington-Pollock

30.1 Introduction

Two contexts for this chapter are provided within the edited volumes, *A Census User's Handbook* (Rhind, 1983) and *The Census Data System* (Rees et al., 2002). First, the chapter by Norris and Mounsey (1983) on analysing change over time informs us that the long time series of censuses is potentially valuable for studying change over time but this is hampered by various problems. These problems include: changes to the questions asked and of the definitions used; changes in the variable combinations used in area output tables; and changes in the boundary systems used for which census data are disseminated, most particularly at small area level. Second, the chapter by Senior (2002) on 'Deprivation indicators' details the derivation of composite measures which combine information from separate census variables, each of which is thought to indicate a dimension of deprivation.

This chapter will first outline some background to deprivation measures including the variety of uses of deprivation indices and why measuring change in deprivation over time is useful in research and policy-related applications. Then the steps to calculate a time comparable deprivation change measure at small area level in Great Britain (GB) for the five decennial censuses from 1971 to 2011 will be documented. The outputs of this process will then be used to show how deprivation has changed over this time period in selected case study districts. Finally, the future of deprivation measures will be considered in the light of changes recommended by the National Statistician regarding the 2021 Census and for the increased usage of administrative data sources (ONS, 2014).

30.2 Measuring deprivation

Inequalities in society may be at an individual level and/or have spatial elements. Social inequalities refer to how people, by characteristics such as age, gender, social class and ethnicity, may have different ease of access to social goods and services, including labour force participation and income, housing tenure, education, health care and political representation. Spatial inequality between different locations may relate to aggregate measures of the unequal distribution of income, resources and environmental amenities (Stillwell et al., 2010; Mitchell et al., 2015).

Deprivation is generally taken to be a state of demonstrable disadvantage relative to the local community, wider society or the nation to which an individual, family or group belongs (Townsend, 1987). Deprivation covers a wide range of facets since people may be deprived of good education and employment, adequate housing and good health as well as sufficient income (Dorling, 1999). Multiple deprivation at area level reflects concentrations of these aspects (Holman, 1978) and is the basis for work using the 1971 Census by Holtermann (1975) on urban deprivation. Since then, Senior (2002, p. 124) notes a "remarkable, and somewhat bewildering, growth in the use of deprivation measures". These deprivation measures have been influential for the allocation of public resources (Simpson, 1996; Brennan et al., 1999; Blackman, 2006) and are regularly used in relation to outcomes such as health (Senior et al., 2000; Boyle et al., 2002; Norman et al., 2005), educational achievement (Higgs et al., 1997) and crime (Kawachi et al., 1999).

The census-based deprivation measures are a composite single figure index which summarises information from several indicator variables, at small area level. A variety of indicators are used but ubiquitous is unemployment (Haynes et al., 1996) and commonly used are variables about household tenure, household overcrowding, access to a car and social class. Some schemes also use indicators such as quality of housing amenities (e.g. presence of central heating) and household structure (lone parents). A comprehensive list is provided by Senior (2002, Table 9.2, pp. 127–128) containing various composite indices which include those developed within the UK by Jarman (1983), Townsend (1987) and Carstairs (Carstairs and Morris, 1989). Recently in the UK, there has been a move away from census-based indices to the use of administrative data for the Indices of Multiple Deprivation (IMD) (Noble et al., 2006). Underpinned by census data but using other sources where appropriate, deprivation indices have also been developed in Australia, Canada, France, New Zealand, the US and elsewhere (Eroğlu, 2007; Bell et al., 2007; Havard et al., 2008; Pornet et al., 2012; Fu et al., 2015; Norman et al., 2016).

In the main, the UK census-based schemes have been at electoral ward scale and the more recent IMDs for statistical geographies (e.g. lower super output areas (LSOAs) in England, see DCLG, 2015). Although there is a lack of consensus on some of the technical aspects including geographical scale (Carr-Hill and Rice, 1995; Mackenzie et al., 1998), census-based index construction is well understood and transparent to practitioners, though this may not be the case for the IMDs because of their multiple inputs and complex methodologies (Adams and White, 2006; Morelli and Seaman, 2007). Despite differences in the detail of the construction of different indices, strong correlation between the schemes has invariably been found (Morris and Carstairs, 1991; Mackenzie et al., 1998; Hoare, 2003; Ajebon and Norman, 2016; D'Silva and Norman, 2015).

Deprivation indices are calculated cross-sectionally using data for a census year (or for administrative data from a quarter or for a calendar year) and thereby show the level of deprivation of an area relative to other areas and/or at national level (Fu et al., 2015). This means that a scheme is time point specific and this includes aspects such as the variable definitions and the geography used. Since deprivation measures are cross-sectional, this means they cannot directly be used to assess change over time and this is something of a drawback. Where changing deprivation has been calculated (Norman 2010a; Exeter et al., 2011; Norman et al., 2016), whether areas become more or less deprived over time has been shown to relate to the impact on population size of net migration (Norman, 2010b; Norman et al., 2016), to differences in cancer incidence and survival (Basta et al., 2014; Blakey et al., 2014; McNally et al., 2012, 2014a, 2014b, 2015) and to inequities in environmental issues (Mitchell and Norman, 2012; Mitchell et al., 2015).

In the main, as areas become less deprived, good health and other benefits accrue and *vice versa*. If change in specific places can be tracked then the effect of a policy such as regeneration can be assessed for its success.

30.3   Calculating changing deprivation

Specifying the deprivation measure

The specification of the measure of changing deprivation reported here is as follows:

- timeframe: 1971, 1981, 1991, 2001 and 2011 Censuses;
- geography: GB by the 2011 definitions of LSOAs in England and Wales and data zones (DZs) in Scotland;
- deprivation indicator variables: the inputs for the Townsend scheme:
  - unemployed as a percentage of persons economically active and seeking work;
  - non-home owners as a percentage of all households;
  - no access to a car or van as a percentage of all households; and
  - overcrowded households (more persons than rooms) as a percentage of all households.

The timeframe here dates from 1971, the first census for which a wide range of questions was asked and for which detailed outputs were available in tables for 100 per cent of the population (Denham and Rhind, 1983). The 2011 LSOAs and DZs are used so that the deprivation data are relevant to a contemporary setting. The coverage is GB, since a UK-wide index that also includes Northern Ireland is challenging because data were only available there in grid cell format prior to 1991 (Rees, 1995) (though see ongoing work by Lloyd, 2016).

The deprivation measure here is based on the Townsend index. This is not to claim the Townsend index is necessarily 'right' as a measure but the scheme has had widespread use in academic work and public health reports (Higgs et al., 1998). The Townsend index has just the four input variables listed above that are available in 100 per cent population tables from all censuses since 1971. 'Low social class', as used in the Carstairs index, would be of interest but is only available for a 10 per cent sample in 1981 and 1991. Further, this dimension is inconsistently defined across the 1971 to 2011 Censuses. In terms of the warnings by Norris and Mounsey (1983) regarding the analysis of change, the input variables for the Townsend index are close enough in their definitions to be comparable over time.

Adjusting for changing geographical boundaries

As noted above, analysing change over time is hampered by changes in the boundary systems used for the dissemination of census data, particularly at small area level. Unless the same areal extent is being compared, even if places have the same name (Norman and Riva, 2012), it is not possible to determine whether the change in a social indicator is real or due to a boundary change (Norman et al., 2003). The ward boundaries commonly used in deprivation schemes are regularly revised as part of electoral processes and are the geography of representation within local government (Norman et al., 2007). The smallest geographies of census data release (known as Enumeration Districts or Output Areas at different censuses, variously in England, Wales and Scotland) have their boundaries defined in ways which ensure the areas are as small as possible for geographically focused studies but have sufficient population and households to ensure personal data cannot be revealed (Martin, 1998; Cockings et al., 2013).

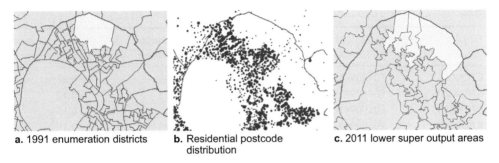

a. 1991 enumeration districts   **b.** Residential postcode   **c.** 2011 lower super output areas
distribution

Figure 30.1   Establishing conversions between census geographies: Birmingham, 1991 to 2011

Source: Boundary data from ONS and postcode data from Royal Mail.

If two boundary systems are different, the size of the area of overlap can readily be calculated using Geographical Information Systems (GIS) software. Since the population is not evenly distributed across space, this method is inappropriate as a means to apportion population counts between the zonal systems. To address this, several approaches to data conversion between geographies are used, including areal interpolation and dasymetric mapping (Gregory, 2002; Mennis and Hultgren 2006). The use of postcode (centroid) point distributions within areas is a proxy to represent both uneven population distribution (to inform the interpolation of a value across space) and also where people do and do not live (the principle of dasymetrics) (Figure 30.1b). This is the approach underpinning *GeoConvert*, an online tool for users to convert their own data (Simpson and Yu, 2003). Since the conversions needed for the work here are not available in *GeoConvert*, the process detailed by Norman et al. (2003) and then implemented for converting deprivation indicators at ward level (Norman, 2010a) is used. For specifics about the census area terminologies in different years and data conversions, see Norman and Riva (2012).

To explain the operationalisation of the conversion process, Figure 30.1a shows the 1991 enumeration district (ED) boundaries for part of Birmingham. This is the 'source' geography for which the data exist, with an ED '07CNL05' highlighted in yellow. Figure 30.1b illustrates the residential postcode distribution for that year, which indicates that this ED is sparsely populated. Only the outer parts of the ED away from the Birmingham boundary (in red) are populated. The different sized symbols relate to the number of addresses at each postcode. Figure 30.1c shows the 2011 LSOA boundaries. This is the 'target' geography for which the data are needed: the LSOAs which are overlapped by the highlighted ED are also highlighted. The postcodes can be associated using GIS functionality to both the 1991 source and 2011 target geographies. An estimate of the proportion of the population in the intersection between each source and target unit can be derived using the linked postcodes and the address counts.

Table 30.1 reports the number of addresses which fall in the intersection between ED 07CNL05 and four 2011 LSOAs. Since there are 223 addresses in the ED in total, apportionment weights are calculated as the intersection divided by the total. There were 45 persons counted as unemployed in the ED in 1991 and applying the weights to this apportions the unemployed to the 2011 geography. Since the 2011 LSOAs which are highlighted also overlap EDs other than 07CNL05, the counts of unemployed in these other EDs will also be apportioned accordingly. This approach ensures that the process is 'exhaustive'. No data are lost in the process so that, across all areas, the sum of the original data is preserved (Simpson, 2002).

Linkages have been made, equivalent to those illustrated in Figure 30.1, between the smallest geographies of the 1971, 1981, 1991 and 2001 censuses (EDs and output areas (OAs) as appropriate by country and year) and the 2011 LSOAs/DZs. All numerators and denominators needed as

Table 30.1 Converting data between geographies: Birmingham, 1991 to 2011

ED 1991	LSOA 2011	Addresses in intersection	Addresses in source	Apportionment weight
07CNGL05	E01009415	126	223	0.57
07CNGL05	E01009416	27	223	0.12
07CNGL05	E01009418	30	223	0.13
07CNGL05	E01009423	40	223	0.18

ED 1991	Unemployed		LSOA 2011	Unemployed
07CNGL05	45		E01009415	25.43
			E01009416	5.45
			E01009418	6.05
			E01009423	8.07

inputs for the Townsend index have been accessed from *Casweb* (for the 1971 to 2001 Censuses) and converted to the 2011 geography (as detailed above in Table 30.1). Simpson (2002) terms this a 'geographical conversion table'. The 2011 Census data have been downloaded from *Nomis*.

Calculating comparable deprivation over time

The steps to derive Townsend scores, as discussed in Senior (2002), for a specific census year, are:

1 Calculate percentages of unemployment, non-home ownership, no car access and household overcrowding and then log transform the unemployment and overcrowding variables to (near) normal distributions.
2 Standardise the four inputs to z-scores.
3 Sum the z-scores to be the deprivation score.

To make the index comparable over time, the calculation of the z-scores varies from the approach used if the index is cross-sectional. In a cross-sectional measure, a z-score is calculated as the observation value for an area minus the mean across all areas divided by the standard deviation. The resulting z-score across all areas then has a mean of zero and a standard deviation of one. Thus, if a z-score for a specific area is positive, this area has a higher value than average and if the value is negative, it has a lower than average value. In a deprivation index, the sum of the z-scores can be interpreted as positive index scores representing more deprived areas and negative index scores less deprived.

In Table 30.2a, the (hypothetical) values of percentage unemployment for the area reduce progressively over time. However, because the mean across all areas changes differently, the relative position as indicated by the z-score does not reflect that this area's unemployment is improving. In Table 30.2b, the mean and standard deviation across all areas and years are calculated and used for the calculation of the z-scores. In this way, the gradual change from above to below average unemployment has the change from a positive to a negative z-score. To identify changing small area deprivation, this approach has been utilised for the UK, Scotland and Australia (Norman, 2010a; Exeter et al., 2011; Norman et al., 2016).

It is useful to categorise deprivation scores into quintiles. So that quintiles are comparable over time, population-weighted quintiles have been calculated here by ranking all areas by their level of deprivation across all years and then dividing the rank order into five categories of equal

Table 30.2 Cross-sectional and time comparable z-scores

a. Cross-sectional	1971	1981	1991	2001	2011
Area unemployment	8	7	6	5	4
Mean: all areas	3.6	9.6	9.5	3.2	4.5
Standard deviation	2.3	6.2	6.6	2.1	2.5
z-score	1.91	-0.42	-0.53	0.86	-0.20

b. Time comparable	1971	1981	1991	2001	2011
Area unemployment	8	7	6	5	4
Mean: all areas and time			6.08		
Standard deviation			3.94		
z-score	0.49	0.23	-0.02	-0.27	-0.53

population size. Thus, from one time point to the next, if an area's score increased or decreased, the relative level of deprivation worsened or improved. Similarly, if an area changed quintile, the change can be interpreted accordingly.

30.4   Investigating changing deprivation

After a summary of change across GB for 1971 to 2011, various locations will have their changing deprivation circumstances outlined. These case studies are local government areas and their constituent LSOAs. Each has been selected for its distinctive circumstances to thereby represent elements of urban, rural, historic, new town and multi-ethnic situations. These are just examples and any local government areas in GB can have their deprivation circumstances presented in the same way.

Figure 30.2a illustrates the mean level of deprivation across the 41,729 LSOAs and DZs from 1971 to 2011. This shows a similar level in both 1971 and 1981 with the mean above zero; GB was on average more deprived than in 1991. By 2001, deprivation reduced further, but in 2011 had increased a little. Figure 30.2b shows that percentage non-home ownership and no car reduced over time but the reduction is curtailed between 2001 and 2011. Household overcrowding reduces from 1971 to 1991 and has stayed at low levels since. It is the unemployed percentage which displays a different pattern. From a relatively low level, unemployment rises between

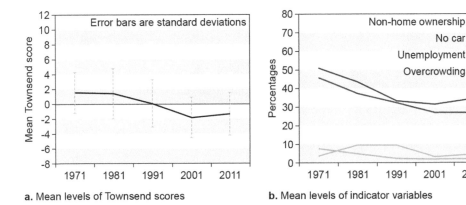

a. Mean levels of Townsend scores   b. Mean levels of indicator variables

Figure 30.2   Changing deprivation and indicator variables: GB, 1971 to 2011

1971 and 1981, and is high in 1991. There is a reduction to 2001 but a small rise by 2011. Given fluctuations in the economy, unemployment is likely to be more volatile than the other variables.

County Durham

Durham, famous for its cathedral and university, is a typical university city whose economy and demography are driven by the educational institution it hosts. Located in the north-east of England, Durham's Norman castle and cathedral earned the city its status as a UNESCO World Heritage Site. The city also has many listed buildings, which resulted in Durham's city centre being recognised as a conservation area in the late 1960s. Whilst Durham City is relatively advantaged compared to other areas in County Durham, Durham City Council (2011) notes marked inequalities within the city.

Figure 30.3a shows that, on average, County Durham is a little above average in its level of deprivation at all census time points between 1971 and 2001 and its deprivation has

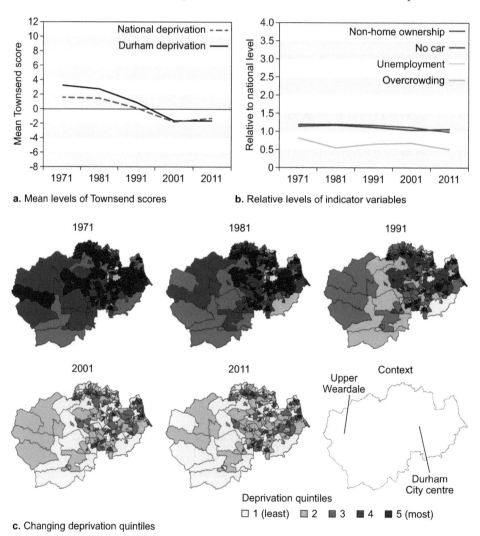

a. Mean levels of Townsend scores   **b.** Relative levels of indicator variables

c. Changing deprivation quintiles

Figure 30.3 Changing deprivation and indicator variables: County Durham, 1971 to 2011

improved decade on decade. By 2011 County Durham is marginally less deprived than GB as a whole. Figure 30.3b shows the mean levels of each indicator variable relative to the national average. The changing deprivation situation is driven by all bar the overcrowding variables being higher than GB levels, although the differences are slight from 2001 (Figure 30.3b). Figure 30.3c maps the changing quintiles of deprivation with the darker colours representing the more deprived areas. In 1971, the more rural areas to the west of County Durham are relatively deprived (and include the old coal and lead mining areas in Upper Weardale). Over time, the more rural areas become less deprived, as do the locations near Durham City centre. The periphery of the city remains deprived; also a legacy of the coal mining industry.

North Kesteven

Located just south of Lincoln and east of Newark-on-Trent, North Kesteven is a non-metropolitan district in the East Midlands with large parts of the district comprising green space, including agricultural land and open spaces. This geography shapes the local economy with the area characterised by large areas of arable farmland, small market towns and rural settlements. In addition, there are three RAF stations within the district (RAF Cranwell, RAF Waddington and RAF Digby). Despite the rurality of this area, it has good levels of social cohesion with a recent study from Ipsos Mori (2010) finding that 82 per cent of residents believe that people from different backgrounds can get along well together there.

Figures 30.4a and 30.4b reveal North Kesteven to have relatively low levels of deprivation with all indicators below national levels. There is a small rise in non-home ownership between 2001 and 2011. The mapped distributions of quintiles (Figure 30.4c) show that the more extensive polygons are the more deprived areas and these will be the more rural areas. The more urban areas in the north of the district are on the outskirts of Lincoln and appear to be quite well off. These relative positions persist over time whilst deprivation eases, in the main. Sleaford in the south of the district has parts of the town which are somewhat deprived.

Lewisham

Lewisham is an ethnically diverse area of south-west London, containing a population that speaks over 170 languages (Lewisham, 2015a). The relative deprivation of Lewisham has likely prompted the council's targeted regeneration of the town centre, including schemes to improve the housing supply and local amenities. For example, despite Lewisham boasting one of the largest commercial areas in south-west London, the local council reported that fewer than 8 per cent of residents in Lewisham town centre's catchment area would choose it as their first choice to shop (Lewisham, 2015b). Housing costs also reflect a significant problem for local residents (van Lohuizen, 2015) and stocks of local social housing are declining (Longden, 2015).

Figure 30.5 demonstrates high levels of persisting deprivation over time, which, given that deprivation is generally easing in most areas, places Lewisham at a relative disadvantage. The deprivation indicator variables (Figure 30.5b) are around or above national average in all years, and between 2001 and 2011 there are relative rises for no car, non-home ownership and particularly for household overcrowding suggesting a worsening deprivation experience for Lewisham's residents. There are more areas in quintile 5, the most deprived areas, by 2011 and many locations have been at this level of deprivation since 1971 (Figure 30.5c).

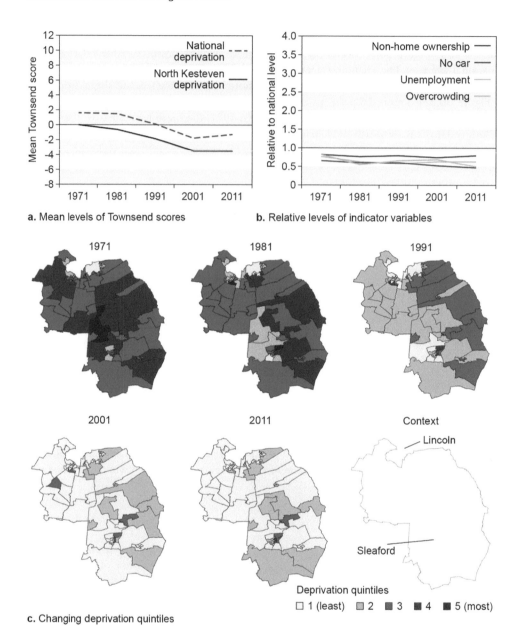

a. Mean levels of Townsend scores

b. Relative levels of indicator variables

c. Changing deprivation quintiles

Deprivation quintiles
☐ 1 (least) ☐ 2 ■ 3 ■ 4 ■ 5 (most)

Figure 30.4 Changing deprivation and indicator variables: North Kesteven, 1971 to 2011

Milton Keynes

Located in Buckinghamshire, Milton Keynes became a designated new town in 1967 with the primary aim of relieving housing pressure in London. Geographically centrally positioned between London, Birmingham, Leicester, Oxford and Cambridge, the town was expected to become a new major regional centre. Built according to the grid systems typical of the USA, the geography of Milton Keynes varies from most other towns and cities in the UK and is also characterised by

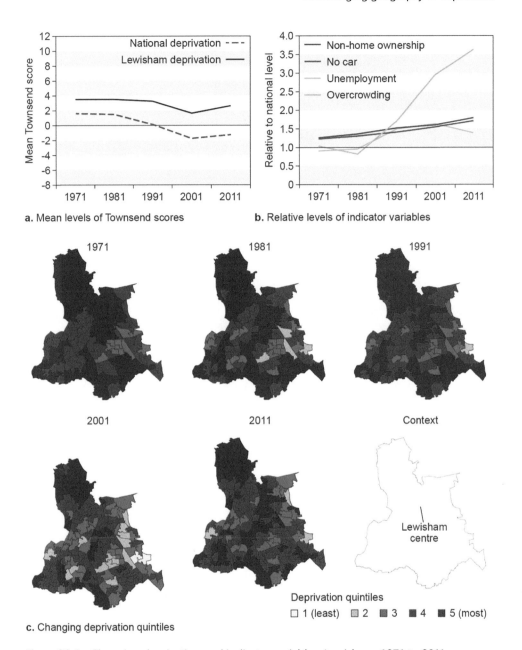

Figure 30.5   Changing deprivation and indicator variables: Lewisham, 1971 to 2011

a business and shopping district alongside varied tenures and housing types. Between 1997 and 2011, Milton Keynes had the fastest growing economy outside of London (*The Economist*, 2013).

Although designated in 1967, almost no building had started by the time of the 1971 Census so the geography of deprivation at that time represents the rural villages and farmland on which the new town was then built (Figure 30.6c). The areas which become the urban centre are at that time relatively non-deprived and the more extensive rural locations somewhat more deprived. As

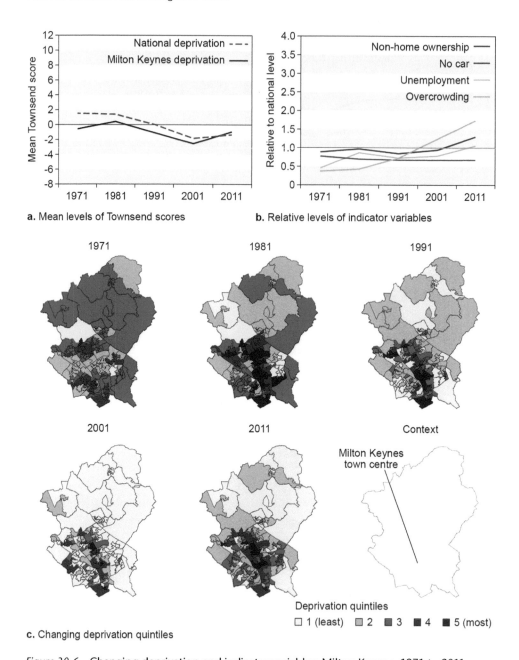

Figure 30.6 Changing deprivation and indicator variables: Milton Keynes, 1971 to 2011

the new town develops through 1981 onwards, the deprivation gradually becomes more focused in the south of the district. It is interesting how the geography of deprivation has developed a typical urban–rural gradient from when the residential building work started. Despite the growing economy noted above, there is a slight rise in deprivation by 2011, due to relative rises in non-home ownership, overcrowding and unemployment (Figure 30.6b).

The case studies reported above serve to illustrate that changing deprivation and indicator variables can be used to reveal how a particular place has changed over time. The reporting here is descriptive and would benefit from more rigorous measures to classify changes. In a similar way to a geodemographic classification, the input variables could be used to group locations by their 1971 to 2011 trajectories. That way, areas which are improving their position relative to others could more readily be identified.

30.5   Ongoing measures of deprivation

Here we have used a time series from the decennial censuses from 1971, the year when detailed small area data became more readily available, to 2011, the most recent census. The future of traditional census-taking is being questioned around the world due to high costs and assertions that a periodic snapshot is not providing the data which contemporary users demand (Dugmore et al., 2011; Yacyshyn and Swanson, 2011; Norman, 2013). This was the context in which the Office for National Statistics (ONS) set up the Beyond 2011 Programme to assess the feasibility of administrative data sources as an alternative to the census in the future (ONS, 2011). Informed by Beyond 2011, the National Statistician recommended to the UK Government that there *will* be a 2021 Census (predominantly online) but that the country's statistical system should be *enhanced* by greater use of administrative data (ONS, 2014).

Administrative data are those collected by government departments and other organisations as part of their day-to-day activities but which are not data specifically collected for practical population applications or social science research (D'Silva and Norman, 2015). Both ONS and others (e.g. Harper and Mayhew, 2012) have made good progress in demonstrating that administrative data have sufficient utility to aid in counting the population. However, administrative data as a source to inform on population characteristics did not receive the same level of focus in Beyond 2011. Nevertheless, administrative data underpin the production of the Index of Multiple Deprivation (IMD) (Noble et al., 2006). This is a sophisticated scheme which has multiple data inputs to seven 'domains' of deprivation which themselves are combined into an indication of an area's relative level of deprivation. There have been a series of IMDs produced separately in the UK's constituent countries at various time points from the late 1990s onwards and they are regularly used in local government applications and academic research.

There are, however, elements about the IMDs which users need to bear in mind. The sophistication of the IMDs means that it is unlikely that practitioners can recreate the indices. Fortunately, most people versed in census data usage could produce an administrative index based on the same approach as Townsend/Carstairs and indices using administrative data in this way have been found to correlate closely with the IMD (Ajebon and Norman, 2016; D'Silva and Norman, 2015). There is a risk of circularity if a variable used as an input to the IMD is highly related to the outcome of interest (Adams and White, 2006). Having separate IMDs has applicability within each country but means that it is inappropriate to carry out a GB- or UK-wide study because the values and meanings of deprivation are not necessarily comparable (ONS, 2010; Morelli and Seaman, 2007; Whynes, 2008). Comparisons of values for the same location over time are not advisable because different inputs and details of methods are used, and because the year of release of an IMD may not be the time point of the input data files used. In addition, although the national statistics agencies consult users on the content of census data, there is no such control by researchers on the specification of administrative data collected nor the tables disseminated (D'Silva and Norman, 2015). A change in policy can create differences in qualification for a benefit which would mean that successive data downloads may not be exactly comparable. Whilst in a cross-sectional deprivation composite this may not matter, any difference in definition over time will render a change measurement less valid. The use of administrative data to track deprivation change does have potential, though, as demonstrated by Gambaro et al. (2015) and Hinks (2015).

A perennial problem to be overcome is that of boundary change (Norman et al., 2003). One of the advances in data supply from the 2000s onwards has been the ready availability of an increasing variety of administrative data. These data have largely been disseminated for super output areas (SOAs) and DZs. These geographies were originally delineated using information from the 2001 Census and are then frozen over the decade to enable time-series analysis. There were some boundary changes as a result of population change since 2001 identified using the 2011 Census (ONS, 2012), but these revisions were not as widespread as changes at small area level for the earlier censuses. Since the IMD uses administrative data inputs and is made available for LSOAs in England and in Wales, DZs in Scotland and SOAs in Northern Ireland, this means that the IMDs are not independent of the census. Whilst a 2021 Census will give the opportunity to re-set the boundaries of these geographies, in the face of ongoing population change, if that census is the last, then another method will be needed for the definition of boundaries to be used for data dissemination. This could be achieved by applying the same approach as Cockings et al. (2011) but using administrative data such as dwelling counts to provide thresholds to maintain confidentiality whilst giving geographic detail and council tax bands to inform within area homogeneity.

30.6 Conclusion

A major use of small area census data has been to construct deprivation measures as a composite of a range of input variables. These deprivation indices have been used in practical circumstances to inform policy on regeneration and to explain occurrences of other phenomena such as health outcomes, education achievement and levels of crime. Appraising how changes in deprivation are related to other change is a very useful activity but the production of time comparable deprivation measures is hampered by some of the drawbacks of data drawn from successive censuses. Drawbacks exist because the questions and dissemination of data may not be close in definition from one time point to the next and the small areas for which data are made available are altered in their areal extents. Creating measures so that like-with-like is being compared is challenging.

An index based on the Townsend scheme which has four variable inputs is achievable over time, since these particular variables are available in closely equivalent definitions for all censuses from 1971 to 2011 for small areas across GB. Those small areas do change but methods to reliably adjust data from one boundary to another are well established and have been used here to convert from the original census geographies to the LSOAs used for the dissemination of 2011 Census data in England and Wales and the DZs as used in Scotland. A deprivation index has then been calculated which allows change to be measured over time. A set of case studies has been used to show how places have changed over time *vis-à-vis* the national average. In the main, deprivation is shown to ease, given that more people own their homes and have cars over the time period. However, the improvements seem to have slowed or even slightly reversed between 2001 and 2011.

We should reflect on the applicability of the input variables to the Townsend index. Although unemployment is a highly relevant indicator reflecting changing economic conditions, non-home ownership might better be replaced by social housing. This would exclude the private rental market, a tenure in which a wide range of people may live. High levels of non-car ownership were assumed to reflect lack of resources through which to buy a car but perhaps now relate more to accessibility. This may have different societal impacts in urban and rural areas where there are different provisions of public transport. Household overcrowding is at such low levels to thereby warrant a low weighting in an index, although as a standalone variable geographic variations still relate to health.

Following the 2021 Census we can update the work here to show deprivation change from 1971 and experiment with revisions to the input variables. Despite various caveats noted above, the future of deprivation measures lies in the use of administrative data which can pick up change on an annual (or more frequent basis) rather than at the beginning of each decade. The IMDs have great value but to have versatility more equivalent to the census-based measures, developments which are needed include: an understanding of how geographies of data release can be defined using administrative data, enabling deprivation to be calculated at a variety of geographic scales; a set of inputs which are applicable across the UK's constituent countries; and a method of combining these inputs into a composite measure which users can readily recreate themselves.

Acknowledgements

The research used census data, the National Statistics Postcode Directory and GIS boundary data obtained via MIMAS, *Casweb* and EDINA UKBORDERS, which are academic services supported by the Economic and Social Research Council (ESRC) and Jisc. The Census, official Mid-Year Estimates and National Statistics Postcode Directory for England, Wales and Scotland have been provided by the Office for National Statistics and National Records of Scotland and the digital boundary data by Ordnance Survey. These data are copyright of the Crown and are reproduced with permission of the Controller of HMSO.

References

Adams, J. and White, M. (2006) Removing the health domain from the Index of Multiple Deprivation 2004 – effect on measured inequalities in census measure of health. *Journal of Public Health*, 28(4): 379–383.

Ajebon, M. and Norman, P. (2016) Beyond the census: A spatial analysis of health and deprivation in England. *GeoJournal*, 81(3): 395–410.

Basta, N.O., James, P.W., Gómez Pozo, B., Craft, A.W., Norman, P.D. and McNally, R.J.Q. (2014) Survival from teenage and young adult cancer in northern England, 1968–2008. *Pediatric Blood & Cancer*, 61(5): 901–6.

Bell, N., Schuurman, N. and Hayes, M.V. (2007) Using GIS-based methods of multicriteria analysis to construct socioeconomic deprivation indices. *International Journal of Health Geographics*, 6: 17.

Blackman, T. (2006) *Placing Health. Neighbourhood Renewal, Health Improvement and Complexity*. The Policy Press, Bristol.

Blakey, K., Feltbower, R.G., Parslow, R.C., James, P.W., Gómez Pozo, B., Stiller, C., Vincent, T.J., Norman, P.D., McKinney, P.A., Murphy, M.F., Craft, A.W. and McNally, R.J.Q. (2014) Is fluoride a risk factor for bone cancer? Small area analysis of osteosarcoma and Ewing sarcoma diagnosed among 0–49 year olds in Great Britain, 1980–2005. *International Journal of Epidemiology*, 43(1): 224–234.

Boyle, P., Norman, P. and Rees, P. (2002) Does migration exaggerate the relationship between deprivation and limiting long-term illness? A Scottish analysis. *Social Science & Medicine*, 55(1): 21–31.

Brennan, A., Rhodes, J. and Tyler, P. (1999) The distribution of SRB challenge fund expenditure in relation to local-area need in England. *Urban Studies*, 36(12): 2069–2084.

Carr-Hill, R. and Rice, N. (1995) Is enumeration district level an improvement on ward level analysis in studies of deprivation and health? *Journal of Epidemiology and Community Health*, 49(2): S28-S29.

Carstairs, V. and Morris, R. (1989) Deprivation: Explaining differences in mortality between Scotland, England and Wales. *British Medical Journal*, 299: 886–889.

Cockings, S., Harfoot, A., Martin, D. and Hornby, D. (2011) Maintaining existing zoning systems using automated zone-design techniques: Methods for creating the 2011 Census output geographies for England and Wales. *Environment and Planning A*, 43(10): 2399–2418.

Cockings, S., Harfoot, A., Martin, D. and Hornby, D. (2013) Getting the foundations right: Spatial building blocks for official population statistics. *Environment and Planning A*, 45(6): 1403–1420.

DCLG (2015) English indices of deprivation. Department for Communities and Local Government. Available at: www.gov.uk/government/collections/english-indices-of-deprivation.

Denham, C.J. and Rhind, D.W. (1983) The 1981 Census and its results. In Rhind, D. (ed.) *A Census User's Handbook*. Methuen, London, pp. 17–88.

Dorling, D. (1999) Who's afraid of income inequality? *Environment and Planning A*, 31(4): 571–574.

D'Silva, S. and Norman, P. (2015) Impacts of mine closure in Doncaster: An index of social stress. *Radical Statistics*, 112: 23–33.

Dugmore, K., Furness, P., Leventhal, B. and Moy, C. (2011) Beyond the 2011 Census in the United Kingdom. *International Journal of Market Research*, 53(5): 619–650.

Durham City Council (2011) 2010 Index of Deprivation Report. Available at: replace with: http://www.countydurhampartnership.co.uk/media/12828/Index-of-Deprivation-2010-Report/pdf/ID2010SummaryWebsite.pdf.

Eroğlu, S. (2007) Developing an index of deprivation which integrates objective and subjective dimensions: Extending the work of Townsend, Mack and Lansley and Halleröd. *Social Indicators Research*, 80(3): 493–510.

Exeter, D.J., Boyle, P.J. and Norman, P. (2011) Deprivation (im)mobility and cause-specific premature mortality in Scotland. *Social Science & Medicine*, 72(3): 389–397.

Fu, M., Exeter, D.J. and Anderson, A. (2015) 'So, is that your "relative" or mine?' A political-ecological critique of census-based area deprivation indices. *Social Science & Medicine*, 142: 27–36.

Gambaro, L., Joshi, H., Lupton, R., Fenton, A. and Lennon, M.C. (2015) Developing better measures of neighbourhood characteristics and change for use in studies of residential mobility: A case study of Britain in the early 2000s. *Applied Spatial Analysis and Policy*, 9(4): 569–590.

Gregory, I.N. (2002) The accuracy of areal interpolation techniques: Standardising 19th and 20th century census data to allow long-term comparisons. *Computers, Environment and Urban Systems*, 26(4): 293–314.

Harper, G. and Mayhew, L. (2012) Applications of population counts based on administrative data at local level. *Applied Spatial Analysis and Policy*, 5(3): 183–209.

Havard, S., Deguen, S., Bodin, J., Louis, K., Laurent, O. and Bard, D. (2008) A small-area index of socioeconomic deprivation to capture health inequalities in France. *Social Science & Medicine*, 67(12): 2007–2016.

Haynes, R., Gale, S., Lovett, A. and Bentham, G. (1996) Unemployment rate as an updatable health needs indicator for small areas. *Journal of Public Health Medicine*, 18(1): 27–32.

Higgs, G., Bellin, W., Farrell, S. and White, S. (1997) Educational attainment and social disadvantage: Contextualizing school league tables. *Regional Studies*, 31(8): 775–789.

Higgs, G., Senior, M.L. and Williams, H.C.W.L. (1998) Spatial and temporal variation of mortality and deprivation 1: Widening health inequalities. *Environment and Planning A*, 30(9): 1661–1682.

Hinks, S. (2015) Deprived neighbourhoods in transition: Divergent pathways of change in the Greater Manchester city-region. *Urban Studies*. DOI: 10.1177/0042098015619142.

Hoare, J. (2003) Comparison of area-based inequality measures and disease morbidity in England, 1994–1998. *Health Statistics Quarterly*, 18: 18–24.

Holman, R. (1978) *Poverty: Explanations of Social Deprivation*. Martin Robertson, London.

Holtermann, S. (1975) Areas of urban deprivation in Great Britain: An analysis of 1971 Census data. *Social Trends*, 6: 33–45.

Ipsos Mori (2010) Mind the gap: Frontiers of performance in local government V. Available at: https://www.ipsos.com/ipsos-mori/en-uk/mind-gap-frontiers-performance-local-government-v?language_content_entity=en-uk.

Jarman, B. (1983) Identification of underprivileged areas. *British Medical Journal*, 286: 1705–1709.

Kawachi, I., Kennedy, B.P. and Wilkinson, R.G. (1999) Crime: Social disorganization and relative deprivation. *Social Science & Medicine*, 48(6): 719–731.

Lewisham (2015a) Regenerating Lewisham town centre. Available at: www.lewisham.gov.uk/inmyarea/regeneration/lewishamtowncentre/Pages/default.aspx.

Lewisham (2015b) Lewisham's joint strategic needs assessment. Demography. Available at: www.lewishamjsna.org.uk/a-profile-of-lewisham/demography.

Lloyd, C. (2016) Population change and geographic inequalities in the UK, 1971–2011. ESRC grant ES/L014769/1.

Longden, H. (2015) The cost of a home: Council housing tenants face being pushed out of east London. *EastLondonLines*. Available at: www.eastlondonlines.co.uk/2015/12/the-cost-of-a-home-council-housing-tenants-face-being-pushed-out-of-east-london/.

Mackenzie, I.F., Nelder, R., Maconachie, M. and Radford, G. (1998) My ward is more deprived than yours. *Journal of Public Health Medicine*, 20: 186–90.

Martin, D.J. (1998) 2001 Census output zones: From concept to prototype. *Population Trends*, 94: 19–24.

McNally, R.J.Q., James, P.W., Ducker, S., Norman, P.D. and James, O.F.W. (2014a) No rise in incidence but geographical heterogeneity in the occurrence of primary biliary cirrhosis in northeast England. *American Journal of Epidemiology*, 179(4): 492–498.

McNally, R.J.Q., Basta, N.O., Errington, S., James, P.W., Norman, P.D. and Craft, A.W. (2014b) Socio-economic patterning in the incidence and survival of children and young people diagnosed with malignant melanoma in northern England. *Journal of Investigative Dermatology*, 134(11): 2703–2708.

McNally, R.J., Basta, N.O., Errington, S., James, P.W., Norman, P.D., Hale, J.P. and Pearce, M.S. (2015) Socio-economic patterning in the incidence and survival of boys and young men diagnosed with testicular cancer in northern England. *Urologic Oncology: Seminars and Original Investigations*, 33(2): 506.e9–506.e14.

McNally, R.J.Q., Blakey, K., Parslow, R.C., James, P.W., Gómez Pozo, B., Stiller, C., Vincent, T.J., Norman, P., McKinney, P.A., Murphy, M.F., Craft, A.W. and Feltbower, R.G. (2012) Small area analyses of bone cancer diagnosed in Great Britain provide clues to etiology. *BMC Cancer*, 12: 270.

Mennis, J. and Hultgren, T. (2006) Intelligent dasymetric mapping and its application to areal interpolation. *Cartography and Geographic Information Science*, 33(3): 179–194.

Mitchell, G. and Norman, P. (2012) Longitudinal environmental justice analysis: Co-evolution of environmental quality and deprivation in England, 1960–2007. *Geoforum*, 43: 44–57.

Mitchell, G., Norman, P. and Mullin, K. (2015). Who benefits from environmental policy? An environmental justice analysis of air quality change in Britain, 2001–2011. *Environmental Research Letters*, 10. DOI: 10.1088/1748-9326/10/10/105009.

Morelli, C. and Seaman, P. (2007) Devolution and inequality: A failure to create a community of equals? *Transactions of the Institute of British Geographers*, 32(4): 523–538.

Morris, R. and Carstairs, V. (1991) Which deprivation? A comparison of selected deprivation indices. *Journal of Public Health Medicine*, 13(4): 318–326.

Noble, M., Wright, G., Smith, G. and Dibben, C. (2006) Measuring multiple deprivation at the small-area level. *Environment and Planning A*, 38: 168–185.

Norman, P. (2010a) Identifying change over time in small area socio-economic deprivation. *Applied Spatial Analysis and Policy*, 3(2–3): 107–138.

Norman, P. (2010b) Demographic and deprivation change in the UK, 1991–2001. In Stillwell, J., Norman, P., Thomas, C. and Surridge, P. (eds.) *Understanding Population Trends and Processes Volume 2: Spatial and Social Disparities.* Springer, Dordrecht, pp. 17–35.

Norman, P. (2013) Whither/wither the census? *Radical Statistics*, 106: 13–17.

Norman, P. and Riva, M. (2012) Population health across space and time: The geographical harmonisation of the ONS Longitudinal Study for England and Wales. *Population, Space & Place* 18(5): 483–502.

Norman, P., Boyle, P. and Rees, P. (2005) Selective migration, health and deprivation: A longitudinal analysis. *Social Science & Medicine*, 60(12): 2755–2771.

Norman, P., Charles-Edwards, E. and Wilson, T. (2016) Relationships between population change, deprivation change and health change at small area level: Australia 2001–2011. In Wilson, T., Charles-Edwards, E. and Bell, M. (eds.) *Demography for Planning and Policy: Australian Case Studies.* Springer, Dordrecht, pp. 197–214.

Norman, P., Purdam, K., Tajar, A. and Simpson, L. (2007) Representation and local democracy: Geographical variations in elector to councillor ratios. *Political Geography*, 26(1): 57–77.

Norman, P., Rees, P. and Boyle, P. (2003) Achieving data compatibility over space and time: Creating consistent geographical zones. *International Journal of Population Geography*, 9(5): 365–386.

Norris, P. and Mounsey, H.M. (1983) Analysing change through time. In Rhind, D. (ed.) *A Census User's Handbook.* Methuen, London, pp. 265–286.

ONS (2010) Comparing across countries' indices of deprivation: Guidance paper. ONS.

ONS (2011) Beyond the 2011 Census Project. Available at: www.ons.gov.uk/ons/about-ons/what-we-do/programmes---projects/beyond-2011/index.html.

ONS (2012) Changes to output areas and super output areas in England and Wales, 2001 to 2011. Available at: www.ons.gov.uk/ons/guide-method/geography/beginner-s-guide/census/super-output-areas--soas-/index.html.

ONS (2014) The census and future provision of population statistics in England and Wales: Recommendation from the National Statistician and Chief Executive of the UK Statistics Authority. Available at: www.ons.gov.uk/ons/about-ons/what-we-do/programmes---projects/beyond-2011/index.html.

Pornet, C., Delpierre, C., Dejardin, O., Grosclaude, P., Launay, L., Guittet, L., and Launoy, G. (2012) Construction of an adaptable European transnational ecological deprivation index: The French version. *Journal of Epidemiology and Community Health*, 66(11): 982–989.

Rees, P. (1995) Putting the census on the researcher's desk. In Openshaw, S. (ed.) *Census Users' Handbook*, GeoInformation International, Cambridge, pp. 27–81.

Rees, P., Martin, D. and Williamson, P. (2002) *The Census Data System*. Wiley, Chichester.

Rhind, D. (1983) *A Census User's Handbook*, Methuen, London.

Senior, M. (2002) Deprivation indicators. In Rees, P., Martin, D. and Williamson, P. (eds.) *The Census Data System*. Wiley, Chichester, pp. 123–139.

Senior, M., Williams, H. and Higgs, G. (2000) Urban–rural mortality differentials: Controlling for material deprivation. *Social Science & Medicine*, 51 (2): 289–305.

Simpson, L. (1996) Resource allocation by measures of relative social need in geographical areas: The relevance of the signed chi-square, the percentage, and the raw count. *Environment & Planning A*, 28: 537–554.

Simpson, L. (2002) Geography conversion tables: A framework for conversion of data between geographical units. *International Journal of Population Geography*, 8(1): 69–82.

Simpson, L. and Yu, A. (2003) Public access to conversion of data between geographies, with multiple look up tables derived from a postal directory. *Computers, Environment and Urban Systems*, 27(3): 283–307.

Stillwell, J., Norman, P., Thomas, C. and Surridge, P. (2010) Spatial and social disparities. In Stillwell, J., Norman, P., Thomas, C. and Surridge, P. (eds.) *Understanding Population Trends and Processes Volume 2: Spatial and Social Disparities*. Springer, Dordrecht, pp. 1–15.

The Economist (2013) Paradise lost, Britain's new towns. 3 August. Available at: www.economist.com/news/britain/21582559-britains-new-towns-illustrate-value-cheap-land-and-good-infrastructure-paradise-lost.

Townsend, P. (1987) Deprivation. *Journal of Social Policy*, 16: 125–46.

van Lohuizen, A. (2015) House prices to rise by a third – why planning reforms must deliver more affordable housing. Shelter policy blog. Available at: http://blog.shelter.org.uk/2015/07/house-prices-to-rise-by-a-third-why-planning-reforms-must-deliver-more-affordable-housing/.

Whynes, D.K. (2008) Deprivation and self-reported health: Are there 'Scottish effects' in England and Wales? *Journal of Public Health*, 31(1): 147–153.

Yacyshyn, A.M. and Swanson, D.A. (2011) The costs of conducting a national census: Rationale for re-designing current census methodology in Canada and the United States. Available at: http://cssd.ucr.edu/Papers/PDFs/Yacyshyn_Swanson_JOS_Aug26_2011.pdf.

Scale, geographic inequalities and the North–South divide in England and Wales, 2001–11

Christopher D. Lloyd

31.1 Introduction

The chapter explores changes in the spatial concentration of members of population sub-groups in England and Wales between 2001 and 2011. Spatial concentrations relate to spatial inequalities and where, for example, social deprivation is shown to be increasingly concentrated in particular areas, spatial inequalities have increased. In this chapter, spatial clustering in population sub-groups is measured using Moran's *I* spatial autocorrelation coefficient. In addition, variograms are used to measure the magnitude and scale of population concentrations, thus providing a summary of spatial inequalities. Variograms can be used to measure the spatial structure of population sub-groups in different directions and thus they characterise inequalities between the North and the South or the East and the West, or indeed any direction of interest. Examining changes in variograms allows an assessment of the ways in which inequalities have changed, including the directions which correspond to most or least change. In other words, it is possible to determine if, for example, the North and South are growing apart while the East and West are becoming more alike.

Studies of geographic inequalities in the UK have focused on differences between the north and the south of the country and between urban and rural areas; health and wealth have provided key foci. Doran et al. (2004) found evidence for a North West–South East divide in social class inequalities in Britain in 2001 and they showed that Wales, the North East and North West regions of England had higher rates of poor health than elsewhere, but that the widest health gaps were between social classes in Scotland and London. A response from Bland (2004) contested these views and argued that the differences between socio-economic groups do not vary markedly between regions. Over a decade ago, Dorling and Thomas (2004) provided evidence for growing regional inequalities in educational qualifications between London and the South East and the rest of the country.

Several other studies have observed North–South divides; these include all-cause mortality (Hacking et al., 2011) and of the growth in the British economy (Gardiner et al., 2013). Other research has provided evidence for growing inequalities between regions within the UK; Riva et al. (2011) considered how residential mobility might explain geographic inequalities in all-cause mortality between urban and rural areas. For those in social housing in England, Tunstall

(2011) shows that some dimensions of social exclusion (income, employment, neighbourhood quality) had reduced by a small amount between 2000 and 2011, but there was evidence of an increase in concentrations of disability.

The present study takes a different approach to previous research in that it does not seek to provide prior definitions of regions such as the North and South. Instead, the chapter explores directional trends in the data flexibly with output areas (OAs) as inputs and using variogram maps to characterise spatial variation in all directions simultaneously. The analysis is based on counts by age, ethnic group, housing tenure, car or van availability, qualifications, employment, limiting long-term illness (LLTI) and National Statistics Socio-economic Classification (NS-SeC). The analysis of spatial concentrations of members of population sub-groups links to studies of social, economic and ethnic segregation (Lloyd et al., 2014). There is a large number of studies of segregation which have focused on urban areas and have been concerned with both ethnic segregation (e.g. Johnston et al., 2007; Catney, 2016; 2017) and socio-economic segregation (e.g. Quillian and Lagrange, 2013; Musterd, 2005). The focus in this chapter is on characterising how particular population sub-groups are spatially concentrated over different spatial scales. For example, how large (on average) are differences within urban areas as opposed to differences between urban areas?

This chapter builds on research conducted by Voas and Williamson (2000), Dorling and Rees (2003) and Lloyd (2015). Voas and Williamson (2000) assessed unevenness (using the index of dissimilarity, D) in population groups using 1991 Census data for England and Wales. Using data for districts, wards and enumeration districts, they measured the proportion of unevenness associated with each of the three scales. Changes in spatial divisions across Britain between 1971 and 2001 were the focus of Dorling and Rees (2003). Lloyd (2015) analysed the spatial structure of population sub-groups in England and Wales in 2001 and 2011 and the results suggested that differences between regions had reduced between 2001 and 2011 for most of the population sub-groups included. Other UK-based studies have explored changes by ethnicity (e.g. Catney, 2016), and demographic and deprivation change (Norman, 2010). In Chapter 25 of this book, Dorling identifies changes at the local authority district scale in a selection of variables. The present study systematically assesses population spatial concentrations at multiple spatial scales and in multiple directions, providing a novel overview of the geographies of population sub-groups in England and Wales in 2001 and 2011.

31.2 Data and methods

Data and data transformation

The analysis makes use of data from the 2001 and 2011 Censuses of England and Wales. Given the concern with the exploration of spatial scale in population concentrations, the smallest available area data were required, thus output areas (OAs) provide the basis of the analysis. There were 175,434 OAs in England and Wales in 2001 (with a mean population of 297), while in 2011 there were 181,408 (mean population of 309). The two sets of OAs were considered to be comparable on the basis that only some 2.6 per cent of the 2001 OAs were changed as a result of the 2011 Census (Office for National Statistics (ONS), 2012). OAs were constructed using clusters of adjacent unit postcodes; they were intended to have similar population sizes and to be as socially homogenous as possible given tenure of household and dwelling type (Martin et al., 2001). The data used come from the Key Statistics tables and they are summarised in Table 31.1. The choice of variables was motivated by a desire to explore demographic, social and economic

Table 31.1 Key Statistics census tables with derived variables and counts

2001	2011	Table	Variable	Abbreviation	Definition	2001 count	2011 count
KS001	KS102	Age structure	Age 0 to 15	A0to15	Persons aged 0 to 15	10,488,736	10,579,132
			Age 16 to 29	A16to29	Persons aged 16 to 29	9,112,810	10,495,245
			Age 30 to 64	A30to64	Persons aged 30 to 64	24,127,596	25,778,462
			Age 65 plus	A65plus	Persons aged 65 plus	8,31,2774	9,223,073
KS006	KS201	Ethnic group	All Whites	White	White persons	47,520,866	48,209,395
			Non-Whites	Non-White	Non-White persons	4,521,050	7,866,517
KS018	KS402	Housing tenure	Owner occupied	OwnOcc	Owner occupied HH	14,916,465	15,031,914
			Social rented	SocRent	Social rented HH	4,157,251	4,118,461
			Private rented	PrivRent	Private rented HH	2,586,759	4,215,669
KS017	KS404	Cars and vans	Cars or vans	CarsVans	HH with cars or vans	15,858,292	17,376,274
			No cars or vans	NoCarsVans	HH with no cars or vans	5,802,183	5,989,770
KS013	KS501	Qualifications and students	Qualifications	Qual	Persons with qualifications	26,670,396	35,189,453
			No qualifications	NoQual	Persons with no qualifications	10,937,042	10,307,327
KS09A	KS601	Economic activity – all persons (aged 16–74)	Employed economically active	EAEmploy	EA employed persons	22,795,520	25,449,863
			Unemployed economically active	EAUnemp	EA unemployed persons	1,261,343	1,799,536
KS14A	KS611	National Statistics Socio-economic Classification (NS-SeC; aged 16–74)*	NS-SeC 1, 2	NSSEC12	NS-SeC 1,2	10,172,697	12,792,224
			NS-SeC 3 to 7	NSSEC37	NS-SeC 3–7	16,650,975	22,324,839
			NS-SeC 8	NSSEC8	NS-SeC 8	1,404,188	2,301,614
KS008	KS301	Health and provision of unpaid care	No LLTI	No LLTI	Persons with no LLTI	42,557,060	46,027,471
			LLTI	LLTI	Persons with an LLTI	9,484,856	10,048,441

Notes: Counts for 2001 are England and Wales level counts and they differ from the sums of the OA-level counts because of small cell adjustment. * The NS-SeC classes are as follows: NS-SeC 1, 2: Managerial, administrative and professional occupations; NS-SeC 3–7: Intermediate, routine and manual occupations; NS-SeC 8: Never worked and long-term unemployed. HH are households; EA is economically active; LLTI is limiting long-term illness. PrivRent for 2001 includes 'Private landlord or letting agency' and 'Other'; PrivRent for 2011 includes 'Private rented: Private landlord or letting agency', 'Private rented: Other' and 'Living rent free'. Qual and NoQual figures for 2001 and 2011 (in italics) use 16–74 and 16 plus population bases respectively, and so should not be directly compared; NS-SeC counts for 2011 (in italics) include imputed persons.

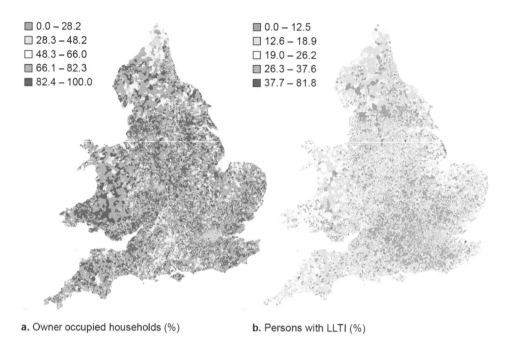

a. Owner occupied households (%)   **b.** Persons with LLTI (%)

Figure 31.1   Percentage of a. owner occupied households and b. persons with an LLTI by OAs for 2011

Note: Contains National Statistics data © Crown copyright and database right 2012. Contains Ordnance Survey data © Crown copyright and database right 2012.

variables which are key components of population profiles in small areas and which have different spatial distributions.

Figure 31.1 includes maps of the percentages of a. owner occupied households and b. persons with a limiting long-term illness (LLTI) by OAs in 2011. These illustrate contrasting spatial patterns with clear urban–rural differences in the map of owner occupied households. The LLTI map indicates that there are lower levels of LLTI in the South than in the North and West. The locations of OAs were represented using population weighted centroids (the median centre of OAs based on household locations and populations; the units are British National Grid co-ordinates in metres).

Sets of values that sum to some fixed value (e.g. in the case of percentages, 100) are termed compositions. Statistical analysis of compositional data is problematic for several reasons. One issue arises from the fact that compositional data such as percentages are closed (i.e. they have a lower and an upper limit) and an analysis of relationships between parts of compositions will result in spurious correlations; at least one covariance/correlation coefficient computed from the components must be negative (Pawlowsky-Glahn and Egozcue, 2006). A fuller summary of the reasons why raw percentages should not be analysed directly is provided by Lloyd et al. (2012). A solution to this problem is to compute log-ratios (Aitchison, 1986). This analysis makes use of balances, a form of isometric-log-ratios (ilr) (Egozcue et al., 2003; Egozcue and Pawlowsky-Glahn, 2005, 2006). The compositions (i.e. sets of percentages) used in this study are two-part (for ethnicity, cars and vans, qualifications, employment and LLTI all comprise two sets of percentages), three-part (for housing tenure and NS-SeC they comprise three sets of percentages) and four-part (for the four age groups used); the log-ratios were computed as outlined below (Table 31.1 defines the input variables).

There are five two-part compositions whose log-ratios are defined as:

$$\text{Ethnicity} = \sqrt{\frac{1}{2}} \ln \frac{\text{White}}{\text{Non-White}} \tag{31.1}$$

$$\text{CarsVans} = \sqrt{\frac{1}{2}} \ln \frac{\text{NoCarsVans}}{\text{CarsVans}} \tag{31.2}$$

$$\text{Qual} = \sqrt{\frac{1}{2}} \ln \frac{\text{NoQual}}{\text{Qual}} \tag{31.3}$$

$$\text{Employ} = \sqrt{\frac{1}{2}} \ln \frac{\text{EAEmploy}}{\text{EAUnemp}} \tag{31.4}$$

$$\text{LLTI} = \sqrt{\frac{1}{2}} \ln \frac{\text{With LLTI}}{\text{No LLTI}} \tag{31.5}$$

The three-part compositions are defined as:

$$\text{Tenure1} = \sqrt{\frac{2}{3}} \ln \frac{(\text{OwnOcc} \times \text{PrivRent})^{\frac{1}{2}}}{\text{SocRent}} \tag{31.6}$$

$$\text{Tenure2} = \sqrt{\frac{1}{2}} \ln \frac{\text{OwnOcc}}{\text{PrivRent}} \tag{31.7}$$

$$\text{NSSEC1} = \sqrt{\frac{2}{3}} \ln \frac{(\text{NSSEC12} \times \text{NSSEC37})^{\frac{1}{2}}}{\text{NSSEC8}} \tag{31.8}$$

$$\text{NSSEC2} = \sqrt{\frac{1}{2}} \ln \frac{\text{NSSEC12}}{\text{NSSEC37}} \tag{31.9}$$

and the four-part compositions are defined for age groups as:

$$\text{Age1} = \sqrt{\frac{3}{4}} \ln \frac{(\text{A0to15} \times \text{A16to29} \times \text{A30to64})^{\frac{1}{3}}}{\text{A65plus}} \tag{31.10}$$

$$\text{Age2} = \sqrt{\frac{2}{3}} \ln \frac{(\text{A0to15} \times \text{A16to29})^{\frac{1}{2}}}{\text{A30to64}} \tag{31.11}$$

$$\text{Age3} = \sqrt{\frac{1}{2}} \ln \frac{\text{A0to15}}{\text{A16to29}} \tag{31.12}$$

The percentages were calculated from counts of population sub-groups $x_1, x_2, x_3, \ldots$ (e.g. persons aged 0–15, aged 16–29, aged 30–64 . . .) with $x_1 + 1, x_2 + 1, x_3 + 1 \ldots$ since some counts are zeros (division by zero is undefined and logs cannot be computed from zeros). The elements to the left-hand side of the logs are normalising constants and justification for their inclusion is provided by Pawlowsky-Glahn and Egozcue (2006).

Christopher D. Lloyd

Moran's I and variograms for measuring spatial variation

The first stage of the analysis entails the measurement of clustering in log-ratios using the Moran's I autocorrelation coefficient (Moran, 1950; Cliff and Ord, 1973). Moran's I, with weights (w_{ij}) between OA centroid locations s_i and s_j row-standardised (the weights for each i sum to one), is given by:

$$I = \frac{\sum_{i=1}^{n}\sum_{j=1}^{n} w_{ij}(z(s_i) - \bar{z})(z(s_j) - \bar{z})}{\sum_{i=1}^{n}(z(s_i) - \bar{z})^2} \tag{31.13}$$

where OA centroids are denoted with i and j, and the values $z(s_i)$ have the mean $\bar{z}$. Positive values of I indicate positive spatial autocorrelation (spatial dependence), while negative values indicate negative spatial autocorrelation. In this study, the weights are equal for the ten nearest neighbours of each OA.

The main part of the analysis is based on the use of the variogram as a means of assessing how the population sub-groups are structured over multiple spatial scales. Variograms provide useful summaries of the magnitude and scale of spatial variation and variograms estimated from population variables which have strong spatial patterning (e.g. large values in urban areas and small values in rural areas) will take very different forms to variograms estimated from properties which do not exhibit clear spatial trends. Thus, variograms can be used to determine the spatial scale of concentrations of population sub-groups and to chart how these have changed over time. The variogram, $\gamma(h)$, relates half the average of the squared differences (that is, the semivariances) between zone centroids to the distances (in lags, or bins) separating them. Taking the example of log-ratios attached to OA centroids, the variogram can be computed through a series of steps as follows:

1   Each OA centroid is compared to every other OA centroid and the squared difference between the paired log-ratio values, as well as the distance between the OA centroids, is stored.
2   Each of the squared differences and distances is then binned such that the average squared difference between all log-ratios separated by a given range of distances (e.g. 0–2 km) is computed.
3   The resulting average squared differences are then multiplied by 0.5 and plotted against the average distance within the distance band (alternatives are possible, such as the middle of the distance band).

The procedure outlined above does not account for the alignment of paired OA centroids. This information can be stored as well so that, for each pair of OA centroids, the squared difference, distance and relative direction is stored. For example, one OA centroid may be orientated in a north-easterly direction with respect to another OA centroid and thus a value of 45° (clockwise from north, with north indicated as 0°) would be stored. The directions could then be binned (e.g. 0°>22.5°, 22.5°>45°, . . .) and variograms can then be constructed for each directional bin. Directional variograms computed in this way allow for assessment of the scale of variation in population sub-groups by direction. For example, there may be little difference in a selected population sub-group when the East and West of England and Wales are compared, while there is a considerable difference in the same sub-group when areas in the North and the South are compared. Thus, we could talk of a North–South divide, but not an East–West divide.

Formally, the experimental variogram is estimated for the $p(h)$ paired observations (log-ratios in the present study), $z(s_i)$, $z(s_i + h)$, $i = 1, 2, \ldots, p(h)$ with:

$$\hat{\gamma}(h) = \frac{1}{2p(h)} \sum_{i=1}^{p(h)} \{z(s_i) - z(s_i + h)\}^2 \tag{31.14}$$

where h is the lag (distance and direction) by which two observations are separated. With directional variograms (as described above), semivariances for selected directions only are shown; a more flexible alternative is the variogram map. The variogram map is centred on 0,0 and the semivariances are binned into the grid cell in which the lag h is located. In other words, the map shows for each cell the average semivariance for the distance and direction represented by the cell (Bivand et al., 2008). Thus, variogram maps depict half the average squared difference between OAs separated by a specific distance and direction. Detailed interpretation of the examples included in the analysis is provided below to aid understanding.

Models are often fitted to variograms and these provide a useful summary of the structure of variograms (Webster and Oliver, 2007). The coefficients of the fitted models capture the amount of variation (the variance) and the spatial scale(s) over which a population sub-group is distributed and they differentiate variables which vary gradually over the study area from those which vary markedly across small areas. In this study, models are fitted to the variograms estimated from OA data. The models fitted in this analysis comprise a nugget effect and one or two spherical model components (defined below). A spherical model (see Webster and Oliver, 2007) component is defined by the range (denoted by a, representing the spatial scale of variation) and the structured component (c, representing spatially correlated variation). With variograms with more complex forms, more than one spherical model component is fitted. The nugget effect, c_0, represents measurement error and variation at a distance smaller than that represented by the sample spacing. The nugget effect plus the structured component(s) is the total sill (the *a priori* variance). The range represents the spatial *scale* of variation while the nugget effect and structured component(s) indicate the *magnitude* of variation. The variograms were estimated using the *Gstat* software package (Pebesma and Wesseling, 1998). The models were fitted in *Gstat* using weighted least squares (the weights are a function of the number of paired observations at each lag). Understanding of the model components is best facilitated by examples, and the variogram models obtained in the present study are described below.

31.3 Analysis

The analysis begins by assessing spatial concentrations in all directions; that is, no account is taken of, for example, East–West or North–South differences. Next, spatial variation in different directions is considered allowing the exploration of how far there are regional trends in population sub-group concentrations and how far these trends have changed between 2001 and 2011.

Moran's I

Table 31.2 includes Moran's I values for each of the log-ratios, computed using ten nearest neighbours (each neighbouring OA is given a weight of 1 and then these are standardised (1/10)). The largest values of I (10N), for both 2001 and 2011, are for *Ethnicity*, followed by *NSSEC2* and *CarsVans*. These reflect distinct urban–rural contrasts in these variables (log-ratios). Small values for the age log-ratios, *Tenure1* and *LLTI* indicate a lack of spatial structure (e.g. no clear urban–rural or larger regional trends). Large values of Moran's I indicate that

Table 31.2 Moran's *I* and standard deviations of log-ratios; variogram model coefficients

Year	Variable	I (10N)	SD	c_0	c_1	Model	a_1	c_2	Model	a_2
2001	Age1	0.335	0.692	0.352	0.033	sph	3679.43			
2011	Age1	0.436	0.731	0.344	0.087	sph	4263.96	0.113	sph	86839.5
2001	Age2	0.381	0.301	0.058	0.024	sph	3243.12			
2011	Age2	0.428	0.294	0.048	0.030	sph	2621.18			
2001	Age3	0.424	0.411	0.098	0.102	sph	2560.00			
2011	Age3	0.516	0.418	0.089	0.129	sph	2773.67			
2001	Ethnicity	0.751	1.168	0.275	0.266	sph	6101.16	0.766	sph	51013.3
2011	Ethnicity	0.838	1.133	0.172	0.35	sph	6262.94	0.686	sph	56303.7
2001	Tenure1	0.381	1.409	1.304	0.271	sph	3576.57	0.395	sph	24080.3
2011	Tenure1	0.392	1.309	1.251	0.412	sph	4014.59	0.087	sph	12878.5
2001	Tenure2	0.466	0.832	0.282	0.163	sph	2073.48	0.284	sph	23406.4
2011	Tenure2	0.569	0.739	0.189	0.196	sph	2369.04	0.214	sph	24782.4
2001	CarsVans	0.574	0.776	0.264	0.378	sph	30782.00			
2011	CarsVans	0.634	0.776	0.243	0.421	sph	26846.50			
2001	Qual	0.591	0.493	0.119	0.113	sph	4740.47			
2011	Qual	0.556	0.507	0.145	0.124	sph	4011.36			
2001	Employ	0.440	0.602	0.196	0.058	sph	4247.50	0.118	sph	41530.9
2011	Employ	0.468	0.522	0.155	0.076	sph	4538.00	0.039	sph	63928.9
2001	NSSEC1	0.485	0.939	0.444	0.157	sph	4438.73	0.325	sph	40446.5
2011	NSSEC1	0.585	0.771	0.265	0.194	sph	4645.67	0.159	sph	48765.6
2001	NSSEC2	0.658	0.499	0.097	0.088	sph	4424.60	0.035	sph	13018.8
2011	NSSEC2	0.691	0.479	0.089	0.096	sph	4242.29	0.049	sph	10735.9
2001	LLTI	0.391	0.371	0.094	0.014	sph	3009.74			
2011	LLTI	0.403	0.388	0.102	0.031	sph	2621.18			

Notes: 10N is ten nearest neighbours; using a random permutation test, all values of *I* have a pseudo *p* value of 0.001; SD is standard deviation; sph is spherical.

neighbouring values are, on average, similar. If the standard deviation is also large then this suggests that there are marked spatial concentrations of the group concerned. Increased *I* and decreased standard deviation suggests, therefore, a reduction in distinct spatial concentrations of the group. Moran's *I* is here computed using only a single neighbourhood and, while it could be computed for multiple scales, the variogram is preferred instead to maintain consistency with previous research.

Omnidirectional variograms

Variograms provide summaries of the spatial distribution of population sub-groups. There are many studies which describe patterns visible in maps of the proportions of people belonging to different groups in the UK. For example, Dorling and Thomas (2004) provide maps and descriptions of spatial patterns in population variables in 1991 and 2001. Regional differences in population groupings are apparent in terms of age, ethnicity, housing tenure, access to cars and vans, (un)employment and occupational grouping (here represented by NS-SeC). Taking examples, the ethnic geographies of England and Wales are marked by an urban–rural contrast in Wales (see Catney and Simpson, 2010), while rates of poor health have been shown to be low in the South East (again, with the exception of London) with high rates in the Welsh valleys (the former coalfields of South Wales) and urban areas of northern England.

Variograms were computed from each of the log-ratios (using 2 km lags with a maximum lag of 80 km) for OAs for 2001 and 2011. Figure 31.2 shows four selected omnidirectional variograms with fitted models (where the model does not extend to the maximum lag, this indicates it was only fitted to a subset of the semivariances). The variograms selected were for *Ethnicity*, *LLTI*, *Tenure1* and *Tenure2*. The log-ratios have different spatial structures and also contrasting directional variation (discussed below). Table 31.2 details the nugget effects, structured components and the ranges of the models fitted to the variograms for all 12 sets of log-ratios. Taking the variogram model for *Ethnicity* in 2011 as an example: it comprises a nugget effect (c_0) of 0.172, a structured component (c_1) of 0.35 and a range (a_1) of 6,263 m with a second spherical component with a structured component (c_2) of 0.69 and a range (a_2) of 56,304 m. The range values represent dominant spatial features at a local scale (a_1 – approximately 6 km) and a larger 'regional' scale (a_2 – approximately 56 km). Smaller range figures indicate more localised patterns while larger range figures correspond to larger-scale spatial trends.

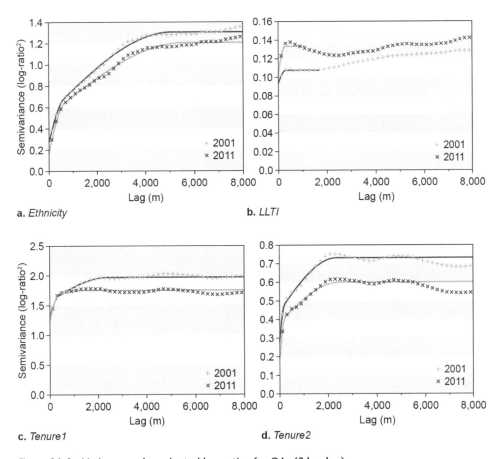

Figure 31.2 Variograms for selected log-ratios for OAs (2 km lag)

The variograms for *Age2*, *Age3*, *Qual* and *LLTI* (model coefficients in Table 31.2) exhibit short-range spatial variation while those for *Ethnicity* and *CarsVans* have much longer-range spatial variation and they highlight the contrast between urban and rural areas. Examination of maps of *Qual* (not shown) and *LLTI* (Figure 31.1b) suggests that there are very large-scale trends, but there is little indication of distinct regional-scale concentrations. The variograms provide a powerful means of comparatively assessing the geography of different variables at different time points.

The variogram models for *Employ* and *NSSEC1* indicate large proportional changes in sill values between 2001 and 2011 (see c_0, c_1, c_2 in Table 31.2). Note that the two sets of log-ratios have similar spatial structures, reflecting the fact that *NSSEC1* relates those with an occupational classification to those who have never worked or are long-term unemployed. For both *Employ* and *NSSEC1*, the sills (the magnitude of variation) have decreased. This could be considered to indicate decreases in geographic inequalities.

The global recession of 2008–09 might have been expected to increase geographical inequalities, rather than reduce them and one possible interpretation is that during (or after) a recession it might appear that inequality has decreased because unemployment has risen in areas where previously it had been very low. There was also a reduction in the *Tenure2* variogram sill between 2001 and 2011. This indicates that there was less spatial variation in the ratio of owner occupied households to private rented households in 2011 than in 2001. This change is partly a function of high house prices, low wage growth and tighter lending requirements (ONS, 2013) and, in combination, these seem to have acted to differentially increase private renting rates with the effect of making owner occupation rates more similar (on average) between areas. The *Ethnicity* variograms also exhibit reduced variation between 2001 and 2011; this probably reflects, at least in part, the destinations of migrants from the 2004 European Union accession countries (Lloyd, 2015), in addition to dispersal from immigrant settlement areas by members of ethnic minority groups (see Simpson and Finney, 2009; Catney and Simpson, 2010). This is consistent with the Moran's *I* results and the standard deviations of the log-ratios, which, together, indicate reduced differences within neighbourhoods (Moran's *I* has increased) and between regions (the standard deviation has decreased).

Variogram maps

Omnidirectional variograms characterise the spatial scale of variation and the amount of variation in all directions simultaneously. Use of directional variograms or variogram maps allows for the exploration of how population characteristics vary by direction. Taking an example, directional variograms make it possible to determine if differences in levels of self-reported ill-health in areas of the East and West of England and Wales tend to be smaller or larger than equivalent differences between areas in the North and South and if these differences have changed through time. Balabdaoui et al. (2001) use directional variograms to explore changes in spatial patterns in a fertility index in India.

Figure 31.3 shows variogram maps for 2011 and variogram map values for 2001–11 for *Tenure2* and *LLTI*. To clarify, the values in the variogram maps indicate half the average squared differences (semivariances) between OAs separated by a particular distance and direction. Taking the example of *LLTI* in 2011 (Figure 31.3c), the centre of the variogram indicates the zero point (the origin of the variogram). Each cell represents a lag of 10 km. From the centre to the top of the map is the North direction (0°), from the centre to the right is the East direction (90°), centre to the bottom is South (180°) and from the centre to the left is West (270°). So, drawing a line 90° from the centre to the right-hand side of the variogram map indicates semivariances

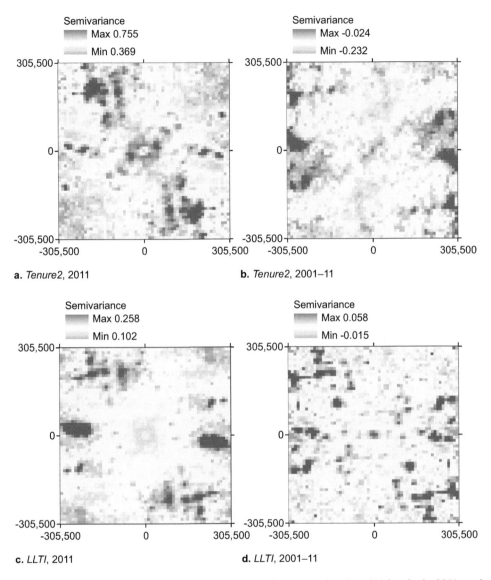

Semivariance
Max 0.755
Min 0.369

Semivariance
Max -0.024
Min -0.232

Semivariance
Max 0.258
Min 0.102

Semivariance
Max 0.058
Min -0.015

a. *Tenure2*, 2011

b. *Tenure2*, 2001–11

c. *LLTI*, 2011

d. *LLTI*, 2001–11

Figure 31.3 Variogram maps for *Tenure2* and *LLTI* log-ratios for OAs (10 km lag), 2011 and 2001–11

for OAs aligned in a West to East direction. For that direction, the semivariances are small for small distances (i.e. cells close to the centre of the variogram map), but they tend to increase with distance and semivariances for cells in the middle of the right-hand side of the variogram map (thus a direction of 90° and lags of > 200 km) are large, indicating that average differences in *LLTI* values for OAs aligned West to East and separated by >200 km are very large. As a further clarifying example, drawing a line 45° from the centre to the top right hand corner of the variogram map indicates semivariances for OAs aligned in a South West to North East direction. Note that, for example, the semivariances for 45° and 225° are the same, as the variogram map is symmetric. The variogram difference maps are the 2011 variogram map values minus the equivalents for

2001. Thus, positive values indicate an increase in semivariances between 2001 and 2011 while negative values indicate a decrease in semivariances.

In the variogram map for *LLTI* in 2011, the largest semivariances are for approximately the 90°/270° (West–East) and 135°/315° (North West–South East) degree directions, while the semivariances for the 45°/225° direction (North East–South West) are much smaller. Taking *LLTI* as an example, differences have increased on average. However, they have increased more in some directions than in others and there was a large increase in the 135°/315° direction. This change suggests that the South and East and the North and West are becoming more dissimilar in terms of *LLTI*. A first order polynomial trend fitted to the *LLTI* log-ratios and visual interpretation of the map of percentages (Figures 31.3c and d) indicates that the *LLTI* proportions were larger in the North and West than in the South and East in both 2001 and 2011. For other log-ratios there has been an increase in differences over most directions and distances, but the differences between (for example) the North West and the South East have remained relatively stable compared to the differences between the North East and the South West. In the case of *Tenure2*, the variogram maps (Figures 31.3a and b) indicate decreasing differences between all directions. The variogram maps for *Qual* (not shown) do not suggest an increase in differences between regions.

31.4 Conclusions

The chapter used omnidirectional variograms and variogram maps to show quantitatively that there are distinct spatial patterns for the selected log-ratios. The population by age is shown to be fairly geographically even (as found by Voas and Williamson, 2000). In contrast, the distinct spatial structure of the population by ethnicity is clear, as demonstrated by the form of the variograms. Log-ratios which exhibit distinct urban–rural patterns (for example, *CarsVans* and *Ethnicity*) have variograms which reflect these structures, while for other log-ratios (the age log-ratios, *Qual* and *LLTI*, for example), there is less obvious patterning over the spatial scales considered. Using variogram maps, for all log-ratios there is evidence of directional variation but it is particularly pronounced for *Qual*, *NSSEC2*, *LLTI* and *Ethnicity*.

The results suggest that, on average, differences between areas reduced between 2001 and 2011, with respect to most of the log-ratios selected. However, there are directional variations both in terms of the geographies of the log-ratios (the *scale* of variation) in each of the two census years, but also in terms of the *magnitude* of change. The larger semivariances for the variogram maps for *LLTI* in the North West to South East direction and in the East to West direction than in the North East to South West direction support the visual assessment in that smaller percentages are found in the South East than in the North and West. The variogram maps indicate that there was an increase in differences between the North and West and the South and East, with a proportionally larger increase in semivariances between 2001 and 2011 at larger lags in the North West to South East direction than there was in the perpendicular direction. In terms of ill-health, therefore, the North–South (or North West–South East) divide in England and Wales appears to be growing.

For *Qual*, the variogram maps (not shown) provide support for the comments of Dorling and Thomas (2004), who argue that there is a North–South divide in educational attainment. However, the North–South (or, North West–South East in some cases) differences between 2001 and 2011 are proportionally similar to differences in the perpendicular direction. Geographical inequalities in qualifications, for example, have increased in all directions (by a relatively small amount) but differences between the North and South (or North West and South East) have not increased relative to differences in other directions.

The results indicate that there has been increased growth in differences between the North West and the South East with respect to self-reported ill-health, with stable North–South (or North West–South East) differences in the case of other log-ratios. The analyses could be extended by expanding the set of variables assessed as well as increasing the timeframe of the analysis. Another possibility would be an assessment of geographical differences between the most wealthy and 'the rest' (e.g. houses with multiple cars or small households in large houses). Local variograms (Lloyd, 2012) could be used to further assess spatial scales of variation in population sub-groups. Expanding the analysis to include the whole of the UK would allow for a fuller assessment of the changing relationships between London and the South East and the rest of the country. This study profiles geographical inequalities; an obvious next step is to move beyond description of patterns to better understand the determinants of inequalities and, ultimately, to create a more geographically equal society.

Acknowledgements

The ONS is thanked for provision of the data on which the analyses were based: Office for National Statistics, 2001 and 2011 Census: Digitised Boundary Data (England and Wales) [computer file]. ESRC/Jisc Census Programme, Census Geography Data Unit (UKBORDERS), EDINA (University of Edinburgh)/Census Dissemination Unit. Census output is Crown copyright and is reproduced with the permission of the Controller of HMSO and the Queen's Printer for Scotland. Part of the research on which this paper was based was supported by the Economic and Social Research Council (grant ES/L014769/1) and this is acknowledged gratefully. John Stillwell is thanked for his very helpful comments on a draft of this chapter.

References

Aitchison, J. (1986) *The Statistical Analysis of Compositional Data*. Chapman and Hall, London.

Balabdaoui, F., Bocquet-Appel, J.-P., Lajaunie, C. and Irudaya Rajan, S. (2001) Space–time evolution of the fertility transition in India, 1961–1991. *International Journal of Population Geography*, 7: 129–148.

Bivand, R.S., Pebesma, E.J. and Gómez-Rubio, V. (2008) *Applied Spatial Data Analysis with R*. Springer, New York.

Bland, J.M. (2004) North–south divide in social inequalities in Great Britain: Divide in social class inequalities may exist but is small. *British Medical Journal*, 329: 52.

Catney, G. (2016) Exploring a decade of small area ethnic (de-)segregation in England and Wales. *Urban Studies*, 53: 1691–1709.

Catney, G. (2017) Towards an enhanced understanding of ethnic group geographies using measures of clustering and unevenness. *The Geographical Journal*, 183: 71–83.

Catney, G. and Simpson, L. (2010) Settlement area migration in England and Wales: Assessing evidence for a social gradient. *Transactions of the Institute of British Geographers*, 35: 571–584.

Cliff, A.D. and Ord, J.K. (1973) *Spatial Autocorrelation*. Pion, London.

Doran, T., Drever, F. and Whitehead, M. (2004) Is there a north–south divide in social class inequalities in health in Great Britain? Cross sectional study using data from the 2001 census. *British Medical Journal*, 328: 1043–1045.

Dorling, D. and Rees, P. (2003) A nation still dividing: The British census and social polarisation 1971–2001. *Environment and Planning A*, 35: 1287–1313.

Dorling, D. and Thomas, B. (2004) *People and Places: A 2001 Census Atlas of the UK*. The Policy Press, Bristol.

Egozcue, J.J. and Pawlowsky-Glahn, V. (2005) Groups of parts and their balances in compositional data analysis. *Mathematical Geology*, 37: 795–828.

Egozcue, J.J. and Pawlowsky-Glahn, V. (2006) Simplicial geometry for compositional data. In Buccianti, A., Mateu-Figueras, G. and Pawlowsky-Glahn, V. (eds.) *Compositional Data Analysis in the Geosciences: From Theory to Practice*. Geological Society Special Publications No. 264, Geological Society, London, pp. 145–160.

Egozcue, J.J., Pawlowsky-Glahn, V., Mateu-Figueras, G. and Barcelo-Vidal, C. (2003) Isometric logratio transformations for compositional data analysis. *Mathematical Geology*, 35: 279–300.

Gardiner, B., Martin, R., Sunley, P. and Tyler, P. (2013) Spatially unbalanced growth in the British economy. *Journal of Economic Geography*, 13: 889–928.

Hacking, J.M., Muller, S. and Buchan, I.E. (2011) Trends in mortality from 1965 to 2008 across the English north-south divide: Comparative observational study. *British Medical Journal*, 342: d508.

Johnston, R., Poulsen, M. and Forrest, J. (2007) The geography of ethnic residential segregation: A comparative study of five countries. *Annals of the Association of American Geographers*, 97: 713–738.

Lloyd, C.D. (2012) Analysing the spatial scale of population concentrations in Northern Ireland using global and local variograms. *International Journal of Geographical Information Science*, 26: 57–73.

Lloyd, C.D. (2015) Assessing the spatial structure of population variables in England and Wales. *Transactions of the Institute of British Geographers*, 40: 28–43.

Lloyd, C.D., Pawlowsky-Glahn, V. and Egozcue, J.J. (2012) Compositional data analysis for population studies. *Annals of the Association of American Geographers*, 102: 1251–1266.

Lloyd, C.D., Shuttleworth, I.G. and Wong, D.W. (eds.) (2014) *Social–Spatial Segregation: Concepts, Processes and Outcomes*. Policy Press, Bristol.

Martin, D., Nolan, A. and Tranmer, M. (2001) The application of zone-design methodology in the 2001 UK Census. *Environment and Planning A*, 33(11): 1949–1962.

Moran, P.A.P. (1950) Notes on continuous stochastic phenomena. *Biometrika*, 37: 17–23.

Musterd, S. (2005) Social and ethnic segregation in Europe: Levels, causes, and effects. *Journal of Urban Affairs*, 27: 331–348.

Norman, P. (2010) Demographic and deprivation change in the UK, 1991–2001. In Stillwell, J., Norman, P., Thomas, C. and Surridge, P. (eds.) *Spatial and Social Disparities. Understanding Population Trends and Processes Volume 2*. Springer, Dordrecht, pp. 17–35.

ONS (2012) 2011 Census, population and household estimates for wards and output areas in England and Wales. November. Available at: www.ons.gov.uk/ons/rel/census/2011-census/population-and-household-estimates-for-wards-and-output-areas-in-england-and-wales/index.html.

ONS (2013) 2011 Census analysis, a century of home ownership and renting in England and Wales. Available at: www.ons.gov.uk/ons/rel/census/2011-census-analysis/a-century-of-home-ownership-and-renting-in-england-and-wales/index.html.

Pawlowsky-Glahn, V. and Egozcue, J.J. (2006) Compositional data analysis: An introduction. In Buccianti, A., Mateu-Figueras, G. and Pawlowsky-Glahn, V. (eds.) *Compositional Data Analysis in the Geosciences: From Theory to Practice*. Geological Society Special Publications No. 264, Geological Society, London, pp. 1–10.

Pebesma, E.J. and Wesseling, C.G. (1998) Gstat, a program for geostatistical modelling, prediction and simulation. *Computers and Geosciences*, 24: 17–31.

Quillian, L. and Lagrange, H. (2013) Socio-economic segregation in large cities in France and the United States. *WP-13–24, Working Paper Series*, Institute for Policy Research, Northwestern University.

Riva, M., Curtis, S. and Norman, P. (2011) Residential mobility within England and urban–rural inequalities in mortality. *Social Science and Medicine*, 73: 1698–1706.

Simpson, L. and Finney, N. (2009) Spatial patterns of internal migration: Evidence for ethnic groups in Britain. *Population, Space and Place*, 15: 37–56.

Tunstall, R. (2011) Social housing and social exclusion 2000–2011. *CASE Paper 153*, Centre for Analysis of Social Exclusion, The London School of Economics and Political Science, London.

Voas, D. and Williamson, P. (2000) The scale of dissimilarity: Concepts, measurement and an application to socio-economic variation across England and Wales. *Transactions of the Institute of British Geographers New Series*, 25: 465–481.

Webster, R. and Oliver, M.A. (2007) *Geostatistics for Environmental Scientists*. Second edition. Wiley, Chichester.

32

Using contemporary and historical census data to explore micro-scale population change in parts of London

Nigel Walford

32.1   Introduction

Most chapters in this book focus on the British 2011 Census or on similar but not identical enumerations that took place over recent decades (see, for example, Chapter 30). Others look forward to an unfamiliar statistical landscape prevailing after a mainly online census in 2021 and a world in which a range of administrative and commercial data sources may, alone or in conjunction with a survey 'not totally unlike' the traditional census, hold sway (see, for example, Chapter 33). It will not have escaped the reader's attention that, before the 2021 Census takes place, which may be the last such full-scale enumeration conducted with the comprehensiveness and completeness familiar to many census analysts, we will celebrate the hundredth anniversary of the 1920 Census Act. This established the legal framework for subsequent enumerations during the twentieth and early twenty-first centuries and still constitutes the primary legislation governing modern censuses.

The present chapter, in contrast to most others in this book, unashamedly steps further back in time to the last two British censuses that were held before the 1920 Census Act came into force and combines information from these with contemporary census data from the 2001 and 2011 enumerations. This serves not only to enable investigation of the persistence or otherwise of socio-demographic characteristics but also to reveal different types of analysis that can be carried out when georeferenced household and individual census data are available. The Censuses in 1901 and 1911 may be viewed as the culmination of incremental development in the administrative formality, topical coverage and recording accuracy of the British census during the nineteenth century. Censuses in the early decades of the 1800s were essentially head counts, which evolved into increasingly more comprehensive enumerators' books from 1841 onwards as the value of statistics for an expanding bureaucracy, particularly towards the end of the nineteenth century, became more apparent. The release of the historical records from the 1901 and 1911 Censuses under the 100 years rule by The National Archives (TNA) has fuelled enthusiasm for genealogical research by individuals seeking to explore their ancestry. Information from these sources has also contributed to popular television programming, helping to find potential beneficiaries from people dying intestate and for portraying social and economic history through the ancestry of celebrities. The access route to these historical census records initially entailed consultation of the original paper documents, although more recently TNA has established partnerships with commercial organisations to provide access to scanned digital images and transcriptions of the sources.

Researchers of economic and social history have not neglected this rich source of information about the conditions of households and individuals during the Victorian and Edwardian eras and followed the lead set by Anderson's early work transcribing a national one-stage cluster sample comprising a stratified 2 per cent systematic sample from the enumerators' records of the 1851 Census (Anderson and Collins, 1973; Anderson, Collins and Stott, 1977; Anderson, 1987), yielding some 415,000 sampled individuals in 945 clusters. Enumerators' books were the sample units, selected on a one in 50 basis, except for settlements in England and Wales with a population less than 2,000, which were sampled as a whole. This initiative set a course navigated by subsequent researchers, who have captured cross-sectional census records for single settlements; for example, Tilley and French (1997) in respect of linking the 1851, 1861, 1871 and 1891 Census enumerators' returns for Kingston upon Thames, and others who have focused on comparing urban or rural areas (Hinde, 1985). More recently the Integrated Census Microdata project at the University of Essex (Higgs et al., 2013) in conjunction with a commercial partner findmypast. co.uk (part of brightsolid) has created a resource of all historical British census records covering the enumerations from 1851 to 1911.

Historians have not surprisingly paid rather less attention to the geographical component of nineteenth- and early twentieth-century censuses than geographers when undertaking research with these data sources. Geographers have pursued two main directions in their research. The first, reflecting the growing importance of Geographical Information Systems and Science (GISS) in the visualisation and analysis of contemporary census statistics, has focused on capturing and harnessing historical data, especially boundaries and aggregate statistics, in order to examine long-term changes in population characteristics (see, for example, Gregory, 2002; Gregory et al., 2001, 2002). The second has pursued the assembly of different types of data, including not only statistical information, but also pictorial images, historical maps, personal narratives, newspaper accounts and administrative records, into a multimedia database. Perhaps the most notable example of such a rich source of historical information in the British Isles is the *Vision of Britain*,[1] created from the Great Britain Historical GIS Project (Southall, 2003, 2006, 2014).

In the context of the present book, this chapter seeks to provide an overview of the data sources, methods and some indicative findings that may be obtained from combining and co-analysing historical and contemporary census data with reference to a case study locality. The following two sections first review the census and other data sources used in this research and then proceed to outline the method used to georeference the early twentieth-century census records. The fourth section presents some indicative results of the analysis before a concluding section giving consideration to the wider implications of the work. In particular, the potential difficulties that may arise for future researchers are considered should something akin to the familiar census records not be available in 100 years' time.

32.2 Data sources

The essence of the census research application outlined in this chapter involved the combination or integration of historical and contemporary geospatial and census data sources within a GISS framework. The aim was to georeference household and individual census records to the addresses where people were enumerated in the 1901 and 1911 enumerations. This corresponds to the population present household base used until the 1971 Census, which counted people where they happened to be on census night rather than where they usually lived. Georeferencing of these occupied 1901 and 1911 Census addresses would then offer opportunities to undertake enhanced analyses of the census data that incorporated a significant geographical component as well as allowing other historical sources to be incorporated in order to enrich our understanding of spatial variation in demographic, economic and social conditions. It must be acknowledged that the topical scope of the early twentieth-century censuses was considerably more limited than those undertaken in 2001 and 2011. Nevertheless, some core variables or attributes are present, albeit defined in different ways, in both pairs of censuses at the start of each century.

There has been a notable increase in the range and scope of datasets available to researchers through services supported by higher education and research funding, such as the UK Data Service and its predecessors, by governmental and public sector bodies, by commercial organisations, such as findmypast.co.uk in respect of a range of historical sources, and by groups, for example local history societies, and occasionally individuals working on a formal or informal basis to capture the past and present conditions of their local communities. Data quality remains an important issue for academic researchers, especially when using information from what might be referred to as 'voluntary sector maintained' internet sites, although cross-referencing or triangulation between different sources can provide reassurance over quality. The following subsections review data sources used in this research application under two headings in relation to the time periods to which they relate: late nineteenth and early twentieth century; and early twenty-first century. Discussion focuses on data relating to England in view of the location of the selected case study areas (see Section 32.3).

Late nineteenth and early twentieth century

The Census held on 31 March/1 April 1901, the eleventh decennial census in England and Wales, introduced a number of important changes to preceding enumerations, especially with regard to procedures for collecting data from households and from people residing in institutions or on vessels, and with respect to the publication of statistics, notably the addition of county reports including tables for parishes, boroughs, sanitary divisions and registration districts. Ten years later, the Census on 2/3 April 1911 collected data on a number of additional matters relating to issues that were topical at the time. Table 32.1 details topics included in each of these censuses and reveals close agreement in a number of respects, but also differences including a survey of married women's fertility, the nationality of people born outside Britain and industry of employment, which was likely to be connected with implementing measures under the 1911 National Insurance Act that provided unemployment benefits.

The documents from these (and earlier) censuses were passed for preservation to the Public Records Office, the predecessor of TNA, and became available for public consultation 100 years after the census. The preservation of such historical documents raises issues over their curation and storage and a proportion of records in any given collection may suffer a degree of deterioration as time passes, for example as a result of water or fire damage leading to the paper becoming increasingly fragile. Expansion of the internet acted as a stimulus to the establishment of organisations offering subscription services allowing access to scanned images of these public documents and transcriptions of the recorded information. The scanned records offer the possibility of arresting further deterioration of the paper records provided adequate measures for their retention, preservation and continued accessibility are implemented. Table 32.2 provides transcriptions of the census records for a family that lived in the same house at the time of the 1901 and 1911 Censuses. A second child was born during this decade and the head of the household remained working in the same occupation as a tapestry weaver. This example illustrates the potential for

Table 32.1 Topics included in 1901 and 1911 British Censuses in England

Addressed to:	1901	1911
Households	Address	Address
	Number of rooms if less than 5	Number of rooms
		Building type
Individuals	Name	Name
	Relationship to head of family	Relationship to head of family
	Marital status	Marital status
	Age	Age
	Sex	Sex
	Occupation	Occupation
	Birthplace (level of geographical detail sought depended on whether birthplace was in England, Wales, Scotland/Ireland, British colony or dependency or a foreign country)	Birthplace (level of geographical detail sought depended on whether birthplace was in England, Wales, Scotland/Ireland, British colony or dependency or a foreign country)
	Medical disabilities (deaf, deaf/dumb, blind, lunatic, imbeciles and the 'feeble-minded')	Medical infirmities (deaf, deaf/dumb, blind, lunatic, imbeciles and the 'feeble-minded')
		Marital fertility (total live births to women in their present marriage, number still alive and number who had died)
		Age at marriage
		Nationality of people born outside of the country
		Employment status
		Whether working at home
		Industry or service of employment

enhancing the information recorded in the 1901 Census documents, because it may be inferred the husband and wife had been married four years at that time, they occupied five or more rooms and the head of household was a worker.

The Ordnance Survey (OS) embarked on producing the County Series of topographic maps at the 6-inch mapping scale (1:10,560) in 1840 and in 1854 expanded this to include mapping at 1:2,500 (approximately 25 inches to the mile) in order to meet the need for greater detail. The first editions of maps at both scales for the whole country were published by the 1890s and surveying for the first revision commenced in 1891. This was completed shortly before the First World War and re-surveying for a third or second revision started in 1907 and continued until the 1940s, although it was never completed and only areas with significant changes were revised (Harley, 1975). The OS and Landmark Information Group created scanned digital copies of these historical maps, which are available through EDINA. The OS archive of historical maps was 93 per cent complete when scanning started in 1995 at 300 dpi and both organisations sought to source the missing maps in order to achieve as complete coverage as possible. The scanned images were processed to conform to the National Grid and to create a tiled seamless mosaic across the country.

Table 32.2 Transcriptions of 1901 and 1911 Census records for linked household and individuals

1901

Schedule number	House number/name, road/street name	Number of rooms if less than 5	Name and surname	Relationship to head of family	Condition as to marriage	Sex	Age last birthday	Profession or occupation	Birthplace	If: (1) deaf and dumb, (2) blind, (3) lunatic or imbecile, (4) feeble-minded
61	Ivy Cottage, Park Road	–	John Martin	Head	Married	Male	29	Aras Tapestry weaver	Middx: Hampton Wick	–
			Louisa Martin	Wife	Married	Female	26	–	London: Stepney	–
			William Martin	Son	–	Male	3	–	Middx: Hampton Wick	–

1911

Schedule number	House number/ name, road/ street name	Number of rooms	Name and surname	Relationship to head of family	Age and sex		Particulars of marriage					Profession or occupation				Birthplace	Nationality for persons born in foreign country	If: (1) deaf and dumb, (2) blind, (3) lunatic or imbecile, (4) feeble-minded
					Male	Female	Marital status	Completed years of present marriage	Total children born alive	Children still living	Children who have died	Personal occupation	Industry or service	Employment status	Whether working at home			
176	Ivy Cottage, Park Road	5	John Martin	Head	39		Married	–	–	–	–	Aras Tapestry weaver	–	Worker	No	Middlesex: Hampton Wick	–	–
			Louisa Martin	Wife		36	Married	14	2	2	–	–	–	–	–	London: Stepney	–	–
			William Martin	Son	13		–	–	–	–	–	–	–	–	–	Middlesex: Hampton Wick	–	–
			Doris Martin	Daughter		4	–	–	–	–	–	–	–	–	–	Middlesex: Hampton Wick	–	–

Source: The National Archives Registrar General 13/671/174/9; Registrar General 14/3568/357/1.

Note: dashes (-) denote blank entries on the census record indicating not applicable; household level data have not been duplicated for individuals.

These scanned registered raster images provide background mapping that aids the process of georeferencing 1901 and 1911 Census addresses. Figure 32.1 provides an illustration of the land cover changes that were taking place at the turn of the twentieth century using the example of an area in Deptford, south-east London. This area was part of Kent on the OS First Edition County Series map published in 1895 and shows that a few metres of Arica Road and St Norbert Road had been constructed and a few residential properties had been built along each (right of centre on upper map image). However, by 1916, when the Second Edition map was published, these roads had been extended southwards and other roads (Aspinall, Avignon, Dundalk, Finland, Revelon and St Asaph) had been built over what were formerly open fields (note field number and areas on First Edition map) and seemingly were lined with mainly terraced residential properties. Each map shows some buildings were named, but neither provides details of the house numbers or names, which were included in the census records. The address of the transcribed census records in Table 32.2, Ivy Cottage, Park Road, (not shown in Figure 32.1) is not a named building on either the First Edition County Series sheet published in 1896 or the Second Edition published in 1913, although other properties on the road are named, for example Thatched Cottage and a Vicarage.

One further historical data source deserves mention before switching to examine the early twenty-first-century mapping and census data sources that contributed to the research. Charles Booth's 'poverty map' was not used to assist with the georeferencing process, but is an example of an additional data source that can enhance analysis of the historical census records. Booth carried out a survey of people in London between 1886 and 1903 and he published some of the results of his inquiry as an innovative example of social mapping in 1889 (Booth, 1902). Such was the significance and interest in this first map that it was comprehensively revised ten years later with a new classification of the surveyed streets. Known as the *Map Descriptive of London Poverty, 1898–9*, the mapping comprises 12 map sheets at the 1:10,560 scale covering the area of the County of London from Hammersmith to Greenwich and Hampstead to Clapham. The maps were digitised and rendered searchable by the Charles Booth Online Archive project.[2] The Booth maps have been digitised so that they conform to modern standards of georeferencing and they were stitched together to create a single seamless image.

Early twenty-first century

One of the main differences between the historical and contemporary census data, at least from the users' perspective, is that the former are available as paper records or digitised images of the original documents from which data about individuals and households present at addresses can be captured. In contrast, contemporary census data, apart from the special case of the Samples of Anonymised Records, are cross-tabulated counts of one or more characteristics aggregated to spatial units of different sizes within nested hierarchies. Although these are generated dynamically with respect to the 2011 Census, users will not be able to access the complete set of household and individual records for a little under 100 years according to current legislation. However, as will become apparent in due course, one of the outputs from the research reported here is similar but inevitably less comprehensive aggregate statistics for both small historical units, for example 1901 and 1911 enumeration districts, and historical census data aggregated to modern spatial units.

Contemporary aggregate census data are accompanied by digital geospatial data corresponding to the boundaries of the spatial units enabling the statistical data to be mapped and visualised thematically. It is inappropriate to provide detailed information about the characteristics of the 2011 Census datasets, as these are contained elsewhere in this volume or in equivalent texts published following the last three censuses (Rhind, 1983; Openshaw, 1995; Rees et al., 2002).

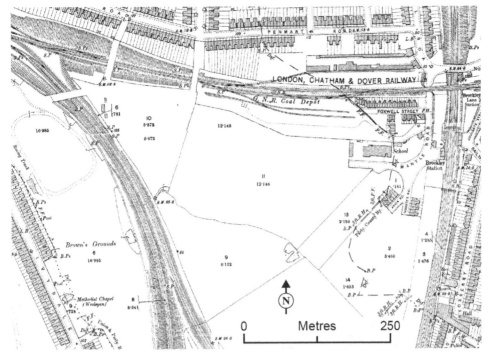

a. County Series, 1:2,500, First Edition, 1895, Kent

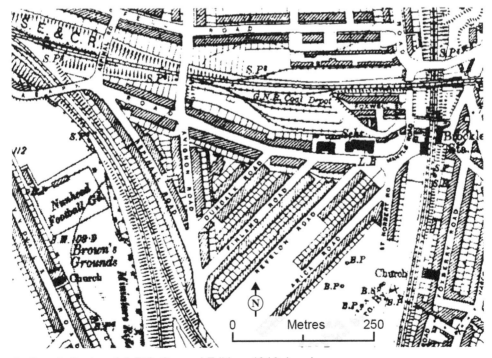

b. County Series, 1:2,500, Second Edition, 1916, London

Figure 32.1 Comparison of Ordnance Survey base topographic mapping for an area in Deptford, south-east London

Source: Landmark Information Group, EDINA.

However, one point worth mentioning at this stage is that the high degree of consistency between the output area geography of the 2001 and 2011 Censuses (over 95 per cent) means that it will be possible to aggregate the georeferenced historical census data to spatial units that are also reasonably consistent between 1901 and 1911.

Modern topographic mapping comparable with the digitised images of historical OS maps is not relevant to georeferencing the 1901 and 1911 Census addresses. However, one of the most important data sources assisting with this process is the Address Layer 2 of the contemporary *MasterMap* data produced by the OS. Having been launched in 2001 following re-engineering of the *LandLine* data, *OS MasterMap* has become the organisation's main digital topographic mapping product, replacing *LandLine* in 2008. *MasterMap* consists of a series of layers (topography, integrated transport, address and imagery), some of which are subdivided into themes. Some of these data are available for research and teaching purposes to UK Higher Education institutions through the Digimap service at the University of Edinburgh. The locations of addresses, both postal and non-postal, are held in *OS MasterMap* Address Layer 2 (MMAL2). It includes not only the georeference of each address, but also fields relating to the postcode, building number and name, thoroughfare and unique Topographic Identifier (TOID).

32.3 Methods and analysis

The various historical and contemporary geospatial and census data sources examined in the previous section are, for the most part, available nationally in a reasonably consistent format and structure. However, embarking on developing a method for georeferencing addresses in the 1901 and 1911 Censuses was not something that could feasibly be undertaken on a national basis with the resources available. It was necessary to carry out a pilot study and to select areas that would potentially demonstrate the feasibility of the method and provide some rewarding insights into demographic, economic and social conditions of Britain in the early twentieth century that could be compared and contrasted with the situation 100 years later.

The nineteenth century is often characterised as the period in which Britain's population transitioned from a largely land-based economy and society located for the most part in rural areas to an urban one centred on the burgeoning cities and towns founded on the 'factory system'. The population of England and Wales grew from 8.9 million in 1801 to 32.5 million in 1901 and by 1851 half of the population was regarded as 'urban'. Although much of this urban industrialisation was focused on northern England and South Wales, these industrialising and urbanising processes permeated the length and breadth of the country and London was emerging as a centre of administrative and political, if not yet perhaps economic, power. London's population exceeded 4.5 million in 1901 with 0.8 million in the neighbouring county of Middlesex. By the close of the twentieth century, these areas had for the most part been absorbed into the Greater London Authority, which additionally had expanded eastwards and southwards to include areas in the former counties of Essex, Kent and Surrey (Figure 32.2). Estimating the 2011 population of the area approximately equivalent to Middlesex and London Counties suggests the 1911 total of 5,609,150 had increased to 5,646,558, which represents 69.1 per cent of people in the GLA area as a whole, which now includes London Boroughs south and east of the two historical counties.

The specific details of the local government areas in place and the statutory functions for which they were responsible at the start of the twentieth century are very different from what exists today, although the principle of larger units containing smaller ones (counties and their constituent Metropolitan Boroughs (MB), County Boroughs, Urban Districts (UD) and Rural

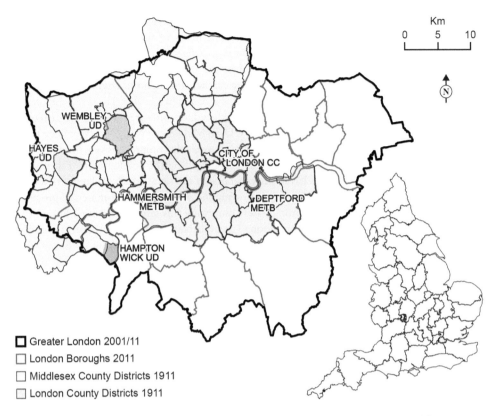

Figure 32.2 Case study boroughs and urban districts in former London and Middlesex counties in relation to present day Greater London Area

Source: Ordnance Survey, EDINA.

Districts) was already in place. Although a two-tier hierarchy currently occurs over much of England, restructuring over recent decades has eroded its existence everywhere. A group of 1911 MBs and UDs, three each from London and Middlesex counties, were selected for the purpose of developing and testing a method for georeferencing 1901 and 1911 Census addresses. These areas were selected to illustrate contrasting trends of population change 1901–11, to reflect different locations in relation to current inner and outer London and localities north and south of the River Thames. They were also chosen on the basis that the number of addresses to be georeferenced would be feasible with the resources available.

The six areas were the City of London MB, Deptford MB and Hammersmith MB in London County and Hampton Wick UD, Hayes UD and Wembley UD in Middlesex County, which together had total populations of 258,821 in 1901 and 268,048 in 1911 (respectively 4.9 and 4.8 per cent of their total person counts); and the equivalent figures for households were 57,823 and 61,369 (1901 and 1911) and for addresses 37,423 and 38,694 (again 4.9 and 4.8 per cent of the total in both cases). This chapter for the most part focuses on the results for the 1911 district of Hammersmith, with reference to the other five case study areas for the purpose of comparison. Hammersmith experienced population growth of 8.3 per cent between 1901 and 1911, which compares with a modest fall of -0.3 per cent across London County as a whole, making it the borough with the fourth highest upward growth over the decade.

It is obvious with the passage of more than 100 years between when the present research started in 2013 and the 1901 and 1911 Censuses that many of the addresses recorded in the historical censuses would not be present in modern georeferenced address data. Similarly, many contemporary addresses would not have existed in the first decade of the twentieth century. The MMAL2 data are not routinely available to researchers through the Digimap service at the University of Edinburgh in large part because of their commercial value. However, they can be used under special licence. Street gazetteers were produced for some areas around the turn of the twentieth century, including one of the sampled UDs in Middlesex (Wembley) (Wembley Urban District Council, 1906), but such information was not available comprehensively. Similarly, there are some local history societies that have created websites where the development of their settlement has been charted. Again, this has occurred in one of the sampled areas, Hampton Wick in Middlesex, where the website[3] includes maps showing the residential and commercial properties on the relatively small number of thoroughfares comprising this settlement with a link to their historical census record and other information. The availability of such local sources on only a piecemeal, patchy basis prompted use of the MMAL2 as an initial starting point for georeferencing historical census addresses for the sampled areas, whilst recognising its limitations.

The method developed to match and georeference these addresses can be summarised as a four-stage process:

1 Match addresses from transcribed 1901 and 1911 Census records and MMAL2 data; geocode joined spatial and attribute data; and create point features for addresses.

2 Examine visualisation of geocoded census addresses alongside MMAL2 addresses; and identify and correct anomalies (e.g. through road name changes).

3 Identify non-geocoded addresses from stages 1 and 2 and locate on historic map images; and manually digitise new points and transfer addresses from 1901 and 1911 Census data to additional points.

4 Create a unified file of matched spatial and census data for each set of households and individuals in 1901 and 1911.

Despite the increase in residential addresses between the 1901 and 1911 Censuses (see example in Deptford MB in Figure 32.1), the number of consistent addresses present in both sets of records was reasonably high. The process of matching census addresses to the MMAL2 records therefore started with 1911 and then worked backwards to 1901 to maximise the advantage arising from those addresses that were identical.

Figure 32.3 shows the different stages of the process and illustrates some of the issues that complicate the simple matching of MMAL2 data with 1911 addresses in Hammersmith. The residential number or name captured from the census records and held in the MMAL2 data were converted into a standard format (e.g. 42 High Street or Oaklea Park Road) as an address field in each database before being matched using standard procedures. Approximately 57 per cent of the 1911 Census addresses in Hammersmith were successfully matched at this stage representing 55 per cent of individuals and 57 per cent of households (Figure 32.3a).

Stage 2 entailed careful examination of the mapped points overlain on the historical topographic map images in order to identify streets or groups of properties that seemed likely to be residential where matching with the MMAL2 address field had not occurred. This scrutiny was able to account for a proportion of the unmatched addresses either as a result of incorrectly captured census record addresses or changes in the name of the thoroughfare. Figure 32.3b shows the example of Jeddo Road, which appeared on the map image and in the MMAL2 data, but was incorrectly transcribed as Jedds Road. The second example in Figure 32.3c shows where

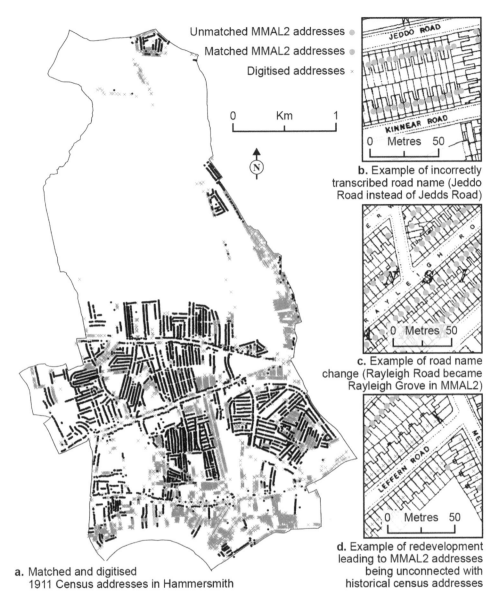

Unmatched MMAL2 addresses ●
Matched MMAL2 addresses ●
Digitised addresses ×

b. Example of incorrectly transcribed road name (Jeddo Road instead of Jedds Road)

c. Example of road name change (Rayleigh Road became Rayleigh Grove in MMAL2)

d. Example of redevelopment leading to MMAL2 addresses being unconnected with historical census addresses

a. Matched and digitised 1911 Census addresses in Hammersmith

Figure 32.3 Application of address matching between 1911 Census addresses in Hammersmith and digitisation of unmatched addresses

Source: Ordnance Survey, EDINA. © Crown copyright and/or database right 2016 OS.

mismatches arose from a change in the name of the thoroughfare from Rayleigh Road in the 1911 Census to Rayleigh Grove in MMAL2. Data editing and reapplication of standard field comparison procedures enabled these addresses to be successfully matched and raised the proportion of georeferenced 1911 addresses in Hammersmith to 63 per cent.

Stage 3 completed the georeferencing process by manually digitising the addresses that were still missing grid co-ordinates (Figure 32.3d). The location of these was achieved by a number of different means including use of local street gazetteers and websites (see examples previously mentioned), by examining the schedule numbers in the census records, which revealed that these generally run in a regular sequence along individual thoroughfares, and by searching for property names on the historical topographic map images. There were 5,944 or 26 per cent of 1911 addresses in Hammersmith that had to be identified and digitised in this way. At the end of stage 3, the grid co-ordinates for all census addresses in each of the six case study areas had been obtained. Matching with the MMAL2 meant that as well as transferring the grid co-ordinates the unit postcode could also be allocated to the historical address. Addresses that were not matched in this way were assigned the postcode of the nearest known MMAL2 address. Further processing of these georeferenced address records enabled these grid references to be attached to the transcribed census data records of individuals and households. This allowed the analysis to proceed for both types of social unit and to aggregate these records to historical and contemporary spatial units (e.g. 1901 and 1911 enumeration districts, thoroughfares, and 2001 and 2011 output areas).

32.4   Then and now

This section offers an introductory selection of results from analysing the georeferenced historical census data on their own and, perhaps more importantly in the context of this volume's focus on the 2011 Census, by aggregation to contemporary spatial units. Two aspects of the 1911 Census information for Hammersmith are explored in Figure 32.4. Georeferencing of the addresses

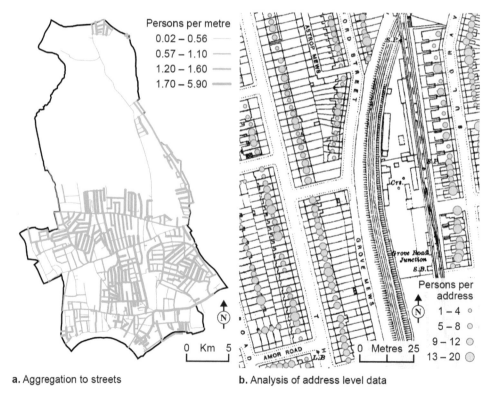

a. Aggregation to streets   b. Analysis of address level data

Figure 32.4   Examples of analyses of 1911 Census for streets and addresses in Hammersmith

Source: Ordnance Survey, EDINA. © Crown copyright and/or database right 2016 OS.

enables the data to be aggregated to thoroughfares and for measures such as the number of persons per metre to be calculated (Figure 32.4a). Viewing Hammersmith Borough as a whole, it is clear that some streets were more densely populated than others. This type of micro-scale demographic analysis is taken a stage further in Figure 32.4b where the numbers of persons per address in a small part of the borough are shown as graduated symbols superimposed on the historical topographic mapping. Both parts reveal some considerable variation in the density of population in 1911 in Hammersmith with dwellings of the same physical size and layout on individual streets having different numbers of people. Further analysis will explore the demographic and socio-economic composition of these streets and addresses in terms of the number of separate households, family structure, occupation, place of birth and childrearing.

The changing age structure of populations in developed countries has become an important issue for researchers and policy makers in recent years and is most clearly articulated in concerns over population ageing and the consequent challenges; for example, in relation to health care and pension provision, for governments faced with a rising proportion of older people in national populations. Exacerbating the issues associated with an ageing population in many countries is the now well-established trend for low fertility in some cases at sub-replacement level. Figure 32.5 offers an introductory exploration of changes in the percentage of two age groups (16–19 and 65 and over) in Hammersmith during the first decades of the twentieth and twenty-first centuries. The 1901 and 1911 individual-level census data were aggregated to the unaltered census output areas in Hammersmith in 2001 and 2011. Each map shows the difference in the percentage of one of the age groups either between 1901–11 or 2001–11; the maps in Figures 32.5a and 32.5c relate to the first period and the maps in Figures 32.5b and 32.5d to the second. The maps for 2001–11 include nine output areas not shown on the 1901–11 maps, which were unpopulated in either 1901 or 1911. Each of the maps uses quantile classification with dark blue shading indicating highest decrease in the age group and dark red greatest increase. In each case the spatial distribution of these changes shows a degree of randomness, although further spatial autocorrelation analysis using Local Moran's I does reveal clustering of high–high and low–low values in some locations in both decades (Anselin, 1995).

32.5 Conclusions

The Census held on 27 March 2011 was the twenty-first complete enumeration of the population in Britain and throughout more than a 200-year history of census-taking changes have been introduced between each successive survey. One perspective on this history is that every census is a unique event reflecting both the economic and social conditions and the administrative and technological norms of its time. However, an alternative view is that the rigour and diligence applied to each enumeration argues for exploring the rich history of demographic, economic and social change that can be charted through careful exploration of different points in this longstanding series of cross-sectional population data. It would be foolhardy to argue that any census or perhaps even alternative assemblage of administrative and survey data, no matter how rigorously conducted, would be capable of achieving 100 per cent accuracy in its count of people and a selection of their attributes. At the extremes of people's lifecourse, there will always be some births and deaths that are missed on the reference date, let alone failures in capturing all in- or out-migration.

The research outlined in this chapter has advanced one stage further the former historical investigations of census data by delving below the level of aggregate statistics to explore the opportunities arising from analysis of household and individual-level historical census records georeferenced to the addresses where people were enumerated. These data reveal the geo-analytical possibilities offered

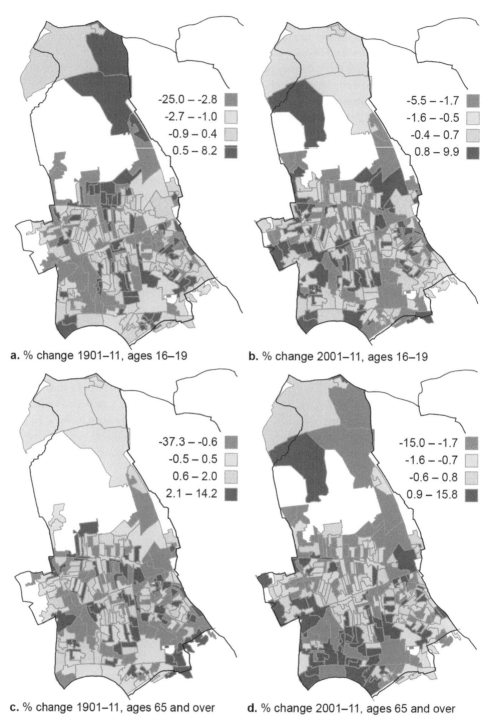

a. % change 1901–11, ages 16–19

-25.0 – -2.8
-2.7 – -1.0
-0.9 – 0.4
0.5 – 8.2

b. % change 2001–11, ages 16–19

-5.5 – -1.7
-1.6 – -0.5
-0.4 – 0.7
0.8 – 9.9

c. % change 1901–11, ages 65 and over

-37.3 – -0.6
-0.5 – 0.5
0.6 – 2.0
2.1 – 14.2

d. % change 2001–11, ages 65 and over

-15.0 – -1.7
-1.6 – -0.7
-0.6 – 0.8
0.9 – 15.8

Figure 32.5 Examples of analyses involving aggregation of 1911 Census addresses in Hammersmith to modern output areas

Source: Ordnance Survey, EDINA.

Note: The maps for 2001–11 include nine output areas not shown on the 1901–11 maps, which were unpopulated at that time. The quantile classification method has been used for each map.

to researchers from having available such a comprehensive database that are impractical with contemporary census data and from being able to explore the persistence or discontinuity in aggregate spatial differentiation over time. The method for georeferencing the historical census records outlined here has shown that it is feasible to use modern address databases to attach grid co-ordinates to historical addresses provided that sufficient supplementary information sources are available to help 'fill in the gaps'. One difference between the London County MB census addresses and those in the UDs in Middlesex when attempting to match with the MMAL2 is that a relatively high proportion of residential addresses on thoroughfares in Hampton Wick, Hayes and Wembley were identified by means of their names rather than by numbers, although the actual property may be the same.

At a time when the future of census-taking is uncertain and despite the commitment to holding a mainly online census in 2021, it is important not to discard the possibility of undertaking long-term analysis and data matching of the type outlined here as we progress through the twenty-first century and beyond. It is to be hoped that academic and genealogical researchers 100 years from now will not be denied the opportunity to explore their family histories and the detailed spatial variations that, for reasons of confidentiality, remain concealed from us as we investigate the demographic, economic and social conditions of the early twenty-first century.

Acknowledgements

Office for National Statistics, 2011 Census: Aggregate data (England and Wales) [computer file]. UK Data Service Census Support. Downloaded from: http://infuse.mimas.ac.uk. These data are licensed under the terms of the Open Government Licence (www.nationalarchives.gov.uk/doc/open-government-licence/version/2).

Office for National Statistics, 2001 Census: Aggregate data (England and Wales) [computer file]. UK Data Service Census Support. Downloaded from: http://infuse.mimas.ac.uk. These data are licensed under the terms of the Open Government Licence (www.nationalarchives.gov.uk/doc/open-government-licence/version/2).

This work is based on data provided through EDINA UKBORDERS with the support of the Economic and Social Research Council (ESRC) and Jisc and uses boundary material which is copyright of the Crown. It also uses data provided under licence by the Ordnance Survey.

The work on which this chapter is based was carried out under a British Academy, Senior Research Fellowship entitled The Geography of London's Population: Populating Places with Historical Census Statistics.

Notes

1 www.visionofbritain.org.uk/.
2 http://booth.lse.ac.uk/.
3 www.brickbybrick.org.

References

Anderson, M. (1987) *National Sample from the 1851 Census of Great Britain: Introductory User Guide*. Department of Economic and Social History, University of Edinburgh, Edinburgh.

Anderson, M. and Collins, B. (1973) National Sample from the 1851 Census of Great Britain: Sample Procedures. *Background Paper 2*, Department of Sociology, University of Edinburgh, Edinburgh.

Anderson, M., Collins, B. and Stott, C. (1977) The national sample from the 1851 Census of Great Britain: Sampling and data handling procedures. *Urban History*, 4: 55–9.

Anselin, L. (1995) Local indicators of spatial association – LISA. *Geographical Analysis*, 27: 93–115.

Booth, C. (1902) *Life and Labour of the People in London.* Macmillan, London.

Gregory, I.N. (2002) The accuracy of areal interpolation techniques: Standardizing 19th and 20th century census data to allow long-term comparisons. *Computers, Environment and Urban Systems,* 26(4): 293–314.

Gregory, I.N., Bennett, C., Gilham, V.L. and Southall, H.R. (2002) The Great Britain Historical GIS Project: From maps to changing human geography. *The Cartographic Journal,* 39(10): 37–49.

Gregory, I.N., Dorling, D. and Southall, H.R. (2001) A century of inequality in England and Wales using standardized geographical units. *Area,* 33(3): 297–311.

Harley, J.B. (1975) *Ordnance Survey Maps: A Descriptive Manual.* Ordnance Survey, Southampton.

Higgs, E., Jones, R., Schürer, K. and Wilkinson, A. (2013) *Integrated Census Microdata (I-CEM) Guide.* Available at: www.essex.ac.uk/history/research/ICeM/documents/icem_guide.pdf.

Hinde, P.R.A. (1985) Household structure, marriage and the institution of service in nineteenth-century rural England. *Local Population Studies,* 35: 43–51.

Openshaw, S. (ed.) (1995) *Census Users' Handbook.* GeoInformation International, Cambridge.

Rees, P., Martin, D. and Williamson, P. (eds.) (2002) *The Census Data System.* Wiley, Chichester.

Rhind, D.W. (ed.) (1983) *A Census User's Handbook.* Methuen, London.

Southall, H.R. (2003) A vision of Britain through time: Making long-run statistics of inequality accessible to all. *Radical Statistics,* 82: 26–43.

Southall, H.R. (2006) Electronic resources for local population studies. A vision of Britain through time: Making sense of 200 years of census reports. *Local Population Studies,* 76: 76–89.

Southall, H.R. (2014) Rebuilding the Great Britain Historical GIS, Part 3: Integrating qualitative content for a sense of place. *Historical Methods: A Journal of Quantitative and Interdisciplinary History,* 47(1): 31–44.

Tilley, P. and French, C. (1997) Record linkage for nineteenth-century census returns: Automatic or computer aided? *History and Computing,* 9(1, 2 and 3): 122–133.

Wembley Urban District Council (1906) *Wembley Directory and Almanack for 1906.* Wembley Urban District Council, Wembley, pp. 58–77.

Part VI

Looking forward and beyond 2021

33

Towards 2021 and beyond

Andy Teague, Lara Phelan, Meghan Elkin
and Garnett Compton

33.1 Introduction

This chapter provides an overview of the work carried out by the Office for National Statistics (ONS) on the Beyond 2011 Programme leading up to the National Statistician's recommendation on the future approach to census and population statistics in March 2014. It then goes on to describe the new Census Transformation Programme (CTP) and the three key objectives: to have a predominantly online census of all households and communal establishments in 2021, with special care taken to support those who are unable to complete the census online; to produce improved and expanded population statistics through increased use of administrative data and surveys; and to establish evidence to enable a decision about the future provision of population statistics after 2021 – and current thinking on the design of the 2021 Census. Notably, there is the move to an online-first approach for the public to complete census questionnaires and to the greater use of administrative data to improve the efficiency, effectiveness and quality of the census. Finally, we describe the Administrative Data Census project which is focused on research into the potential of using administrative data and surveys for replacing the ten-yearly census beyond 2021.

33.2 The Beyond 2011 Programme and the National Statistician's recommendation on the future of census and population statistics

The ONS has periodically reviewed its methods for collecting population data and the effectiveness of potential alternatives. It did so particularly in the wake of the 1991 and 2001 Censuses as part of planning for the next census. The ONS post-2011 review needed to address the requirements of users for more frequent population data, and concerns in the media and the public about perceptions of greater intrusiveness and the increasing cost to taxpayers. This review also re-assessed the alternative ways of collecting census-type information that had been examined in a previous review in 2003 to see if any were now viable.

In parallel with similar initiatives at the National Records of Scotland (NRS) and the Northern Ireland Statistics and Research Agency (NISRA), ONS set up the Beyond 2011 Programme to examine these issues, and test new models for gathering population and socio-demographic statistics. Improvements in technology and in government data sources offered the opportunity either to modernise the census or to develop an alternative approach based on re-using the administrative data that the public has already provided to the Government.

Review and user consultation

The Beyond 2011 Programme covered: identification and prioritisation of all options; review of international practices; identification of alternative data sources, including surveys, that could be used to meet user requirements; and testing and evaluation of the main options. Stakeholder engagement and communication plans were designed to ensure that users, stakeholders and all those with an interest in the Programme clearly understood the work being done, its research findings, the evaluation results and the decision-making processes and procedures. All the options considered were carefully researched and tested by ONS, and then assessed transparently using an agreed set of criteria (ONS, 2014b) to ensure that they could meet users' requirements, provide population and socio-demographic statistics of the required quality, and be acceptable to the public.

Consultation with users was a key component of the Beyond 2011 Programme, helping to inform the assessment and evaluation of options as well as the final recommendation. The initial public consultation, between October 2011 and January 2012, sought information from users about their information requirements and priorities, and their views on the relative importance of accuracy, frequency and geography in the production of population and socio-demographic statistics. Detailed results are published on the ONS website (ONS, 2014c).

All sectors of users were represented in the responses to the consultation; most were from local authorities (44 per cent), from genealogists and family historians. Although aware of the particular genealogical interest in the census, the consultation was primarily designed to capture the views of 'statistical' users on: (i) the current and future requirements for population and socio-demographic statistics on different topics; and (ii) the trade-off between accuracy, geography and the frequency of statistics. A shortlist of six options was then reviewed (four different approaches with variants) as follows:

1   *Full census*: to be carried out decennially as at present, but modernising the methodological approach by, for example, putting more emphasis on data collection via the internet (similar to the approach in Canada).
2   *Rolling census*: an annual enumeration of up to 10 per cent of the population, carried out in different areas each time so that, over ten years, the whole country is covered (similar to the approach in France).
3   *Short-form census with an annual national sample (4 per cent) survey*: in which a short form is delivered to every household every ten years, supplemented by an annual survey using a long form to collect the full range of census characteristics (similar to the approach in the USA).
4   *Annual data linkage with a decennial national sample (10 per cent) survey*: where administrative data are linked to produce population estimates, supplemented by a decennial large-sample long-form survey to derive the necessary population characteristics.
5   *Annual data linkage with an annual national sample (4 per cent) survey*: similar to the previous option but with a smaller annual survey, which could produce more frequent statistics.

6 *Annual data linkage with a decennial national sample (40 per cent) survey*: similar to the fourth option above but with a much larger a smple survey that would allow small-area statistics to be produced.

Of these options, two clear front runners emerged: the full online census option and the administrative data linkage option with a 4 per cent annual survey. ONS identified clear pros and cons to these two approaches in terms of quality, frequency and the nature of outputs, and the different risks they carried. An online census would produce the wealth of small-area data and detailed cross-tabulations that have traditionally come from the census, but only every ten years. On the other hand, an administrative data solution would deliver statistics much more frequently (annually for many key topics). This had the potential to be more responsive to user needs, but would not provide the level of detail provided by the traditional census option for the smallest areas or smallest population groups.

The second public consultation ran from 23 September to 13 December 2013. This presented the pros and cons of the two options in a detailed consultation document, and invited views via an online questionnaire on which approach would best fit user needs. The results would enable the National Statistician to make a recommendation for the collection of future population statistics. The three-month public consultation resulted in more than 700 responses from the Government, local authorities, public bodies, commercial organisations, charities, academics and genealogists. Two-thirds were from individual citizens and users, while a third were from organisations representing users. The report of the public consultation was published in March 2014 (ONS, 2014a) and its key messages were as follows:

- Population statistics were highly valued by a range of national and local users across England and Wales.
- There was continuing demand from the Government, local authorities, public bodies, business, the voluntary sector and individual citizens for the detailed information about small areas and small populations offered by the decennial census, whether online or paper-based; such statistics were regarded as essential to local decision making, policy making and diversity monitoring in fulfilment of legally binding public duties.
- Most users recognised the value of making greater use of administrative data to produce more frequent population statistics. There was a strong concern that the proposed use of an annual survey of 4 per cent of households (to support the use of existing administrative data) would not meet these needs, nor deliver the required small-area and small-population statistics offered by the decennial census.
- While the methods using administrative data and surveys showed considerable potential, and the more frequent statistics they could provide between censuses would be welcome, there was concern that these were not yet mature enough or of sufficient statistical quality to replace the decennial census.
- Many respondents noted that other countries have taken decades to develop replacement systems, and some stated that it would be 'reckless' to move too fast in that direction.
- Many respondents proposed a hybrid approach, making the best of both approaches, with an online census in 2021, enhanced by administrative data and household surveys.
- Many individual users acknowledged that their primary interest was in the census as a historical source and urged continuation of the historic series.

The great majority of respondents valued the decennial census and particularly the small-area and small-population data at its heart. However, they also saw the potential benefits of using administrative data to increase frequency and potential range of population statistics. Support for one approach clearly did not preclude support for the other.

Public acceptability

ONS also carried out a programme of research into public attitudes about the different approaches to providing population statistics and the use of administrative data. This programme of research was conducted from 2009 to 2014 and employed both quantitative (surveys) and qualitative (focus groups) methods. The research insights showed:

- The public are generally positive towards the ten-yearly census as a means of gathering information about the population. Census-taking is widely understood by the public but there can be misconceptions about its purpose. The majority of people are happy to provide sensitive personal information on a census.
- The majority of the public do not object to data held by other government departments being shared with ONS, but public benefits of data sharing are not widely understood; there is a general assumption that data are already shared between government departments.
- Concerns over information generally centre on personal privacy, data security, unauthorised access and general objections to the amount/type of data being shared.
- The majority of the public trust ONS to protect the confidentiality of their data.
- When provided with reassurance with regard to security and privacy, public support improves for ONS re-using administrative data to produce statistics.

In general, the public's concerns encompass the following themes and these will need to be managed carefully by ONS: security and confidentiality; privacy and anonymity; transparency, control, consent and trust; governance and regulations, and lack of understanding of public and personal benefits.

Independent review

ONS commissioned an independent review of the methodologies of the two options (ONS, 2014d). Led by Chris Skinner, Professor of Statistics at the London School of Economics, the review sought in particular to: (i) assess the methodological research and evidence reported by ONS as the basis of its evaluation of the alternative options; (ii) identify the main risks with the two front-running options, and to identify areas where further work is required to mitigate these risks; and (iii) enable a sound assessment of methodology issues relevant to the decision on how to proceed.

The review team gave more attention to the administrative data option, because it represented a more radical change in methodology than the online census option. The review noted that the online census would represent a natural evolution of the traditional decennial census and would mirror lines of development in some other countries, for example Canada, where 54 per cent completed the census online in 2011. It also emphasised that the administrative data option would represent "the most substantial change in the production of statistics for over one and a half centuries" (ONS, 2014d, p. 4).

In order to progress the administrative data option for the future provision of population statistics, the review proposed further research by ONS and the statistical community. Furthermore, the review team noted that: "a key requirement is that there is suitable data sharing legislation between the statistical office and the authorities with control over the administrative systems so that these data will meet the statistical needs over time" (ONS, 2014d, p. 5).

Professor Skinner and his colleagues came to the conclusion that the online census option was relatively low risk and "would represent a natural evolution of the traditional census, drawing on technological innovations and developments in best practice for census taking around the world" (ONS, 2014d, p. 4). The review team stressed the importance of maintaining the compulsory nature of the census and following up any households where there was no online return. They had no hesitation in saying that an online census represented "a methodologically sound basis for replacing Census 2011" (ONS, 2014d, p. 8) but were not prepared to say the same for the administrative data option at the current stage of development in England and Wales.

National Statistician's recommendation

Following discussions with the Registrars General for Scotland and Northern Ireland, and with the Chief Statistician for Wales, the National Statistician recommended (ONS, 2014e) on 27 March 2014 that the UK Statistics Authority (UKSA) should make the best use of all sources, using data from an online census in 2021 *and* administrative data and surveys. This would include:

- an online census of all households and communal establishments in England and Wales in 2021, as a modern successor to the traditional, paper-based decennial census, taking special care to support those who are unable to complete the census online; and
- increased use of administrative data and surveys in order to enhance statistics from the 2021 Census and improve statistics between censuses.

The recommendation added that this approach would provide the population statistics which the nation requires for the next decade and offer a springboard to the greater use of administrative data and annual surveys. This approach may offer a future Government and Parliament the possibility of moving further away from the traditional decennial census to annual population statistics provided by the use of administrative data and annual surveys. It was noted that further research would be required to determine the optimal blend of methods and data sources. It was made clear that the future development of the administrative data approach would depend on public consent, as expressed through Parliament, and it was recognised that data sharing legislation would be required to maximise the benefits of using administrative data for statistical purposes.

The National Statistician's recommendation was commended to the Government in a letter from Sir Andrew Dilnot, Chair of the UKSA, to the Rt Hon Francis Maude MP, Minister for the Cabinet Office, on 27 March 2014. On 18 July 2014, the Minister for the Cabinet Office wrote to Sir Andrew Dilnot (Cabinet Office, 2014) endorsing the National Statistician's recommendation for a predominantly online census in 2021 supplemented by further use of administrative and survey data. The Government's response referred to the: "[Government's] ambition . . . that censuses after 2021 will be conducted using other sources of data . . . In the period up to 2021, UKSA's plans should include ensuring that adequate research into the use of administrative data and surveys is carried out to enable a decision about the future methodology for capturing population and census data" (Cabinet Office, 2014, pp. 1–2).

33.3   The Census Transformation Programme

The ONS established the CTP in January 2015 to take forward the National Statistician's recommendation on the future of the census and population statistics. The recommendation to make the best use of all available data in the production of population statistics is reflected in the three major objectives of the Programme, to deliver:

- a predominantly online census of all households and communal establishments with special care taken to support those who are unable to complete the census online;
- improved and expanded population statistics through increased use of administrative data and surveys; and
- evidence to enable a decision about the future provision of population statistics after 2021.

The Programme has been segmented into three strands to enable ONS to deliver the objectives and to ensure that the development of administrative-based statistics is a priority for ONS in both delivering the 2021 Census and enabling change after 2021.

Strand 1: 2021 Census data collection operations

This involves research, development and operation of a 2021 Online Census. The census operation is a large complex activity that requires years of planning. Most of the operation related to data collection occurs in a relatively short period around census day. Once the data collection operation is concluded, a complex set of processes to convert responses into statistics begins. ONS is investigating how administrative data can be used to target follow-up and support services and to help understand and model non-response. This includes looking to add additional information to the address register to provide contextual information about individual addresses. With more information about each individual address there are opportunities to improve plans for census and survey operations, target follow-up and support services, and help understand and model non-response.

Strand 2: integrated population statistics outputs

This objective involves developing and implementing methods for enhancing 2021 outputs; for example, using administrative data to help estimate non-response or to collect characteristics of the population not historically included on the census (for example, using information from the Department for Work and Pensions (DWP) and Her Majesty's Revenue and Customs (HMRC) to produce statistics on household and personal income).

Strand 3: administrative Data Census

Building on the work carried out during the Beyond 2011 Programme (see Section 33.2), this objective involves research into the potential to switch to an administrative data census after 2021. Key to this will be assessment criteria that will need to be met to provide evidence to move away from the ten-yearly census approach after 2021, including comparing administrative and survey-based outputs with those from the 2021 Census. This will culminate in a recommendation about the future provision of population statistics in 2023.

33.4   2021 Census design

This section describes current thinking on the 2021 Census design. It considers the impact on the design of a move to an online-first approach as a means for the public to complete census questionnaires, and the greater use of administrative data to improve the efficiency, effectiveness and quality of the census. The design will evolve through iterative testing as ONS works through a test in 2017 and a rehearsal in 2019, before the census in 2021.

According to this design, a census is defined as the collection of information from all people about their personal and household characteristics in reference to a particular date, census day; and the subsequent production and publication of population statistics. This description is confined to the design of the 2021 Census in England and Wales, for which ONS is responsible.

Drivers for change and design principles for 2021 Census design

Every census is different due to changes in society impacting what users require to be collected and how the census should be conducted. The initial high-level design for the 2021 Census is driven by the lessons learned from the 2011 Census and international census-taking, changes in technology, improvements in administrative data sources (such as the patient register and tax and benefits data), and the requirement to continue to deliver value for money.

ONS has defined a set of principles that will guide the development of the design for 2021. ONS will:

- utilise the elements of the 2011 Census that worked well and are still relevant;
- embrace new technologies and methods wherever possible;
- design the operational and statistical processes for online first;
- make it as easy as possible for the public to respond without assistance, while ensuring that the design complies with recognised standards on equality and accessibility;
- seek to minimise the respondent burden on the public – the length of time to complete the questionnaire (online or paper if applicable) should be about the same as, or less than, that in 2011;
- protect, and be seen to protect, confidential personal data;
- attempt to get a response from every person and household in England and Wales;
- the operation and statistical methods employed should be developed to deliver the highest quality for the population estimates by age and sex at the local authority level; and
- maximise the use of administrative data in all areas of the operation.

Main areas of change in the 2021 Census

The main change for 2021 is the move to a predominantly online census. While the move to online has many opportunities for all aspects of the 2021 Census, from the design of the questionnaire to the processing and production of outputs, it also introduces new challenges such as minimising digital exclusion by assisting respondents to complete online, providing support through the field force, and community liaison. This section provides an overview of the main design features and, where relevant, lists some of the opportunities and challenges that will be considered when finalising the design for the 2021 Census. For the most recent plans, see the CTP pages on the ONS website.

Accessing census questionnaires

In previous censuses, a paper questionnaire was delivered to every household. For the 2021 Census, the way householders are introduced to their census questionnaire will be different, encouraging them to go online first. The 2011 Census online questionnaire was accessed using a 20-character authentication code printed on the paper questionnaire that was posted to households; this provided a unique link between a response and the address. International experience shows that sending just a letter with an invitation to complete results raises online response. Work is underway to understand the most effective way of contacting respondents in an 'online first' solution, and to investigate alternative methods of authentication.

To encourage online response, ONS will ensure that respondents who would like to complete online but are unable to, are supported. Understanding respondents and how they wish to interact with a census collection exercise, based on an understanding of interactions with other government services, is vital to achieving the required response rate. Current research takes account of assisted digital requirements to meet Government Digital Service (GDS) guidelines.

Target populations and digital inclusion

A primarily online census introduces new challenges. There will always be hard-to-count populations who are at risk of low levels of engagement or response. These are by their nature specific and require special attention. These groups were identified and prioritised in the 2011 Census. ONS will build on this successful strategy for the 2021 Census, ensuring that any emerging groups are included and engaged with at an early stage to understand their needs and explore appropriate ways to achieve higher levels of response.

In addition, new digital services need to meet the Government 'digital by default' standard and provide services that are straightforward and convenient to use so that all those who can use digital services will choose to do so, while those who cannot are not excluded. An estimated 10 per cent of the adult population (not households) may never be able to gain basic digital capabilities because of disabilities or basic literacy skills (Cabinet Office and Government Digital Service, 2014). Digital exclusion typically affects some of the most vulnerable and disadvantaged groups in society. The design and operation of the census will therefore need to take particular account of the requirements of these individuals.

Follow-up of non-responding householders

As with previous censuses, ONS will use different strategies to contact non-responding households and encourage them to take part. The precise timing and combination of follow-up modes is under development, but it will include the use of reminder letters and household visits from census collectors. Use of telephone follow-up is also under consideration. As in 2011, there will be support available in local venues such as libraries, drop-in centres and religious centres.

Collectors will make multiple visits to non-responding households to persuade non-responders and offer support where it is needed. If an address has been incorrectly included, they will provide verification before removing it from the system. As time progresses, the emphasis would shift from support and encouragement to the legal requirement for completing the census and the penalty for not doing so. Later on, a small, separate field force will manage more persistent non-compliance.

Operational management

It is expected that census collectors and their managers will be provided with up-to-date information about which households have not completed a questionnaire. As in 2011, this will enable managers to effectively refine and target the follow-up operation, flexibly deploying staff to the areas with the lower response rates. Online collection will also enable up-to-date information about the types of populations (e.g. students) that have not responded. This could be used to further target publicity or community engagement activities. The prioritisation of collector resource, publicity and community engagement will be important factors in minimising variability in response rates between areas and also between population sub-groups.

Data processing

The aim to have a high proportion of online completions will also provide opportunities to improve the quality of processing of the information. As shown in the 2011 Census, the quality of online completions was significantly better than that of paper questionnaires. There were substantially lower rates of question non-response for internet returns (in comparison with the Census Quality Survey), and internet responses were generally more accurately completed than paper responses. Given that only 16.4 per cent of household responses were completed online in 2011, there is an opportunity to increase the quality of the information captured. A substantial online response also provides opportunities to speed up the processing of census information, compared with a predominantly paper operation. ONS aims to build on the early availability of data and publish results earlier.

Census outputs

ONS will continue to seek ways of making outputs available sooner, while maintaining the high levels of quality and confidentiality of the information provided by the census. ONS will be reviewing the way that it processes, protects and disseminates a range of census outputs to ensure that they continue to meet evolving user needs.

Key elements of the design

As with the 2011 Census, there will be a number of key elements that help to deliver the 2021 Census.

Address register

As with the 2011 Census, a high-quality address list will be essential for delivery and linking of internet access codes to addresses, to allow targeted follow-up of non-responding households and to underpin estimation and the production of outputs following the census. The core of this register is likely to be the National Address Gazetteer (AddressBase) that is maintained on behalf of Government by Geoplace. This incorporates address information supplied by local government, Royal Mail and Ordnance Survey (the main sources used in 2011).

ONS aims to include the use of more address-level intelligence from administrative, commercial and open data sources. By knowing more about addresses, it may be possible to better target resources by, for example, identifying those areas where we are likely to find vacant properties, second residences, or encounter problems with access. A register of communal establishments, such as university accommodation, prisons, caravan parks and army camps, will form an integral part of the address register – again drawing on the best sources to ensure the proper enumeration of residents. ONS will work with data suppliers to ensure improvements to, and improved links between, address sources. A main part of the work will be to assess the quality of such sources. This will probably include some field address checking.

As well as being central to the census operation, address-level data are increasingly important in the linking of administrative and other sources to produce statistics. Work on the address register will underpin research on the reuse of administrative data for 2021 outputs and beyond. The quality of administrative sources would be improved if all owners of such data used a harmonised address framework.

The online questionnaire

Building on the successful 2011 online questionnaire design, research is being undertaken to understand how to transfer the questions to mobile and electronic devices. In addition, the online questionnaire will be designed independently from the design of any paper questionnaire to maximise online take-up and data quality. Opportunities that will be considered and tested include: contextual help to help complete questions; use of detailed drop-down boxes to reduce, or eliminate altogether, the amount of coding for more detailed questions such as occupation, industry or country of birth; more comprehensive validation within and between questions; and design of questions to fit smaller screens.

Stakeholder engagement

Partnerships with, and support from, stakeholders helped to make the 2011 Census such a success. The ONS' aim is to build on these strong relationships to help to design and run the 2021 Census, and to provide users with the census outputs they need. Based on experiences from the 2011 Census, ONS knows that local authorities' knowledge and understanding of their local areas and communities significantly contributed to the success of the census. The need for early engagement with local authorities is a recommendation for the 2021 Census. The aims of the planned 2021 Census local authority engagement programme will be: to raise awareness and understanding of the 2021 Census; to explain to local authorities the role that they can play in participating in and supporting the 2021 Census; and to build confidence and trust in the 2021 Census methodology and outputs.

Previous censuses have shown that certain population groups are less likely to complete their census questionnaires for a number of sometimes complex reasons. These include young men, students, certain Black and minority ethnic groups (BME), the very elderly, low-income families, non-English speakers, and people with disabilities. The 2021 Census will give support to these groups (as was done in 2011). Using an approach similar to the 2011 Census (see Chapter 3 of this book), ONS' planned 2021 Census community engagement programme will aim to encourage strong partnerships with community groups, charities and voluntary organisations, and universities.

Publicising the census

Every census presents a unique publicity challenge, communicating with and motivating every household to fill in their census questionnaire. The target audience is effectively 'everyone'. As in 2011, a paid-for publicity campaign – including TV, radio, outdoor and digital advertising – will be needed in 2021 to raise awareness of the census and to communicate a clear call-to-action to the entire population at the same time. With the greater diversity of the population, the fragmentation of media channels, and the proliferation of digital media channels, it is impossible to reach everyone through one channel. On the upside, particular channels can still be very effective in reaching specific audiences. Consequently the 2021 Census will use a combination of channels to make contact with every household in England and Wales. Digital channels have become the default communication channel used by government to engage with the public and will play a more prominent role in 2021. Social media will be more important than it was in 2011.

In 2021, ONS will need to encourage some audiences more than others to fill in their census questionnaires. Learning from the 2011 Census, the publicity campaign in 2021 will have an additional focus on specific audiences such as young people, students and Black and minority ethnic (BME) groups. Similarly, some audiences such as the elderly will require more help to fill in their questionnaires online, and the publicity campaign will need to steer people to the support available.

Supporting respondents

As in the 2011 Census, ONS will offer a comprehensive support structure for respondents. This will range from initial contact and requests for questionnaire access through to the supply of translation services and ensuring support is provided to those who cannot use online government services on their own (meeting all the requirements of 'assisted digital'). ONS will provide a telephone helpline in 2021 and, as in 2011, will provide online help, Interactive Voice Response (IVR), translation services (both 'live' and via documentation) and services for the visually or hearing impaired. For 2021, consideration is being given to extending the support services available.

Field force management

The objective of the field design will be to maximise overall response, while achieving sufficient response in all local authorities to enable successful estimation of the population. To achieve this, ONS needs to track responses to understand response patterns and, using area-based response targets identified by the hard-to-count work, allow us to direct follow-up activities. The 2021 Census will be primarily online, while recognising that maximising response is an overriding objective. While encouraging the vast majority of respondents to complete online, ONS will need to minimise digital exclusion, assisting respondents to complete online, providing support through the field force and community liaison, or by providing paper questionnaires. To achieve this objective ONS will require a large field force (expected to be a similar number to that used in 2011).

There are several design changes being explored that would facilitate changes to the field force number, structure and workload, as follows:

- *Designing for more cost-effective methods of non-respondent follow-up*: extensive use of follow-up letters (and possibly emails), reducing the need for census collectors to be 'on the ground' in areas with higher initial response rates.
- *Investigating the use of administrative data to inform follow-up*: reducing wasted visits by understanding the type of resident (for example, those who are out all day), and giving us a better understanding of the type of property (high turnover of residents, or second home).
- *Early analysis of online responses*: comparing distributions of main characteristics of responding households and people with the expected distributions; this will enable flexible deployment of the field force in areas where distributions are not as expected.
- *Equipping the field force with hand-held technology*: to enable automatic direction of census collectors to follow-up addresses in priority areas; route planning; assistance with training; communication and tracking of hours and expenses. This would also reduce the census co-ordinators' workload, which in 2011 concentrated on allocation of work. This is subject to considerations of security/privacy and cost/ benefit.

Communal establishments

In reviewing the 2011 Census, ONS identified the enumeration of communal establishments as an area where improvements could be made for 2021. As a result of having a predominantly online census in 2021, and a greater use of administrative data, there are opportunities to improve the enumeration of communal establishments. A main focus of the 2021 design will be to tailor the enumeration approach to the type of establishment, but in full consideration of the administrative data available for that type. Extra support will be given for those communal establishments where residents may have difficulty completing a questionnaire online, including the option of completing on paper.

Coverage assessment

Every effort is made to ensure everyone is counted in a census. However, no census is perfect and some people are missed. This under-enumeration does not usually occur uniformly across all geographical areas, or across other sub-groups of the population such as age and sex groups. In order to achieve the objectives of the CTP, coverage assessment and adjustment is required. This is achieved by conducting a Census Coverage Survey (CCS), independent of the census. This is a large, focused survey undertaken shortly after the census, collected using a method different from the census. The current expectation is that it will be similar to the survey conducted in 2011.

Use of administrative data in the census design

A significant change from 2011 will be the greater use of administrative data in the design and conduct of the 2021 Census. Administrative data may be able to improve the efficiency, effectiveness and quality of the 2021 Census through prioritisation and operation improvements and through improving statistical quality.

Prioritisation and operational improvements

For 2021, the strategic aim is the same as 2011 – to maximise overall response but also minimise the variability in non-response. While initial collector resource will still be targeted at areas where the initial returns are predicted to be lowest, there are likely to be data that could support lower-level targeting, potentially at address level. This could enable a more efficient targeting of resources, provided that predictions of likely non-response patterns are accurate. Research is underway to identify such data and assess their ability to predict non-response patterns. However, it is worth noting that predicting the patterns of non-response may be more of a challenge than in 2011 as a result of the online-first approach.

Administrative data may help us in better understanding the status of addresses and therefore enable us to prioritise field resources. For example, administrative data may be able to identify households that are vacant or second residences. This would enable field resources to be prioritised to those households which are harder to contact or need further encouragement for completing their questionnaire.

Statistical quality

Administrative data could be used to improve the quality of the census estimates. In the 2011 Northern Ireland Census, high-quality administrative data were added to the census data for some households that did not respond. This improved the quality of the resulting statistics. Such an approach would depend on the quality and availability of suitable administrative data and the outcome of further research to understand the implications and quality gains of such an approach.

Outputs

ONS currently assumes that it will continue to use output areas as the base geography to provide data for small areas, comparability with previous censuses, and as building blocks for additional and changing geographies. Given the success of the additional population bases in 2011, ONS will look to retain these and consider additional bases for 2021. ONS will be seeking to integrate additional administrative data within the main census outputs as much as possible, and build a platform for the provision of small-area statistics beyond 2021.

33.5 Administrative Data Census

As outlined in Section 33.2, the third objective of the CTP is about delivering evidence to enable a decision about the approach to provision of census and population statistics after 2021. The evidence will be crucial to determining whether the Government's ambition "that censuses after 2021 will be conducted using other [than traditional census] sources of data" (Cabinet Office, 2014, pp. 1–2) is possible. ONS' aim is to replicate the information collected through the census with administrative data already held by government, supplemented by surveys. The goal is to be able to compare outputs based on administrative data and targeted surveys against the 2021 Census to demonstrate to government and other users that the alternative can produce high-quality information at a lower cost, and can do so on a more regular basis. To track progress, ONS intends to publish an annual assessment of progress towards an Administrative Data Census after 2021 starting in Spring 2016. This will culminate in a recommendation, at the time of writing, in 2023.

ONS outlines that, in order to produce the type of information that is collected by a ten-yearly census (on housing, households and people), an Administrative Data Census will require a combination of:

- access to record-level administrative data held by government;
- a population coverage survey; and
- a population characteristics survey.

ONS has identified that the following five criteria would need to be met to enable ONS to move to an Administrative Data Census:

1 rapid access to existing and new data sources;
2 an ability to link data efficiently and accurately;
3 an ability to produce statistical outputs that meet priority information needs of users;
4 to be acceptable to key stakeholders (users, data suppliers, public and Parliament); and
5 that an Administrative Data Census would be value for money.

Further details will be published as part of the annual assessments on the CTP pages on the ONS website. This will include the evaluation criteria which are used to assess the five criteria.

The annual assessment will be supported by ONS' research into administrative data, which will include the regular production of administrative data-based research outputs. Through publishing updates on administrative data research, ONS aims to: demonstrate progress; seek feedback from stakeholders about quality, how methods could be improved and how new data sources could be used to improve quality in the future; operationalise procedures to ensure they can provide the accompanying infrastructure to produce and publish statistics using integrated data sources; and derive early benefits from the CTP.

The first research outputs were published by ONS in October 2015. This first publication (ONS, 2015) included:

- the methodology used to produce the research outputs, with consideration for how the statistical population dataset (SPD) methodology differs from the existing method used for the production of mid-year population estimates (MYE);
- a summary of national performance of the SPD and some of the coverage issues associated with the use of administrative data;
- a more detailed analysis of SPD performance at local authority level by sex and five-year age groups; and
- a series of detailed case studies that highlight some of the particular quality issues relating to the accuracy of administrative data, including timeliness of updates and operational processes.

ONS will expand the range of topics included in the research outputs each year, depending on the availability of data and their quality. At the time of writing, research outputs being explored include personal or household income, ethnicity, qualifications, as well as the number, size and composition of households.

The 2015 Research Outputs release (ONS, 2015) contained the first release of estimates of the size of the population that have been produced by pseudonymously linking multiple administrative datasets to produce a Statistical Population Dataset (SPD). Pseudonymisation is a procedure by which identifying fields (i.e. names, dates of birth and addresses) within a data record are replaced by one or more artificial identifiers to protect the privacy of individuals. As shown in Figure 33.1, the administrative datasets used to construct the SPD include the National Health

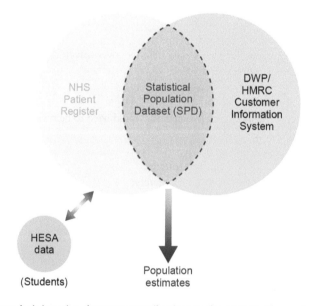

Figure 33.1   The administrative datasets contributing to the statistical population dataset

Service (NHS) Patient Register, Department for Work and Pensions (DWP) Customer Information System, and data from the Higher Education Statistics Agency (HESA).

The latest administrative data research outputs can be found on the CTP pages of the ONS website (ONS, 2015). As noted during the research carried out in the Beyond 2011 Programme, administrative data alone are unlikely to produce population statistics that are robust and of the necessary quality. Surveys are likely to be needed to help assess and adjust for coverage error in the administrative data and to provide some of the information traditionally collected in a census that is currently not available in administrative data.

The integration of administrative and survey data to provide the detailed multivariate small-area statistics provided by a census is a significant methodological challenge, compounded by the fact that all other countries that have switched to a register or administrative data-based approach have population registers. There is no doubt that administrative data have a major role to play. There are significant benefits in terms of reduced costs to the taxpayer and in the potential to provide more frequent small-area statistics. Meeting the criteria outlined above will be essential if and when we make the biggest change to the production of statistics on our population in over 200 years.

33.6   Conclusion

The census has always been about both continuity and change. It is about continuity because government and other users want to be able to compare the results between censuses to trace large-scale social change (e.g. the ageing population; the changing nature of immigration). But every census is also about change – the census evolves to meet social expectations and technological change.

The 2021 census is undergoing probably the most radical transformation since the introduction of punched card technology in 1911. This transformation includes, for instance:

- a digital by default census, with 75 per cent of responses (nearly 19 million returns) expected online – the highest online response rate target in the world;
- the online census will be the largest public facing government digital service in 2021;
- development of alternative census estimates based on administrative data;
- benchmarking of administrative data-based estimates to the 2021 Census;
- the integration of census and administrative data (e.g. on income) producing new outputs; and
- the move from paper to mobile technology to manage the field force workload.

ONS' aspiration is to be able to readily access administrative data held by government and be able to replicate, as far as possible, the information provided by the census in 2021. The prize sought is to be able to benchmark administrative data and targeted surveys (as not all topics are covered by administrative data) against the 2021 Census to prove to government and other users that an administrative data census can produce high-quality information at a lower cost, and can do so every year. The Government can then make a decision, after 2021, about the future of the census and whether the most significant change in over 200 years is made to the production of official population and social statistics.

This chapter was written in April 2016. For the latest information about the 2021 Census and Administrative Data Census please visit: www.ons.gov.uk/census.

References

Cabinet Office (2014) Government's response to the National Statistician's recommendation. Letter from Rt Hon Francis Maude to Sir Andrew Dilnot. Available at: www.statisticsauthority.gov.uk/wp-content/uploads/2015/12/letterfromrthonfrancismaudemptosirandrewdilnot18071_tcm97-43946.pdf.

Cabinet Office and Government Digital Service (2014) Government Digital Inclusion Strategy. Policy Paper. December. Available at: www.gov.uk/government/publications/government-digital-inclusion-strategy/government-digital-inclusion-strategy.

ONS (2014a) The census and future provision of population statistics in England and Wales: Report on the public consultation. March. Available at: https://www.ons.gov.uk/census/censustransformation programme/beyond2011censustransformationprogramme/reportsandpublications.

ONS (2014b) Beyond 2011: Final options report (O4). April. Available at: https://www.ons.gov.uk/census/censustransformationprogramme/beyond2011censustransformationprogramme/reports andpublications.

ONS (2014c) Beyond 2011: Public consultation on user requirements: Report. August. Available at: www.ons.gov.uk/ons/about-ons/who-ons-are/programmes-and-projects/beyond-2011/reports-and-publications/consultation-reports/beyond-2011-user-requirements-consultation-report.pdf.

ONS (2014d) Beyond 2011: Independent review of methodology. Available at: www.ons.gov.uk/census/censustransformationprogramme/beyond2011censustransformationprogramme/independentreview ofmethodology.

ONS (2014e) The census and future provision of population statistics in England and Wales: Recommendation from the National Statistician and Chief Executive of the UK Statistics Authority. Available at: www.ons.gov.uk/census/censustransformationprogramme/beyond2011censustransformationprogramme/thecensusandfutureprovisionofpopulationstatisticsinenglandandwalesrecommendationfromthenational statisticianandchiefexecutiveoftheukstatisticsauthorityandthegovernmentsresponse.

ONS (2015) Administrative data research report: 2015. October 2015. Available at: http://www.ons.gov.uk/census/censustransformationprogramme/administrativedatacensusproject under 'Administrative Data Census Research Outputs'.

Index

T - #0164 - 111024 - C512 - 246/174/24 - PB - 9780367660031 - Gloss Lamination